Introduction to IONIC LIQUIDS

Introduction to IONIC LIQUIDS

Hannah Foster
Patrick Robertson

Kruger Brentt
Publishers
2026

Kruger Brentt Publishers UK. LTD.
Company Number 9728962

Regd. Office: 68 St Margarets Road, Edgware, Middlesex HA8 9UU

ISBN: 978-1-78715-666-1

For information on all our publications visit our website at http://krugerbrentt.com/

PREFACE

Ionic liquids have emerged as a revolutionary class of solvents with unique physicochemical properties, including negligible vapor pressure, high thermal stability, tunable solubility, and remarkable electrochemical characteristics. These attributes have positioned ionic liquids at the forefront of green chemistry, catalysis, electrochemistry, and materials science. Their ability to replace conventional volatile organic solvents while offering enhanced performance has opened new avenues in both academic research and industrial applications.

This book, Introduction to Ionic Liquids, provides a comprehensive overview of the fundamental principles, properties, and applications of ionic liquids. It is designed for students, researchers, and professionals in chemistry, chemical engineering, materials science, and related fields who wish to understand the science and potential of these versatile compounds.

The initial chapters introduce the basic concepts, classification, and structural features of ionic liquids, explaining how their unique ionic interactions contribute to their remarkable properties. Subsequent sections explore synthesis methods, characterization techniques, and physicochemical behavior, offering a clear understanding of how these liquids can be designed and tailored for specific applications.

Applications of ionic liquids are covered extensively, ranging from green solvents in chemical reactions and catalysis to electrolytes in batteries, supercapacitors, and fuel cells. Special attention is given to their role in extraction processes, separation technologies, and advanced materials development. The book also addresses environmental and safety considerations, emphasizing the potential of ionic liquids to contribute to sustainable and eco-friendly processes.

By combining theoretical concepts with practical applications and contemporary research findings, Introduction to Ionic Liquids equips readers with the knowledge necessary to explore, design, and apply ionic liquids effectively in both academic and industrial contexts.

Hannah Foster

Patrick Robertson

CONTENTS

Chapter 1 SOPHISTICATED UTILIZATIONS OF IONIC LIQUIDS IN POLYMER SCIENCE

1. INTRODUCTION: AN OVERVIEW

In recent years, ionic liquids have garnered significant attention as "green and designer" solvents for chemical reactions. Room-temperature ionic liquids have surfaced as a viable alternative to organic solvents in catalytic processes at both laboratory and industrial levels (Holbrey & Seddon, 1999b). Literature documents a diverse array of reactions, encompassing advancements in alkylation reactions (Earle et al., 1998), Diels-Alder cyclizations (Earle et al., 1999; Jaeger & Tucker, 1989), and the establishment of commercially viable processes for the dimerization, oligomerization, and polymerization of olefins (Abdul-Sada et al., 1995a; 1995b; Ambler et al., 1996; Chauvin et al., 1988; 1989). Ionic liquids possess a distinct array of chemical and physical properties (Chauvin, 1996; Chauvin & Mussmann, 1995; Seddon, 1997) and exhibit no observable vapor pressure, making them suitable substitutes for volatile organic solvents. The extensive and easily obtainable array of room-temperature ionic liquids, exhibiting diverse physical properties due to straightforward structural modifications of cations (Gordon et al., 1998; Holbrey & Seddon, 1999a) or alterations in anions (Bonho^te et al., 1996; Wilkes & Zaworotko, 1992), presents the potential to engineer an ionic liquid-solvent system tailored for specific processes. Consequently, these ionic liquids may be regarded as "designer solvents" (Freemantle, 1998).

The utilization of ionic liquids as solvents in polymerization processes has been extensively examined in the literature (Kubisa, 2004; Shen & Ding, 2004; Lua et al., 2009). Ionic liquids have been employed in polymer science primarily as media for various polymerization processes, including conventional free radical polymerization (Sarbu & Matyjaszewski, 2001), living/controlled radical polymerizations such as atom transfer radical polymerizations (ATRP) (Ding et al., 2005; Shen & Ding, 2004; Biedron & Kubisa, 2001; Biedron & Kubisa, 2002; Biedron

& Kubisa, 2003), reversible addition-fragmentation transfer (RAFT) (Perrier & Davis, 2002), and in ionic and coordination polymerizations (Chiefari et al., 1998; Vijayaraghavan & MacFarlane, 2004). Radical polymerizations performed in ionic liquids typically exhibit a marked increase in the kp/kt ratio compared to those conducted in other polar or coordinating solvents. Ionic liquids serve as solvents for ATRP and RAFT, enabling the separation of the polymer from residual catalysts and minimizing the occurrence of side reactions.

The application of ionic liquids in polymer research extends beyond their role as solvents. Ionic liquids are utilized as additives, such as plasticizers, constituents of polymer electrolytes, and porogenic agents in polymers. Recent investigations focus on the characteristics of polymers with chemically bonded ionic liquid moieties (polymeric ionic liquids) and the potential applications thereof. Ionic liquids are examined as constituents of polymeric matrices (e.g., polymer gels), templates for porous polymers, and innovative electrolytes for electrochemical polymerizations (Przemysław & Kubisa, 2009). This chapter examines new advancements and accomplishments derived from the use of ionic liquids in the synthesis of functional polymers and the alteration of polymer characteristics induced by ionic liquids.

2. RADICAL POLYMERIZATION

2.1 Radical polymerization (ATRP) of acrylates in ionic liquids

It is known, that properties of ILs commonly depend on the structure of the cation (the symmetry and the length of alkyl substituents, the presence of hydrophobic groups, etc.) as well as on the degree of anion charge delocalization. To elucidate the influence of the nature of ILs on the yield and molecular weight of vinyl polymers, the radical polymerization of appropriate monomers (MMA) in different ionic liquids has been studied (Yakov et. al, 2007).For the MMA polymerization, the dependence between polymer ηinh and the length of alkyl substitute in asymmetrical 1-methyl-3-alkylimidazolium ILs was examined (Table 1). The results presented in Table 1 (entries 1–9) show, that increase of carbon chain length of imidazolium ILs leads to the reduction of polymer molecular weight, especially in the case of tetrafluoroborate ILs (Table 1, entries 5–9). Table 1 (entries 10, 11) shows, that a polymer having relatively high ηinh value was obtained in both tetrafluoroborate IL: [1,3-Bu2Im]BF4 and [1-Bu-3-(iso-Bu)Im]BF4, but in the case of 1,3-di-n- butylimidazolium tetrafluoroborate (entry 10) the ηinh value of the polymer is higher (3.13 and 2.30 dl g^{-1}, respectively).

The influence of the anion nature on the radical polymerization of MMA can be revealed by comparison of the data, obtained in ILs with ordinary used [1-Me-3-BuIm] cation (Table 1, High values of polymer molar mass and an increase in the rate of free radical polymerization of MMA in ILs can be assigned to the strong effect of ionic media on the chain propagation (activation energy decrease) and chain termination, for the reason of high viscosity of the reaction system, the so-called gel effect. (Yakov et. al, 2007).

2.2 Polymerized ionic liquids: synthesis and applications

Recently, much attention has been paid to polymerized ionic liquids or polymeric ionic liquids, which are macromolecules obtained from polymerizing ionic liquid monomers (Lu et. al, 2009). Their potential applications involve polymeric electrolytes (Galiński et. al, 2006; Ricks-Laskoski& Snow, 2006; Sato et. al, 2007; Susan et. al, 2005), catalytic membranes (Carlin & Fuller 1997), ionic conductive materials (Hirao et. al, 2000; Washiro et. al, 2004; Matsumi et. al, 2006), CO2 absorbing materials (Tang et. al, 2005a; Tang et. al, 2005b; Tang et. al, 2005c; Tang et. al, 2005d), microwave absorbing materials (Tang et. al, 2008; Amajjahe, & Ritter 2009), and porous materials (Yan & Texter, 2006; Yan et. al, 2007).

Table 1. IL's nature effect upon the radical polymerization of MMA

	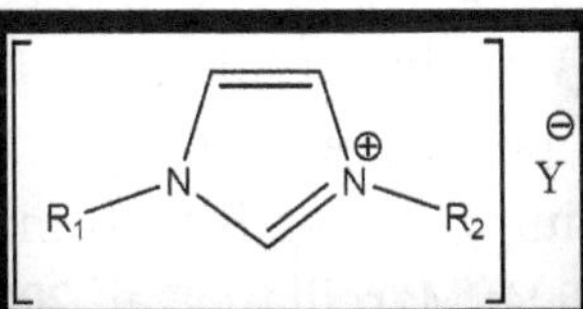			PMMA	
Entry[a]	R_1	R_2	Y	Yield (%)	ηin[b] (dlg^{-1})
1	CH_3	C_2H_5	$(CF_3SO_2)_2N$	98	3.06
2	CH_3	C_4H_9	$(CF_3SO_2)_2N$	98	2.81
3	CH_3	C_2H_5	CF_3SO_3	96	4.09c
4	CH_3	C_4H_9	CF_3SO_3	94	2.97
5	CH_3	C_3H_7	BF_4	93	3.00
6	CH_3	C_4H_9	BF_4	92	2.70
7	CH_3	C_5H_{11}	BF_4	91	2.86
8	CH_3	C_6H_{13}	BF_4	90	1.58
9	CH_3	C_7H_{15}	BF_4	95	1.55
10	C_4H_9	C_4H_9	BF_4	92	3.13
11	C_4H_9	*i*-C_4H_9	BF_4	96	2.30
12	CH_3	C_4H_9	PF_6	97	3.89
13	CH_3	C_4H_9	SbF_6	98	2.92

14	CH_3	C_6H_{13}	$(CF_3CF_2)_3PF_6$	98	3.29
15		$[P^+(C_6H_{13})C_{14}H_{29}]BF_4^-$		96	3.01
16		$[P^+(C_6H_{13})C_{14}H_{29}]PF_6$		98	3.31
17		$P^+(C_6H_{13})C_{14}H_{29}]C$		98	3.48 [d]
18		$[P^+(C_8H_{17})_4]B^-$		95	3.32
19		Benzene [e]		41	0.36

[a] Polymerization parameters: [AIBN]= 0.5 wt%, [MMA] =50 wt%, reaction time = 4 hr, reaction temperature, T = 60 °C.

[b] For the solutions of 0.05 g of PMMA in 10.0 ml of $CHCl_3$ at 25.0 °C.

[c] M_w = 5,770,000 g/mol (determined by static light scattering in acetone).

[d] Mw = 4,100,000 g/mol (determined by static light scattering in acetone).

[e] For comparison. (Reproduced from Vygodskii1, et. al (2007) *Polym. Adv. Technol.* 3, 18, 50–63, Copyright (2007), with permeation from John Wiely & Sons)

A variety of polymers having imidazolium moieties in the side chains have been reported, including poly (meth) acrylate (Washiro et. al, 2004; Ding et. al, 2004; Nakashima et. al, 2007; Juger et. al, 2009), polystyrene (Tang et. al, 2005c; Tang et. al, 2005e), and poly(N- vinylimidazolium) derivatives (Amajjahe, & Ritter 2009; Leddet et. al, 2001; Marcilla et. al, 2004; Marcilla et. al, 2005), and most of these poly(ionic liquid)s were prepared by conventional radical polymerizations. Free radical polymerizations of various N- vinylimidazolium derivatives were reported to proceed in the presence of conventional radical initiator, and various copolymers involving the imidazolium group were also synthesized by this method (Mu, et. al, 2005; Sugimura et. al, 2007).

2.2.1 crowave-absorbing ionic liquid polymer

Microwave-absorbing materials are applicable to reducing electromagnetic interference from personal computers, stealth aircraft technology, microwave cookware, and microwave darkroom protection (Petrov & Gagulin, 2001; Bregar, 2004; Yoshihiro et. al, 2002; Saib et. al, 2006; Zou et. al, 2006). Microwave absorption is usually achieved by combining dielectric and magnetic loss. Most microwave absorbing materials are polymer composites with conductive fillers, such as graphite, carbon black, and metals, or magnetic fillers, such as ferrites and carbonyl iron powders. These fillers make the microwave-absorbing materials black and hard to fabricate, for example, into precise parts or thin films (Peng et. al, 2005; Bosman et. al, 2003). Conductive polymers such as polyaniline, polypyrrole, polyalkylthiophenes, and poly(4,4'-diphenylene diphenylvinylene) were also reported as microwave-absorbing materials (Truong et. al, 2003; Chandrasekhar & Naishadham, 1999; Olmedo et. al, 1995; Wan, et. al, 2001;

Phang, et. al, 2005). The structures of poly(ionic liquid)s studied by Tang and coworker are shown in Fig. 1.

Since the poly(ionic liquid)s exhibit no magnetic loss and are insulators (the reported ionic conductivity of a poly(ionic liquid) with a similar structure to the polymers reported here is about 10^{-8}s m^{-1}[34]), their microwave absorptions are solely due to the dielectric loss.

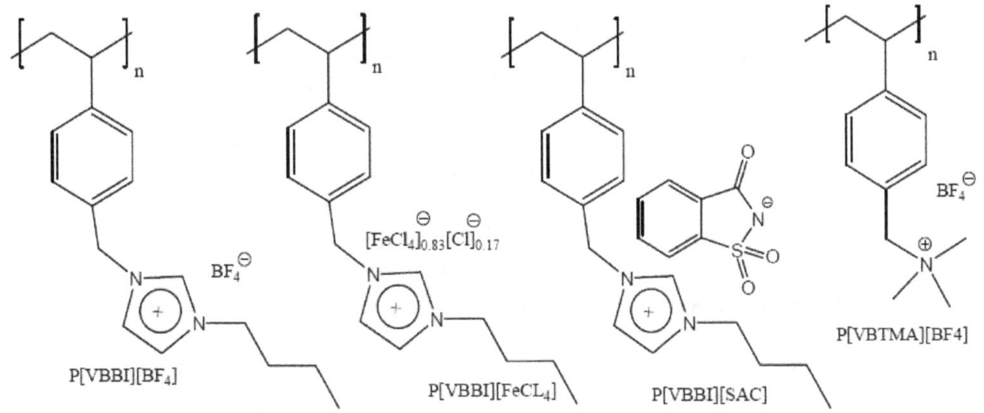

Fig. 1. Structures of poly(ionic liquid)s: poly[1-(*p*-vinylbenzyl)-3-butylimidazolium tetrafluoroborate] (P[VBBI][BF_4]), poly[1-(*p*-vinylbenzyl)- 3-butylimidazolium tetrachloroferrate] (P[VBBI][$FeCl_4$]), poly[1-(*p*-vinylbenzyl)-3-butylimidazolium o-benzoicsulphimide] (P[VBBI][Sac]), and poly[*p*-vinylbenzyltrimethylammonium tetrafluoroborate] (P[VBTMA][BF_4]).

(Reproduced from Tang et. al, (2008) *Macromolecules*, 41, 2, 493-496, Copyright (2008) with permeation from American Chemical Society)

The dielectric constant έ is shown in Fig. 2 as a function of frequency. At frequencies lower than 1 GHz, the έ decreases with increasing frequency. At higher frequencies, the έ remains essentially constant. The poly(ionic liquid)s with the imidazolium cations have similar dielectric constants (έ ≈ 4). In contrast, P[VBTMA][BF_4] with the ammonium cations has a higher dielectric constant (έ ≈ 5.2). This is not surprising because this material is more polar than poly(ionic liquid)s with imidazolium cations, which is in line with results of other ionic liquids (Tokuda et. al, 2006).

To probe the effect of anion on the dielectric loss, the dielectric losses of three poly(ionic liquid)s, P[VBBI][Sac], P[VBBI][BF_4], and P[VBBI][$FeCl_4$], with the same backbone and cations but different anions are compared. Sac^- is a mostly used organic anion while BF_4^- is a widely used inorganic anion in ionic liquids. A poly(ionic liquid) with $FeCl_4^-$ anions containing transition metal ions Fe^{3+} is also synthesized to test whether such kinds of anions can further increase the loss factor (Tang et. al, 2008).

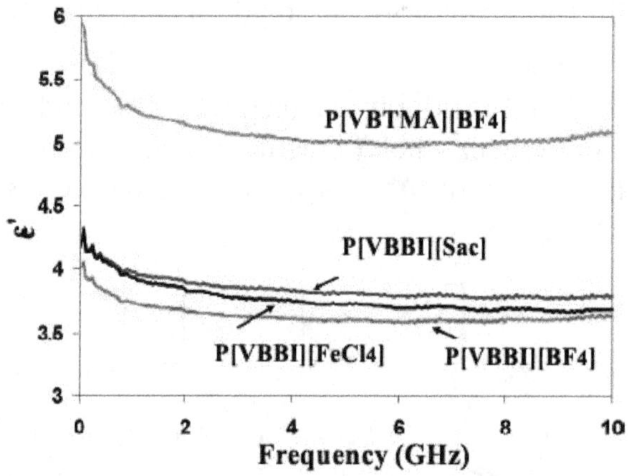

Fig. 2. Dielectric constant ἐ (the real part of complex permittivity) as a function of frequency (Reproduced from Tang et. al, (2008) *Macromolecules*, 41, 2, 493-496, Copyright (2008), with permeation from American Chemical Society)

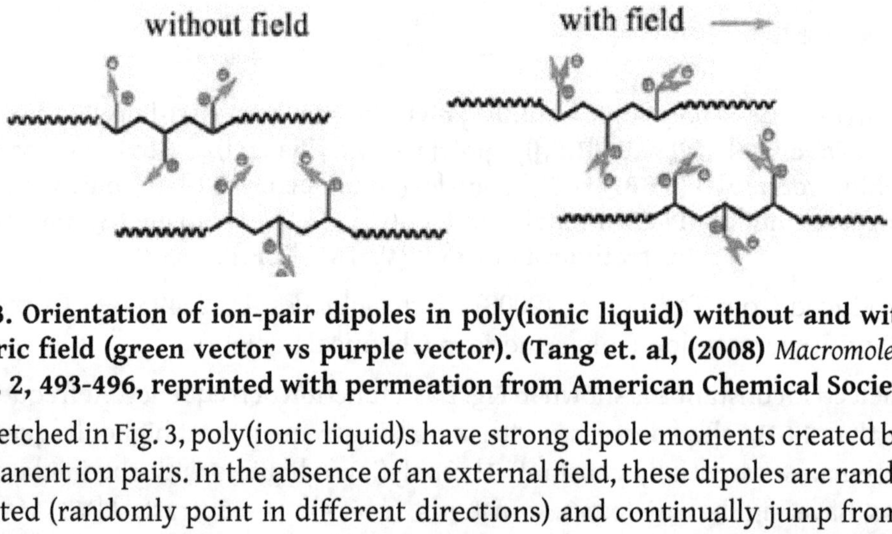

Fig. 3. Orientation of ion-pair dipoles in poly(ionic liquid) without and with an electric field (green vector vs purple vector). (Tang et. al, (2008) *Macromolecules*, 41, 2, 493-496, reprinted with permeation from American Chemical Society)

As sketched in Fig. 3, poly(ionic liquid)s have strong dipole moments created by the permanent ion pairs. In the absence of an external field, these dipoles are randomly oriented (randomly point in different directions) and continually jump from one orientation to another as a result of thermal agitation. In an external field, these dipoles orient themselves in the direction of the applied field (Fig. 3). (Tang et. al, 2008)

2.2.2 ectrowetting of a new ionic liquid monomer and polymer system

Synthesis and electrowetting of a new ionic liquid monomer and polymer system was reported by Holly and coworkers (Holly et. al, 2008). The formation of the monomeric ionic liquid salt and its polymerization is depicted in Fig. 4.

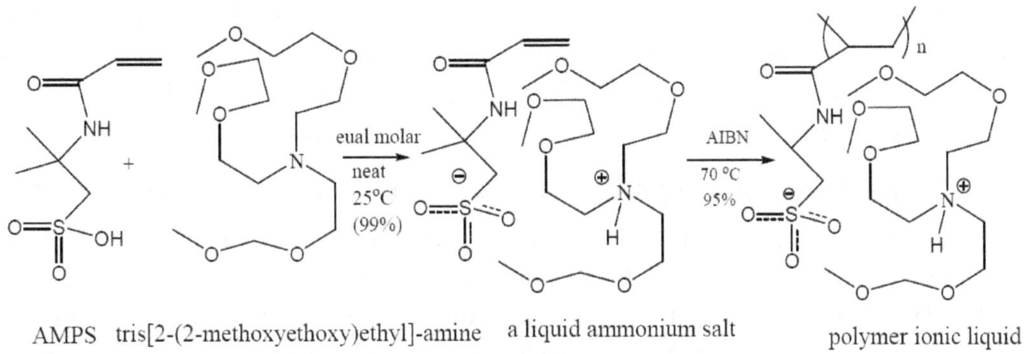

Fig. 4. Reaction Scheme for AMPS oxyethylene ammonium salt monomer and polymer (Reproduced from Ricks-Laskoski & Snow (2006) *J. Am. Chem. Soc.* 128, 38, 12402-12403, Copyright (2006), with permeation from American Chemical Society)

The uniqueness of the oxyethylene amine in the formation of the ammonium cationic species contributes to both the ionic and liquid nature of the monomer and polymer. Even more remarkable is the ability of this polymer to maintain its liquid nature as a macromolecule and to wet a substrate, showing preference for one polarity based upon the makeup of the ionic backbone of the polymer formed (Ricks-Laskoski & Snow, 2006).

Polymerizable ionic liquids and their actuation in an electric field are a combination of material and properties with unique potential to display structural and fluid dynamics above that found for small molecule ionic liquids.

Electrowetting is an electrostatically driven surface effect where a liquid droplet's spreading on a hydrophobic surface is modulated by application of a voltage to the droplet and an underlying conducting substrate (Quilliet &. Berge 2001). A schematic of this effect is illustrated in Figure 5.

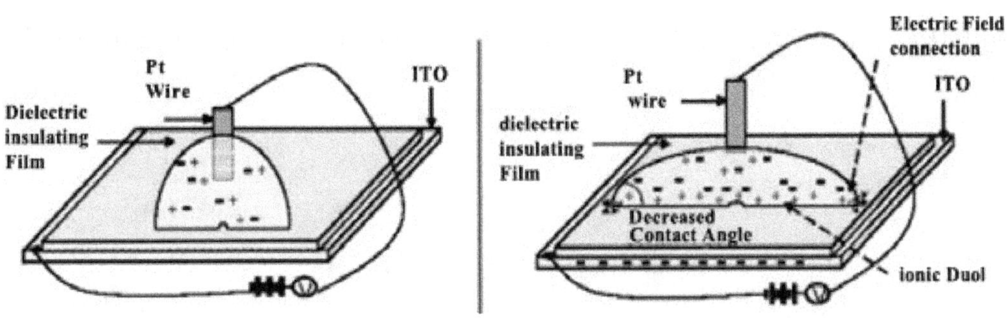

Fig. 5. Depiction of an electrowetting actuation electrode setup with (right) and without (left) an induced electric field (Reproduced from Ricks-Laskoski & Snow (2006) *J. Am. Chem. Soc.* 128, 38, 12402-12403, Copyright (2006), with permeation from American Chemical Society)

The droplet rests on a very thin low-dielectric insulating film (Teflon AF) which is supported on a conducting substrate and is contacted at the top by a very fine wire contact. Application of a voltage builds up a layer of charge on both sides of the interface with the dielectric film and decreases the interfacial energy.

2.2.3 Polymerized ionic liquids: solution properties and electrospinning

The solution properties and electrospinning of a polymerized ionic liquid was explored by Chen and Elabd (Chen & Elabd, 2009) Polymerized ionic liquids are synthesized from polymerizing ionic liquid monomers, where ionic liquids are of great interest due to their unique physiochemical properties. Compared to other polyelectrolyte solutions, this polymerized ionic liquid solution exhibits similar viscosity scaling relationships in the semidilute unentangled and semidilute entangled regimes. However, the electrospraying- electrospinning transition occurs at similar polymer solution concentrations compared to neutral polymers, where electrospinning produced beaded fibers and defect free fibers at ~1.25 and ~2 times the entanglement concentration, respectively. Due to high solution conductivities, electrospinning produces fibers approximately an order of magnitude smaller than neutral polymers at equivalent normalized solution concentrations. In addition, a high ionic conductivity of the solid-state fiber mat was observed under dry conditions and even higher conductivities were observed for polyelectrolyte fiber mats produced from electrospinning polyelectrolyte-ionic liquid solutions, where both anion and cation are mobile species. Structure of polymerized iIonic liquid poly(MEBIm-BF_4) is shown in Fig. 6. Fig. 7 shows the morphology of the fibers at various ionic liquid contents. Instead of reduced fiber sizes, the existence of ionic liquid results in larger fibers with a ribbon structure. This can be attributed to the nonvolatility of ionic liquid that hinders the solidification of fibers to smaller sizes. Moreover, the liquid component in the fiber collapses the fiber into a ribbon structure.

With the increase of ionic liquid content, more ribbons were observed in the fiber mat (Fig 7).

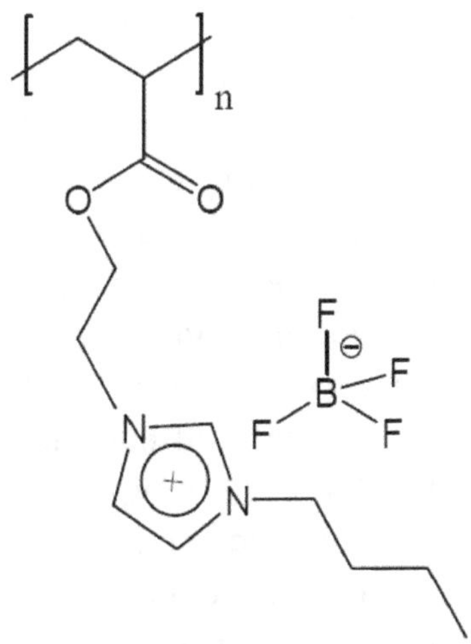

Fig. 6. Structure of polymerized ionic liquid poly(MEBIm-BF_4) (Reproduced from Chen et al., (2009) *Macromolecules* 42, 3368-3373, Copyright (2009), with permeation from American Chemical Society)

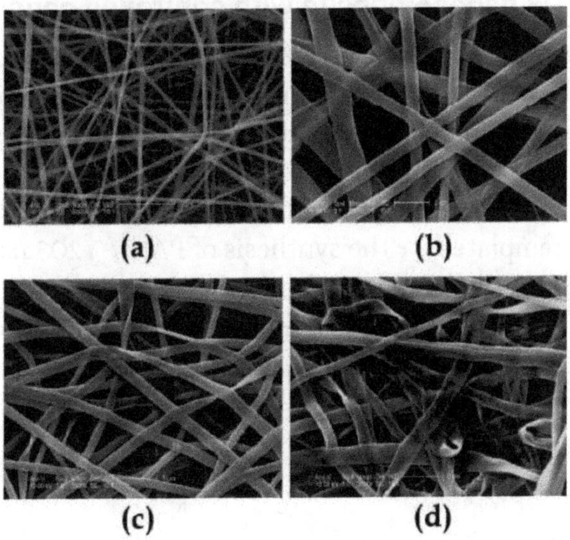

Fig. 7. Field emission scanning electron microscope images of electrospun Nafion-PAA- BMIm-BF4 blend at ionic liquid weight fraction of (a) 0%, (b) 10%, (c) 20%, (d) 30%. The weight ratio of Nafion:PAA is 3:2 at a 10 wt % total polymer concentration (Reproduced from Chen et al., (2009) *Macromolecules* 42, 3368-3373, Copyright (2009), with permeation from American Chemical Society)

2.3 Ultrasound and ionic-liquid-assisted synthesis method

Currently, the study of physical and chemical effects of ultrasound irradiation is a rapidly growing research area. When liquids are irradiated with high-intensity ultrasound irradiation, acoustic cavitations (the formation, growth, and implosive collapse of bubbles) provide the primary mechanism for sonochemical effects. During cavitation, bubble collapse produces intense local heating, high pressures, and extremely rapid cooling rates (Suslick et.al 1991, Suslick 1988). These transient, localized hot spots can drive many chemical reactions, such as oxidation, reduction, dissolution, decomposition, and promotion of polymerization (Suslick 1988). One of the most important recent aspects of sonochemistry has been its application in the synthesis of nanodimensional materials (Suslick & Price, 1999). Ultrasound irradiation offers a very attractive method for the preparation of novel materials with unusual properties and has shown very rapid growth in its application in materials science due to its unique reaction effects and ability to induce the formation of particles of much smaller sizes (Suslick 1988, Suslick & Price, 1999). The advantages of the sonochemical method include a rapid reaction rate, controllable reaction conditions, and the ability to form nanoparticles with uniform shapes, narrow size distributions, and high purity.

2.3.1 Ultrasound and ionic-liquid-assisted synthesis and characterization of polyaniline/Y2O3 nanocomposite with controlled conductivity

A sonochemical method has been employed to prepare polyaniline-Y_2O_3 (PANI/Y_2O_3) nanocomposite with controlled conductivity with the assistance of an ionic liquid by Kowsari and Faraghi (Kowsari & Faraghi, 2010) Ultrasound energy and the ionic liquid replace conventional oxidants and metal complexes in promoting the polymerization of aniline monomer. Ionic liquids (with unique properties) can act as morphology templates for the synthesis of PANI/ Y2O3 nanocomposite with novel or improved properties. Here task specific acidic ionic liquids with different counter ions induce different template and different PANI morphology. The structures of ionic liquids studied by Kowsari Faraghi are shown in Fig. 8.

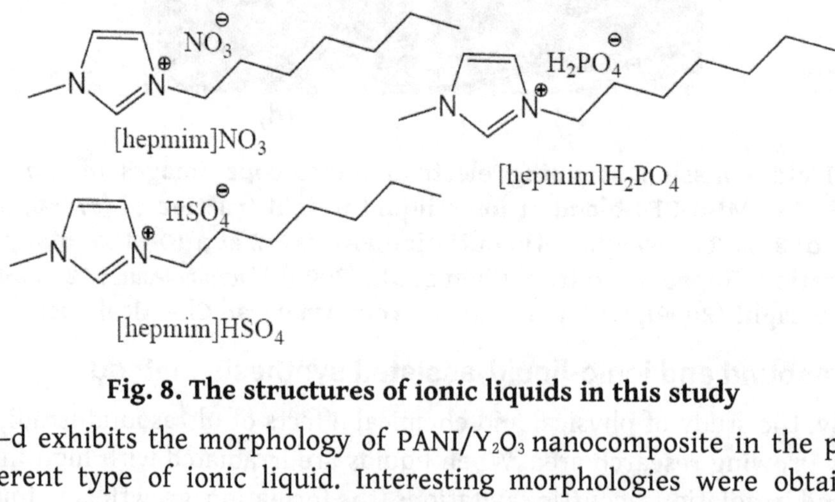

Fig. 8. The structures of ionic liquids in this study

Fig. 9a–d exhibits the morphology of PANI/Y_2O_3 nanocomposite in the presence of different type of ionic liquid. Interesting morphologies were obtained for PANI/Y_2O_3 nanocomposites. To compare the effects of ionic liquid additives on the properties of the resulting PANI/ Y_2O_3 nanocomposite, the monomer concentration ratio was kept constant. The differences in the structure of PANI/ Y_2O_3nanocomposite prepared, in the presence of ionic liquids are clearly visible. At the same magnification PANI/ Y_2O_3 nanocomposite (with [hepmim]. H_2PO_4 ionic liquid) reveals an interesting ribbon structure (Fig. 9a). The nanosheet structure of PANI/ Y_2O_3 in the presence of [hepmim] HSO_4, ionic liquid is shown in Fig. 9b. In Fig. 9b the smooth surface of PANI/ Y_2O_3 is visible. Smooth surfaces of it can be a reason for the better conductivities of these samples compared to Y_2O_3-free PANI. The SEM study shows that the presence of ionic liquid additives in polymerization strongly affects the morphology of PANI/ Y_2O_3. ionic liquids play a key role in tailoring the resultant conducting PANI/ Y_2O_3 structures. It was found that in the presence of [hepmim] NO_3, the products were regular solid microspheres covered with some nanoparticles.(Fig 9c) .TEM image of Y_2O_3 is shown in Fig 9d.

The influence of three different ionic liquid counter anions, namely HSO_4, H_2PO_4 and NO_3 on conductivity was investigated by preparing composites in the presence of these anions. It was observed that the conductivity was strongly influenced by the type of anion, although the yields of the respective composites were subject to variation. Since the PANI in the composite is in its emeraldine salt form irrespective of the acid used, the variation in conductivity most probably stems from the differences in the size and nature of the dopant anions.

As can be seen in Fig. 10, the conductivities may be classified in the decreasing order $NO_3^- < HSO_4^- < H_2PO_4^-$

The effect of the concentration of ionic liquid = [hepmim]HSO_4) on PANI-Y_2O_3 morphology is shown is Fig 11. As is vivid, at 0.2 M concentration morphology is a mixture of fibers and plates. When the concentration increases to 0.4, fibers gradually disappear and change into plates. Also, when the concentration reaches 0.6, the plates become thin in terms of thickness. As for both mentioned ionic liquids, increase in concentration brings about higher conductivity and yield, as well.

In the present study, when there is an increase in a frequency from 20 to 40 kHz, the morphology evolves as sphere shapes. In additions, the yield would be 55 wt %, 76 wt %, and 80 wt % at 20, 30, and 40 kHz for frequency.

Ultrasonic irradiation at 40 kHz, yields constantly higher degradation efficiencies compared with that at 20 kHz for all ionic liquids. Since, in the present study, the ionic liquid replaces conventional oxidant and metal complexes for polymerization, the increase of ionic liquids degradation leads to the increase of the concentration of alkyl radicals and the increase of aniline polymerization and, therefore, the increase of yield of product thickness.

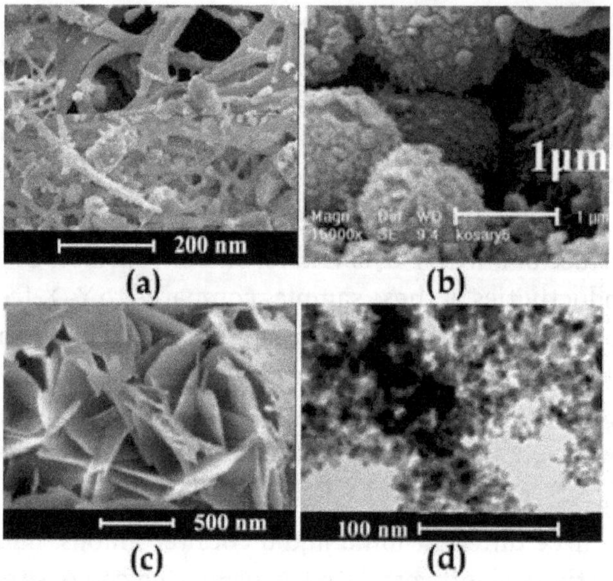

Fig. 9. SEM images of PANI/Y_2O_3 composite at different type of IL: (a) PANI/ Y_2O_3 composite (Y2O3 = 30%, aniline = 0.2 M, and IL = [hepmim]•H_2PO_4 = 0.6 M), (b) PANI/ Y_2O_3 composite (Y_2O_3 = 30%, aniline = 0.2 M, and IL = [hepmim]•HSO_4 = 0.6 M), (c) PANI/ Y_2O_3 composite (Y_2O_3= 30%, aniline = 0.2 M, and IL = [hepmim]•NO_3 = 0.6 M). (d) TEM image of Y_2O_3 (Reproduced from Kowsari & Faraghi (2010) *Ultrason. Sonochem*. 17, 4, 718- 725, Copyright (2010), with permeation from Elsevier)

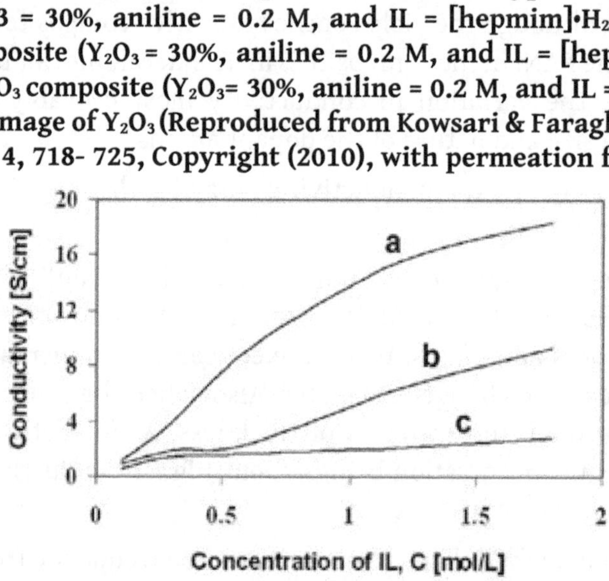

Fig. 10. The conductivities of PANI/Y_2O_3 composites at different concentrations of ILs: (a) PANI/Y_2O_3 composite (Y_2O_3 = 30%, aniline = 0.2 M, and IL = [hepmim]·H_2PO_4 = 0.6 M), (b) PANI/Y_2O_3 composite (Y_2O_3 = 30%, aniline = 0.2 M, and IL = [hepmim]·HSO_4 = 0.6 M), (c) PANI/Y_2O_3 composite (Y_2O_3 = 30%, aniline = 0.2 M, and IL = [hepmim]·NO_3 = 0.6 M). (Reproduced from Kowsari & Faraghi (2010) *Ultrason. Sonochem*. 17, 4, 718-725, Copyright 2010), with permeation from Elsevier)

As for both mentioned ionic liquids, increase in concentration brings about higher conductivity and yield, as well.

In the present study, when there is an increase in a frequency from 20 to 40 kHz, the morphology evolves as sphere shapes. In additions, the yield would be 55 wt %, 76 wt %, and 80 wt % at 20, 30, and 40 kHz for frequency.

Ultrasonic irradiation at 40 kHz, yields constantly higher degradation efficiencies compared with that at 20 kHz for all ionic liquids. Since, in the present study, the ionic liquid replaces conventional oxidant and metal complexes for polymerization, the increase of ionic liquids degradation leads to the increase of the concentration of alkyl radicals and the increase of aniline polymerization and, therefore, the increase of yield of product.

2.4 Ziegler-Natta polymerisation of ethylene

Ziegler-Natta polymerisation is used extensively for the polymerisation of simple olefins (e.g. ethylene, propene and 1-butene) and is the focus of much academic attention, as even small improvements to a commercial process operated on this scale can be important. Ziegler-Natta catalyst systems, which in general are early transition metal compounds used in conjunction with alkylaluminium compounds, lend themselves to study in the chloroaluminate(III) ionic liquids, especially the ones with an acidic composition.

Fig. 11. SEM images of PANI/Y_2O_3 composite (Y_2O_3 = 30%, aniline = 0.2 M, and IL = [hepmim]·HSO_4) at different concentrations of ILs (a) [IL] = 0.2 M, (b) [IL] = 0.4 M, (c) [IL] = 0.6 M (Reproduced from Kowsari & Faraghi (2010) *Ultrason. Sonochem.* 17, 4, 718-725, Copyright (2010), with permeation from Elsevier)

During studies into the behaviour of titanium(IV) chloride in chloroaluminate(III) ionic liquids Carlin *et al* carried out a brief study to investigate if Ziegler-Natta polymerisation was possible in an ionic liquid (Carlin et. al 1990)

2.5 Microemulsion polymerization by ionic liquids

Microemulsions are thermodynamically stable dispersions containing two immiscible liquids stabilized by surfactants at the liquid-liquid interface. Compared with "classic" water-inoil microemulsions, ionic liquid-in-oil microemulsions, in which an ionic liquid dispersed in an oil-continuous phase by suitable surfactants, are of great interest due to the unique features of both ionic liquid and microemulsion system (Eastoe et al., 2005; Binks et al., 2003; He et al., 2006; Atkin et al., 2007; Gao et al., 2004; Li et al., 2007). Microemulsions composed of a room temperature ionic liquid, 1-butyl-3-methylimidazolium tetrafluoroborate ([Bmim][BF_4]), cyclohexane, and a surfactant, Triton X-100, have been recently studied by Han (Gao et al., 2004) and Eastoe (Eastoe et al., 2005). Small angle neutron scattering data clearly demonstrated the formation of ionic liquid nanodroplets dispersed in cyclohexane (Eastoe et al., 2005). More recently, microemulsions composed of an ionic liquid dispersed in a variety of organic oils, such as toluene, (Li et al., 2007; Gao et al., 2006) xylene (Gao et al., 2006b) and benzene, (Gao et al., 2007) have been investigated by several groups. These microemulsions behaviors are consistent with water-in-oil microemulsions and provide nanosized ionic liquid domains as reaction or extraction media which should avoid contacting with water. However, potential uses of these ionic liquid - based microemulsions have not been intensively studied so far.

2.5.1 Microemulsions with an ionic liquid surfactant

The formation of amphiphilic association structures in and with ionic liquids, such as micelles, vesicles, microemulsions and liquid crystalline phases has been reviewed recently (Hao & Zemb, 2007, Qiu, & Texter, 2008). The formation of micelles formed by common surfactants in EAN was first documented in 1982 by Evans and coworkers (Evans et al, 1982, 1983), whereas the formation of liquid crystals of lipids in ethyl ammonium nitrate (EAN) (Evans, & Kaler, 1983) as well as lyotropic liquid crystals of non-ionic surfactants were observed in the same room temperature ionic liquid (Araos & Warr, 2005). The formation of liquid crystalline phases of binary mixtures of [C_{16}mim][Cl] and EAN has also been described lately (Zhao et al, 2009). The self-aggregation of common ionic and non-ionic surfactants in imidazolium based RTIL was also reported. (Anderson et al, 2003; Patrascu et al, 2006; Hao et al, 2005 As imidazolium based ionic liquids (ILs) with long-chain hydrocarbon residues exhibit surfactant properties in water (Thomaier& Kunz, 2007; (Bowers et al, 2004; Miskolczy et al, 2004; Sirieix-Plenet et al, 2004; Kaper & Smarsly, 2006; Seoud et al, 2007), Thomaier and coworkers could recently demonstrate that the surfactant like ionic liquid (SLIL) 1- hexadecyl-3-methylimidazolium chloride ([C_{16}mim][Cl]) forms colloidal structures in EAN as well. They found a critical aggregation concentration (cac) approximately ten times higher than the critical micelle concentration (cmc) in the corresponding aqueous system. (Thomaier & Kunz, 2007) This gap is in accordance with the results of Evans *et al.*, who found that the cacs of classical surfactants in EAN are between 5 and 10 times higher than in water. (Evans et al, 1982)

2.5.2 Polymerization of ionic liquid-based microemulsions: The synthesis of polymer electrolytes

S. Yu and coworkers (Yu et al, 2008) reported the first example of polymerization of microemulsions comprising surfactant stabilized ionic liquid nanodomains. Polymerization of these ionic liquid-based microemulsions yielded free-standing, flexible, and transparent polymer electrolytes even though the resulting vinyl polymers are incompatible with ionic liquid cores. The obtained ionic liquid/polymer composites show high conductivity at both room temperature and elevated temperature. This facile yet and effective method provides a versatile platform for the preparation of semisolid polymer electrolytes containing ionic liquids. Although They only demonstrated this method with vinyl monomers of styrene, methyl methacrylate, vinyl acetate, and *N,N*-dimethylacrylamide and only one surfactant as examples, they believe that this method should be extendable to other liquid monomers with similar polarity and ionic liquids with desirable properties. The elucidation of interaction between the polymeric matrix and ionic liquids need to be further explored in future work.

2.6 Aqueous/ionic liquid interfacial polymerization

Interfacial polymerization (IP) involves step polymerization of two reactive monomers or agents, which are dissolved respectively in two immiscible phases and the reaction takes place at the interface of the two liquids. IP has been used to prepare various polymers, such as poly(urea), poly(amide), and poly(ester) capsules.

It allows the synthesis of polymers at low temperature with limited side reactions, and can avoid the use of catalysts or phase transfer agents. The relative ease of IP has made it the preferred technique in many fields, ranging from microencapsulation of pharmaceutical products to preparing conducting polymers. IP is commonly done with volatile organic solvents as the organic phase, such as benzene (Chu et al, 2002), chloroform (Ashgarian et al, 1999), and toluene (Lu et al, 2002). The nonvolatile nature of ionic liquids gives them significant advantage in minimizing solvent consumption. They are good solvents for organic, inorganic, and polymeric compounds, and have been used as media for chemical reactions (Dupont et al, 2002), including polymerization (Harrisson et al, 2003) and ionic liquid/CO_2 biphasic reactions. Poly aniline (PANI) has been synthesized with water/ionic liquid IP by Kowsari and Yavari (Kowsari & Yavari,

2009). This method has some obvious advantages, such as PANI nanofiber can be fabricated and both solvents used are environmentally benign.

3. POLYCONDENSATION PROCESSES IN IONIC LIQUIDS

Polycondensation is typically conducted at relatively high temperature thus non-volatile and thermally stable ionic liquids seem to be suitable solvents for polycondensation processes. Research in this area has mainly been directed towards synthesis of polyamides, polyimides and polyesters. On the other hand, in some studies of polycondesation catalytic effect of ionic liquids was observed. More recently, two step procedures (involving post polycondensation in ionic liquid) was applied to polycondensation of sebacic, adipic, and succinic acid with aliphatic diols. Aliphatic polyesters with Mw up to 6×104 were obtained and once more it was noted that solubility of polyesters in ionic liquid was a limiting factor. Solubility depends on structure of IL (nature of cation and anion) and correlation was found between the miscibility of aliphatic polyester/ionic liquid system and the extent to which their solubility parameters matched (Kubisa, 2009).

3.1 Ionic liquids as novel solvents and catalysts for the direct polycondensation

The direct polycondensation of *N, N'*-(4,4'-oxydiphthaloyl)- bis-L-phenylalanine diacid with various aromatic diamines was performed in ionic liquid media by Mallakpour and Kowsari (Mallakpour & Kowsari, 2005). The influence of various reaction parameters, including the nature of the ionic liquid cations and anions, the monomer structures, the reaction temperature, and the reaction time, on the yields and inherent viscosities of the resulting optically active poly(amide imide) s (PAIs) were investigated. Direct polycondensation preceded in ionic liquids and triphenyl phosphite (a condensing agent) without any additional extra components, such as LiCl and pyridine, which are used in similar reactions in ordinary molecular solvents. Therefore, ionic liquids can act as both solvents and catalysts. Various high-molecular weights, optically active PAIs were obtained in high yields with inherent viscosities ranging from 0.54 to 0.88 dL/g. This method was also compared with three other classical methods for the polycondensation of the aforementioned monomers.

3.2 Polycondensation processes in ionic liquid under microwave irradiation

Application of ionic liquids as media to microwave-assisted reactions offers several advantages. Typical organic solvents are frequently flammable and volatile, which is a safety hazard for high-temperature and closed-vessel applications using microwaves. In contrast, ionic liquids have high boiling-points, low vapor pressures, and high thermal stabilities. In addition, typical ionic liquids have moderately high dielectric constants (in the range of 10–15), and relatively low heat capacities (in the range of 1–2 J/g K). This combination allows ionic liquids to absorb microwaves efficiently. Owing to these advantages, ionic liquids have been investigated as solvents in a number of microwave- mediated reactions. Microwave heating in ionic liquids was used also for polycondensation reactions leading to polyamides (Mallakpour& Kowsari, 2006). Certain advantages have been indicated, but until now only slight improvement of reactions conditions (more efficient heating, higher rates) has been achieved.

4. IONIC POLYMERIZATION

4.1 Cationic polymerization

Ability of ionic liquids to dissolve wide range of inorganic compounds was exploited in the study in which organoborate acids (HBOB) (bisoxalatoboric acid, bissuccinatoboric and bisglutaratoboric acids) were used as initiators of the cationic polymerization of styrene in pyrollidonium, imidazolium and phosphonium bis (trifluromethanesulfonyl) amide ionic liquids. In another study, cationic polymerization of styrene initiated with $AlCl_3$ in ionic liquid ([bmim][PF6]), supercritical CO_2 and organic solvent (CH_2Cl_2) was investigated. The only conclusion was that in ionic liquids rates and molecular weights are higher than in organic solvent (Kubisa, 2009).

Studies on the dimerisation and hydrogenation of olefins with transition metal catalysts in acidic chloroaluminate(III) ionic liquids report the formation of higher molecular weight fractions consistent with cationic initiation (Chauvin et. al 1990; Ellis et. al 1999). These studies ascribe the occurrence of the undesired side reaction to both the Lewis acid and the proton catalysed routes. Their attempts to avoid these side reactions led to the preparation of alkylchloroaluminate(III) ionic liquids and buffered chloroaluminate(III) ionic liquids (Chauvin et. al 1990; Ellis et. al 1999).

Attempts to bring the benefits of ionic liquid technology by drawing on the inherent ability of the chloroaluminate(III) ionic liquids to catalyse cationic polymerisation reactions, as opposed to minimising them, were patented by Ambler et al of BP Chemicals Ltd in 1993 (Ambler, 1993). They used acidic [EMIM][Cl-$AlCl_3$] ($X(AlCl_3)$ = 0.67) for the polymerisation of butene to give products that find application as lubricants. The polymerisation can be carried out by bubbling butene through the ionic liquid. The product formed a separate layer that floats upon the ionic liquid and was isolated by a simple process. Alternatively, the polymerisation was carried out by injecting the ionic liquid into a vessel charged with butene. After a suitable settling period the poly(butene) was isolated in a similar fashion. The products from these reactions must be best described as oligomers as opposed to polymers as the product is still in the liquid form. Chain transfer to impurities, ionic liquid, monomer and polymer will terminate the propagation reaction resulting in the low-mass products.

4.1.1 Cationic polymerization of styrene in $scCO_2$ and [bmim][PF6]

Bueno and coworkers (Bueno et. al 2009) presented a study on the cationic polymerization of styrene in different reaction media, supercritical CO_2 ($scCO_2$), [bmim][PF_6], dichloromethane, and $scCO_2$ plus [bmim][PF_6] using $AlCl_3$ as initiator at temperatures from 273 to 333 K. The reactions were analyzed in relation to the monomer conversion rate, polymer structures, average molecular weights and molecular weight distribution. In all cases, the styrene polymerization produced oligomers. Reactions using ionic liquid as a solvent led to higher molecular weight and monomer conversion rate. Monomer conversion rates of about 100% and weight average molecular weight (Mw) of 2400 at 298 K were obtained. Reactions with $scCO_2$ as a solvent yielded low monomer conversion rates (around 50%) and a *Mw* = 2000 at 298 K. The oligolystyrenes presented rr syndiotatic-rich sequences in the microstructure in all reaction conditions. The use of ionic liquid and $scCO_2$ results in better yields than the use of the other organic solvents.

4.2 Anionic polymerization

4.2.1 Anionic polymerization of methyl methacrylate in an ionic liquid

Anionic polymerization reactions of methyl methacrylate (MMA) in ionic liquids were carried out by utilizing alkyl lithium initiators such as n-butyl lithium (n-BuLi) and diphenylhexyl lithium (DPHLi) by H. Kokubo and Watanabe (Kokubo & Watanabe, 2008). The polymerization in ionic liquids having bis(trifluoromethyl sulfonyl)amide ([NTf_2]) anion did not yield poly(MMA) (PMMA), because of the deactivation of the initiator due to an attack on the trifluoromethyl group. By using 1-butyl-3-methylimidazolium hexafluorophosphate ([C_4mim][PF_6]), the polymerization reactions proceeded. The PMMA prepared in [C_4mim][PF_6] gave low yields (5–9%), as compared with those (15–62%) obtained forPMMAprepared in tetrahydrofuran (THF), and the prepared PMMA had large polydispersity indices ($\approx$ 2.0). The results obtained can be attributed to the high reaction temperature (0 ºC), in comparison to the common anionic polymerization temperature of MMA, -78 ºC, and the reaction between the initiator and the imidazolium cation. The initiator was considered to be deactivated because the hydrogen atom at the 2-position of the imidazolium ring was withdrawn by the alkyl lithium initiator. The tacticity of the obtained PMMA, prepared in [C_4mim][PF_6] by utilizing DPHLi, was rich in *mm* triads, similar to that polymerized in toluene

5. MISCELLANEOUS APPLICATION OF IONIC LIQUIDS IN POLYMER CHEMISTRY

5.1 Polymer electrolytes: Compatible system between polymers and ionic liquids (Ion gels)

Polymer electrolytes containing room temperature ionic liquids were first reported on such a system by Noda & Watanabe (Noda & Watanabe, 2000). Successively, the study was expanded to polymer electrolytes containing chemically stable ionic liquids (Noda & Watanabe, 2000; Watanabe et. al 1993; Watanabe et. al 1995; Watanabe & Mizumura 1996; Ogata et. al 1995; Susan et. al 2005). Watanabe and coworkers found that common vinyl monomers were widely soluble in common ionic liquids and that they could be polymerized by free radical polymerization (Noda & Watanabe 2000; Susan et. al 2005). In certain cases, surprisingly good compatibility of the resulting polymers with the ionic liquids could be achieved irrespective of the polymer concentration and temperature (Fig. 12).

Typically, the polymerization of methyl methacrylate (MMA) in a common ionic liquid, 1- ethyl-3-methyl imidazolium bis(trifluoromethane sulfone)imide ([C_2mim][NTf_2]) in the presence of a small amount of a cross-linker gives self standing, flexible, and transparent polymer gels. The polymer and ionic liquid composite gels, which they term "ion gels", show a single T_g for a given range of [C_2mim][NTf_2] composition, and the T_g decreases with an increase in the mole fraction of [C2mim][NTf2] (Fig 13) (Susan et. al 2005; Ueki &.Watanabe 2008).

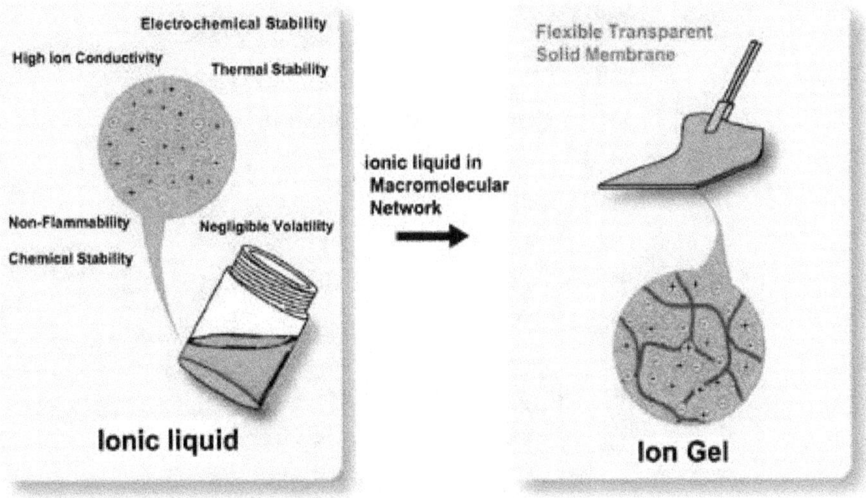

Fig. 12. Preparation of ion gels (compatible binary systems between polymer networks and ionic liquids) by a variety of methods such as in situ radical polymerization of vinyl monomers in ionic liquids and sol-gel transition of macromolecules in ILs.

The flexibility of molecular design of ionic liquids can facilitate various interesting applications of ion gels, particularly as ion-conducting polymer electrolytes

(Reproduced from Ueki & Watanabe (2008) *Macromolecules*, 41, 11, 3739-3748, Copyright (2008) ,with permeation from American Chemical Society)

5.2 Ion gel gated polymer thin-film transistors

Lee and coworkers (Lee et. al 2007) demonstrate that a gel electrolyte (a so-called "ion gel") based on a mixture of an ionic liquid and a block copolymer can provide both large specific capacitance (>10 μF/cm2) and greatly improved polarization response times (1 ms) when used as the gate dielectric in a polymer TFT (thin-film transistor). The improved properties allow transistor operation at frequencies greater than 100 Hz, significantly faster than what has been demonstrated previously and opening the door to a broader range of applications. Furthermore, the ion gel material is solution processible, making it potentially compatible with high throughput patterning methods (e.g., ink jet printing). Unlike traditional solid polymer electrolytes, which are obtained by the dissolution of salts by ion-coordinating polymers, the ion gel electrolyte is formed by gelation of poly(styrene-block-ethylene oxide- block-styrene) → (SOS) → triblock → copolymer in 1-butyl-3-methylimidazolium hexafluorophosphate ([BMIM][PF6]), a room temperature ionic liquid; the structures of these materials are shown in Fig. 14.

6. CONCISE SUMMARY

Advancements in the domain of ionic liquids, particularly regarding polymerization mechanisms, are unavoidable.

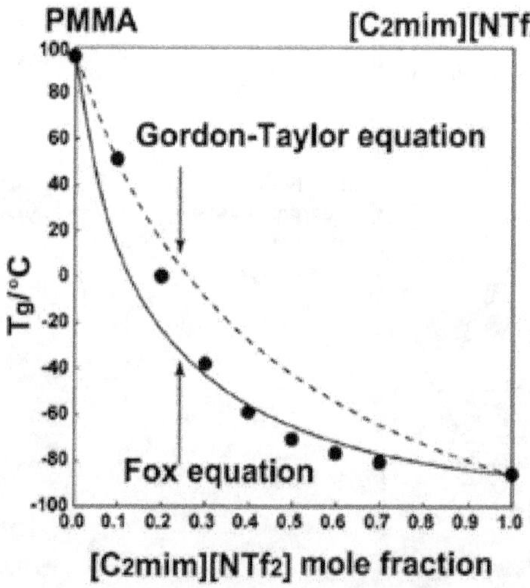

Fig. 13. Glass transition temperatures (*Tg*) of PMMA network polymers with dissolved [C_2mim][NTf_2] as a function of [C_2mim][NTf_2] mole fraction. The points denote experimental *Tg* results, the broken lines denote the fitted profiles by the Gordon-Taylor equation, and the solid lines denote those by the Fox equation. (Reproduced from Susan et. al (2005) *J Am. Chem. Soc.* 2005, 127, 4976, Copyright (2005), with permeation from American Chemical Society)

Researchers have deemed the study of ionic liquids, owing to its unique characteristics and environmentally friendly applications in industry, as both intriguing and promising. Promising initial applications for the integration of ionic liquid monomers into polymers have been documented; nonetheless, the diversity of potential ionic liquid architectures remains significantly underexplored. It is essential to meticulously select the systems examined where genuine synthetic benefits of utilizing ionic liquids are presented or novel insights into polymerization mechanisms are revealed. These study requirements contrast with simply mentions, albeit being novel, of processes that can be conducted about ionic liquids. In this manner, accomplishments in the domain might be anticipated.

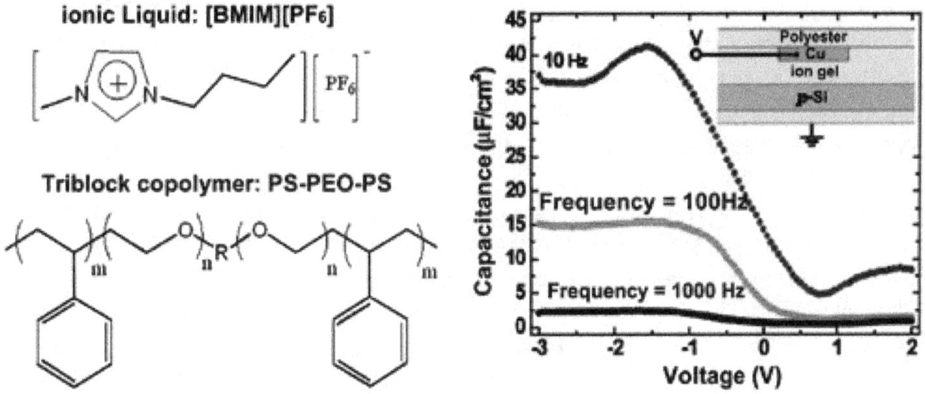

Fig. 14. Structure of the ionic liquid and triblock copolymer ion gel components (left) and C- V characteristics of a p-Si/ion gel/Cu test structure (see inset) at three frequencies (right).

The C-V curves indicate large frequency-dependent hole accumulation in the Si at negative bias on the top contact (Reproduced from Lee et. al (2007) *J. Am. Chem.* Soc. 129, 4532-4533, Copyright (2007), with permeation from American Chemical Society).

REFERENCES

Abdul-Sada, A. K.; Ambler, P.W.; Hodgson, P. K. G., Seddon, K. R. & Stewart N. J. (1995a). Ionic liquids. *World patent*, WO95/21871.

Abdul-Sada, A. K.; Atkins, M. P.; Ellis, B.; Hodgson, P. K. G.; Morgan, M. L. M.& Seddon K. R. (1995b). Alkylation process. *World patent*, WO95/21806.

Amajjahe, S. & Ritter, H. (2009). Microwave-Sensitive foamable poly(ionic liquids) bearing tert-butyl ester groups: Influence of counterions on the ester pyrolysis. *Macromol. Rapid Commun*, 30, 2, 94–98, 1521-3927

Ambler, P. W.; Hodgson, P.K. G & Stewart, N.J. (1993) *European Patent*, EP, 0558187.

Ambler, P. W.; Hodgson, P. K. G. & Stewart, N. J. (1996). Butene polymers. *European patent application*, EP/0558187A0558181.

Anderson, J. L.; Pino, V.; Hagberg, E. C.; Sheares, V. V. & Armstrong, D. W. (2003). Surfactant solvation effects and micelle formation in ionic liquids. *Chem. Commun*, 19 2444-2445, 1359-7345

Araos, M. U. & Warr, G. G. (2005). Self-assembly of nonionic surfactants into lyotropic liquid crystals in ethylammonium nitrate, a room-temperature ionic liquid. *J. Phys. Chem. B*, 109, 30, 14275-14277, 1520-6106

Ashgarian, B.; Cadenhead, D. A. (1999). nterfacial Compression and Polymerization of 4- and 9-((Butoxycarbonyl)methyl)urethane-Diacetylene Monomolecular Films: A Fluorescence Microscopy Study. *Macromolecules*, 32, 17, 5575-5580, 0024-9297

Atkin, R.; Warr, G. G. (2007). Phase behavior and microstructure of microemulsions with a room-temperature ionic liquid as the polar phase. *J. Phys. Chem. B*, 111, 31, 9309- 9316, 1520-6106

Biedron, T. & Kubisa, P. (2001). Atom-transfer radical polymerization of acrylates in an ionic liquid. *Macromol Rap Commun*, 22, 15, 1237–1242, 1521-3927

Biedron, T. & Kubisa P. (2002). P. Atom transfer radical polymerization of acrylates in an ionic liquid: synthesis of block copolymers. *J Polym Sci Part A Polym Chem*, 40, 16, 2799–809, 1099-0518

Biedron, T. & Kubisa, P. (2003) Ionic liquids as reaction media for polymerization processes: atom transfer radical polymerization (ATRP) of acrylates in ionic liquids. *Polym Int*, 52, 10, 1584–8, 1097-0126

Binks, B. P.; Dyab, A. K. F.; Fletcher, P. D. I. (2003) Novel emulsions of ionic liquids stabilised solely by silica nanoparticles. *Chem. Commun.*, 20, 2540–2541, 1359-7345

Bosman, Y.; Lau, Y.; Gilgenbach, R. M. (2003) Microwave absorption on a thin film. *Appl. Phys. Lett.*, *82*, 9, 1353-1355, 0003-6951

Bowers, J.; Butts, C. P.; Martin, P. J.; Vergara-Gutierrez, M. C. & Heenan, R. K. (2004). Aggregation behavior of aqueous solutions of ionic liquids. *Langmuir*, 20, 6, 2191- 2198, 0743-7463

Bregar, V. B. (2004). Advantages of ferromagnetic nanoparticle composites in microwave absorbers. *IEEE Trans. Magn.*, 40, 3, 1679-1684, 0018-9464

Bueno, C.; Cabral,V. F.; Cardozo-Filho, L.; Dias, M.L. & Antunes.O. A. C. (2009). Cationic polymerization of styrene in $scCO_2$ and [bmim][PF_6], *J. of Supercritical Fluids*, 48, 2, 183–187, 0896-8446

Carlin, R. T.; Osteryoung, R. A.; Wilkes, J. S. & Rovang, J. (1990). Studies of titanium(IV) chloride in a strongly Lewis acidic molten salt: electrochemistry and titanium NMR and electronic spectroscopy. *Inorg. Chem*, 29, 16, 3003-3009, 0020-1669

Carlin, R. T. & Fuller, J. (1997) Ionic liquid–polymer gel catalytic membrane. *Chem. Commun.*, 15, 1345–1346, 1359-7345

Chandrasekhar, P.; Naishadham, K. (1999). Broadband microwave absorption and shielding properties of a poly(aniline). *Synth. Met*, 105, 2, 115- 120, 0379-6779

Chauvin, Y. (1996). Two-phase catalysis in nonaqueous ionic liquids. *Actualite chimique* v, 44–46, 0151-9093

Chauvin, Y.; Commereuc, D.; Hirschauer, A.; Hugues, F. & Saussine, L. (1988). Process and catalyst for the dimeriation or codimerization of olefins. *French Patent, FR* 2,611,700.

Chauvin, Y.; Commereuc, D.; Hirschauer, A.; Hughes, F. & Saussine, L. (1989). Process and catalyst for the alkylation of isoparaffins with alkenes. *French Patent, FR* 2,611,572.

Chauvin, Y.; Gilbert, B. & Guibard, I. (1990) Catalytic dimerization of alkenes by nickel complexes in organochloroaluminate molten salts. *J. Chem. Soc., Chem. Commun*, 23, 1715-1716, 0022-4936

Chauvin, Y.; Mussmann, L.& Olivier, H. (1995). A novel class of versatile solvents for two- phase catalysts: Hydrogenation, isomerization, and hydroformylation of alkenes catalyzed by rhodium complexes in liquid 1,3-dialkylimidazolium salts. *Angew Chem Int Ed Engl*, 34, 23-24, 2698–2700, 1521-3773

Chen, H.; Yossef A. Elabd,Y. A.(2009). Polymerized ionic liquids: solution properties and electrospinning. *Macromolecules*, *42*,9, 3368-3373, 0024-9297

Chiefari, J.; Chong, Y. K.; Ercole, F., Krstina, J.; Jeffery, L.; Mayadunne, R. T. A. (1998). Living free-radical polymerization by reversible additionfragmentation chain transfer: the RAFT process. *Macromolecules*, 31, 16, 5559–5562, 0024-9297

Chu, L. Y.; Park, S. H.; Yamaguchi, T.; Nakao, S. I. (2002). Preparation of micron-sized monodispersed thermoresponsive core-shell microcapsules. *Langmuir,* 18, 5, 1856-1864, 0743-7463

Ding, S.; Tang, H.; Radosz, M. & Shen, Y. (2004). Atom transfer radical polymerization of ionic liquid 2-(1- butylimidazolium-3-yl) ethyl methacrylate tetrafluoroborate. *J. Polym. Sci., Part A: Polym. Chem,.* 42, 22, 5794–5801, 1099-0518

Ding, S.; Radosz, M. & Shen, Y. (2005). Ionic liquid catalyst for biphasic atom transfer radical polymerization of methyl methacrylate. *Macromolecules,* 38, 14, 5921–5928, 0024-9297

Dupont, J.; de Souza, R. F.; Suarez P. A. Z. (2002). Review ionic liquid (Molten Salt) phase organometallic catalysis. *Chem Rev*, 102, 10, 3667-3692, 0009-2665

Earle, M. J.; McCormac, P. B. & Seddon, K. R. (1999). Diels-Alder reactions in ionic liquids. *Green Chem*, 1, 23–25, 1463-9262

Eastoe, S.; Gold, S. E.; Rogers, A.; Paul, T.; Welton, R. K.; Heenan, I. G.(2005). Ionic liquid-in- oil microemulsions. *J. Am. Chem. Soc*, 127, 20, 7302–7303, 0002-7863

Ellis,B.; Keim,W. & Wasserscheid, P.(1999). Linear dimerisation of but-1-ene in biphasic mode using buffered chloroaluminate ionic liquid solvents. *J. Chem. Soc., Chem. Commun*, 4, 337-338, 0022-4936

Elseoud, O. A.; Pires, P. A. R.; Abdel-Moghny, T.; Bastos E. L. (2007). Synthesis and micellar properties of surface-active ionic liquids: 1-alkyl-3- methylimidazolium chlorides. *J. Colloid Interface Sci.*, 313, 1, 296-304, 0021-9797

Evans, D. F.; Yamauchi, A.; Roman R.& Casassa E. Z. (1982). Micelle formation in ethylammonium nitrate, a low-melting fused salt. *J. Colloid Interface Sci.*, 88, 1, 89-96, 0021-9797

Evans,D. F.; Kaler, E. W. & Benton; W. J. (1983). Liquid crystals in a fused salt: distearoylphosphatidylcholine in *N*-ethylammonium nitrate *J. Phys Chem*, 87, 18, 533-535

Evans, D. F.; A. Yamauchi, A.; Wei, G. J.; Bloomfield, V. A. (1983). Micelle size in ethylammonium nitrate as determined by classical and quasi-elastic light scattering. A, *J. Phys. Chem. 87*, 18, 3537-3541

Flannigan, J. D; Hopkins, D. S. D.; Suslick, K. S. (2005). Sonochemistry and sonoluminescence in ionic liquids, molten salts, and concentrated electrolyte solutions. *J. Organomet. Chem,* 690,15, 3513–3517, 0022-328X

Freemantle, M. (1998). Designer solvents—Ionic liquids may boost clean technology development. *Chem Eng News*, 76, 32–37, 0009-2347

Galiński, M.; Lewandowski, A. & Ste-pniak, I. (2006) Ionic liquids as electrolytes. *Electrochim. Acta*, 51, 26, 5567–5580, 0013-4686

Gao, H. X.; Li, J. C.; Han, B. X.; Chen, W. N.; Zhang, J. L.; Zhang, R.; Yan, D. D. (2004) Microemulsions with ionic liquid polar domains. *Phys. Chem. Chem. Phys.* 6, 2914– 2916, 1463-9076

Gao, Y. A.; Wang, S. Q.; Zheng, L. Q.; Han, S. B.; Zhang, X.; Lu, D. M.; Yu, L.; Ji, Y. Q.; Zhang, G. Y. (2006) Microregion detection of ionic liquid microemulsions. *J. Colloid Interface Sci.*, 301, 2, 612–616, 0021-9797

Gao, Y. A.; Zhang, J.; Xu, H. Y.; Zhao, X. Y.; Zheng, L. Q.; Li, X. W.; Yu, L. (2006). Structural Studies of 1-Butyl-3-methylimidazolium Tetrafluoroborate/TX-100/ *p*-Xylene Ionic Liquid Microemulsions. *Chem Phys Chem* , 7, 7, 1554–1561, 1439-7641

Gao, Y.; Li, N.; Zheng, L.; Bai, X.; Yu, L.; Zhao, X.; Zhang, J.; Zhao, M.; Li, Z. (2007). Role of solubilized water in the reverse ionic liquid microemulsion of 1-Butyl-3- methylimidazolium Tetrafluoroborate/ TX-100/Benzene. *J. Phys. Chem. B,* 111, 10, 2506–2513, 1520-6106

Gordon, C. M., Holbrey, J. D.; Kennedy, A. & Seddon, K. R. (1998). Ionic liquid crystals: Hexafluorophosphate salts. J Mater Chem, 8, 2627–2636, 0959-9428

Hansen, F. K. & Rødsrud, G. (1991). Surface tension by pendant drop: I. A fast standard instrument using computer image analysis. *J. Colloid Interface Sci.*, 141,1, 1-9, 0021- 9797

Hao, J.; Song, A.; Wang, J.; Chen, X.; Zhuang, W.; Shi, F.; Zhou, F. & Liu, W. (2005). Self- assembled structures in room-temperature ionic liquids, *Chem. Eur. J.*, 11,13, 3936- 3940, 1521-3765

Hao, J.; Zemb,T. (2007). Self-assembled structures and chemical reactions in room- temperature ionic liquids, *Curr. Opin. Colloid Interface Sci.*, *12*, 3, 129-137, 1359-0294

Harrisson, S., Mackenzie, S. R.; Haddleton, D. M. (2003). Pulsed laser polymerization in an ionic liquid: Strong solvent effects on propagation and termination of methyl methacrylate. *Macromolecules*, 36, 14, 5072-5075, 0024-9297

He, Y.; Li, Z.; Simone, P.; Lodge, T. P. (2006). Self-assembly of block copolymer micelles in an ionic liquid. *J. Am. Chem. Soc.*, 128, 8, 2745–2750, 0002-7863

Hirao, M.; Ito, K. &; Ohno, H. (2000). Preparation and polymerization of new organic molten salts; *N*-alkylimidazolium salt derivatives. *Electrochim. Acta*, 45, 8-9, 1291–1294. 0013-4686

Jaeger, D. A. & Tucker, C. E. (1989). Diels-Alder reactions in ethylammonium nitrate, a low- melting fused salt. *Tetrahedron Lett*, 30, 1785–1788, 0040-4039

Holbrey, J. D. & Seddon, K. R. (1999a). The phase behaviour of 1-alkyl-3- methylimidazolium tetrafluoroborates: ionic liquid and ionic liquid crystals. *J Chem Soc Dalton Trans*, 13, 2133–2139, 1472-7773

Holbrey, J. D. & Seddon, K. R. (1999b). Ionic liquids. *Clean Prod Proc*, 1, 223–237, 1435-2974 Bonho^te, P., Dias, A. P.; Papageorgiou, N., Kalyanasundaram, K. & Gra¨tzel, M. (1996). Hydrophobic, highly conductive ambient-temperature moltensalts. *Inorg Chem*, 35, 5, 1168–1178, 0020-1669

Holly L. Ricks-Laskoski, and Arthur W. Snow (2006). Synthesis and electric field actuation of an ionic liquid polymer. *J. Am. Chem. Soc.*, 128, 38, 12402-12403, 0002-7863

Juger, J.; Meyer, F.; Vidal, F.; Chevrot, C. & Teyssie, D. (2009). Synthesis, polymerization and conducting properties of an ionic liquid-type anionic monomer. *Tetrahedron Lett.*, 50, 1, 128–131, 0040-4039

Kaper, H. & SmarslyB. (2006). Templating and phase behavior of the long chain ionic liquid C_{16}mimCl,. *Phys. Chem*, *220*, 10, 1455-1471, 0942-9352

Kato, Y.; Satoshi, S.; Sugimoto, S.; Shinohara, K., Tezuka, N., Kagotani, T. & Inomata, K. (2002). Magnetic Properties and Microwave Absorption Properties of Polymer- Protected Cobalt Nanoparticles. *Mater. Trans*, 43, 3, 406-409, 1345-9678

Kokubo, H. & Watanabe M. (2008). Anionic polymerization of methyl methacrylate in an ionic liquid. *Polym. Adv. Technol*, 19,10, 1441- 1444, 1099-1581

Ku, C. C.; Liepins, R. (1987). *Electrical Properties of Polymers: Chemical Principles*, Hanser Publishers, 978-3446142800, New York

Kubisa, P. (2004). Application of ionic liquids as solvents for polymerization processes. *Prog Polym Sci*, 29, 1, 3–12, 0079-6700

Kubisa, P. (2009) Ionic liquids as solvents for polymerization processes—progress and challenges. *Prog Polym Sci* , 34, 12, 1333–1347, 0079-6700

Kowsari, E.; Yavari, I. (2009) Proceedings of the Polymer Processing Society Europe/Africa Regional Meeting ~ PPS-2009 ~ October 18-21, 2009 Larnaca (Cyprus)

Kowsari, E.; Faraghi, G. (2010). Ultrasound and ionic-liquid-assisted synthesis and characterization of polyaniline/Y_2O_3 nanocomposite with controlled conductivity. *Ultrason. Sonochem*, 17, 4, 718-725, 1350-4177

Leddet, C. & Fischer, A.; Brembilla, A.; Lochon, P. (2001). Influence of the alkyl-chain size on the amphiphilic behaviour of poly(3-alkyl-1-vinylimidazolium) bromide in aqueous medium . *Polym. Bull.*, 46, 1 , 75–82, 0170-0839

Lee, J; Panzer, M. J.; He, Y.; Lodge, T. P & Daniel Frisbie, C. (2007) Ion gel gated polymer thin-film transistors, *J. Am. Chem. Soc.* , 129, 4532-4533, 0002-7863

Li, N.; Gao, Y.; Zheng, L.; Zhang, J.; Yu, L.; Li, X. (2007) Studies on the micropolarities of bmimBF_4/ TX-100/Toluene ionic liquid microemulsions and their behaviors characterized by UV-Visible spectroscopy. *Langmuir*, 23, 3, 1091–1097, 0743-7463

Lu, X. F.; Bian, X. K., Shi, L. Q. (2002) Preparation and characterization of NF composite membrane. *J. Membr Sci* 210, 1, 3-11, 0376-7388

Lu, J.; Yan, F. & Texter, J. (2009) Advanced applications of ionic liquids in polymer science *Prog. Polym. Sci.*, 34, 5, 431–448, 0079-6700

Mallakpour, S., Kowsari, E. (2005). Ionic liquids as novel solvents and catalysts for the direct polycondensation of *N,N'*-(4,4'-oxydiphthaloyl)-bis-L-phenylalanine diacid with various aromatic diamines. *J. Polym. Sci. Part A: Polym. Chem*, 43, 24, 6545-6553, 1099-0518

Mallakpour, S; Kowsari, E. (2006). Microwave heating in conjunction with ionic liquid as a novel method for the fast synthesis of optically active poly(amide-imide)s derived from *N,N'-(4,4'*-hexafluorois opropylidenediphthaloyl bis-*L*methionine and various aromatic diamines. *Iranian Polym., J.* 15, 3, 239-247, 1026-1265

Marcilla, R.; Blazquez, J. A.; Rodriguez, J.; Pomposo, J. A. & Mecerreyes, D. *J.* (2004). Tuning the solubility of polymerized ionic liquids by simple anion-exchange reactions. *Polym. Sci., Part A: Polym. Chem.*, 42, 1, 208–212, 1099-0518

Marcilla, R.; Blazquez, J. A.; Fernandez, R.; Grande, H.; Rodriguez, J.; Pomposo, J. A. & Mecerreyes, D. (2005) Synthesis of novel polycations using the chemistry of ionic liquids. *Macromol. Chem. Phys.*, 206, 2, 299–304, 1521-3935

Matsumi, N.; Sugai, K.; Miyake, M. & Ohno, H. (2006). Polymerized ionic liquids via hydroboration polymerization as single ion conductive polymer electrolytes. *Macromolecules*, 39, 20, 6924–6927, 0024-9297

Miskolczy, Z.; Sebok-Nagy, K.; Biczok, L. & Goektuerk, S. (2004). Aggregation and micelle formation of ionic liquids in aqueous solution. *Chem. Phys. Lett,. 400*, 4-6, 296-300, 0009-2614

Mu, X. d.; Meng, J. q.; Li, Z. C.; Kou, Y. (2005). Rhodium nanoparticles stabilized by ionic copolymers in ionic liquids: long lifetime nanocluster catalysts for benzene hydrogenation. *J. Am. Chem. Soc,.* 127, 27, 9694–9695, 0002-7863

Nakashima, T.; Sakashita, M.; Nonoguchi, Y. & Kawai, T. (2007). Sensitized photopolymerization of an ionic liquid-based monomer by using CdTe nanocrystals. *Macromolecules,* 40, 18, 6540–6544, 0024-9297

Noda, A. & Watanabe, M. (2000). Highly conductive polymer electrolytes prepared by in situ polymerization of vinyl monomers in room temperature molten salts. *Electrochim. Acta,,* 45, 8-9, 1265-1270, 0013-4686

Ogata, N.; Sanui, K.; Rikukawa, M.; Yamada, S.;Watanabe, M.; Mizumura, T. (1996). *Solid State Ionics* 353, 86.

Olmedo, L.; Hourquebie, P.; Jousse, F. (1995). Microwave properties of conductive polymers. *Synth. Met.*, 69, 1-3, 205- 208, 0379-6779

Patrascu, C.; Gauffre, F.; Nallet, F.; Bordes, R.; Oberdisse, J.; de Lauth-Viguerie, N. & Mingotaud, C. (2006). Micelles in ionic liquids: aggregation behavior of alkyl poly(ethyleneglycol)- ethers in 1-butyl-3-methyl-imidazolium type ionic liquids, *Chem. Phys. Chem,.* 7, 99- 101, 1439-7641, 1439-7641

Peng, C. H.; Hwang, C. C.; Wan, J.; Tsai, J. S.; Chen, S. Y. (2005). Microwave-absorbing characteristics for the composites of thermal-plastic polyurethane (TPU)-bonded NiZn-ferrites prepared by combustion synthesis method. *Mater. Sci. Eng., B* , 117,1, 27-36, 0921-5107

Perrier, S.; Davis, T. P.; Carmichael, A. J.; Haddleton, D. M. (2002). First report of reversible addition-fragmentation chain transfer (RAFT) polymerization in room temperature ionic liquids. *Chem Commun*, 19, 2226–2227, 1359-7345

Petrov, V. M. & Gagulin, V. V. (2001). Microwave absorbing materials. *Inorg. Mater.*, 37,2, 93- 98, 0020-1685

Phang, S.W.; Daik, R; Abdullah, M. H. (2005). Poly(4,4'-diphenylene diphenylvinylene) as a non-magnetic microwave absorbing conjugated polymer. *Thin Solid Films*, 477, 1-2, 125- 130, 0040-6090

Qiu, Z. & Texter, J. (2008). Ionic liquids in microemulsions, *Curr. Opin. Colloid Interface Sci*, 13, 4, 252-262, 1359-0294

Quilliet, C. & Berge, B. (2001). Electrowetting: a recent outbreak. *Curr. Opin. Colloid Interface Sci.*, 6, 1, 34-39, 1359-0294

Ricks-Laskoski, H. L. & Snow, A. W. (2006). Synthesis and electric field actuation of an ionic liquid polymer. *J. Am. Chem. Soc.*, 128, 38, 12402–12403, 0002-7863.

Saib, A.; Bednarz, L.; Daussin, R.; Bailly, C.; Lou, X.; Thomassin, J. M.; Pagnoulle, C.; Detrembleur, C.; Jerome, R. & Huynen, I. (2006). Carbon nanotube composites for broadband microwave absorbing materials. *IEEE Trans. Microwave Theor. Tech*, 54, 6, 2745-2754, 0018-9480

Sarbu, T. & Matyjaszewski, K. (2001). ATRP of methyl methacrylate in the presence of ionic liquids with ferrous and cuprous anions. *Macromol Chem Phys*; 202, 17, 3379–91, 1022-1352

Sato, T.; Marukane, S.; Narutomi, T. & Akao, T. (2007). High rate performance of a lithium polymer battery using a novel ionic liquid polymer composite. *J. Power Sources*, 164,1, 390–396. 0378-7753

Seddon, K. R. (1997). Ionic liquids for clean technology. *J Chemical Tech Biotech* 68, 351–356, 0268-2575

Shen, Y. & Ding, S. (2004). Catalyst separation in atom transfer radical polymerization. *Prog Polym Sci*, 29, 10, 1053–1078, 0079-6700

Sirieix-Plenet, J.; Gaillon, L. & Letellier, P. (2004). Behavior of a binary solvent mixture constituted by an amphiphilic ionic liquid, 1-decyl-3-methylimidazolium bromide and water. Potentiometric and conductimetric studies, *Talanta*, *63*, 4, 979-986, 0039-9140

Sugimura, R.; Qiao, K.; Tomida, D.; Kume, Y. & Yokoyama, C. (2007) immobilization of palladium acetate on ionic liquid copolymerized polystyrene: A way to eliminate inhibiting effect of imidazolium chloride and enhance catalytic performance. *Chem. Lett*,. 36, 7, 874–875, 0366-7022

Susan, M. A. B. H.; Kaneko, T.; Noda, A.; Watanabe, M. (2005) Ion gels prepared by in situ radical polymerization of vinyl monomers in an ionic liquid and their characterization as polymer electrolytes. *J. Am. Chem. Soc*. 127, 13, 4976-4983, 0002-7863

Suslick, K. S. (1988). *Ultrasound*, VCH, Weinhein, 0-89573-328-5, Germany

Suslick, K. S.; Choe, S. B.; Cichowlas, A. A.; Grinstaff, M. W. (1991). Sonochemical synthesis of amorphous iron. *Nature*, 353, 6343, 414-416, 0028-0836

Suslick, K. S.; Price G. J. (1999) Application of ultrasound to materials chemistry. *Ann. Rev. Mater. Sci*,. 29, 295-326, 0084-6600

Tang, J.; Tang, H.; Sun, W.; Plancher, H.; Radosz, M.; Shen, Y. (2005). Poly(ionic liquid)s: a new material with enhanced and fast CO_2 absorption. *Chem. Commun*, 26, 3325– 3327. 1359-7345

Tang, H.; Tang, J.; Ding, S.; Radosz, M. & Shen, Y. (2005). Atom transfer radical polymerization of styrenic ionic liquid monomers and carbon dioxide absorption of the polymerized iIonic liquids. *J. Polym. Sci., Part A: Polym. Chem*,. 43, 7, 1432–1443, 1099-0518

Tang, J.; Sun, W.; Tang, H.; Radosz, M.; Shen, Y. (2005). Enhanced CO_2 absorption of poly(ionic liquid)s. *Macromolecules*, 38,6, 2037–2039, 0024-9297

Tang, J.; Tang, H.; Sun, W.; Radosz, M. & Shen, Y. (2005) Low-pressure CO_2 sorption in ammonium-based poly(ionic liquid)s .*Polymer*, 46, 26, 12460–12467, 0032-3861, 0024-9297

Tang, J.; Tang, H.; Sun, W.; Radosz, M. & Shen, Y. (2005). Poly(ionic liquid)s as new materials for CO_2 Absorption. *J. Polym. Sci., Part A: Polym. Chem*,. 43, 22, 5477–5489, 1099-0518

Tang, J.; Radosz, M.; Shen, Y. (2008) Poly(ionic liquid)s as optically transparent microwave- absorbing materials. *Macromolecules*, 41, 2, 493– 496, 0024-9297

Thomaier, S. & Kunz, W. (2007) Aggregates in mixtures of ionic liquids, *J. Mol. Liq*, 130, 1-3, 104-107, 0167-7322

Tokuda, H.; Tsuzuki, S.; Hayamizu, M. K.; Watanabe, M. (2006). How ionic are room- temperature ionic liquids? An indicator of the physicochemical properties. J. Phys. Chem. B, 110, 39, 19593-19600, 1520-6106

Truong, V. T.; Riddell, S. Z.; Muscat, R. F. (1998). Polypyrrole based microwave absorbers. *J. Mater. Sci*, 33, 20, 4971-4976, 0022-2461

Ueki, T. & Watanabe, M. (2008). Macromolecules in ionic liquids: progress, challenges, and opportunities. *Macromolecules* , 41, 11, 3739-3748, 0024-9297

Vijayaraghavan, R. & MacFarlane, D. R. (2004) Living cationic polymerisation of styrene in an ionic liquid. *Chem. Commun*,6,700-701

Vijayaraghavan, R.; Pringle, J. M. & MacFarlane D. R. (2008). Anionic polymerization of styrene in ionic liquids. *European Polymer Journal*, 44 , 6,1758–1762, 0014-3057

Vygodskii1,Y. S.; Mel'nik1,O. A.; Lozinskaya1, E. I.; Shaplov1, A. S.; I. A. Malyshkina, I. A.; Gavrilova,n. D.; Lyssenko1,k. A.; Antipin1,M. Y. Golovanov, D. G.; G.; Korlyukov1, A. A.; Ignat'ev, N .& Welz-Biermann, U. (2007). The influence of ionic liquid's nature on free radical polymerization of vinyl monomers and ionic conductivity of the obtained polymeric materials. *Polym. Adv. Technol*, 3, 18, 50–63, 1099-1581

Wan, M. X.; Li, J. C.; Li, S. Z. (2001). Microtubules of polyaniline as new microwave absorbent materials. *Polym. Adv. Technol.*, 12, 11-12, 651- 657, 1099-1581

Wang, Y. Y.; Li,W. & Dai, L. Y. (2007). Cationic ring-opening polymerization of 3,3- bis(chloromethyl) oxacyclobutane in ionic liquids, *Chinese Chemical Letters*, 18, 10, 1187–1190, 1001-8417

Washiro, S.; Yoshizawa, M.; Nakajima, H. & Ohno, H. (2004). Highly ion conductive flexible films composed of network polymers based on polymerizable ionic liquids. *Polymer*, 45, 5 1577–1582, 0032-3861

Watanabe, M.; Yamada, S.; Sanui, K.; Ogata, N. J. (1993). High ionic conductivity of new polymer electrolytes consisting of polypyridinium, pyridinium and aluminium chloride. *J Chem. Soc Chem. Commun.* , 11, 929-931, 0022-4936

Watanabe, M.; Yamada, S. & Ogata, N. (1995) Ionic conductivity of polymer electrolytes containing room temperature molten salts based on pyridinium halide and aluminium chloride. *Electrochim. Acta* , 40,13-14, 2285-2288, *0013-4686*

Wilkes, J. S. & Zaworotko, M. J. (1992). Air and water stable 1-ethyl-3-methylimidazolium based ionic liquids. *J Chem Soc Chem Commun*, 13, 965–967, 0022-4936

Yan, F. & Texter, J. (2006). Surfactant ionic liquid based microemulsions for polymerizatiom. *Chem. Commun*, 2696–2698. 1359-7345

Yan, F. & Texte J. (2007) Solvent-reversible poration in ionic liquid copolymers. *Angew. Chem., Int. Ed*, 46, 14, 2440–2443, 1521-3773

Yu, S.; Yan, F.; Zhang, X.; You, J.; Wu,P.; Lu, J.; Xu, Q.; Xia, X. & Ma, G. (2008). Polymerization of ionic liquid-based microemulsions: A versatile method for the synthesis of polymer electrolytes. *Macromolecules*, 41, 10, 3389-3392, 0024-9297

Zhao, Y.; Chen, X.; WengX. (2009). Liquid crystalline phases self-organized from a surfactant-like ionic liquid C_{16}mimCl in ethylammonium nitrate, *J. Phys. Chem B*, 113,7, 2024-2030, 1520-6106

Zou, Y. H.; Liu, H. B.; Yang, L.; Chen, Z. Z. (2006). The influence of temperature on magnetic and microwave absorption properties of Fe/graphite oxide nanocomposites. *J. Magn. Magn. Mater*, 302, 2, 343-347, 0304-8853

2 Chapter

IONIC LIQUIDS IN BIPHASIC ETHYLENE POLYMERIZATION

1. INTRODUCTION: AN OVERVIEW

In recent years, there has been a significant rise in the interest in ionic liquids as multifunctional reaction media for diverse synthetic processes. The appeal of ionic liquids is primarily attributed to their distinctive physical and chemical features, which may be customized through the careful selection of cations and anions. These characteristics enable the utilization of ionic liquids as solvents for a wide array of organic, inorganic, and organometallic compounds, encompassing catalysts and co-catalysts, across diverse reaction types. The polar yet weakly coordinating nature of ionic liquids allows their use as media in catalysis for immobilizing diverse catalyst precursors containing transition metals. Ionic liquids have recently been seen as an environmentally friendly alternative to traditional organic solvents. (Olivier-Bourbigou, 2002; Wasserscheid et al., 2004; Wilkes, 2004; Welton, 2004; Kubisa, 2004; Jain, 2005 ; Parvulescu et al., 2007; Chowdhury et al., 2007; Ochędzan-Siodłak, 2009).

Noteworthy among the several processes conducted in an ionic liquids milieu are the dimerization and oligomerization reactions catalyzed by nickel complexes. In most instances, enhancements in the activity and selectivity of the processes, together with the mitigation of unwanted side reactions, were attained. The excellent solubility of transition metal compounds in the ionic liquid phase, coupled with the limited miscibility of the resultant olefins, typically necessitates the reaction to be conducted in a biphasic manner. The result forms a distinct phase above the ionic liquid phase that contains the dissolved catalyst. Consequently, it becomes feasible to efficiently isolate the product from the reaction mixture, reduce the waste of the costly transition metal complex, and reutilize the catalyst (recycling). (Chauvin et al.,1990; 1995; 1997; Einloft et al. 1996; Simon et al., 1998; Ellis et al., 1999; Pinheiro et al., 2001; Wasserscheid et al., 2001; Bernando-Gusmão et al., 2003; Wasserscheid et al., 2004).

Prominent among the several processes occurring in an ionic liquid environment are the dimerization and oligomerization events mediated by nickel complexes. In most cases, improvements in the activity and selectivity of the processes, together with the reduction of undesirable side reactions, were achieved. The superior solubility of transition metal compounds in the ionic liquid phase, along with the restricted miscibility of the resulting olefins, generally requires the reaction to be performed in a biphasic manner. The outcome creates a separate phase above the ionic liquid phase that encompasses the dissolved catalyst. Thus, it becomes possible to effectively separate the product from the reaction mixture, minimize the waste of the expensive transition metal complex, and recycle the catalyst. (MAO). Furthermore, the homogenous systems reveal low stabilities during the polymerisation reactions, reaction fouling, and a lack of control of the polymer morphology (Kaminsky, Laban, 2001; Ewen, 2000).

To overcome these undesirable features, the metallocene systems were subjected to an Attachment on a stable substrate, such as SiO2, Al2O3, MgCl2, and other polymeric substances. The achieved heterogenization of the metallocene systems enables control over product morphology, eliminates reactor fouling, and reduces the quantity of the MAO activator needed. Regrettably, the attachment of the catalyst to the solid support leads to a reduction in activity (10-60%) relative to the homogeneous counterpart. The process of heterogenization is technically challenging, time-consuming, and results in a polymer product with significant inhomogeneity. (Chen, 2000; Hlatky, 2000; Severn et al., 2005; Ochędzan-Siodłak, Nowakowska, 2005).

Consequently, it is imperative to explore novel, unconventional methodologies for olefin polymerization utilizing metallocene catalysts. The immobilization of the metallocene catalyst in ionic liquids may serve as a viable alternative to its heterogenization on a solid support.

2. ETHYLENE POLYMERISATION

The ethylene polymerisation performed using the [C_2-mim][$AlCl_4$] ionic liquid as a medium of the $TiCl_4$ catalyst activated by the $AlEtCl_2$ alkylaluminium compound was described for the first time by Carlin and Osteryoung. The polymerisation was carried out in the single phase mode (in the ionic liquid). Although only small amounts of the polyethylene product were obtained (0.23 kgPE/(molTi · 10 min)), the results indicated that ionic liquids can be applied to olefin polymerisation (Carlin et al., 1990). It was further found that amongst the Cp_2MCl_2 metallocene catalysts (M = Ti, Hf, Zr) activated by the $AlCl_{3-x}R_x$ alylaluminium compounds (R = Me, Et), only the titanocene catalyst turned out to be active (Carlin, Wilkes, 1990).

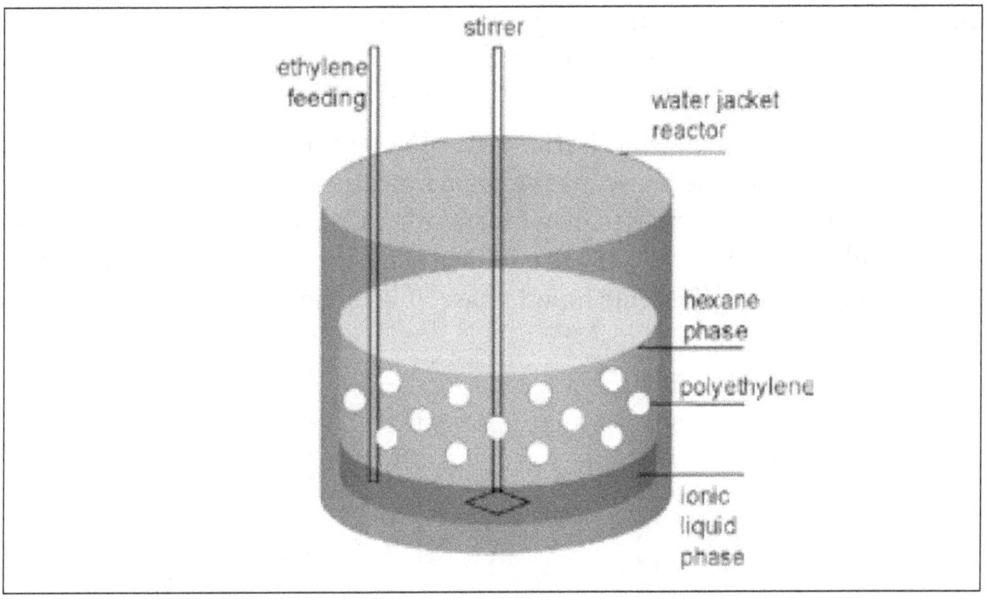

Fig. 1. The schematic representation of the studied biphasic ionic liquid/hexane polymerisation.

In our work, the polymerisation reaction was performed in the biphasic mode (Figure 1). The polymerisation of the simplest olefin, ethylene, was investigated as a model. The lower phase consisted of the ionic liquid, in which the Cp_2TiCl_2 catalyst and the alkylaluminium activators were dissolved. Hexane constituted the upper phase. For the ionic liquids, 1-*n*-alkyl-3-methylimidazolium and 1-*n*-alkyl-4-methylpyridinium chloroaluminates substituted by alkyl chains of various length (n = butyl, hexyl, octyl) were applied (Ochędzan-Siodłak et al., 2007; 2008; 2009).

2.1 Biphasic process

For all biphasic ionic liquid/hexane polymerisation experiments performed, the following facts have been observed. Two phases, hexane and ionic liquid, are clearly visible from the beginning to the end of the reaction. At the beginning of the reaction the ionic liquid phase becomes white and swells considerably as the polyethylene appears. At the same time, the hexane phase remains colourless and transparent. After 5-10 minutes, the hexane phase becomes a white suspension as the polyethylene is progressively transferred from the ionic liquid phase. The polyethylene is obtained in form of a powder. It is white when obtained from the hexane phase and off-white when obtained from the ionic liquid phase. The polyethylene from the hexane phase can be easily isolated and subjected to analysis.

It is very important for the biphasic catalytic process to maintain the catalyst in the selected phase. This has been shown by experiments where the ionic liquid phase contained the Cp_2TiCl_2 catalyst and the alkylaluminium compound was washed with hexane. Catalyst leakage was not detected, which indicates that the titanocene is firmly immobilised in the ionic liquid phase and remains during the entire duration of the polymerisation (Ochędzan- Siodłak, Pawelska, 2008).

During the course of the polymerisation reaction, the polyethylene is progressively transferred to the hexane phase. At the same time, the amount of polyethylene in the ionic liquid phase decreases, however, the total polyethylene yield increases with the polymerisation time (Figure 2). This confirms that the polymerisation reaction takes place in the ionic liquid phase. It also indicates that the catalyst is stable.

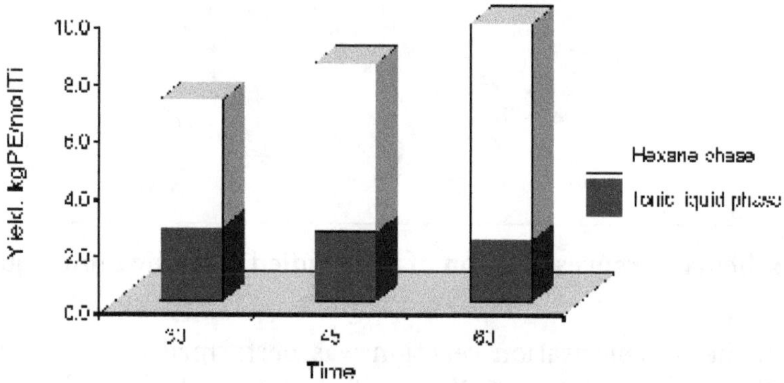

Fig. 2. The influence of the reaction time on the PE yield in the biphasic ethylene polymerisation. Polymerisation conditions: ionic liquid – [C_4-mim][$AlCl_4$], catalyst – Cp_2TiCl_2, activator–$AlEt_2Cl$, Al/Ti = 33, standard pressure, room temperature.

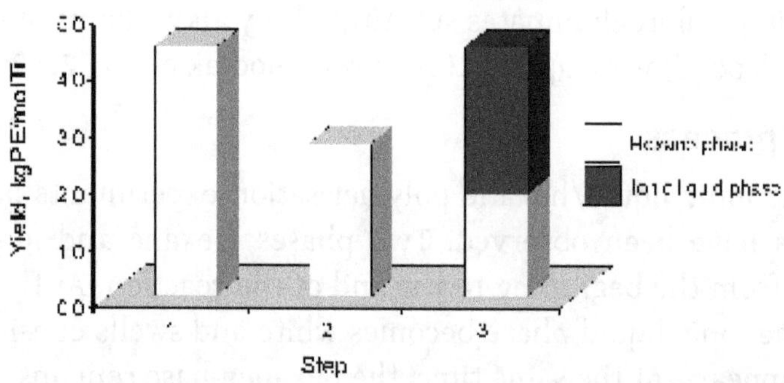

Fig. 3. The 3-step catalyst recycling in the biphasic ethylene polymerisation. Polymerisation conditions: ionic liquid – [C_4-mim][$AlCl_4$], catalyst – Cp_2TiCl_2, activator – $AlEt_2Cl$, Al/Ti = 33, standard pressure, room temperature, reaction time – 30 min.

The stability of the catalyst was also proved by the re-use of the ionic liquid phase containing the catalyst and the activators in the consecutive polymerisation reactions (Figure 3). The experiments consisted of three cycles, which were successfully performed, although the polymerisation yield gradually decreased.

The catalyst was maintained in the ionic liquid phase whereas the polyethylene product was removed after each cycle with the hexane phase. Thus, catalyst recycling is possible. Also, it seems to be a very important way to increase catalyst performance in the biphasic processes (Ochędzan-Siodłak et al., 2008).

2.2 Activators

The great advantage of ionic liquids over classic solvent is the fact that their properties can be tailored by the choice of suitable cations and anions. Applying a titanocene catalyst to the ethylene polymerisation limits the choice to the chloroaluminate ionic liquids because so far it is only this anion which makes the use of the alkylaluminium compounds as necessary catalyst activators possible. It should be noted that introduction of the alkylaluminium compound inevitably influences the anionic part of the ionic liquid. The kind and the amount of the alkylaluminium compound used are important because a range of mixtures of the chloroaluminate anions can be created. Also, the Lewis acidity of the ionic liquid is changed. This has considerable influence on the course and the performance of the ethylene polymerisation carried out in the biphasic mode.

The following alkylaluminium compounds: $AlEtCl_2$, $AlEt_2Cl$, $AlEt_3$, and MAO were investigated as activators of the Cp_2TiCl_2 catalyst (Figure 4). The best results were obtained for the compounds containing the chlorine atom, $AlEtCl_2$ and $AlEt_2Cl$. The more chlorine atom the activator contains, the higher the total yield, with the greater amount of polyethylene product transferred to the hexane phase. Interestingly, the worst results were obtained using methylaluminoxane (MAO), which is the best activator for the homogeneous metallocene catalysts. Therefore, the application of the ionic liquid to the ethylene polymerisation eliminates the need for the expensive MAO as the activator, which can be superseded by traditional, cheaper alkylaluminium compounds (Ochędzan-Siodłak et al., 2007; 2008).

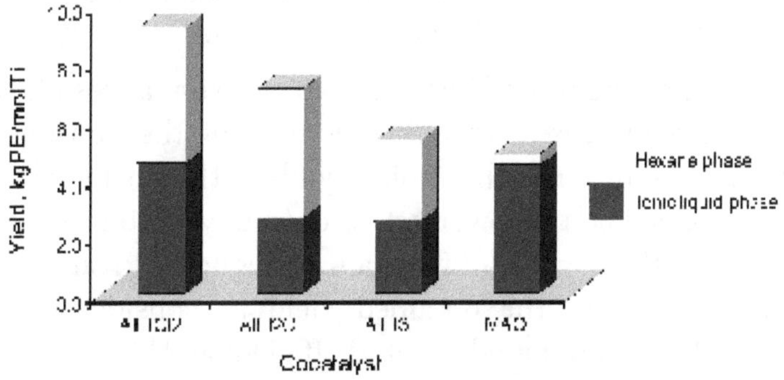

Fig. 4. The influence of various kinds of alkylaluminium activators on the performance of the biphasic ethylene polymerisation. Polymerisation conditions: ionic liquid - [C_4- mim][$AlCl_4$], catalyst - Cp_2TiCl_2, Al/Ti = 33, standard pressure, room temperature, reaction time - 30 min.

2.3 Imidazolium ionic liquids

The further investigations focused mainly on the modification of the cationic part of the ionic liquid. Substituting the imidazolium and pirydinium cations by alkyl chains of various lengths makes it possible to tailor some of the physical properties of the ionic liquids. The two most effective activators, $AlEtCl_2$ and $AlEt_2Cl$, were applied (Ochędzan-Siodłak et al., 2007; 2008; 2009). The study performed using 1-*n*-alkyl-3-methylimidazolium ionic liquids, $[C_n\text{-mim}][AlCl_4]$, show that the length of the alkyl chain at the ionic liquid cation and the concentration of the alkylaluminium compound influence the performance of the biphasic polymerisation reaction, i.e. the total polymerisation yield, the amount of the polyethylene product gathered in the hexane phase, and the phase separation (Figure 5). For each studied ionic liquid, there is an optimal concentration of the alkylaluminium compound, in which the yield of the reaction is the highest and the majority of the product transfers to the hexane phase. For $[C_4\text{-mim}][AlCl_4]$ and $[C_6\text{-mim}][AlCl_4]$, ionic liquids with shorter alkyl chains, the activator/catalyst molar ratio (Al/Ti) reaches optimum at the value 100. A higher ratio results in a decrease of the yield and the majority of the product remains in the ionic liquid phase. For $[C_8\text{-mim}][AlCl_4]$, the ionic liquid with a longer alkyl chain, the optimal activator/catalyst molar ratio is higher (Al/Ti = 133) resulting in a total PE yield 120 kgPE/molTi·h. An even higher yield can be obtained at the activator/catalyst molar ratios 167 and 200. However, the majority of the product remains in the ionic liquid phase, which is disadvantageous to the biphasic process. Therefore, if two criteria for the performance of the biphasic process are the amount of the product gathered in the hexane phase and the ratio between the amount of product in both phases, the ionic liquid with the longest alkyl chain is the better medium for the titanocene catalyst as it results in the greatest amount of the pure polymer product from the hexane phase. It was also found that prolongation of the reaction time (up to 120 min.) improves the reaction yield. This confirms the stability of the catalyst immobilised in the ionic liquid and makes it possible to obtain the greatest amount of the product from the hexane phase.

When the $AlEt_2Cl$ compound is used instead of $AlEtCl_2$, similar results were obtained (Figure 6). The yield increases with the increase of the concentration of the alkylaluminium compound. For the ionic liquids with shorter alkyl chains, the best results are obtained at the lower activator/catalyst molar ratio, whereas for the ionic liquid with the longer alkyl chain, a higher activator/catalyst molar ratio is required. However, the obtained yield is considerably lower, which indicates that the $AlEt_2Cl$ compound is not as efficient as $AlEtCl_2$.

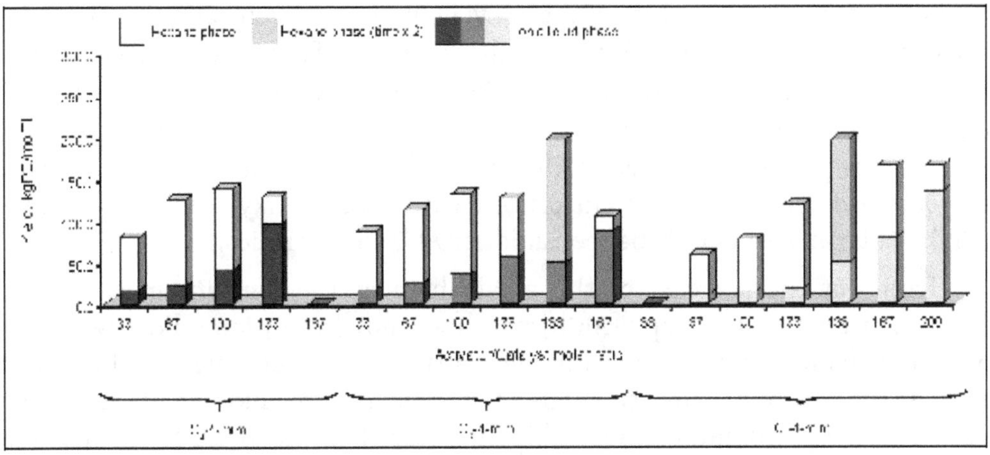

Fig. 5. The influence of the alkyl chain at the cation of the ionic liquids and the concentration of the $AlEtCl_2$ alkylaluminium activator on the performance of the biphasic ethylene polymerisation. Polymerisation conditions: ionic liquid - [C_n-mim][$AlCl_4$] (n = 4,6,8), catalyst - Cp_2TiCl_2, activator -$AlEtCl_2$, pressure - 0.5MPa, room temperature, reaction time - 60 min.

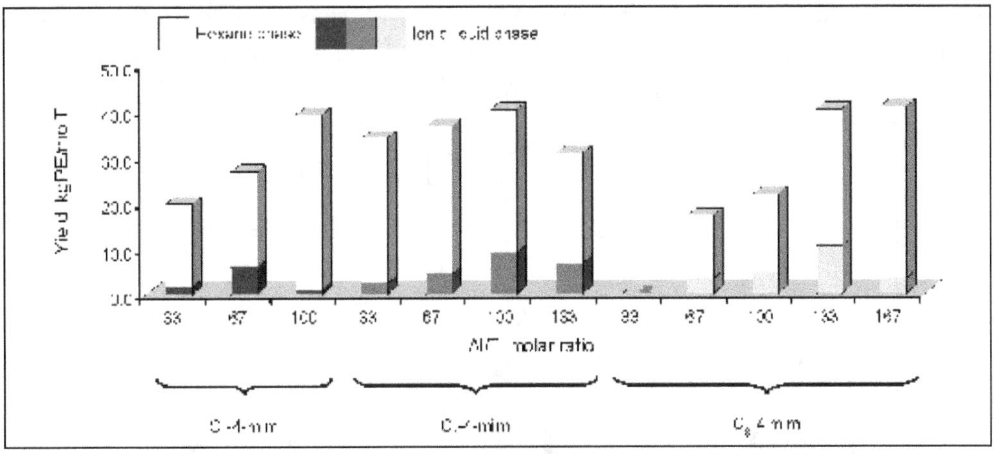

Fig. 6. The influence of the alkyl chain at the cation of the ionic liquids and the concentration of the $AlEt_2Cl$ alkylaluminium activator on the performance of the biphasic ethylene polymerisation. Polymerisation conditions: ionic liquid - [C_n-mim][$AlCl_4$] (n = 4,6,8), catalyst - Cp_2TiCl_2, activator - $AlEt_2Cl$, pressure - 0.5MPa, room temperature, reaction time - 60 min.

2.4 Pyridinium ionic liquids

Similar studies were carried out for the pirydinium ionic liquids. The pyridinium cation not only differs in its structure in comparison to the imidazolium cation, but also the position of the alkyl chain can be changed. 1-*n*-Alkyl-4-methylpyridinium chloroaluminates [C_n-4- mpy][$AlCl_4$] were investigated at similar polymerisation conditions as for the imidazolium ionic liquids (Figure 7). It was found that the yield increases with the increase of the alkylaluminium concentration. Nevertheless at higher activator/catalyst molar ratios, the majority of the product remains in the ionic liquids, which is a disadvantageous phenomenon. Only at a lower activator/catalyst molar ratio (Al/Ti = 67) is the majority of the polymer product gathered in the hexane phase. In contrast to the imidazolium ionic liquids, the best results were obtained for the pyridinium cations with shorter alkyl chains ([C_4-4-mpy][$AlCl_4$]). This can particularly be seen at lower Al/Ti molar ratios.

The change of the position of the alkyl chain at the pyridinium cation was investigated on the [C_8-4-mpy][$AlCl_4$] and [C_8-3-mpy][$AlCl_4$] ionic liquids. In both cases the polymerisation yield and the transfer of the polymer to the hexane phase is slightly better for the ionic liquid having the alkyl chain at the *para* position at the pyridinium cation. This indicates that the position of the alkyl chain at the ionic liquid cation influences the performance of the biphasic process, although the effect is moderate. Generally, the application of pyridinium ionic liquids to the biphasic polymerisation results in higher yields in comparison to imidazolium analogues at similar reaction conditions.

To improve the performance of the biphasic polymerisation, the reaction time was extended twice (up to 120 min.). Advantageous increases in the amount of the product in the hexane phase as well as in the total polymerisation yield were observed. Again, the best result was obtained for the [C_4-4-mpy][$AlCl_4$] ionic liquid with the shortest alkyl chain.

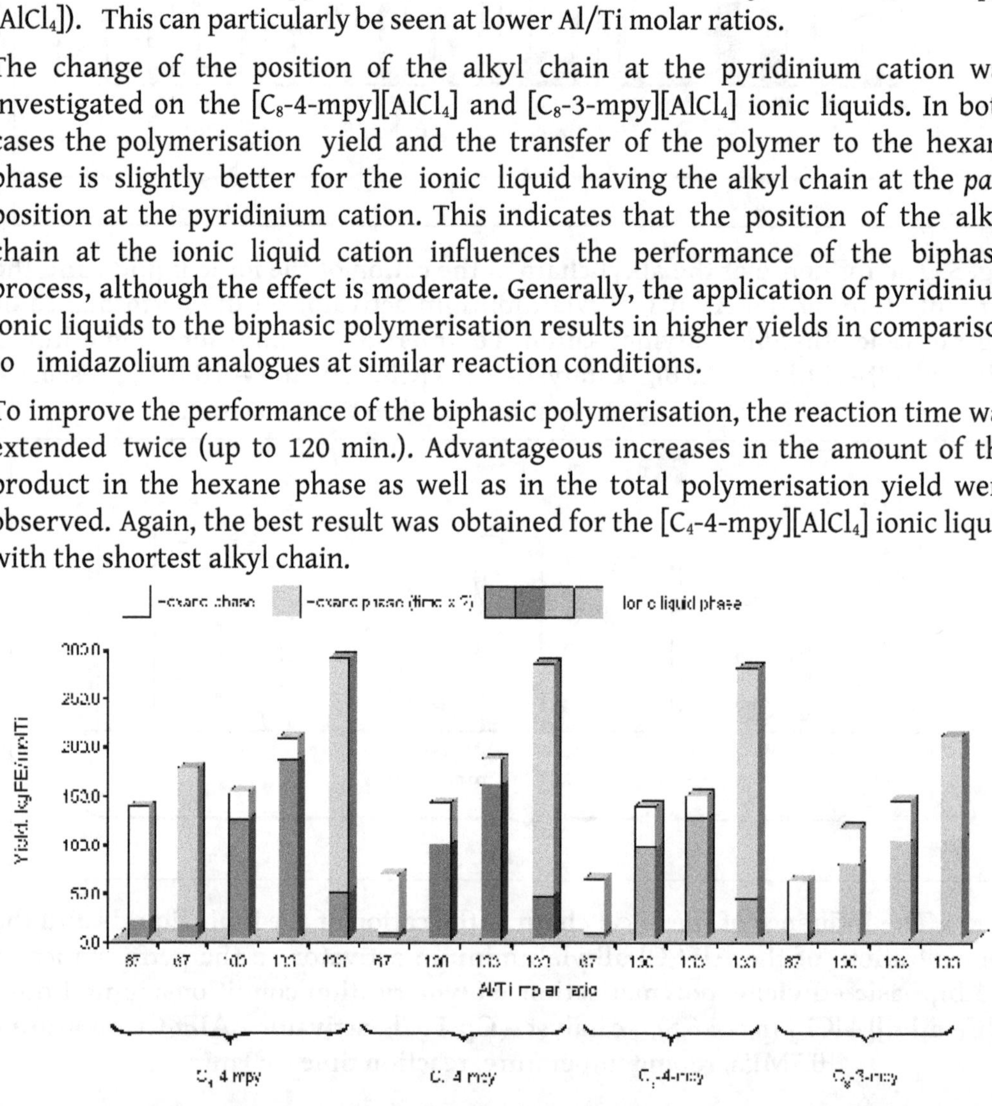

Fig. 7. The influence of the alkyl chain at the pyridinium cation and the concentration of the $AlEtCl_2$ alkylaluminium activator on the performance of the biphasic ethylene polymerisation. Polymerisation conditions: ionic liquids: [C_n-4-mpy][$AlCl_4$] (n = 4,6,8); [C_8-3-mpy][$AlCl_4$], catalyst – Cp_2TiCl_2, activator –$AlEtCl_2$, pressure – 0.5MPa, room temperature, reaction time – 60 min.

By contrast, the length of the alkyl chain at the pyridinium cations in comparison to imidazolium analogues has a negative effect, which can clearly be seen using the $AlEt_2Cl$ activator instead of $AlEtCl_2$ (Figure 8). Similarly to the imidazolium ionic liquids, the application of the $AlEt_2Cl$ activator results in a considerable decrease of the reaction yield. However, it should be noted that almost the entire amount of the polymer product is gathered in the hexane phase. This phenomenon is of great importance because it results in a product of high purity. Furthermore, the negligible amount of the product in the ionic liquid phase makes it possible to apply this phase in consecutive multi-step reactions. Thus, the recycling of both the catalyst and the ionic liquid can easily be done. It seems that the $AlEt_2Cl$ activator could be successfully applied in a continuous flow process, whereas the $AlEtCl_2$ activator is better in a single reaction step.

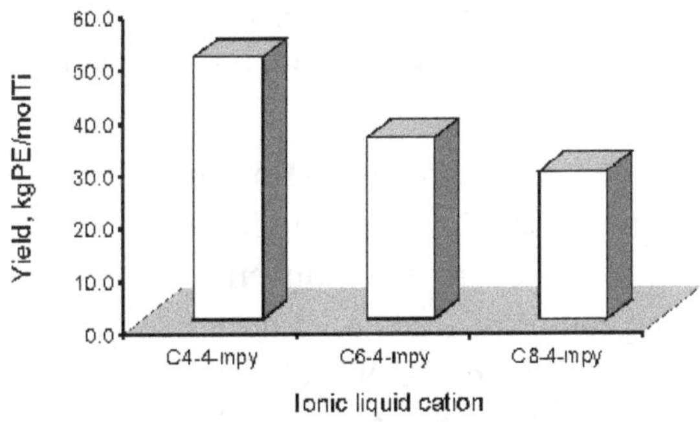

Fig. 8. The influence of the alkyl chain at the pyridinium cation and the concentration of the $AlEt_2Cl$ alkylaluminium activator on the performance of the biphasic ethylene polymerisation. Polymerisation conditions: ionic liquid – $[C_n\text{-4-mpy}][AlCl_4]$ (n = 4,6,8), catalyst – Cp_2TiCl_2, activator – $AlEt_2Cl$, Al/Ti = 133, standard pressure – 0.5MPa, room temperature, reaction time – 60 min.

2.5 Ionic liquids properties

The presented results show that the structure of the ionic liquid cation has a considerable influence on the performance of the biphasic polymerisation. The structure of the cation influences the physical properties of the ionic liquids, such as density and viscosity (Olivier- Bourbigou, 2010).

Generally, an increase in the alkyl chain length leads to a decrease in the density and an increase in the viscosity of the ionic liquid. In biphasic ethylene polymerization the viscosity and the density of the ionic liquid play a key role in the dispersion of both phases, which influences the mass transfer of the product from the ionic liquid phase to the hexane phase. This can be seen the most clearly in the case of the polymerisations carried out in the imidazolium ionic liquids (Figure 5). In the case of the pyridinium analogues, the mass transfer seems to be more hindered, and this effect is much smaller.

For polymerisation performed at lower alkylaluminium (activator) concentration, the majority of the product is gathered in the hexane phase. This can be explained by the presence of a relatively small number of active sites. In consequence, a smaller amount of the polymer is produced, but this does not exceed the amount of product which the system is able to transfer from the ionic liquid phase where the product is created to the hexane phase where the product is gathered. For the polymerisation performed at a higher alkylaluminium concentration, a greater number of active sites appears, which results in a greater amount of the polymer produced than the system is able to transfer to the hexane phase. Thus, a considerabe amount of the product remains in the ionic liquid phase. This phenomenon is observed for all the ionic liquids studied. The polymerisation performed using the $AlEt_2Cl$ compound support this conclusion. $AlEt_2Cl$ is an activator of lower efficiency than $AlEtCl_2$. It creates a smaller number of active sites, and thus, a smaller mass of the polymer product, but this is easily transferred to the hexane phase.

The optimum activator/catalyst molar ratio observed in the case of the polymerisation carried out in the imidazolium ionic liquids is another interesting phenomenon. Too low or too high a concentration of the alkylaluminium compound is disadvantageous and leads to a decrease of the reaction yield. This phenomenon is connected with the acidity of the ionic liquid phase. Carlin and Wilkes proved that only ionic liquids of the Lewis acid character are useful for the ethylene polymerisation reaction. The acid character of the chloroaluminate ionic liquid can be obtained by adding the correct amount of $AlCl_3$. Usually, during the synthesis of the ionic liquids from $AlCl_3$ and the given chloride of the organic cation, an excess of $AlCl_3$ is added $AlCl_3/[C_n\text{-mim}]Cl>1$). However, at a higher concentration of the $AlCl_4^-$ and $Al_2Cl_7^-$ anions, complexation of the titanocene catalyst to either the $Cp_2Ti(AlCl_4)_2$ or $Cp_2TiCl(Al_2Cl_7)$ forms can occur (Carlin, Wilkes, 1990).

$$Cp_2TiCl_2 + Al_2Cl_7 \leftrightarrow Cp_2TiCl(AlCl_4) + AlCl_4 \quad (1)$$

$$Cp_2Ti(AlCl_4) + Al_2Cl_7 \leftrightarrow Cp_2Ti(AlCl_4)_2 + AlCl_4 \quad (2)$$

$$Cp_2Ti(AlCl_4) + Al_2Cl_7 \leftrightarrow Cp_2Ti(Al_2Cl_7) + AlCl_4 \quad (3)$$

The proper acidity of the ionic liquid can be obtained using alkylaluminium compounds containing chlorine atoms. These compounds, such as the $AlEtCl_2$ applied in this work, lead to the creation of various anions in the reaction medium: $AlCl_4$, Al_2Cl_7, Al_2EtCl_6, $AlEtCl_3$, $Al_2Et_2Cl_5$. The optimum parameters of the chloroaluminate ionic liquid applied as a medium of the catalyst for the polymerisation depends on the amount and kinds of the anions. Higher concentrations of the $AlCl_4^-$ and $AlEtCl_3$ anions, of neutral character, considerably decrease the acidity of the chloroaluminate ionic liquid. In contrast, higher concentrations of the Al_2Cl_7

and Al_2EtCl_6 anions result in the ionic liquid being too acidic, which also leads to a decrease of the polymerisation yield. Therefore, the judicious selection of the kind and amount of the alkylaluminium Lewis acid is very important when tailoring the acidity of the ionic liquid. The presented results indicate that the application of the $AlEtCl_2$ compound, which reveals a higher Lewis acidity in comparison to $AlEt_2Cl$, results in a better performance of the ionic liquid in the polymerisation reaction (Chauvin, 1990; Thiele, de Souza, 2007).

The optimum activator/catalyst molar ratio not only enables the reduction of the titanium compound, but also creates the appropriate environment for the polymerisation reaction, which results from the correct proportion of the chloroaluminate anions being present in the ionic liquid medium.

It is known that in the homogeneous metallocene systems, the active site is represented by the cationic form of the metallocene; Cp_2MR^+. Thus, the necessary condition for the polymerisation reaction is:

i. detachment of the $AlCl_4$ group and creation of the unsaturated Cp_2TiR^+, according to the equation:

$$Cp_2TiR(AlCl_4) \leftrightarrow Cp_2TiR^+ + AlCl_4 \quad (4)$$

ii. exchange of the $AlCl_4$ $Cp_2TiR(CH_2CH_2)^+$complex:

group by ethylene molecule and creation of the

$$Cp_2TiR(AlCl_4)+CH_2CH_2 \leftrightarrow Cp_2TiR(CH_2CH_2)^+ +AlCl_4 \quad (5)$$

In both cases, the centre of the polymerisation reaction is the cationic complex, created by the break of the M-Cl bond, which is stronger for Zr and Hf and weaker for Ti. Therefore, the inactivity of the Zr and Hf compounds results from the stronger M-Cl bonds, which make it impossible to create the active cationic form [Cp_2MR^+] (Carlin, Wilkes, 1990). It should be noted, however, that the chloroaluminate ionic liquid participates in the creation of the active centre, which is supported by literature (Chauvin, 1997; Simon, 1998; Bernardo- Gusmao, 2003).

3. POLYETHYLENE PROPERTIES

The presented biphasic polymerisation experiments show that the polyethylene product is present in both the ionic liquid and hexane phases. According to the concept of biphasic polymerisation, the product gathered in the hexane phase is the most important. Therefore only the properties of this polyethylene have been analysed. The polyethylene from the ionic liquid phase is contaminated by the products of decomposition of the catalyst, activator, and ionic liquids, which means that it is impossible to determine the PE properties correctly.

The data presented in Table 1 show the influence of the imidazolium ionic liquids on the properties of the PE produced. The obtained polyethylene was linear in character. On average, 3.0-9.4 methyl groups on the 1000 methylene groups occur. The melting temperature (~133°C) is relatively narrow and typical for commercial HDPE (Vasilie , Pascu, 2005). The high crystallity degree (~86%) should be noted. This value varies depending on the length of the alkyl chain at the ionic liquid cation; the lowest is for [C_4-mim][$AlCl_4$] (~82%), and the highest for [C_8-mim][$AlCl_4$] (~87%). Another advantageous feature of the polyethylene product is a high value of the bulk density, which again varies depending on the ionic liquid applied. Again, the lowest value was obtained for [C_4-mim][$AlCl_4$] (~230 g/dm^3) and the highest for [C_8-mim][$AlCl_4$] (~470 g/dm^3). Such high values of the PE bulk density can be obtained only for the heterogeneous metallocene catalyst (Kang, 1999; Razavi, 2000; Jang, 2003; Wei, 2004; Ochędzan-Siodłak, 2005; Hammawa, 2006). In the case of the homogeneous systems, the bulk density usually does not exceed 100 g/dm^3 (Soares, 2000; Lee, 2000). The high values of the crystallity degree and bulk density of the polyethylene obtained show a high level of the molecular order and indicate good mechanical parameters of the polymer product. Nevertheless, the polyethylene obtained has a relatively small molecular weight (M_w) in the range of 70 000–170 000 g/mol. The molecular weight decreases with the increase of the activator/catalyst molar ratio. Therefore, an increase of the activator concentration results in shorter PE chains. It indicates that the alkylaluminium compound participates, most probably, in the chain termination reaction. The molecular weight distribution is monomodal (MWD from 4.3 to 7.2). This is characteristic for the polymer obtained using supported metallocene catalysts. The polyethylene obtained using the $AlEt_2Cl$ activator has a similar melting temperature, crystallity degree, molecular weight, molecular weight distribution, and linearity to the analogue polymer obtained using the $AlEtCl_2$ activator, regardless of the ionic liquid used. However, the bulk density is lower when $AlEt_2Cl$ is applied.

Table 1. Selected properties of the polyethylene produced using imidazolium ionic liquids in the biphasic polymerisation

Item	Activator/Catalyst molar ratio	T_m (°C)	Crystallinity (DSC) (%)	Mw ·10^{-3} (g/mol)	MWD	Bulk density (g/dm^3)	CH_3/1000CH_2
	[C$_4$mim]$^+$[$AlCl_4$]$^-$						
1	67	134	86.8	159.6	4.4	243	3.7
2	100	133	83.5	86.7	5.8	220	6.3
3	133	137	81.6	71.7	4.3	240	3.0
	[C$_6$mim]$^+$[$AlCl_4$]$^-$						
4	67	133	85.9	121.5	6.2	–	7.6
5	100	132	86.4	110.0	6.8	370	7.3
6	133	132	86.6	45.9	4.8	440	7.0

7[a]	100	133	85.5	116.3	4.3	170	9.4
$[C_8mim]^+[AlCl_4]^-$							
8	67	135	89.2	165.6	7.2	–	–
9	100	134	88.6	129.2	6.8	340	6.2
10	133	133	85.8	82.5	5.3	470	8.3
11[a]	100	134	80.5	77.8	5.4	160	7.0
[a] $AlEt_2Cl$ activator							

The application of the pyridinium ionic liquid in the biphasic polymerisation results in a polyethylene product with slightly different properties than that obtained using the imidazolium ionic liquids (Table 2). It is a linear polyethylene and the number of branches (from 4.3 to 12.9 of the CH_3 groups per 1000 CH_2 groups) is comparable for all the samples analysed. The number of branches increases with an increase in the activator concentration, regardless of the ionic liquid used. The melting temperatures (127-130°C) are lower but the molecular weights (M_w = 100 000–270 000 g/mol) are higher than for the PE samples obtained in the biphasic polymerisation using imidazolium ionic liquids. It should be noted that the lowest M_w have PE samples obtained using $[C_8\text{-3-mpy}][AlCl_4]$. The molecular weight distributions (MWDs) of the studied samples are relatively broad, from 3.1 to 6.9. Again, the distinguishing feature of the PE produced is the high bulk density, which reaches up to a value of 520 g/dm^3, which corresponds to that of the polyethylenes obtained over supported metallocene catalysts. A decrease of the bulk density is observed with the increase of the activator/catalyst molar ratio. The position of the alkyl chain at the pyridinium cation has an influence on the bulk density. The polyethylene produced using the $[C_8\text{-4-mpy}][AlCl_4]$ ionic liquid has a higher bulk density than that obtained using $[C_8\text{-3-mpy}][AlCl_4]$ (Al/Ti=67 and 100).

All the studied PE samples reveal an exceptionally high crystallinity degree, with an average value of 93%, as determined by the DSC method. Such a high crystallinity is atypical for PE obtained using a metallocene catalyst (Prasad, 1999; Wei, 2004) and it corresponds to that of the commercial HDPE (crystallinity = 77-90%) obtained using Ziegler- Natta catalysts (Doak, 1986).

The polyethylene samples obtained using the $AlEt_2Cl$ activator revealed considerable purity due to the efficient mass transfer to the hexane phase of the polymer produced. They reveal much lower molecular weights, bulk densities, and crystallinity – similar to those of the polyethylene obtained in the biphasic process using imidazolium ionic liquids (Ochędzan- Siodłak, 2008). Generally, the polyethylene obtained using $AlEtCl_2$ as the activator is characterised by better properties than that obtained using $AlEt_2Cl$.

Table 2. Selected properties of the polyethylene produced using pyridinium ionic liquids in the biphasic polymerisation

Item	Activator/Catalyst molar ratio	T_m (°C)	Crystallinity (DSC) (%)	Mw ·10-3 (g/mol)	MWD	Bulk density (g/dm3)	CH_3/1000CH_2
[C_4-4-mpy][$AlCl_4$]							
1	67	128	94.0	261.6	5.9	370	5.9
2	100	126	95.0	233.0	6.7	215	8.9
3	133	127	95.1	151.5	5.4	-	9.9
4[a]	133	129	94.2	142.9	6.7	165	9.1
[C_6-4-mpy][$AlCl_4$]							
5	67	129	94.9	260.0	5.5	444	5.2
6	100	128	93.1	229.3	6.7	283	7.1
7	133	126	95.1	150.9	6.1	136	12.9
8[a]	133	127	90.6	130.1	6.9	145	11.1
[C_8-4-mpy][$AlCl_4$]							
9	67	129	96.0	266.8	6.9	520	4.4
10	100	128	95.0	222.9	5.4	395	5.3
11	133	133	93.0	158.6	5.9	210	6.2
12[a]	133	130	87.4	122.5	5.6	137	9.8
[C_8-3-mpy][$AlCl_4$]							
13	67	128	96.0	141.3	3.4	481	5.0
14	100	128	98.0	138.2	3.9	357	6.0
15	133	127	98.0	111.5	3.1	-	8.1

[a] $AlEt_2Cl$ activator

The presented results show that the polyethylene product obtained in the biphasic process joins the properties of the polyethylenes obtained using the homogenous and heterogeneous metallocene catalysts (a low molecular weight as well as broad molecular weight distribution and high bulk density, respectively). Moreover, the product is characterised by a high crystallity degree resulting from the high level of the macromolecular order, which is particularly observed for the samples obtained from the pyridinium ionic liquids. Properties such as molecular weight, bulk density, and crystallity degree can be modified by both the kind of the alkyl chain at the ionic liquid cation and the concentration of the alkylaluminium compound.

4. CONCISE SUMMARY

The findings reported elucidate the biphasic polymerization of ethylene utilizing a titanocene catalyst in an ionic liquid/hexane system. Imidazolium and pyridinium chloroaluminates were utilized as ionic liquids. The catalyst was determined to be stable during the polymerization reaction and securely immobilized within the ionic liquid. Consequently, utilizing ionic liquids as catalytic media presents a compelling alternative to the heterogenization of catalysts on solid supports. The examined biphasic mode facilitates the recycling of both the catalyst and the ionic liquid phase. Conventional alkylaluminium compounds serve as more appropriate activators for the titanocene catalyst compared to methylaluminoxane, which is regarded as the most efficient activator for metallocene catalysts. Polyethylene (PE) generated in the ionic liquid phase is gradually transferred to the hexane phase. Consequently, the polymer exhibits great purity and may be readily isolated from the reaction mixture using a straightforward decantation method.

The performance of biphasic polymerization and the qualities of the polymer product can be influenced by altering the characteristics of the ionic liquid. This can be achieved by altering the cationic and anionic components of the ionic liquid. This study examined 1-n-alkyl-3-methylimidazolium and 1-n-alkyl-4-methylpyridinium cations with alkyl chains of different lengths. The chloroaluminate anions can be altered by selecting the kind and concentration of the alkylaluminium compound, which also serves as the catalytic activator. The alkylaluminium compounds examined were $AlEt_3$, $AlEt_2Cl$, $AlEtCl_2$, and MAO. The results obtained using compounds containing the chlorine atom, AlEtCl2 and AlEt2Cl were better than when methylaluminoxane was applied. The AlEtCl2 compound turned out to be the most suitable for the biphasic ethylene polymerisation as regardless of the ionic liquid cation used, the highest yields were obtained. It should be noted that the reaction yield depends on the concentration of the alkylaluminium compound. This can be explained by the balance of the various chloroaluminate anions created, which influence the acidity of the ionic liquids, which is necessary for the polymerisation reaction. The concentration of the alkylaluminium compound, described in the from of activator/catalyst molar ratio (Al/Ti), determines the amount of the polymer product gathered in the hexane phase. Generally, the lower the Al/Ti molar ratio, the greater amount of product present in the hexane phase and a correspondingly smaller amount remaining in the ionic liquid phase. Relatively low activator/catalyst molar ratios result in a low concentration of the active sites, and the amount of the polymer produced does not exceed the ability of the biphasic system to transfer the product from the ionic liquid to the hexane phase. Therefore, for each ionic liquid an optimum concentration of the alkylaluminium compound can be determined. For the imidazolium ionic liquids, the longer the alkyl chain at the cation the better the mass transfer of the product to the hexane phase. The length of the alkyl chain at the imidazolium cation changes the physical properties of the ionic liquids, particularly density and viscosity, which influence the dispersion and separation of both phases. For the pyridinium ionic liquids such a tendency is not observed. The transfer of the product to the hexane phase is best with a shorter alkyl chain at the pyridinium cation and only at a lower concentration of the alkylaluminium compound. Although the application of pyridinium ionic liquids results in better total yield (from both the ionic liquid and hexane phases), the mass transfer of the product to the hexane phase is hindered at higher $AlEtCl_2$ concentrations. This can be overcome, however, by extending the reaction time.

The analysis of the physical and molecular properties of the polyethylene product obtained in the biphasic process shows that regardless of the ionic liquid used, the product displays unique properties, particularly a non-typical, very high crystallinity degree - up to 95%. It is also possible to control the selected properties of the polymer product, e.g. the melting point, crystallinity degree and bulk density (Ochędzan-Siodłak et al., 2007; 2008; 2009; 2010). The results demonstrate that biphasic polymerization utilizing ionic liquid as a catalyst medium is a compelling and advantageous choice in the context of contemporary green chemistry. This enables the production of a polymer with desirable properties, which can be customized by altering the parameters of the ionic liquid. Subsequent research should concentrate on the eradication of the detrimental characteristics of biphasic polymerization. Enhancing the catalytic efficacy of the system, optimizing product mass transfer, and regulating polymer morphology are of paramount importance. The supported ionic liquid phase (SILP) process, which involves anchoring an ionic liquid containing an organometallic catalyst onto a solid carrier, presents a viable approach to meet these requirements (Valkenberg et al., 2001; 2002; Mehnert, 2005; Riisager et al., 2006). This approach enables simultaneous control of the polymerization reaction and the physical and chemical properties of the polymer product. The quantity of the ionic liquid phase can be significantly minimized, enhancing catalytic performance while preserving catalyst reusability and facilitating separation procedures. (Gu, Li, 2009).

REFERENCES

Bernando-Gusmão, K., Queiroz, L.F.T., de Souza, R.F., Leca, F., Loup, C., Réau, R. (2003). Biphasic oligomerisation of ethylene with nickel-1,2-diiminophosphorane complexes immobilized in 1-n-butyl-3-methylimidazolium organochloroaluminate. *J. Catal.*, Vol., 219, pp. 59-62.

Carlin, R.T., Osteryoung, R.A., Wilkes, J.S., Rovang, J. (1990). Studies of titanium(IV) chloride in a strongly Lewis acidic molten salt: Electrochemistry and titanium NMR and electronic spectroscopy. *Inorg. Chem.*, Vol., 29, pp. 3003-3009.

Carlin, R.T., Wilkes, J.S. (1990). Complexation of Cp_2MCl_2 in a chloroaluminate molten salt: relevance to homogeneous Ziegler–Natta catalysis. *J. Mol. Catal.*, Vol., 63, pp. 125-129.

Chauvin, Y., Einloft, S., Olivier, H. (1995). Catalytic dimerization of propene by nickel- phosphine complexes in 1-butyl-3-methylimidazolium chloride/$AlEt_xCl_{3-x}$ (x = 0, 1) ionic liquids. *Ind. Eng. Chem. Res.*, Vol., 34, pp. 1149-1155.

Chauvin, Y., Gilbert, B., Guibard, I. (1990). Catalytic dimerization of alkenes by nickel complexes in organochloroaluminate molton salts. *J. Chem. Soc. Chem. Commun.*, Vol., 23, pp. 1715-1716.

Chauvin, Y., Olivier, H., Wyrvalski, C.N., Kimon, L.C., Souza, R.F. (1997). Oligomerization of n-butenes catalyzed by nickel complexes dissolved in organochloroaluminate ionic liquids. *J. Catal.*, Vol., 165, pp.275-278.

Chauvin, Y., Olivier, H., Wyrvalski, C.N., Kimon, L.C., Souza, R.F. (1997). Oligomerization of n-butenes catalyzed by nickel complexes dissolved in organochloroaluminate ionic liquids. *J. Catal.*, Vol., 165, pp. 275-278.

Chien, J.C.W. (1999). Supported metallocene polymerization catalysis. *Topics in Catal.*, Vol., 7, pp. 23-36.

Chowdhury, S., Mohan, R.S., Scott, J.L. (2007). Reactivity of ionic liquids. *Tetrahedron*, Vol., 63, pp. 2363-2389.

Doak, K.W. (1986). *Encyclopedia of polymer science and engineering*, vol. 6. J.I. Kroschwitz (Ed.), p. 384, Willey, New York.

Einloft, S., Dietrich, F.K., de Souza, R.F., Dupont, J. (1996). Selective two-phase catalytic ethylenedimerization by Ni complexes/$AlEtCl_2$ dissolved in organoaluminate ionic liquids. *Polyhedron*, Vol., 15, pp. 3257- 3259.

Ellis, B., Keim, W., Wasserscheid, P. (1999). Linear dimerisation of but-1-ene in biphasic mode using buffered chloroaluminate ionic liquid solvents. *Chem. Commun.*, pp. 337-338.

Ewen, J.A. (2000). Metallocene polymerization catalysts: Past, Present and Future, In: *Metallocene-based polyolefins*, J. Scheirs, W. Kaminsky (Ed.), pp. 3-31, John Wiley, Sons, Ltd., Chichester.

Gu, Y., Li, G. (2009). Ionic Liquids-Based Catalysis with Solids: State of the Art. *Adv. Synth. Catal.*, Vol., 351, pp. 817–847.

Hammawa, H., Wanke, S.E. (2006). Gas-phase olefin polymerization over supported metallocene/MAO catalysts: influence of support on activity and polydispersity. *Polym. Int.*, Vol., 55, pp. 426-434.

Hlatky, G.G. (2000). Heterogeneous single-site catalysts for olefin polymerization. *Chem. Rev.*, Vol., 100, pp. 1347-1376.

Jain, N. Kumar, A. Chauchan, S. Chauchan, S.M.S. (2005) Chemical and biochemical transformations in ionic liquids. *Tetrahedron*, Vol., 61, pp. 1015-1060.

Jang, Y-J., Nenov, N., Klapper, M., Muller, K. (2003). Organic nanoparticles with polypropyleneoxide chains as support for metallocene catalysts: ethylene homopolymerization and ethylene α-olefin copolymerization. *Polymer Bull.*, Vol., 50, pp. 343-350.

Kaminsky, W., Laban, A. (2001). Metallocene catalysis. *Appl. Catal. A: Gen.*, Vol., 222, pp. 47-61. Kang, K.K., Oh, J.K., Jeong, Y.T., Shiono, T., Ikeda, T. (1999). Highly active $MgCl_2$-supported $CpMCl_3$ (M= Ti, Zr) catalysts for ethylene polymerization. *Macromol. Rapid Commun.*, Vol., 20, pp. 308-311.

Kubisa, P. (2004). Application of ionic liquids as solvents for polymerization processes. *Prog. Polym. Sci.*, Vol., 29, pp. 3-12.

Lee, B.Y., Oh, J.S. (2000). Preparation of anchored metallocene complexes on dehydroxylated silica and their use in the polymerization of ethylene. *Macromolecules*, 33, 3194-3195.

Mehnert, C.P. (2005). Supported ionic liquid catalysis. *Chem. Eur. J.*, Vol., 11, pp. 50-56. Ochędzan-Siodłak, W. (2009). Ionic liquids in olefins oligomerization and polymerization (in Polish). *Wiadomości Chemiczne* (Wrocław), Vol., 63, pp. 776-801.

Ochędzan-Siodłak, W., Dziubek, K., Czaja, K. (2009). Comparison of imidazolium and pyridinium ionic liquids as mediums of the titanocene catalyst for biphasic ethylene polymerization. *Polimery (Warsow)*, Vol., 54, pp. 501-506.

Ochędzan-Siodłak, W., Dziubek, K., Siodłak, D. (2008). Biphasic ethylene polymerisation using 1-n-alkyl-3-methylimidazolium tetrachloroaluminate ionic liquid as a medium of the Cp_2TiCl_2 titanocene catalyst. *Eur. Polym. J.*, Vol., 44, pp. 3608-3614.

Ochędzan-Siodłak, W., Nowakowska, M. (2005). Heterogeneous zirconocene catalyst on magnesium support $MgCl_2(THF)_2$ modified by $AlEt_2Cl$ for ethylene polymerisation *Eur. Polym. J.*, Vol., 41, pp. 941-947.

Ochędzan-Siodłak, W., Pawelska, P. (2008). Chloroaluminate ionic liquids as a medium of titanocene catalyst activated by alkyl aluminium compounds for ethylene polymerisation. *Polimery (Warsow)*, Vol., 53, pp. 36-41.

Ochędzan-Siodłak, W., Sacher-Majewska, B. (2007). Biphasic ethylene polymerisation using ionic liquid over a titanocene catalyst activated by an alkyl aluminium compound. *Eur. Polym. J.*, Vol., 43, pp. 3688–3694.

Olivier-Bourbigou, H., Magna, L. (2002). Ionic liquids: perspectives for organic and catalytic reactions. *J. Mol. Catal. A: Chem.*, Vol., 182-183, pp. 419-437.

Olivier-Bourbigou, H., Magna, L., Morvan, D. (2010). Ionic liquids and catalysis: Recent progress from knowledge to applications. *Appl. Catal. A: Gen.*, Vol., 373, pp. 1-56.

Parvulescu, V.I., Hardacre, Ch. (2007). Catalysis in ionic liquids. Chem. Rev., 107, 2615–2665.

Pinheiro, M.F., Mauler, R.S., de Souza, R.F. (2001). Biphasic ethylene polymerisation with a diiminenickel catalyst. *Macromol. Rapid. Commun.*, Vol., 22, No., 6, pp. 425-428.

Prasad, A. (1999). *Polymer data handbook.* J.E. Mark (Ed.), p. 531, Oxford University Press: Inc., New York.

Razavi, A.C.R. (2000). Metallocene catalysts technology and environment. *Chemistry*, Vol., 3, pp. 615-625.

Riisager, A., Fehrmann, R., Haumann, M., Wasserscheid, P. (2006). *Topics Catal.*, 40, 91-102. Severn, J.R., Chadwick, J.C., Duchateau, R., Friedrich, S. (2005). "Bound but not gagged" -

Immobilizing single-site α-olefin polymerization catalysts. *Chem. Rev.*, Vol., 105, pp. 4073-4147.

Simon, L.C., Dupont, J., de Souza, R.F. (1998). Two-phase n-butenes dimerization by nickel complexes in molten salt media. *Appl. Catal. A: Gen.*, Vol., 175, pp. 215-220.

Soares, J.B.P., Kim, J.D.J. (2000). Copolymerization of ethylene and α-olefins with combined metallocene catalysts. I. A formal cwiterion for molecular weight bimodality. *J. Polym. Sci. Chem.*, Vol., 38, pp. 1408-16.

Thiele, D., de Souza, R.F. (2007). The role of aluminum species in biphasic butene dimerization catalyzed by nickel complexes. *J. Mol. Catal. A: Chem.*, Vol., 264, pp. 293-298.

Valkenberg, M.H., de Castro, C., Hölderich, W.F. (2001). Immobilisation of chloroaluminate ionic liquids on silica materials. *Topics in Catal.*, Vol., 14, pp. 139-144.

Valkenberg, M.H., de Castro, C., Hölderich, W.F. (2002). Immobilisation of ionic liquids on solid supports. *Green Chem.*, Vol., 4, pp. 88–93.

Vasilie C., Pascu M. (2005). *Practical Guide of Polyethylene*, p. 32, Rapra Technology Limited, ISBN: 1-85957-493-9, Shawbury, UK Wasserscheid, P., Eichmann, M. (2001). Selective dimerisation of 1-butene in biphasic mode using buffered chloroaluminate ionic liquid solvents—design and application of a continuous loop reactor. *Catal. Today*, Vol., 66, pp.309-316.

Wasserscheid, P., Hilgers, C., Keim, W. (2004). Ionic liquids-weakly-coordinating solvents for the biphasic ethylene oligomerisation to α-olefins using cationic Ni-complexes. *J. Mol. Catal. A: Chem.*, Vol., 214, pp.83-90.

Wei, L., Tang, T., Huang, B. (2004). Synthesis and characterization of polyethylene/clay- silica nanocomposites: a montmorillonite/silica-hybrid-supported catalyst and in situ polymerization. *J. Polym. Sci. Part A: Polym. Chem.*, Vol., 42, pp. 941-949.

Welton, J. (2004). Ionic liquids in catalysis. *Coord. Chem. Rev.*, Vol., 248, pp. 2459-2477. Wilkes, J.S. (2004). Properties of ionic liquid solvents for catalysis. *J. Mol. Catal. A: Chem.*, Vol., pp. 214, 11-17.

3 Chapter

SYNTHESIS OF IONIC LIQUIDS, SOLUBILITY IN WOOD, AND ITS APPLICATION IN GRAFT COPOLYMERIZATION WITH ACRYLAMIDE

1. INTRODUCTION: AN OVERVIEW

As awareness of sustainable environmental preservation grows, there is a strong desire to investigate natural, recoverable, and eco-friendly materials [1]. Wood is a commonly available natural resource that is environmentally sustainable and readily accessible. Cellulose, hemicellulose, and lignin constitute the primary components of wood powder that form the porous three-dimensional structures[2]. The limited solubility of wood powder in conventional solvents significantly restricts its use; therefore, it is crucial to identify environmentally safe solvents for wood powder.

The parameters for employing traditional methods to dissolve wood in solvents like phenol or polyhydric alcohols were exceedingly stringent, leading to the degradation of the wood's natural structure and a consequent loss of mechanical qualities [3, 4]. Ionic liquids, recognized as environmentally friendly solvents for cellulose, can partially dissolve wood powder and enhance reactivity due to structural flexibility[10-12]. The applications of ionic liquids encompass chemical synthesis, biological catalysis, and the creation of functional materials, among others [13-19]. Moreover, as a recently discovered approach extensively utilized in diverse chemical fields, microwave application can facilitate the dissolution of wood powder as an innovative and highly efficient method.

Numerous studies on lignin have been conducted over several decades in several countries, including the U.S. and Japan, highlighting the significance of developing environmentally sustainable polymer materials from natural resources. Polyacrylamide, renowned as the "hundred industry assistant," is utilized in diverse sectors. However, the production cost for its application is exceedingly high. The development of multi-functional lignin flocculants from affordable lumber is an unavoidable trend with significant potential[20, 21]. Concurrently, copolymerizations of lignin or cellulose with vinyl monomers in aqueous media have been documented [22-25], exhibiting drawbacks such as low polymerization conversion, unstable end products, and challenges in reaction control. To address the aforementioned drawbacks, we present the synthesis of an innovative flocculant composed of wood and acrylamide (AM) via graft copolymerization (AM-g-wood) in ionic liquids, utilizing wood powder as a substitute for lignin, thereby introducing novel concepts for the utilization of renewable natural resources. Furthermore, the graft copolymer of AM-g-wood, a novel modified flocculant with significantly reduced manufacturing costs in this study, has potential applications in various fields including water treatment, the petrochemical sector, mineral processing, and papermaking. The pertinent research is now underway.

2. EXPERIMENTAL

2.1 Materials

Materials: N-methylimidazole, N-ethyllimidazole, and 1-vinlylimidazole (Purity≥99%) were obtained from Linhai Kaile Chemical Plant; Allyl Chloride and Chloroethanol were from Sinopharm Chemical Reagent Co. Ltd. and Tianjin Guangfu Fine Chemical Research Institute separately; Potassium Persulfate (K2S2O8 or KPS) and acrylamide were received from Sinopharm Chemical Reagent Co. Ltd.; Benzene, Methanol, Acetone, Sodium hydroxide(NaOH), and Diethyl Ether were provide by Shanghai Suyi Chemical Reagent Co. Ltd.; Fir powder was donated by Huangshan Wood Factory. All chemicals and materials were used as received without further purification.

2.2 Synthesis of ionic liquids

Ionic liquids were synthesized from imidazole and allyl chloride (or chloroethanol), and their chemical structures were confirmed by FTIR and 1HNMR. Typical procedure is: into a round bottom flask, imidazole materials and allyl chloride (or chloroethanol) were added. The reaction was conducted under refluxing for 8-48 h. After reaction, the mixture was extracted with ether for several times to remove the residual imidazole. The ionic liquids can be obtained after being dried at 80°C (vacuum: 0.08MPa) for 24h (yield: >90%). The chemical structures of the ionic liquids were shown in Scheme 1.

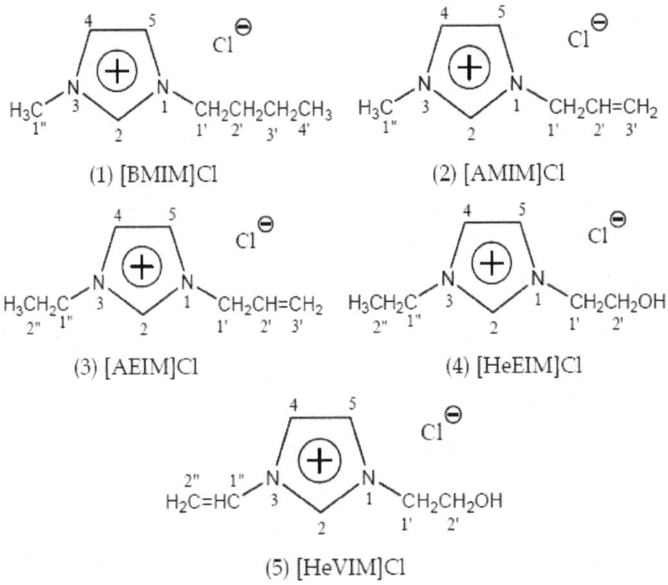

Scheme 1. Chemical structures and abbreviation of the examined ionic liquids

1. 1-butyl-3-methyl imidazolium chloride ([BMIM]Cl) was synthesized according to the general procedure provided by Lu et al.[26] with a slight of modification. FTIR (film): v=3422 (v O-H), 2918 and 2843 (v C-H), 1632 (v C=C), 1575 (v C=N), 1472 (β C-H), 1169 (δ C-H), 718 (δ imidazole ring); 1H NMR (DMSO, 25°C): δ=0.84 (3H, t, H-4'), 1.26 (2H, m, H-3'), 1.76 (2H, m, H-2'), 4.00 (3H, s, H-1"), 4.23 (2H, t, H-1'), 7.42 (1H, d, H-4), 7.58 (1H, d, H-5), 10.26 (1H, s, H-2).

2. 1-allyl-3-methyl imidazolium chloride ([AMIM]Cl) was synthesized according to the general procedure provided by Zhang et al.[9] with a slight of modification. FTIR (film): v=3409 (v O-H), 3083 and 2857 (v C-H), 1645 (v C=C), 1573 (v C=N), 1424 (β C-H), 1167 (δ C-H), 998 (ρ C-H), 950 (ρ C-C), 763 (δ imidazole ring); 1H NMR (DMSO, 25°C): δ=3.88 (3H, s, H-1"), 4.80 (2H, dd, H-1'), 5.22 (2H, m, H-3'), 5.81 (1H, m, H-2'), 7.26 (1H, d, H-4), 7.59 (1H, d, H-5), 10.08 (1H, s, H-2).

3. 1-allyl-3-ethyl-imidazolium chloride ([AEIM]Cl) was synthesized from allyl chloride 25.3g (0.33mol) and N-ethylimdiazole 28.8g (0.3mol). The reaction was conducted under refluxing for 8h at 70°C in water bath with stirring under the protection of N2. FTIR (film): v=3432 (v O-H), 3082 and 2980 (v C-H), 1645 (v C=C), 1563 (v C=N), 1449 (β C- H), 1163 (δ C-H), 997 (ρ C-H), 943 (ρ C-C), 761 (δ imidazole ring); 1H NMR (DMSO, 25°C): δ=1.45 (3H,t, H-2"), 4.20 (2H, q, H-1"), 4.77 (2H, dd, H-1'), 5.34 (2H, m, H-3'), 6.00 (1H, m, H-2'), 7.43 (1H,d, H-4), 7.47 (1H, d, H-5), 8.74 (1H, s, H-2).

4. 1-(2-hydroxylethyl)-3-ethyl imidazolium chloride ([HeEIM]Cl) was synthesized from chloroethanol 16.7mL (0.25mol) and N-ethylimdiazole 28.8g (0.3mol). The reaction was conducted under refluxing for 48h at 80°C in water bath with stirring. FTIR (film): v=3392 (v O-H), 3145 and 2994 (v C-H), 1632 (v C=C), 1566 (v C=N), 1449 (β C-H), 1165 (δ C-H), 1072 (v C-O), 750 (δ imidazole ring); 1H NMR (DMSO, 25°C): δ=1.43 (3H, t, H- 2"), 3.73 (2H, m, H-2'), 4.22 (2H, q, H-1"), 4.26 (2H, t, H-1'), 5.48 (1H, br, H-O-H), 7.81 (1H, d, H-4), 7.86 (1H, d, H-5), 9.39 (1H, s, H-2).
5. 1-(2-hydroxylethyl)-3-ethylene imidazolium chloride ([HeVIM]Cl) was synthesized from chloroethanol 16.7mL (0.25mol) and N-ethylene-imdiazole 26.3g (0.28mol). The reaction was conducted under refluxing for 10h at 60°C in water bath with stirring under the protection of N2. FTIR (film): v=3406 (v O-H), 2994 and 2878 (v C-H), 1651 (v C=C), 1572 (v C=N), 1495 (vas C-H), 1372 (β C-H), 1078 (v C-O), 962 (ρ C-H), 916 (ω O- H), 750 (δ imidazole ring); 1H NMR (DMSO, 25°C): δ=3.60 (2H, m, H-2"), 3.77 (2H, m, H-2'), 4.30 (2H, t, H-1'), 5.47 (1H, br, H-O-H), 7.40 (1H, m, H-1"), 7.97 (1H, d, H-5), 8.32 (1H, d, H-4), 9.80 (1H, s, H-2).

2.3 The pretreatment of wood powder

2.3.1 Extraction of wood powder with benzene and ethanol

Wood powder with particle size at 0.2mm was extracted with a mixture of benzene/ ethanol (volume ratio: 1/1) using soxhlet extractor [27]. Then, the powder was dried at 70°C (vacuum: 0.04MPa) for 12h in Vacuum drying oven.

2.3.2 Treatment of wood powder with 25% NaOH

Wood powder with particle size at 0.2mm was treated with 25% NaOH solution in air for 2h. Alternative conditions are: microwave treatment for 5min at 60°C (360w), or treated MCC in 25% NaOH solution (40mL) at 140°C for 2h in an airtight high pressure pot (60mL), which was placed in Electronic constant temperature oven. The mixture was filtered, neutralized and washed with water several times. Then the fir powder was dried at 70°C (0.04 MPa) for 12h in Vacuum drying oven.

2.4 Dissolution of fir powder by microwave heating

Into a flask, ionic liquid and pretreated fir powder (W0) above (weight ratio: 25:1) were introduced. The flask was placed in microwave oven (NJL 07-3) and heated at 90°C (400w) for 1h with stirring. The parallel experiment was copied. One of them was shifted to polymerization tube for future use. The other mixture was allowed to cool to room temperature followed by the centrifuge separation (2500r/ min, 5min). The finial fir powder was dried at 70°C (0.04Mpa) for 12h in order to calculate dissolution rate [11].

2.5 Synthesis and purification of AM-g-Wood

2.5.1 Synthesis of AM-g-Wood

Into a polymerization tube which was placed in water bath at 45°C, the mixture of fir powder/ionic liquid treated by microwave was added. Then air was purged out by N2 stream before the introduction of initiator K2S2O8 (I/M: 1/1000). AM was added dropwise into the mixture, and the dropping rate was controlled to avoid the homo-polymerization. The polymerization was stopped after 3h, and aged for 24h before purification.

2.5.2 Purification of AM-g-Wood

Typical purification procedure of AM-g-Wood is: the copolymer was washed over ethanol several times to remove the ionic liquid and residual monomer, followed by drying at 60°C under vacuum (0.08 MPa) for 24h. Then, the raw product (W1) was extracted with acetone for 48h using soxhlet extractor to give the refined product, whose mass was W2 after being dried under the same conditions above. The effect of graft reaction was evaluated by grafting degree (D) and grafting efficiency (E) which can be calculated by the following formulas:

$$\text{Grafting degree (\%)} = (W_2 - W_0) / W_2 \times 100\%$$

$$\text{Grafting efficiency (\%)} = (W_2 - W_0) / (W_1 - W_0) \times 100\%$$

W_0-----Total mass of wood powder

W_1-----Mass of raw product of AM-g-Wood

W_2-----Mass of refined product of AM-g-Wood

2.6 Characterizations

MAGNA-IR750 FT-IR Spectrometer and AVANCE AV-400 Super-conducting Fourier Digital NMR Spectrometer was used to analyze the structures of ionic liquids; MAGNA- IR750 FT-IR Spectrometer, D/Max-rB X-ray diffractometer was used to analyze the structures and the crystallinity of pretreated fir powder and the graft copolymer of AM-g- Wood; ShimazuTGA-50H Thermo-gravimetric Analyzer and JSM6700F Field Emission Scanning Electron Microscope were applied to analyze the thermal stability and apparent morphology of AM-g-Wood respectively.

3. RESULTS AND DISCUSSION

3.1 Wood pretreatment and dissolution in ionic liquids

3.1.1 Chemical structures of fir powder pretreated with different methods

Fir power was extracted with a mixture of benzene/ethanol or activated with 25% (mass fraction) NaOH under normal temperature and pressure, microwave and high pressure in order to make it dissolve more easily in ionic liquids. Chemical structures of the treated fir power, including the change of wave number, wave form and the interaction of hydrogen bonding, were examined by FTIR and the results were shown in Fig.1.

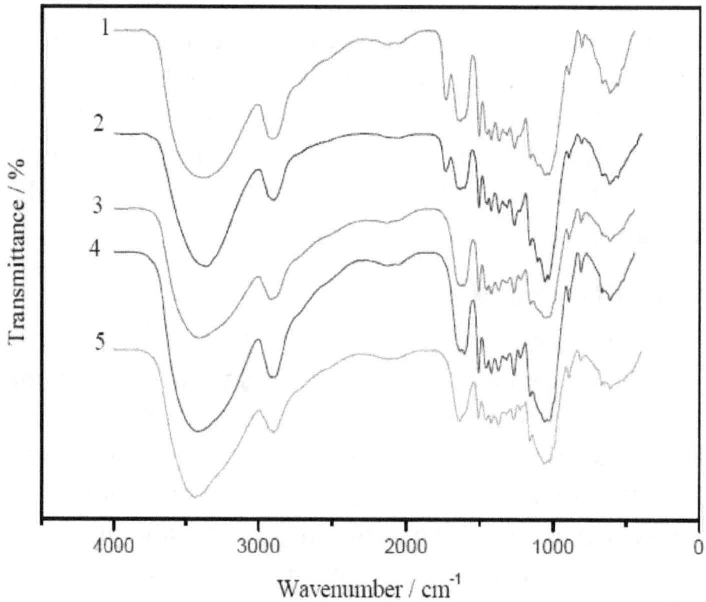

Fig. 1. FTIR spectra of fir powder pretreated with different methods

(1-Original wood; 2-Extracted by benzene/ethanol; 3-Treated with 25% NaOH (mass fraction) at normal temperature and pressure; 4-Treated with 25% NaOH by microwave heating; 5-Treated with 25% NaOH in high pressure pot).

The peak around 3400cm-1 is stretching vibration of O-H in hydrogen bonding. The peak originally ascribe to the absorption of lignocelluloses weakened which suggested the breaking of hydrogen bonding in the wood system after extraction and NaOH treatments, and this peak shifted to high wave number by the greatest degree and the waveform was the most narrow after high pressure treatment (Fig.1), which indicated the hydrogen bonding of sample 5 was broken most seriously. Similarly, the shrinkage of peak around 1730cm-1 indicated that certain amount of fatty acids and acetyl groups or uronic acids of the hemicelluloses have been removed partly by the extraction, and this peak disappeared after NaOH treatment which suggested that the hemicelluloses of wood was degraded and the fatty acids, uronic acids and acetyl groups, etc, were completely dissolved by NaOH solution. However, absorption peaks of wood powder mainly around 1500cm-1, 1268cm-1, 1060cm-1 remained unchanged after extraction and NaOH treatments under different conditions [28, 29]. Followed by the change of wave forms and analysis above, the preliminary conclusion could be reached that it is easier to dissolve fir powder treated with 25% NaOH under high pressure.

3.1.2 Crystallinity and dissolution rate of the pretreated fir powder

Crystallinity of fir powder treated under different technologic conditions were measured by Turley method, which scanned and counted in rule times at diffraction angles of 5°, 18.6°,22.6° and 40°, then wrote down the diffraction peak intensity respectively and calculating the crystallinity. Dissolution rate of pretreated fir powder dissolved in [HeEIM]Cl by microwave heating was studied, and the results were shown in Table 1.

Table 1. Crystallinity and dissolution rate in [HeEIM]Cl of pretreated fir powder

Sample number	Angle and Intensity 5°	18.6°	22.6°	40°	Crystallinity (%)	Dissolution rate (
1	17177	37523	74608	10498	61.0	6.2
2	16614	21780	38468	4949	60.2	8.4
3	14156	45671	54162	11196	19.6	13.6
4	8171	45386	49822	8773	10.6	16.8
5	8914	42504	43507	10450	2.4	18.3

(1-Original wood; 2-Extracted by benzene/ethanol; 3-Treated with 25% NaOH (mass fraction) at normal temperature and pressure; 4-Treated with 25% NaOH by microwave heating; 5-Treated with 25% NaOH in high pressure pot)

Crystallinity of fir powder, which extracted by mixture of benzene/ethanol (the extract content was 6.1%), had little change while the dissolution rate increased (in Table 1). This is because some impurity and small molecule fattiness of wood powder were extracted by the mixture. When the fir powder was pretreated with 25% NaOH, the crystallinity of sample 5 was decreased and the dissolution rate was increased, especially under high pressure condition, the crystallinity decreased from 61.0% to 2.4%, while the dissolution rate increased from 6.2% to 18.3%. There were two reasons for the result: on the one hand, alkali had a certain effect on decreasing crystallinity and degrading hemicellulose of wood; on the other hand, higher temperature and pressure were helpful to destroy the crystal form of fir powder, causing the crystallinity to decrease and dissolution rate to increase. High pressure method, therefore, was used as the best pretreatment method for wood powder.

3.1.3 Solubility of different ionic liquids

Fir powder pretreated with 25% NaOH under high pressure (Sample 5 in Table 1) was dissolved in different ionic liquids by microwave heating, and the results were shown in Fig.2.

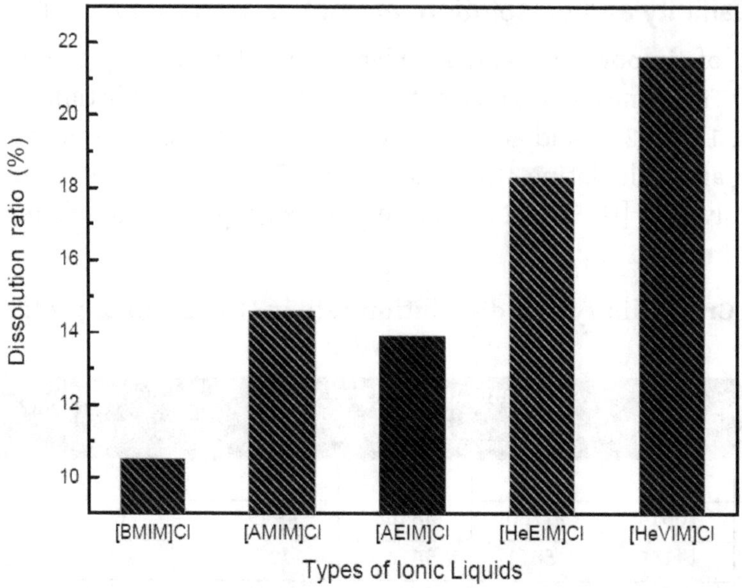

Fig. 2. Influence of different ionic liquids on dissolution rate

Due to the presence of hydroxyl and alkene groups in the side chains which contribute to the destroy of hydrogen bond, the solubility of fir powder using [HeVIM]Cl was the highest (Fig.2). The solubility of ionic liquid with hydroxyl side chain ([HeEIM]Cl) was better than alkene side chain ([AEIM]Cl) and alkyl side chain ([BMIM]Cl) (Fig.2). However, solubility of ionic liquids with shorter side chains ([AMIM]Cl) was better than that of [AEIM]Cl , which is ascribe to the reduced efficient chloride concentration [6]. Ionic liquid of [BMIM]Cl, whose side chain was of the longest alkyl groups, therefore, had the worst solubility. [HeEIM]Cl was used as radium of the following graft copolymerizations instead of [HeVIM]Cl, which had the best solubility, but it was more difficult to be purified.

3.2 Effect of reaction medium on graft copolymerization

AM was polymerized with fir powder (Sample 5 in Table 5) at 45°C using aqueous and [HeEIM]Cl medium respectively. The results were shown in Fig.3.

Graft copolymer of AM-g-Wood could be prepared much more easily in [HeEIM] Cl than in aqueous system (Fig.3). Similarly, grafting degree (D) and grafting efficiency (E) of AM-g- Wood prepared in [HeEIM]Cl were much higher than that in aqueous medium. This is because fir powder can be dissolved in ionic liquid before copolymerization which resulted in the breaking of hydrogen bond and enhancing the reactivity of the powder. In Fig.3, grafting degree and grafting efficiency increased very fast during the beginning of the reaction (≤1h), and became stable in aqueous medium, while increased slowly in [HeEIM]Cl thereafter. This phenomenon can be explained by the typical conventional free radical polymerization mechanism: In the first stage, a great number of free radicals were formed and the reaction was very quick. With the reaction went on, propagation slowed down due to the loss of radicals during the process. The best condition for graft copolymerization, therefore, was used [HeEIM]Cl as the reaction medium with the reaction time of 1h at least.

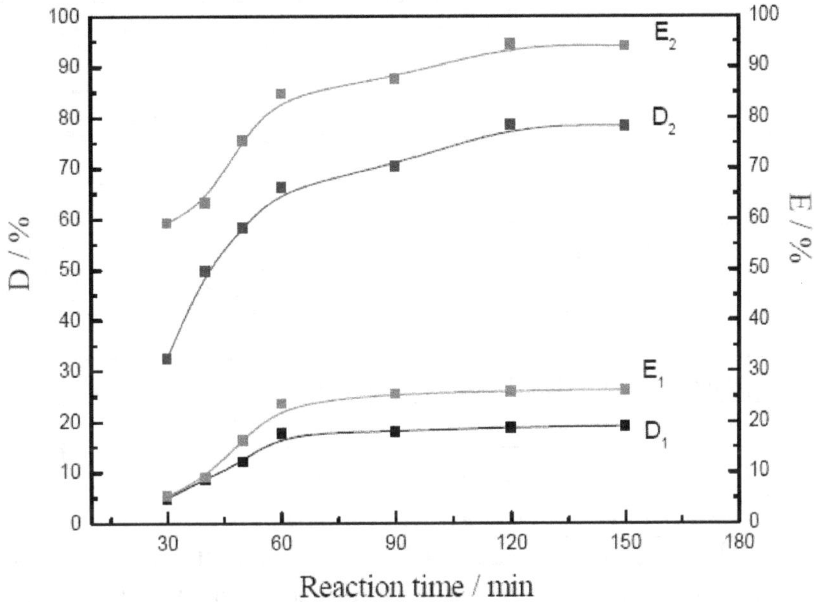

1-reaction medium of H2O; 2- reaction medium of [HeEIM]Cl)

Fig. 3. Influence of reaction medium on grafting degree (D) and grafting efficiency (E)

3.3 Orthogonal test

To optimize the technical conditions, orthogonal test was designed based on three factors and levels (Table 2) separately. The effects of different conditions, such as mass fraction of AM, fluid ratio of [HeEIM]Cl to Wood, and mass proportion of wood to AM on graft copolymerization were carefully studied. Details were shown in Table 2 and 3.

Table 2. Factors and levels of the orthogonal test

Levels	Factors		
	A (Mass fraction of AM)	B (Fluid ratio of [HeEIM]Cl to wood)	C (Mass proportion of wood to AM)
1	20%	25:1	1:3
2	25%	30:1	1:4
3	30%	35:1	1:5

Table 3. Result of orthogonal test

No.	A	B	C	Grafting degree (%)	Grafting efficiency (%)
1	1	1	1	61.2	82.2
2	1	2	1	77.5	89.3
3	1	3	3	70.4	69.5
4	2	1	2	84.1	91.4
5	2	2	3	82.5	75.8
6	2	3	1	78.9	89.5
7	3	1	1	74.8	82.6
8	3	2	1	79.1	94.1
9	3	3	2	74.3	90.5
K1	209.1 / 241.0	220.1 / 256.2	219.2 / 265.8		
K2	245.5 / 256.7	239.1 /259.2	235.9 /271.2	-	-
K3	228.2 /267.2	223.6 /249.5	227.7 /227.9		
R	36.4 / 26.2	19.0 / 9.7	16.7 / 43.3		

The data in Table 3 showed that factors A, B and C had different effect on grafting degree and grafting efficiency of the synthesized products. The affecting order on grafting degree was A>B>C, while on grafting efficiency was C>A>B. The best comprehensive effects came from $A_2B_2C_2$ sample, which means that the mass ratio of AM was 25%, fluid ratio of [HeEIM]Cl to fir powder was 30:1, and mass proportion of wood to AM was 1:4.

3.4 Chemical structures of original wood and AM-g-Wood

Chemical structures of original wood powder and its graft copolymer of AM-g-Wood, which were prepared in [HeEIM]Cl using the best conditions according to the results of orthogonal test, were confirmed using FTIR. The results were shown in Fig.4.

For curve A of Fig.4, peak around 1735cm^{-1} was the stretching vibration of C=O in fatty molecules and acetyl groups or uronic acids of the hemicelluloses of original wood powder, peak around 1638cm^{-1} was characteristic vibration of conjugated carbonyl in lignin of wood powder, and peaks around 1509cm^{-1}, 1160cm^{-1}, and 1058cm^{-1} were typical absorption of aromatic skeletal vibration, phenol-ether group and aliphatic ether vibration in lignin of wood, peak around 1268cm^{-1} was the stretching vibration of C-O-C in celluloses and hemicelluloses [28-30]. For curve B, the peak around 1735cm^{-1} disappeared after grafted, which suggested that fatty molecules and hemicelluloses have been dissolved by ionic liquids. The new stronger peak in 1671cm^{-1} came from the typical stretching vibration of -$CONH_2$. There were still existence of peaks around 1160cm^{-1} and 1058cm^{-1} after grafted polymerization which came from the wood. The wave form of peak around 1268cm^{-1} became unnoticeable after grafted, which suggested that the content of cellulose in the wood was decreased. These analysis strongly suggested that AM has been grafted onto the fir powder.

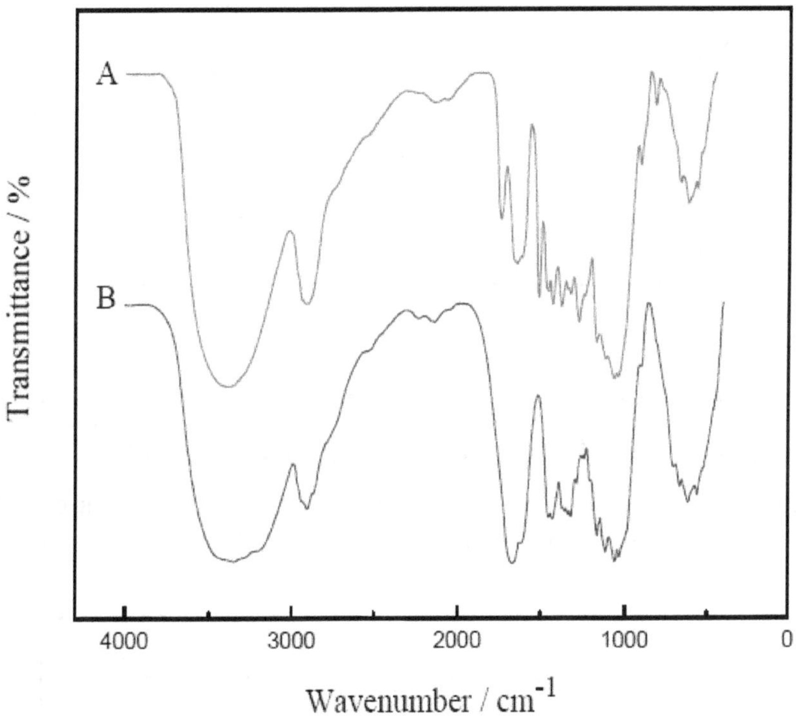

Wavenumber / cm^{-1}

Fig. 4. FTIR spectra of original Wood (A) and graft copolymer AM-g-Wood (B)

3.5 Crystal form of original wood and AM-g-Wood

Crystal forms of original wood powder and its graft copolymer of AM-g-Wood were measured by X-ray diffractometer and the results were shown in Fig.5.

In Fig 5, diffraction peaks appeared at angles of 16.6°, 22.1°, and 34.8° which demonstrate that original wood powder have the same diffraction results as wood fiber. Crystallinity of the original wood powder was 61% based on the Segal formula calculation [31]. The diffraction peaks from wood powder disappeared after grafted polymerization and the crystal area were completely converted into amorphous region, which suggested that the ordered crystal structures have been destroyed. The change of the crystal type also suggested the graft copolymerization occurred between wood powder and acrylamide (Fig 5).

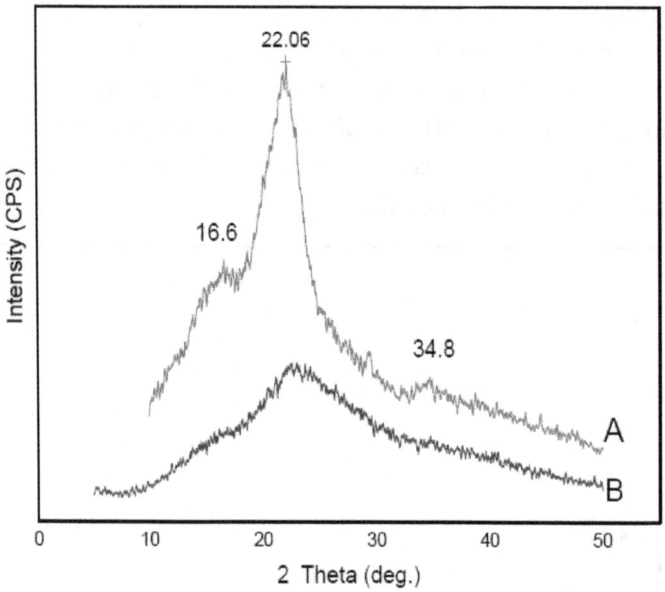

Fig. 5. X-ray spectra of original Wood (A) and graft copolymer AM-g-Wood (B)

3.6 Thermal stability of original wood and AM-g-Wood

Thermal stabilities of original wood powder and its graft copolymer of AM-g-Wood were examined by Thermo-gravimetric analyzer and the results were shown in Fig.6.

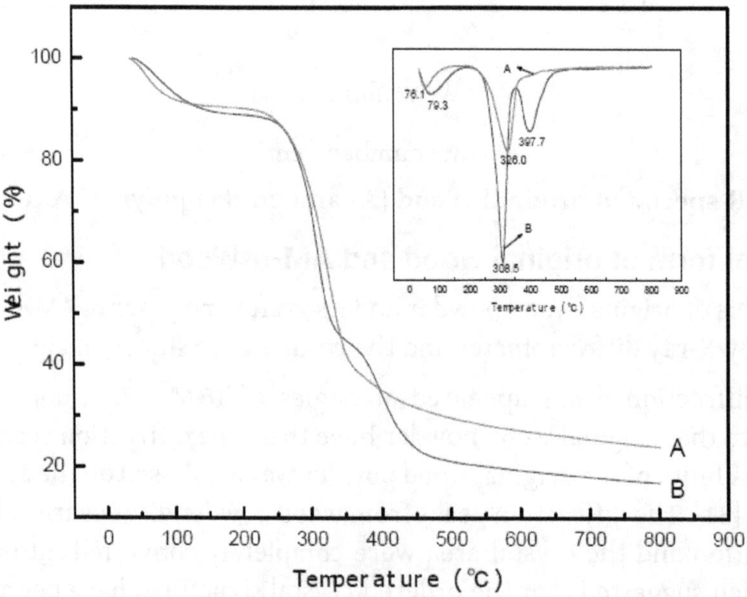

Fig. 6. TGA and DTG curves of original Wood (A) and graft copolymer AM-g-Wood (B)

As shown in Fig. 6, there were two steps for thermal decomposition of original wood: The first one was from 34°C to 175°C, which indicated that wood lost water in this range and the weight loss rate was around 9.5%; The second step was from 175°C to 794°C, responding to the thermal decomposition of wood and the weight loss rate was 66.2%. Actually, the original fir powder began to decompose and the chains of macromolecules were broken at 175°C. The temperature at 326°C refers to the largest weight loss rate. The residual weight rate was 24.4% after the temperature reached 800°C.

However, the thermal decomposition of the graft copolymer AM-g-Wood had three steps: The first step was from 38°C to 189°C, indicating the lost of water from copolymer with weight loss rate around 10.3%. The second step was from 190°C to 343°C, suggesting the weight loss (43.5%) resulted from imino reactions between amide groups of the PAM in the copolymer. The third step was from 343°C to 800°C with weight loss of 28.4%, and this is because the graft chains of PAM were broken during this temperature range. The final residual weight rate was 17.8% after the temperature reached 800°C. As such, the initial thermal decomposition temperature of the grafted wood was 343°C which was much higher than that of the original wood (175°C). Meanwhile, the temperature of the maximum weight loss rate for the grafted wood was 398°C, which was also higher than original wood (326°C). These results demonstrated that AM has been successfully grafted onto fir powder and the thermal stability of the wood after copolymerization was greatly improved.

3.7 Morphology of wood and AM-g-Wood

SEM results of the wood powder before dissolved (A), wood residue after dissolved in [HeEIM]Cl (B) and graft copolymer of AM-g-Wood (C) which was synthesized in [HeEIM]Cl were shown in Fig.7.

Microstructure of the wood powder before dissolved in [HeEIM]Cl was shown in A (Fig.7) which was loose and porous, while looser structure and bigger porous were found in B for wood after dissolved in [HeEIM]Cl. However, the loose and porous structures completely disappeared after polymerization (C) and it was difficult to detect fibril, which is very different from A and B. This difference should be ascribe to the existence of AM-g-wood structure which was synthesized from the graft copolymerization of AM and wood. Followed by the changes of wood appearance, therefore, a graft copolymerization occurred between fir powder and AM, and grafted copolymer of AM-g-Wood should be prepared more easily with wood residue as its looser structure and bigger porous.

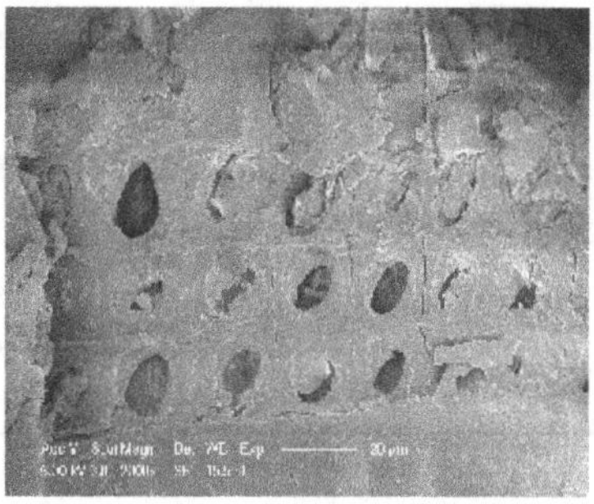

(A) wood before dissolved

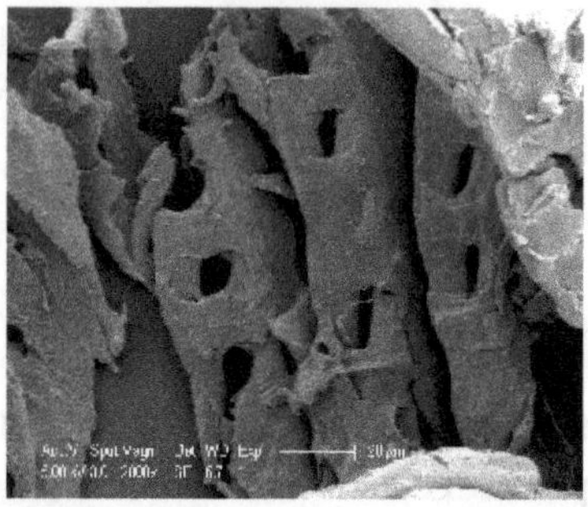

(B) wood residue after dissolved

(C) graft copolymer of AM-g-Wood

Fig. 7. SEM micrographs of wood and its graft copolymer

4. CONCISE SUMMARY

Diverse ionic liquids with distinct imidazole side chain functional groups have been produced, and their solubility for fir powder has been examined. AM-g-Wood was synthesized in [HeEIM]Cl using K2S2O8 as the initiator, and the conditions for graft copolymerization were optimized using an orthogonal test. The precise conclusions were enumerated as follows:

1. The structures of the produced ionic liquids have been validated using FT-IR and 1H NMR analysis. The characteristics of side groups significantly influenced the solubility of the ionic liquids. The ionic liquid [HeVIM]Cl, featuring hydroxyl and alkene groups in its imidazole side chains, proved to be the most effective solvent for fir powder; nonetheless, it posed challenges in purification.
2. The fir powder prepared with a 25% NaOH solution under high pressure exhibits the lowest crystallinity and the maximum dissolving rate.
3. AM-g-Wood could be easily synthesized in [HeEIM]Cl. In comparison to an aqueous medium, AM-g-Wood synthesized in [HeEIM]Cl exhibited advantages including significantly enhanced grafting degree and efficiency, as well as superior yields.

4. The orthogonal test indicated that the optimal conditions for graft copolymerization were a mass fraction of acrylamide (AM) at 25%, a fluid ratio of [HeEIM]Cl to wood at 30:1, and a mass proportion of wood to AM at 1:4.
5. The FT-IR and SEM results indicated that AM has been effectively grafted onto the wood powder. The crystalline structure of the wood has been entirely obliterated following graft copolymerization, and heat stability has been significantly enhanced.

REFERENCE

Wang Y, He J, Lu T, Wu YY, Fan YM (2004) Research on graft copolymerization of cellulose. Fine and Specialty Chemicals 12:18-21

Vogli E, Sieber H, Greil P (2002) Biomorphic SiC-ceramic prepared by Si-vapor phase infiltration of wood. Journal of the Europe Ceramic Society 22: 2663-2668

Lin LZ, Yoshioka M, Yao YG, Shiraishi N (1994) Liquefaction of wood in the presence of phenol using phosphoric acid as a catalyst and the flow properties of the liquefied wood. Journal of Applied Polymer Science 52: 1629-1636

Yamada T, Ono H (2001) Characterization of the products resulting from ethylene glycol liquefaction of cellulose. Journal of Wood Scicence 47: 458-464

Cuissinat C, Navard P, Heinze T (2008) Swelling and dissolution of cellulose. Part IV: Free floating cotton and wood fibres in ionic liquids. Carbohydrate Polymers 72: 590-596

Swatloski RP, Spear SK, Holbrey JD, Rogers RD (2002) Dissolution of cellulose with ionic liquids. J Am Chem Soc 124:4974-4975

LY, Chen HZ (2006) Enzymatic hydrolysis of cellulose materials treated with ionic liquid [BMIM]Cl. Chinese Science Bulletin 51: 2432-2436

Guo LY, Shi TJ, Li Z, Duan YP, Wang YG (2008) Synthesis of novel and functionalized ionic liquid [HeEIM] Cl and its solubility for cotton fibre. Chemical Journal of Chinese Universities 29:1901-1907

Zhang H, Wu J, Zhang J, He JS (2005) 1-allyl-3-methylimidazolium chloride room temperature ionic liquid: A new and powerful nonderivatizing solvent for cellulose. Macromolecules 38: 8272-8277

Fort DA, Remsing RC, Swatloski RP, Moyna P, Moyna G, Rogers RD (2007) Can ionic liquids dissolve wood? Processing and analysis of lignocellulosic materials with 1- n-butyl-3-methylimidazolium chloride. Green Chemistry 9: 63-69

Guo LY, Shi TJ, Li Z, Duan YP (2008) Solubility of two kinds of imidazolium ionic liquids for fir powder. Journal of Chemical Industry and Engineering 59:1299-1304

Kilpeläinen I, Xie HB, King A, Granstrom M, Heikkinen S, Argyropoulos DS (2007) Dissolution of wood in ionic liquids. J Agric Food Chem 55: 9142-9148

Jain N, Kumar A, Chauhan S, Chauhan SMS (2005) Chemical and biochemical transformations in ionic liquids. Tetrahedron 61:1015-1060

Kubisa P (2004) Application of ionic liquids as solvent for polymerization Processes. Progress in Polymer Science 29: 3-12

Scurtoa M, Leitner W (2006) Expanding the useful range of ionic liquids: melting point depression of organic salts with carbon dioxide for biphasic catalytic reactions. Chemistry Communication 35: 3681-3683

Lu XB, Hu JQ, Yao X, Wang ZP, Li JH (2006) Composite system based on chitosan and room-temperature ionic liquid: Derict electrochemistry and electrochemistry and electrocatalysis of hemoglobin. Biomacromolecules 7: 975-980

Weuster-Botz D (2007) Process intersification of whole-cell biocatalysis with ionic liquids. Chem Rec 7: 334-340

Zein El-Abedin S, Endres F (2006) Electrodeposition of metals and semiconductors in air- and water-stable ionic liquids. Chemphyschem 7: 58-61

Lu XB, Zhang Q, Zhang L, Li JH (2006) Direct electron transfer of horseradish peroxidase and its biosensor based on chitosan and room temperature ionic liquid. Electrochemistry Communications 8: 874-878

Mckague AB (1974) Flocculating agent derived from kraft lignin. Appl Chem Biotechnol 24: 607-615

Dizhbite T, Telyshera G (1998) Multifunctional activity of lignin derivatives in pulp/ filler compositions. Adv Lignocellul Chem Ecol Friendly Pulping Bleaching Technol, Eur Workshop lignocellul Pulp 97-100

Liu S, Sun G (2008) Radical graft functional modification of cellulose with allyl monomers: Chemistry and structure characterization. Carbohydrate Polymers 71:614-625

Canché-Escamilla G., Pacheco-Catalán DE, Andrade-Canto SB (2006) Modification of properties of rayon fibre by graft copolymerization with acrylic monomers. J Mater Sci 41:7296-7301

Seven P, Coskun M, Demirelli K (2008) Synthesis and characterization of two-armed graft copolymers prepared with acrylate and methacrylate using atom transfer radical polymerization. Reactive & Functional Polymers 68: 922-930

Yang F, Li G, Ren FX, Yu XQ, Wang GX, Zhu SB (2007) Study on Preparation and Properties of Poly (AM -co- CMC). Polymer Materials Science and Engineering 23:78-81

Lu ZX, Yuan X, Wu J, Wang LJ, LuoHA (2005) Synthesis and spectroscopic characterization of N, N'-dialkylimidazolium-based ionic liquids. Chemical World 46:148-150

YH, Zeng XQ, Lu HM (2006) Purification and application of lignin in urea- formaldehyde resin. Contemporary Chemical Industry 35: 377-379

O, Bremer J, Schmidit O, Stevanovic T (1991) Monitoring of chemical changes in white-rot degraded beech wood by pyrolysis-gas chromatography and Fourier transform infrared spectroscopy. Journal of Analytical and Applied Pyrolysis 21:147–162

Pandey KK, Pitman AJ (2003) FTIR studies of the changes in wood chemistry following decay by brown-rot and white-rot fungi. International Biodeterioration & Biodegradation 52:151-160

O (1988) Practical uses of FTIR spectroscopy in wood science and technology. Mikrochim Acta [Wien] 1:21-25

Xue ZH, Zhao GJ (2007) Influence of different treatments on wood crystal properties.Journal of Northwest Forestry University 22:169-171

4 Chapter

TARGETED DECOMPOSITION OF (LIGNO)CELLULOSE IN IONIC LIQUIDS

1. INTRODUCTION: AN OVERVIEW

Lignocellulose primarily consists of cellulose, lignin, hemicellulose, and extractives, constituting a plentiful carbon-neutral renewable resource. Rising apprehensions regarding global warming and depleting fossil fuel sources have prompted significant endeavors to utilize (ligno)cellulose as a substitute for petroleum in the manufacture of fuels and chemicals. From a sustainability perspective, transitioning society's reliance from fossil-based energy sources to renewable alternatives is a significant contribution to the development of a sustainable and "Low Carbon" economy for all nations. (Ragauskas et al. 2006; Corma et al. 2007). By 2025, it is projected that as much as 30% of raw materials for the chemical industry will originate from renewable sources. To attain this objective, there is a significant demand for novel processing technologies, separation and depolymerization methods, and catalytic conversion systems. Dissolving (ligno)cellulose in its original state is generally challenging due to the three-dimensional cross-linked lignin network and the robust hydrogen bonds within the polymeric matrix. The lack of a dependable solvent for dissolving (ligno)cellulose has significantly obstructed the effective use of lignocellulosic biomass. (Kamm 2007).

The development of ionic liquid-based technology has facilitated an eco-friendly and uniform approach to utilizing (ligno)cellulose. Recent studies have highlighted advancements in biomaterials generated from biopolymers utilizing ionic liquids (Zhu et al. 2006; Pinkert et al. 2009); hence, we will not address this topic here. This chapter primarily aims to present a contemporary overview of the selective degradation of (ligno)cellulose into platform molecules utilizing catalytic strategies in ionic liquids, while also addressing critical issues that connect carbohydrate chemistry and petroleum chemistry through ionic liquids. In this context, studies concerning the chemistry of monosaccharides in ionic liquids will be referenced just when necessary to elucidate matters pertaining to the transformation of (ligno)cellulosic materials.

2. DISSOLUTION OF (LIGNO)CELLULOSE IN IONIC LIQUIDS

For a long time, the dissolution and processing of (ligno)cellulose and cellulose are one of the biggest obstacles for their utilization. Rogers et al (Swatloski et al. 2002) reported that imidazolium-based ionic liquids could dissolve cellulose (up to 25 wt%), and an amorphous cellulose can be produced with addition of anti-solvents (e.g. water, ethanol). Since then, a large library of ionic liquids have been synthesized to develop a more efficient and greener solvent system for dissolution of cellulose (Fukaya et al. 2006; Kilpelainen et al. 2007; Zavrel et al. 2009; Zakrzewska et al. 2010) and lignocellulosic materials (Kilpelainen et al. 2007). Recently, dissolution of cellulose in ionic liquids has been reviewed, and the structures of ionic liquids and their solubility to different types of carbohydrates have been collected (Zakrzewska et al. 2010). Some representative structures of ionic liquids are shown in Figure 1.

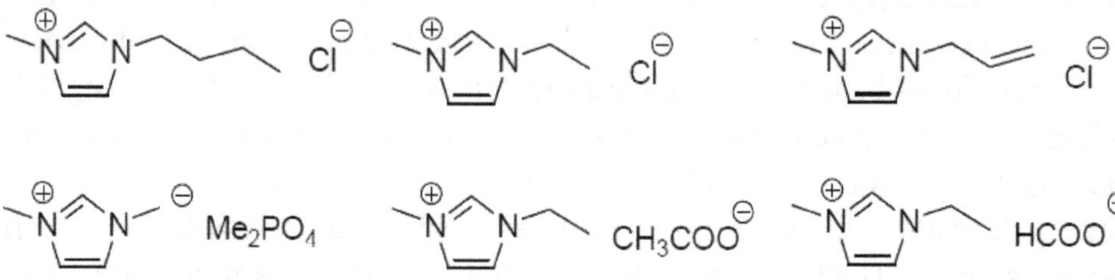

Fig. 1. Representative ionic liquids for biomass dissolution

Despite the increasing number of known ionic liquids capable of dissolving biomass, the mechanism of this dissolution process remains not well understood. There have been a few theoretical and experimental studies, including molecular dynamic studies and NMR analyses. For example, ^{13}C and $^{35/37}Cl$ NMR relaxation measurements have been done for cellulose solution in ionic liquids at varied temperatures and concentrations to investigate the interaction of the chloride ion of the ionic liquids with cellulose, and it was found that the solvation of cellulose by the ionic liquid 1-*n*-butyl-3-methylimidazolium chloride ($[C_4mim]Cl$) involved hydrogen-bonding between the carbohydrate hydroxyl proton and the chloride ion in a 1 : 1 stoichiometry (Remsing et al. 2006). Further multinuclear NMR spectroscopy experiments demonstrated that the anions in ionic liquids were involved in specific interactions with the solutes, and thus controlled the solvation process (Remsing et al. 2007). On the other hand, the ^{13}C relaxation rate of the imidazolium carbons showed no strong correlation with sugar concentration, regardless of the structure of either the cation or the cellulose, indicating that their role was not essential to the solvation process. Yet, molecular dynamics simulations showed that weak hydrogen bonding interactions with ionic liquids cation occurred through the acidic hydrogen at the C(2) position on the imidazolium ring (Youngs et al. 2007).

To provide an in-depth rationale for selecting ionic liquids capable of dissolving cellulose, further molecular dynamics simulations with an all-atom force field were performed on a system consisting of 1-ethyl-3-methyl imidazolium acetate ([Emim][OAc]) (Remsing et al. 2008; Liu et al. 2010). Results showed that the interaction energy between the polysaccharide chain and the ionic liquids was stronger (3 times) than that for either water or methanol. In addition to the anion acetate forming strong hydrogen bonds with hydroxyl groups of cellulose, some of the cations were found to be in close contact with the polysaccharides through hydrophobic interactions. It was also observed that the preferred β-(1,4)-glycosidic linkage conformation of the cellulose was altered when the cellulose was dissolved in [Emim][OAc] as compared to that found in crystalline cellulose dispersed in water. Furthermore, this kind of changes was not reversible during the regeneration process. They hypothesized that this conformational change in the cellulose was one of the key factors in the observed increase in enzymatic hydrolysis rates after ionic liquids pretreatment (Liu et al. 2010). Although the fundamental molecular knowledge of these systems elucidated the role of cation during the dissolution process, recently publications of non-imidazolium based ionic liquids demonstrated good solubility to cellulose (Yang et al. 2010), which argued against the key role of anions to a certain extent. Such molecular level knowledge of ionic liquids system should enable the design and rapid computational screening of a wide range of ionic liquids for biomass pretreatment and facilitate the development of new efficient and economic ionic liquids-based biorefinery technologies (Pinkert et al. 2009).

In the traditional petrochemical industry, crude oil is fractionated and refined to produce various grades of liquid transportation fuels, and hydrocarbon feedstocks are functionalized to produce intermediates and specialty chemicals with advanced catalytic strategies (Gallezot 2008; Cheng et al. 2009; Ong et al. 2010). The concept of biorefinery is similar in terms of the utilization of a renewable source of carbon to produce heat, biomaterials, fuels and value- added chemicals. The overall strategy in the production of hydrocarbon fuels and chemicals from biomass is (1) to de-polymerize biopolymers, (2) to reduce the oxygen content of the parent feedstocks or de-polymerized intermediates and, (3) to create C-C bonds between biomass-derived intermediates to increase the molecular weights of the end products (Alonso et al. 2010). Ionic liquids have been regarded as 'greener' solvents, partially because their properties can be designed through tuning the structures of their constitutive ions (Rogers et al. 2003). The dissolution of (ligno) cellulose in ionic liquids resulted in a full release of all the functional groups and bonds from the well-organized strong hydrogen bonding matrix (Xie et al. 2007; Xie et al. 2009). Thus, biomass dissolved in ionic liquids was more susceptible to chemical attack by external reagents and catalysts. Bearing these in mind, a lot of efforts have been devoted to producing valuable chemicals from biomass *via* hydrolysis, dehydration, hydrogenolysis, etc., taking the advantages of this new homogenous platform.

3. SELECTIVE BREAKDOWN OF (LIGNO)CELLULOSE IN IONIC LIQUIDS

3.1 Hydrolysis of (ligno)cellulose

Hydrolysis of cellulose to fermentable sugars is virtually an essential step in any practical cellulosic biofuel production via a biological route. Furthermore, more chemical processes are seen to convert monosaccharides into fuels and value-added chemicals (Huber et al. 2006; Chheda et al. 2007). Two methods including acid hydrolysis and enzymatic hydrolysis are currently known for biomass hydrolysis. To make cellulosic materials more susceptible to hydrolysis, a pretreatment process is needed to reduce the crystallinity and increase the porosity. Traditional pretreatment technologies using oxidants, organic solvents, lime, or mineral acids, remain imperfect in terms of process greenness and eco- and environmental friendliness (Stephanopoulos 2007). Therefore, the development of advanced pretreatment technologies is one of the main issues in biorefinery. The inherent of dissolution of (ligno)cellulose in ionic liquids provided a new pretreatment technology and an inert homogeneous platform for the hydrolysis of (ligno) cellulose.

3.1.1 Enzymatic hydrolysis of (ligno)cellulose

Presently, two types of research have been carried out to hydrolyze the lignocellulose into fermentable sugars in ionic liquids. One is enzymatic hydrolysis of cellulosic materials regenerated from ionic liquids with the addition of an anti-solvent, and the other is in situ enzymatic hydrolysis in ionic liquids.

Shortly after regeneration of cellulose was demonstrated, efforts have been dedicated to characterizing the regenerated material, finding new applications, and further optimizing the process. It was confirmed that the crystalline structure of cellulose was destroyed (*vide ante*), which should allow a greater accessibility for the hydrolytic enzymes to rapidly penetrate and hydrolyze the (ligno)cellulose. Liu et al firstly investigated the enzymatic hydrolysis of cellulosic samples that had been pretreated with ionic liquids (Liu et al. 2006). It was found that the hydrolysis rates of wheat straw and steam-exploded wheat straw treated with [C_4mim]Cl were improved significantly. For example, wheat straw samples pretreated with [C_4mim]Cl gave a 70% conversion, while those pretreated with water had only a

42% conversion. Further WXRD and morphology study by SEM of the regenerated (ligno)cellulose material showed that diffraction signals from the crystalline regions of spruce sawdust disappeared after the dissolution-regeneration process (Kilpelainen 2007). Those were in an amorphous and porous morphology (Figure 2). For the spruce thermomechanical pulp (TMP) samples regenerated from 1-methyl-3-alkylimidazolium ([Amim]Cl) solution by precipitation upon water addition, glucose yield was 60% by the enzymatic hydrolysis, while it was only 12% for the untreated sample.

Fig. 2. Morphology study of spruce TMP treated by the ionic liquid [Amim]Cl

As noted earlier, such kinds of crystalline structure and morphology transformation, as well as β to α conformational change are anticipated to allow a greater accessibility for the hydrolytic enzymes. An extended research demonstrated an initial clue and potential to produce bioethanol from wheat straw with the ionic liquid pretreatment technology (Li et al. 2009). The fermentability of the enzymatic hydrolysates was evaluated using the yeast *Saccharomyces cerevisiae*, and the ethanol production was 0.43 g/g glucose after 26 h of fermentation. Results indicated that the pretreatment with the ionic liquid 1-methyl-3- ethylimidazolium diethylphosphate ([Emim]DEP) had negligible affects on the downstream ethanol production. The reusability of ionic liquid was also investigated in this study, and the yields of reducing sugars were higher than 52% even though the ionic liquids were recycled for 5 times. Although biomass pretreatment using ionic liquid is technically attractive, issues about process-costs and the ecological and environmental impacts require further scrutinize.

Recently ionic liquids have also been considered as solvents for biocatalysis (Zhao 2010), and examples of bioreactions in ionic liquids were seen with higher selectivity, faster reaction rate and even greater enzyme stability (Moniruzzaman et al. 2010). It is reasonable to perceive direct enzymatic hydrolysis of cellulose using ionic liquids as the reaction media. However, those conventional cellulose-dissolving ILs are composed of anions such as Cl^-, dca$^-$, HCO_2, or CH_3CO_2. While these anions facilitate the dissolving of cellulose through the formation of strong hydrogen bonds, they also tend to damage molecular interaction networks in a protein leading to denaturation of enzyme. In addition, high viscosity of ionic liquids may also limit the effectiveness of enzymatic transformations. All of these limitations hindered the enzymatic conversion of cellulose in ionic liquids. Therefore, the design and synthesis of ionic liquids capable of dissolving cellulose, and enzyme-compatible, especially compatible with cellulase as well, have been paid much attention to pursue an *in situ* enzymatic saccharification of cellulose (Zhao 2010). In 2008, Kamiya et al firstly reported their efforts on this topic (Kamiya et al. 2008). They dissolved cellulose sample in the ionic liquid [Emim]DEP, and then mixed the ionic liquid solution with different volumes of citrate buffer (10 mM, pH 5.0), followed by the addition of cellulase directly at 40 ºC. Little cellulose activity was observed when the volumetric ratio of ionic liquid to water was higher than 3:2. However, decreasing the ratio enhanced enzymatic activity remarkably, and a reducing sugar yield over 70% (composed of 50% glucose + 20% cellobiose) was achieved after 24 h when the ionic liquid to water ratio was 1:4. Importantly, glucose formation in the aqueous-ionic liquid mixture was approximately 2-fold more than that of the aqueous system under identical conditions, indicating that the ionic liquid was truly enzyme-compatible. In comparison, when the ionic liquid [Emim][OAc] was applied under these conditions, cellulase activity was only approximately half of that with [Emim]DEP, suggesting that the anion structure played a key role in terms of the cellulase activity in ionic liquids.

The activity and stability of a commercial cellulase samples in eight ionic liquids have been investigated by optical and calorimetric techniques to gain further insights into the *in situ* enzymatic saccharification process (Bose et al. 2010). Some important issues concerning the cellulase in ionic liquids have been elucidated. Among those ionic liquids, hydrolysis was observed only in 1-hydrogen-3-methylimidazolium chloride and tris-(2-hydroxyethyl) methyl ammonium methylsulfate (HEMA). Interestingly, hydrolysis at 65 ºC had a higher initial rate but reached a plateau after 2 h in the aqueous buffer, whereas the reaction in these two ionic liquids proceeded with a lower initial rate but went monotonically over time. Further investigation of relationship of enzymatic activity and thermal stability with viscosity and reaction temperature demonstrated that this difference in the rate of hydrolysis was largely attributed to two factors, the higher viscosity of the ionic liquids and enzyme stability. The inherent high viscosity of the ionic liquids was a retarding factor for the enzymatic hydrolysis, since it slowed the diffusion of the enzyme to its substrate resulting in a lower observed activity. Reversibility of the unfolding process was studied by gradual cooling of the denatured enzyme from temperature higher than the $T_{1/2}$. It was found that cellulase unfolding was irreversible in buffer and accompanied by precipitation of the enzyme. In HEMA, folding and unfolding processes were reversible even at up to 120 ºC, which indicated that excellent cellulase thermo stability can be achieved in particular ionic liquids. However, in 1-hydrogen-3-methylimidazolium chloride, the unfolding process was completely irreversible. The potential of enzymatic saccharification process based on ionic liquids technology can be envisioned, however, it remains challenging to design an ionic liquid that compromises these opposing properties of dissolving cellulose, retaining the activity of the enzymes, yet having a low viscosity. From an eco- and environment-friendly point of view, efforts should be put on reducing operation costs and synthesis of novel ionic liquids.

3.1.2 Acid hydrolysis of (ligno)cellulose

Many attempts to hydrolyze cellulose into fermentable sugars catalyzed by mineral acids have been reported. These processes were usually done under heterogeneous and harsh conditions in which the (ligno)cellulose raw material was typically not solubilized in the reaction medium. Thus, the supramolecular and crystalline structure of cellulose imposed important restrictions on the kinetics of the hydrolysis. Consequently, traditional acid hydrolysis of (ligno)cellulose was inefficient and cost-intensive. Noting that the full dissolution of cellulose in ionic liquids could break internal and external supramolecular structures among the cellulosic fibers, and we envisioned that it should provide an opportunity to revisit the acid-catalyzed hydrolysis technology. In 2007, we firstly reported the hydrolysis behavior of cellulose in ionic liquids in the presence of mineral acids (Li and Zhao 2007). It was interesting to find that catalytic amounts of acid were sufficient to drive the hydrolysis reaction. For example, when the acid/cellulose mass ratio was set to 0.46, yields of total reducing sugar (TRS) and glucose were 64% and 36%, respectively, after 42 min at 100 ºC. In fact, excess acid loading in the ionic liquid system was detrimental in terms of sugar yields because side reaction tended to occur that consumed the hydrolysis products. Preliminary kinetic study indicated that the cellulose hydrolysis catalyzed by H_2SO_4 followed a consecutive first-order reaction sequence, where k_1 for TRS formation and k_2 for TRS degradation were 0.073 min^{-1} and 0.007 min^{-1}, respectively. Our further study on the hydrolysis behavior of (ligno)cellulose in ionic liquids demonstrated that hydrochloric acid was also an effective catalyst (Li et al. 2008). TRS yields were up to 66%, 74%, 81% and 68% for hydrolysis of corn stalk, rice straw, pine wood and bagasse, respectively, in the presence of only 7 wt% catalyst at 100 ºC under an atmospheric pressure within 60 min. Under those conditions, the constants for k_1 and k_2 were 0.068 min^{-1} and 0.007 min^{-1}, respectively, for the hydrolysis of corn stalk.

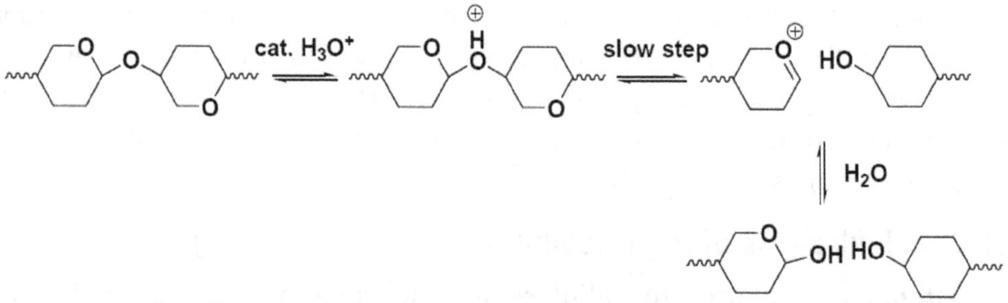

Fig. 3. Proposed mechanism for hydrolysis of cellulose. Hydrogen, hydroxyl, and hydroxymethyl groups are omitted for better clarity (Adopted from the reference (Rinaldi et al. 2010)).

Further study showed that the rates of these two competing reactions varied with acid strength and that for acids with an aqueous pKa below zero (Ka < 1), and the hydrolysis reaction was significantly faster than the degradation of glucose, thus allowing hydrolysis to be performed with a high selectivity in glucose (Vanoye et al. 2009). Although the acidity in ionic liquids is a concept still in its infancy, the relative acidity order of several organic and inorganic acids in ionic liquids was found identical to that in water (MacFarlane et al. 2006). The proposed hydrolysis of cellulose usually consists of three consecutive processes, (1) protonation of the glycosidic oxygen; (2) formation of a cyclic carbocation; (3) nucleophilic attack on the cyclic carbocation species by a water molecule (Figure 3). Although the second one is the rate-limiting step, the protonation of the glycosidic oxygen is difficult in some electron deficient acetals on the basicity of the acetal O-site and the hydroxyl O-sites and the acidic strength. Therefore, although some ionic liquids containing weak basic anions (e.g. acetate, organophosphates) are capable of dissolving cellulose, they are not good solvents for hydrolysis of cellulose, because a relative weak acid formed by addition of a strong acidic catalyst (Rinaldi et al., 2010).

A recent work on pretreatment of wood species in ionic liquids in the presence of mineral acids suggested that a significant amount of lignin was also degraded, together with hydrolysis of cellulose and hemicellulose (Li et al. 2010). This was based on the results of GC-MS and 31P NMR analysis of the ethyl acetate extracted samples produced during the pretreatment process. In fact, authors demonstrated the presence of compounds such as 5- hydroxymethylfurfural, furan-2-carboxylic acid, catechol, methylcatechol, methylguaiacol, acetoguaiacone, and acetol, among which those phenolic compounds and acetol should be originated from lignin.

The facts that the higher the degree of polymerization (DP) value of cellulose, the longer was the reaction time required for a satisfactory glucose yield, and that the longer reaction time favored glucose formation, while the shorter reaction time produced more TRS, implied that cellulose hydrolysis in ionic liquids catalyzed by mineral acids most likely followed a random hydrolysis mechanism, as observed with the concentrated acid system (Li et al. 2007). During the hydrolysis process, both endoglycosidic and exoglycosidic scission occurred, but the endoglycosidic product, oligoglucoses, was the major one in the initial stage, which usually was observed in traditional heterogeneous hydrolytic systems. Such phenomena were later confirmed in the study of solid acid (Amberlyst 15DRY) catalyzed hydrolysis of cellulose and (ligno)cellulose in ionic liquids (Rinaldi et al. 2008; Rinaldi et al. 2010). In those studies, depolymerized cellulose was precipitated and recovered by addition of water to the hydrolytic system, and the DP value was estimated by gel-permeation chromatography. It was found that the size of recovered cellulose fibers became successively smaller over time, resulting in a colloidal dispersion for the material recovered after 5 h. The depolymerization of cellulose proceeded progressively, resulting in the formation of soluble oligosaccharides if the reaction was carried out over a long time. For example, cellooligomers consisted of approximately ten anhydroglucose units (AGU) were seen after 5 h.

There is an intriguing question in the case of hydrolysis of cellulose catalyzed by solid acids in ionic liquids, where does the hydrolysis reaction take place, on the catalyst surface or in the bulk solution? The observation of induction period for the production of glucose and titration results of the ionic liquids separated from a suspension of Amberlyst 15DRY in [Bmim]Cl suggested that proton was progressively released into the bulk liquid within an hour upon through an ion-exchange process involving $[Bmim]^+$ of the ionic liquid and H^+ species of the solid acid. The true catalytic species in the case of Amberlyst 15DRY-catalyzed hydrolysis was the released H^+ species, and an activation process by washing it with a H_2SO_4 solution is required for the reuse of this solid catalyst (Rinaldi et al. 2010).

A full depolymerization of cellulose into fermentable sugar or other derivatives in ionic liquids challenges separation technology, because those ionic liquids *per se* are excellent solvents for monosaccharides, polar, water soluble and unstable under high temperature. Up to now, an efficient and economic method has not been achieved (Binder and Raines 2010; Brennan et al. 2010), which maybe the severe barrier of the ionic liquids based technologies for a practical use point of view. In this regard, the controlled depolymerization of cellulose into water insoluble oligosaccharides followed by an enzymatic hydrolysis seemed the most interesting transformation. For example, cellooligomers with a DP value around 30 were precipitated in 90% yield by the addition of water after microcrystalline cellulose was hydrolyzed in ionic liquids for 1.5 h. Another general strategy is to use ionic liquids as pretreatment agents to facilitate the hydrolysis process (*vide ante*). For example, pretreatment of cellulose with [Bmim]Cl led to loosen crystalline cellulose through partial transformation of cellulose I to cellulose II, which could facilitate the attack of proton during the subsequent hydrolysis over Nafion NR50 (Kim et al. 2010).

Other acids, such as trifluoroacetic acid, Brönsted acid ionic liquids, have also been used and satisfactory results were obtained in hydrolysis of cellulose in ionic liquids (Amarasekara and Owereh 2009; Amarasekara and Owereh 2010). In 2010, a more environment-friendly process was reported for biomass conversion in ionic liquid-water mixture (Zhang et al. 2010). Under relatively mild conditions (≤ 140 °C, 1 atm) and in the absence of acid catalysts typically employed in biomass conversion, the dissolved cellulose in [Emim]Cl can be converted into reducing sugars in up to 97% yield. The combined study of experimental methods and *ab initio* calculations demonstrated the *Kw* value of water in the mixture was up to 3 orders of magnitude higher than the pure water under ambient conditions. Such high *Kw* values are typically achievable under high temperature or subcritical conditions, which is responsible for the remarkable performance without adding acid catalysts. They hypothesized that the increased [H^+] was due to the enhanced water auto ionization by ionic liquids. This process will be affected by the electrostatic environment of the solution, the broad dielectric medium of the solvent and the temperature. Comparative *ab initio* calculations based on the thermodynamic cycle showed that IL-water mixture exhibited higher concentrations of both [H^+] and [OH^-] than pure water, thus enabling the acid- and base-catalyzed reactions.

3.2 Catalytic conversion of (ligno)cellulose into furfurals

Given the chemical differences between biomass feedstocks and conventional fuels and chemicals, biomass should obviously be subjected to deoxygenation/ dehydration in order to reduce the oxygen content. A desired intermediate derived from biomass, usually referred as platform molecules, should afford a great degree of flexibility in downstream processes and applications (Alonso et al. 2010). In 2004, the National Renewable Energy Laboratory proposed twelve building block chemicals that can be readily produced from sugars *via* biological or chemical transformations (Werpy and Petersen 2004). These twelve building blocks can be subsequently converted into a number of high-value chemicals or bulky materials. Among them, 5-hydroxymethylfurfural (HMF), is recognized as a versatile intermediate in biofuel chemistry and the petroleum industry (Figure 4). Production of HMF from fructose by dehydration is relatively easy. Although various conditions were published for this transformation, we will not discuss it. Instead, we will focus on production of HMF from cellulose and lignocellulosic materials. We also cover a few examples where HMF was produced using glucose because it had a tight connection to our main focus.

Although there were early studies on HMF production using glucose as the feedstock, yields were not appreciable. In 2007, a system of metal chloride in ionic liquids was introduced to convert glucose into HMF (Zhao et al. 2007). Specifically, a record high yield of 68% was achieved in the presence of $CrCl_2$ in [Emim]Cl. It was proposed that glucose was isomerized through a $CrCl_2$-associated enediol intermediate to fructose followed by dehydration reaction. It was interesting to note that an addition of strong ligands, 2,2'- bipyridine or glyceraldehydes, essentially shut down the reaction: HMF was receive in less than 2% yield in the presence of 2,2'-bipyridine. The authors concluded that the metal catalyst interacted with the hemiacetal portion of glucopyranose, and that there was little interaction with polyalcohol portion of the sugar. The authors also proposed that the $CrCl_3^-$ anion played an important role in proton transfer, facilitating mutarotation of glucose to fructose.

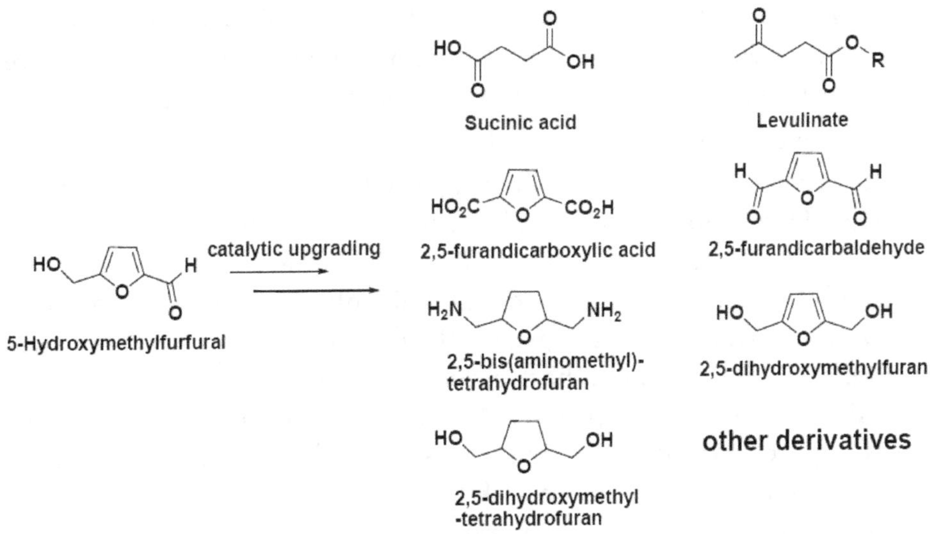

Fig. 4. Versatile derivatization of 5-hydroxymethyl furfural

Given these advancements on production of HMF from monosaccharides, the direct conversion the most abundant of cellulose and (ligno)cellulose into HMF remains challenging. We were the pioneer in the direct transformation of cellulose and (ligno)cellulose into furans (Li 2008). We demonstrated that $CrCl_3$ in ionic liquid was an outstanding system for HMF production from cellulose under microwave irradiation conditions. Thus, treatment of Avicel cellulose (100 mg) and $CrCl_3 \cdot 6H_2O$ (10 mg) in [C_4mim]Cl (2.0 g) under microwave irradiation (400 W) for 2 min gave HMF in 62% yield (Li et al. 2009). Interestingly, when these components were heated at 100 ºC over an oil-bath for 240 min, HMF yield was only 17%, but TRS yield was 45%. These results suggested that microwave irradiation was the key to high yields of HMF and $CrCl_3 \cdot 6H_2O$ was capable of catalyzing hydrolysis of cellulose. It should also mention that this system afforded HMF in over 90% yield when glucose was employed. Following up this work, we also demonstrated that furans production under similar conditions using biomass as the feedstocks (Zhang et al. 2010). Corn stalk, rice straw and pine wood treated under typical reaction conditions produced HMF and furfural in yields of 45–52% and 23–31%, respectively, within 3 min. In a separate study, a pair of metal chlorides ($CuCl_2$ and $CrCl_2$) was used as the catalysts for direct conversion of cellulose into HMF in [Emim]Cl under mild conditions (Su et al. 2009). For example, HMF yields were around 55% when the reactions were held at 80–120 ºC for 8 h in the presence of the paired catalyst $CuCl_2/CrCl_2$ (χ_{CuCl2} = 0.17). Authors also demonstrated that the catalytic performance of recovered [Emim]Cl and the catalysts was maintained in repeated uses once the reaction was done in an extractive separation manor using methylisobutylketone as the extracting agent. It was interesting to note that single metal chlorides at the same total loading showed considerably less activity under similar conditions. This results were in sharp contrast to our work where high HMF yield was achieved using $CrCl_3{\cdot}6H_2O$ along as the catalyst (Li et al. 2009).

In parallel, chemical transformation of lignocellulosic biomass into furans in the DMA/ionic liquids mixture catalyzed by acid catalysts was also developed (Binder and Raines 2009). In this system, the ionic liquids were used as additives. A broad range of substrates including cellulose and (ligno)cellulose were tested, and good results were achievable. For example, in the case of cellulose, HMF yield was 22% in the DMA/LiCl-HCl system in the presence of $CrCl_2$, while the yield reached 54% if ionic liquids were added to 60 wt% concentration. Interestingly, a mineral acid was required to hydrolyze the cellulose into glucose for a satisfactory HMF yield in this mixed solvent system, while in the pure ionic liquid system, $CrCl_2$ acted as a catalyst both for the hydrolysis process and for the muturonation and dehydration processes. In 2010, conversion of cellulose into HMF in up to 89% yield was reported using $CrCl_2$ as a co-catalyst in [Emim]Cl-water mixture under relatively mild conditions (≤ 140 ºC, 1 atm) (Zhang et al. 2010).

Given the significant achievements in the conversion hexoses into HMF in ionic liquid in the presence of metal chlorides, in-depth information is still needed for a creditable mechanism. Back to 1970s, the conversion of glucose into HMF in acidic water was studied (Harris and Feather 1974; Harris and Feather 1975; Farber et al. 1989). Although the yields were low, mechanistic insights were collected from isotopic labeling experiments. It was believed that an 1,2-hydride shift was a major contributor in the process of formation of fructose and HMF from glucose. Inspired by the enzyme-catalyzed process, in which the initial proton transfer is catalyzed by a basic residue (His53) located in the cavity of the enzyme active site, Pidko and co-workers reported the molecular-level details of the unique reactivity of chromium (II) chloride towards selective glucose dehydration in ionic liquids through combining kinetic experiments, in situ X-ray absorption spectroscopy and density functional theory calculations (Pidko et al. 2010). They proposed that the chloride anions of the ionic liquids play the role of basic mediators upon formation of a hydrogen-bonding network with the hydroxyl groups of the carbohydrate, thus facilitating the proton transfer; the rate- limiting H-shift reaction of the open form of the carbohydrate is facilitated by the transient self-organization of the Lewis acidic Cr^{2+} centers into a binuclear complex. This research provided a primary cooperative nature of the Cr complexes and the presence of moderately basic sites in the ionic liquids.

Accordingly, two variations for the mechanism were proposed recently (Binder and Raines 2009; Binder and Raines 2010). The enolization pathway that goes through an enediol(ate) intermediate that is protonated at C-1 to yield a ketose assisted by Cr^{2+}, which is dehydrated into HMF *via* a furanosyl oxocarbenium ion. The hydride shift pathway that goes through a chromium assisted 1,2-hydride shift to form the ketose from the aldose in a single step. Their further study using isotopic labeled glucose demonstrated the ketose formation occurring *via* 1,2-hydride shift, similar to that found in the catalytic mechanism of xylose isomerase. We considered that both in these two pathways, the cooperative action of Cr^{2+} and halide anion contributed the formation of the enediol intermediate and the 1,2-hydride shift to form ketose from aldose. Collectively, a putative mechanism for the chromium- catalyzed conversion of glucose into HMF was presented in Figure 5. It is worthy to mention that although $CrCl_2$ has been very successful in terms of HMF production from glucose and cellulose, $CrCl_3$ or $CrBr_3$ gave similar, even better results for the same chemistry (Binder and Raines 2009; Li et al. 2009; Binder and Raines 2010). We believe that the good performance of $CrCl_2$ might be due to the oxidation of $CrCl_2$ into $CrCl_3$ by residual oxygen in the reaction system. Apart from the metal catalysts for the efficient conversion of glucose into HMF in ionic liquids, heteropoly acids were also very efficient for the dehydration of glucose to HMF, particularly in the presence of acetonitrile as a cosolvent (Chidambaram and Bell 2010). For example, a glucose conversion of 98% with a HMF selectivity of 99% can be achieved using 12-molybdophosphoric acid (12-MPA) as a catalyst in [Emim]Cl and acetonitrile after 3 h at 393 K. The authors proposed that the high selectivity was ascribed to the stabilization of the 1,2-enediol and other intermediates involved in the dehydration of glucose, the avoidance of the 2,3-endiol intermediate leading to furylhydroxymethyl ketone and the suppression of the formation of humans from glucose by the added acetonitrile.

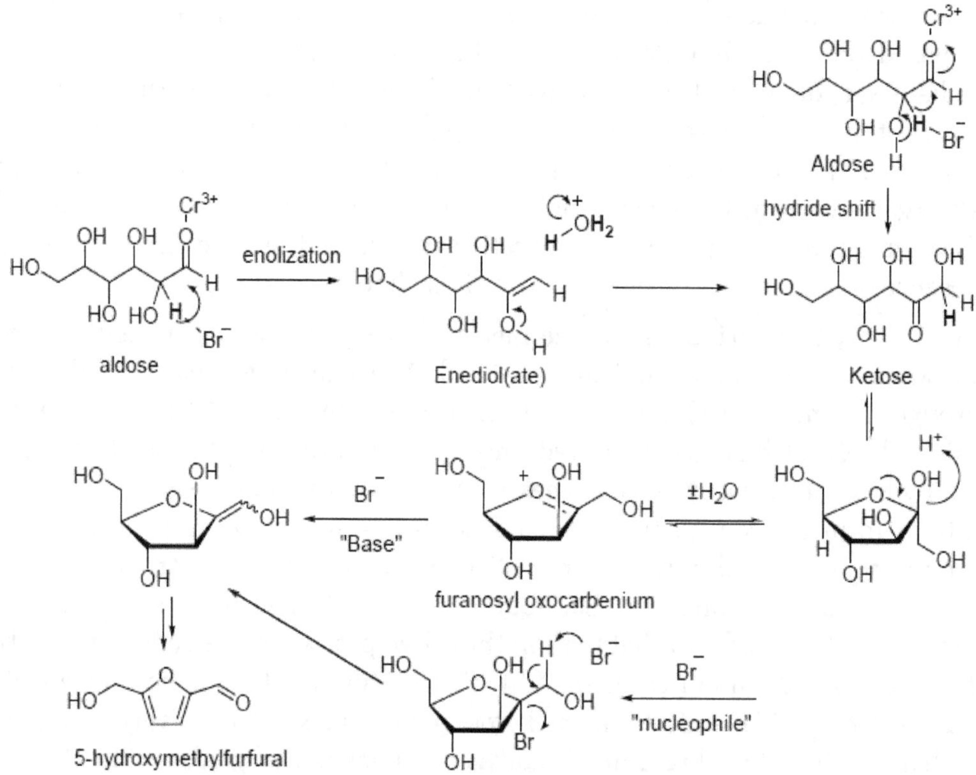

Fig. 5. Possible mechanisms for the dehydration of glucose into HMF in the case of bromide based ionic liquids in the presence of Cr2+ (Formulated and improved from that in references (Binder and Raines 2009; Binder and Raines 2010)).

3.3 Reactions leading to miscellaneous products

As expected, a successful lignocellulosic biorefinery can be realized through a combination of different catalytic technologies and biomass processing strategies in ionic liquids for the flexible production of fuels, chemicals and materials (Ragauskas et al. 2006; Corma et al. 2007; Gallezot 2008; Alonso et al. 2010). Sugar alcohols, a family of important bioproducts, are usually produced from the hydrogenation of monosaccharides. Given that cellulose can be dissolved in ionic liquids, it is logical to develop processes for direct hydrogenolysis of cellulose into sugar alcohols. Reductive depolymerization of cellulose was recently reported in the presence of hydrogen gas catalyzed by the combination of a heterogeneous metal catalyst and a homogeneous ruthenium catalyst (Ignatyev et al. 2010). Authors began their study with the hydrogenation of 1,1-diethoxycyclohexane using a combination of the heterogeneous catalysts Rh/C and Pt/C in the presence of Lewis acid $BF_3{\cdot}Et_2O$ at 20°C under 2 MPa hydrogen gas, and ethoxycyclohexane was obtained in 100% yield. Further study found that the catalytic system was not effective for the hydrogenation of cellobiose. However, a much better result, 43% yield of sorbitol was obtained when the heterogeneous catalysts were replaced by a homogeneous catalyst precursor ($HRuCl(CO)(PPh_3)_3$). Under optimized conditions, sorbitol yield was 51% using cellulose as the feedstock. Additional control experiments confirmed that each component of the catalytic mixture, such as hydrogen, $HRuCl(CO)(PPh_3)_3$, Pt/C, and water, was necessary for the reaction and that the role of the ruthenium complex was to act as a hydrogen transport agent in the ionic liquids *via* formation of hydride compounds, thus supplying Pt/C catalyst with hydrogen.

Transition metal nanoparticles as catalysts for organic transformation has been received extensively attention due to their superior catalytic activities and selectivity relative to their corresponding bulk catalytic materials (Johnson 2003). It is generally received that their small particle size and high surface structure dominated their better performance. Zhu and co-workers (Zhu et al. 2010) designed an ionic liquid capable reversible binding cellulose with the consideration of the ability to stabilize transition metal nanoparticles and the reversible reaction of a boronic acid with sugars (Figure 6). They anticipated that boronic acids could break up the crystal packing of cellulose by reversible binding with the multiple hydroxyl groups, thereby improving solubility and catalytic activity. It was found that in the absence of the designed ionic liquids, a 15% conversion was achieved for cellulose hydrogenation to hexitols using a Ru nanocluster catalyst in [Bmim] Cl. However, the cellulose can be successfully converted into hexitols in high yields ranging from 76% to 93%. It is also interesting to find that cellulose was smoothly hydrolyzed to glucose using the designed ionic liquids with a 95% yield after 5 h at 80 °C. Furthermore, the nanoparticle catalyst was easily recovered from the reaction mixture, and the activity was maintained after 5 runs. Further study by 11B NMR supported the complex formations between the ionic liquid with cellulose, which facilitated the solubilization of cellulose.

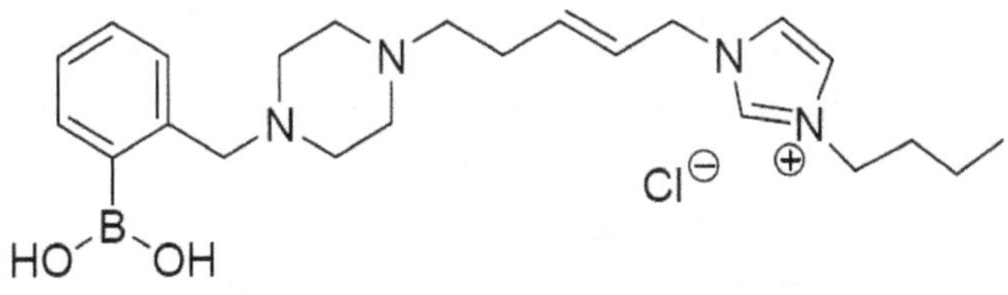

Fig. 6. Boronic acid functionalized ionic liquids

Long-chain alkyl glycosides are no-ionic compounds with excellent surfactant properties, low toxicity and good biodegradability. They have a couple of applications as cosmetics and detergents, food emulsifiers and pharmaceutical dispersing agents. One-pot catalytic conversion of cellulose into biodegradable surfactants was reported in ionic liquids under mild conditions (Villandier and Corma 2010). It was found that a sulfuric resin can catalyze a direct conversion of cellulose into alkyl glucosides with mass yields up to 82% (Figure 7). A lot of alcohols including butanol, hexanol, and octanol were used in the study, and total yields of surfactants were ranged from 7.2% to 91%. The amount of water in the reaction medium played an important role. During the hydrolysis process, water was a reactant inhibiting the formation of HMF as well as hemiacetals between the aldehyde group of the monosaccharide and the alcohols. They found that a removal of water from the system by reducing the reaction pressure could facilitate the glycosidation significantly.

A very similar work was also reported where cellulose was converted into the related alkyl glycosides in [Bmim]Cl in the presence of an acidic resin Amberlyst 15DRY (Ignatyev et al. 2010).

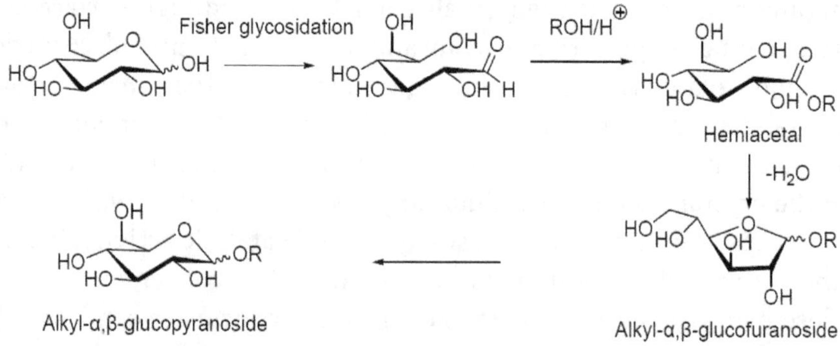

Fig. 7. Conversion of glucose into alkyl-α,β-glycoside surfactants

3.4 Greener pathways for the conversion of (ligno)cellulose in ionic liquids

The development of a sustainable chemical industry requires sustainable raw materials and technologies for the conversion of raw materials into end products. The design of a "green" compound or 'green' technology, whether the role is as a solvent, reagent or catalyst should ideally address issues such as low toxicity and ready biodegradability without the generation of toxic, persistent metabolites (Kamm 2007) . The conversion of biomass into value-added chemicals with the ionic liquids platform already have met two of the 12 green chemistry principles, using green solvents and biorenewable resources as raw materials. With more studies in this field, researchers are also realizing the potential problems, such as potential biotoxicity of ionic liquids, low biodegradability, cost-intensive separation process and energy consumption (Coleman 2010; Thi et al. 2010). With the development of green technology and green chemistry, researchers are also introducing new technologies and concepts into this field (Anastas and Zimmerman 2003) (Figure 8).

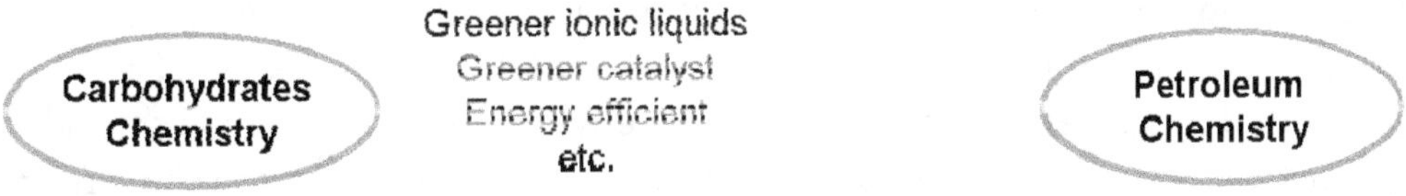

Fig. 8. Greener pathways for the conversion of (ligno)cellulose in ionic liquids

Microwave irradiation is regarded as a more energy efficient heating method and it has been widely used in organic synthesis. In 2009, we demonstrated that microwave irradiation could drastically improve the CrCl3-catalyzed HMF production form cellulose in ionic liquids (Li et al. 2009). In fact, ionic liquids have excellent dielectric properties which can facilitate the energy transformation to heat the reaction rapidly, volumetrically and simultaneously. Other microwave effects, such as lowering activation energy or increasing the preexponential factor in the Arrhenius law due to orientation effect of polar species in an electromagnetic field might also contributed.

With increasing concerns about the environmental effects, the design, synthesis and application of low toxic and biodegradable ionic liquids have been obtained much attention. In this regard, some renewable materials-based ionic liquids have been prepared to investigate their effectiveness in the conversion of fructose into HMF (Hu et al. 2008; Hu et al. 2009). Some ionic liquids prepared entirely from cheap and renewable materials, such as ChoCl/malonic acid, ChoCl/oxalic acid and ChoCl/citric acid were found very effective for the reaction. For example, HMF yield was 83% where the ionic liquid ChoCl/citric acid was used as the solvent. It was also found that HMF was rather stable in these ionic liquids. In order to improve the HMF yield and selectivity in ChoCl/citric acid, and to reuse the ionic liquid, ethyl acetate was added as an extracting solvent to extract HMF out of the system, which gave a HMF yield of 86%, about 9% higher than that in the absence of the extract solvent. The ionic liquids-catalyst mixture was successfully recycled after pouring out the ethyl acetate phase. The most important aspects of this work are, 1) ionic liquids synthesized from biorenewable feedstock, and 2) the reaction performed under continuous reaction- extraction fashion. These are two important factors determining the practical potential of a technology in a large scale. Further study from the same group also showed that the system was also appropriate for the direct conversion of inulin to HMF in 64% yield under mild conditions (Hu et al. 2009).

With the consideration of the toxicity of chromium ion, researchers are devoting to discovering a cheap, non-toxic, easily handled catalysts for the selective conversion of (ligno)cellulose into furans. $SnCl_4$ is a common, cheap, easy handling Lewis acid, and has a much lower toxicity. Hu and co-workers investigated its catalytic activity in several ionic liquids for the conversion of glucose into HMF. The activity mostly varied with the anions in ionic liquids, and a best result of 61%

yield was obtained in [Emim]BF_4. It is interesting to mention that although the Cl^- is essential for the Cr^{2+} catalytic system, it does not work well in this system (Hu et al. 2009). Inspired by a catalytic effect on glucose dehydration by lanthanide chlorides in supercritical water, the catalytic effect of lanthanide catalysts on the conversion of glucose into HMF in ionic liquids was investigated (Stahlberg et al. 2010). However, HMF yields were moderate or relatively low (less than 30%). Germanium (IV) chloride is an environment-friendly Lewis acid. We recently investigated the direct conversion of carbohydrates into HMF in [C_4mim]Cl (Zhang et al. 2010). It was found that monosaccharides D-fructose and D-glucose, disaccharides sucrose and maltose, and even polysaccharides cellulose were successfully converted into HMF with good yields ranging from 30% to 92% under mild conditions. Thus, a HMF yield of 92% was achieved in 5 min in the case of fructose. Interestingly, we noticed that a higher content of water content in the system resulted in a lower HMF yield. However, addition of 5 Å molecular sieves during the dehydration of glucose afforded an increase in HMF yield from 38% to 48%.

4. CONCISE SUMMARY

The conversion of biomass into value-added chemicals using ionic liquids is in its nascent stages, and the intricate challenges will undoubtedly necessitate the integration of sophisticated catalytic and separation technologies. This chapter aims to elucidate the current advancements in this domain, particularly the molecular-level comprehension of biomass dissolving in ionic liquids, methodologies for converting (ligno)cellulose into fermentable sugars and furans, and associated compounds.

Undoubtedly, many milestone-results have been obtained so far. For example, a remarkable TRS yield of 97% was obtained for hydrolysis of cellulose in the water system under mild conditions, which should provide cheaper fermentable feedstocks. If the separation of monosaccharides from ionic liquids were realized economically, the cost of the second generation of bioenergy is expected to be reduced significantly. Indeed, there are at least two references about sugar recovery after the hydrolysis of cellulose in ionic liquids. One is to use ion-exclusion chromatography to simultaneously recover the ionic liquids and the sugars from the reaction mixture, and the obtained sugar solution was proved suitable for the production of ethanol by ethanologenic bacterium or yeast (Binder and Raines 2010). An alternative approach involves the extraction of sugars from ionic liquids utilizing the chemical affinity of sugars for boronates, such phenyl boronic acid and naphthalene-2-boronic acid (Brennan et al. 2010). Boronate complexes were demonstrated to extract up to 90% of mono- and disaccharides from aqueous ionic liquid solutions, pure ionic liquid systems, or hydrolysates of maize stover containing ionic liquids. Both methods produced a solution of fermentable sugars, minimized hazardous byproducts, and offered technological strategies for the complex separation issue in the ionic liquid-based biorefinery. The incorporation of these technologies with conventional fermentable methods is expected to enhance the prospects for producing second-generation biofuels.

Furans, particularly HMF, are considered significant intermediates connecting carbohydrate chemistry and petroleum chemistry. The direct and efficient conversion of cellulose and glucose into HMF has long been unsuccessful. The ionic liquids approach enables the production of HMF in yields exceeding 70% straight from cellulose under mild circumstances. The ongoing advancement of downstream goods derived from furans presents significant challenges. Future endeavors should focus on the synthesis, structural characterisation, and property analysis of novel bioproducts or biofuels derived from these platform molecules. The expanding end product market will foster the development of the associated field.

Despite these substantial advancements, numerous technological and scientific challenges persist, including large-scale economic synthesis, the establishment of recycling pathways, the efficient separation of products and ionic liquids, the development of novel efficient catalysts, the production of new value-added chemicals, biodegradation and bioaccumulation studies, as well as toxicity and handling considerations for ionic liquids. The future is promising, however prospects are invariably accompanied by hurdles in the pursuit of industrial applications. Additional investigation into the selective degradation of (ligno)cellulose, or more broadly, the application of biomass utilizing ionic liquid technology is anticipated as follows.:

1. Advancement of novel catalytic systems for enhanced transformations, such depolymerization, dehydration, hydrogenolysis, or alcoholysis, of (ligno)cellulose and analogous materials;
2. Utilization of innovative chemistry of (ligno)cellulose in ionic liquids or ionic liquid-infused media;
3. Integration of ionic liquid processes with additional chemical or biological methodologies;
4. Design and synthesis of cost-effective, environmentally friendly, and process-compatible ionic liquids;
5. Formulating innovative solutions for the recycling of ionic liquids and the recovery of products;
6. Catalytic transformation of lignin into high-value compounds.

REFERENCES

Amarasekara, A. S. and Owereh, O. S. (2009). Hydrolysis and Decomposition of Cellulose in Bronsted Acidic Ionic Liquids under Mild Conditions. Industrial & Engineering Chemistry Research 48(22): 10152-10155.

Amarasekara, A. S. and Owereh, O. S. (2010). Synthesis of a Sulfonic Acid Functionalized Acidic Ionic Liquid Modified Silica Catalyst and Applications in the Hydrolysis of Cellulose. Catalysis Communications 11(13): 1072-1075.

Alonso, D. M., Bond, J. Q. and Dumesic, J. A. (2010). Catalytic Conversion of Biomass to Biofuels. Green Chemistry, 12(9):1493-1513

Anastas, P. T. and Zimmerman, J. B. (2003). Design through the 12 Principles of Green Engineering. Environmental Science & Technology 37(5): 94A-101A.

Binder, J. B., Cefali, A. V., Blank, J. J. and Raines, R. T. (2010). Mechanistic Insights on the Conversion of Sugars into 5-Hydroxymethylfurfural. Energy & Environmental Science 3(6): 765-771.

Binder, J. B. and Raines, R. T. (2009). Simple Chemical Transformation of Lignocellulosic Biomass into Furans for Fuels and Chemicals. Journal of the American Chemical Society 131(5): 1979-1985.

Binder, J. B. and Raines, R. T. (2010). Fermentable Sugars by Chemical Hydrolysis of Biomass. Proceedings of the National Academy of Sciences of the United States of America 107(10): 4516-4521.

Bose, S., Armstrong, D. W. and Petrich, J. W. (2010). Enzyme-Catalyzed Hydrolysis of Cellulose in Ionic Liquids: A Green Approach toward the Production of Biofuels. Journal of Physical Chemistry B 114(24): 8221-8227.

Brennan, T. C. R., Datta, S., Blanch, H. W., Simmons, B. A. and Holmes, B. M. (2010). Recovery of Sugars from Ionic Liquid Biomass Liquor by Solvent Extraction. Bioenergy Research 3(2): 123-133.

Cheng, S. M. and Zhu, S. D. (2009). Lignocellulosic Feedstock Biorefinery-the Future of the Chemical and Energy Industry. Bioresources 4(2): 456-457.

Chheda, J. N., Huber, G. W. and Dumesic, J. A. (2007). Liquid-Phase Catalytic Processing of Biomass-Derived Oxygenated Hydrocarbons to Fuels and Chemicals. Angewandte Chemie-International Edition 46(38): 7164-7183.

Chidambaram, M. and Bell, A. T. (2010). A Two-Step Approach for the Catalytic Conversion of Glucose to 2,5-Dimethylfuran in Ionic Liquids. Green Chemistry 12(7): 1253-1262. Coleman, D. and Gathergood, N. (2010). Biodegradation Studies of Ionic Liquids. Chemical Society Reviews 39(2): 600-637.

Corma, A., Iborra, S. and Velty, A. (2007). Chemical Routes for the Transformation of Biomass into Chemicals. Chemical Reviews 107(6): 2411-2502.

Farber, G. K., Glasfeld, A., Tiraby, G., Ringe, D. and Petsko, G. A. (1989). Crystallographic Studies of the Mechanism of Xylose Isomerase. Biochemistry 28(18): 7289-7297.

Fukaya, Y., Sugimoto, A. and Ohno, H. (2006). Superior Solubility of Polysaccharides in Low Viscosity, Polar, and Halogen-Free 1,3-Dialkylimidazolium Formates. Biomacromolecules 7(12): 3295-3297.

Gallezot, P. (2008). Catalytic Conversion of Biomass: Challenges and Issues. ChemSusChem 1(8-9): 734-737.

Harris, D. W. and Feather, M. S. (1974). Intramolecular C-2- C-1 Hydrogen Transfer- Reactions During Conversion of Aldoses to 2-Furaldehydes. Journal of Organic Chemistry 39(5): 724-725.

Harris, D. W. and Feather, M. S. (1975). Studies on Mechanism of Interconversion of D- Glucose, D-Mannose, and D-Fructose in Acid Solution. Journal of the American Chemical Society 97(1): 178-182.

Hu, S. Q., Zhang, Z. F., Song, J. L., Zhou, Y. X. and Han, B. X. (2009). Efficient Conversion of Glucose into 5-Hydroxymethylfurfural Catalyzed by a Common Lewis Acid $Sncl_4$ in an Ionic Liquid. Green Chemistry 11(11): 1746-1749.

Hu, S. Q., Zhang, Z. F., Zhou, Y. X., Song, J. L., Fan, H. L. and Han, B. X. (2009). Direct Conversion of Inulin to 5-Hydroxymethylfurfural in Biorenewable Ionic Liquids. Green Chemistry 11(6): 873-877.

Hu, S. Q., Zhang, Z. F., Zhou, Y. X., Han, B. X., Fan, H. L., Li, W. J., Song, J. L. and Xie, Y. (2008). Conversion of Fructose to 5-Hydroxymethylfurfural Using Ionic Liquids Prepared from Renewable Materials. Green Chemistry 10(12): 1280-1283.

Huber, G. W., Iborra, S. and Corma, A. (2006). Synthesis of Transportation Fuels from Biomass: Chemistry, Catalysts, and Engineering. Chemical Reviews 106(9): 4044- 4098.

Ignatyev, I. A., Van Doorslaer, C., Mertens, P. G. N., Binnemans, K. and de Vos, D. E. (2010). Reductive Splitting of Cellulose in the Ionic Liquid 1-Butyl-3-Methylimidazolium Chloride. ChemSusChem 3(1): 91-96.

Ignatyev, I. A., Mertens, P. G. N., Van Doorslaer, C., Binnemans, K. and de Vos, D. E. (2010). Cellulose Conversion into Alkylglycosides in the Ionic Liquid 1-Butyl-3- Methylimidazolium Chloride. Green Chemistry 12(10): 1790-1795.

Johnson, B. F. G. (2003). Nanoparticles in Catalysis. Topics in Catalysis 24(1-4): 147-159.

Kamiya, N., Matsushita, Y., Hanaki, M., Nakashima, K., Narita, M., Goto, M. and Takahashi, H. (2008). Enzymatic *in Situ* Saccharification of Cellulose in Aqueous-Ionic Liquid Media. Biotechnology Letters 30(6): 1037-1040.

Kamm, B. (2007). Production of Platform Chemicals and Synthesis Gas from Biomass. Angewandte Chemie-International Edition 46(27): 5056-5058.

Kilpelainen, I., Xie, H., King, A., Granstrom, M., Heikkinen, S. and Argyropoulos, D. S. (2007). Dissolution of Wood in Ionic Liquids. Journal of Agricultural and Food Chemistry 55: 9142-9148.

Kim, S. J., Dwiatmoko, A. A., Choi, J. W., Suh, Y. W., Suh, D. J. and Oh, M. (2010). Cellulose Pretreatment with 1-n-Butyl-3-Methylimidazolium Chloride for Solid Acid- Catalyzed Hydrolysis. Bioresource Technology 101(21): 8273-8279.

Li, B., Filpponen, I. and Argyropoulos, D. S. (2010). Acidolysis of Wood in Ionic Liquids. Industrial & Engineering Chemistry Research 49(7): 3126-3136.

Li, C. Z. (2008). Application of Ionic Liquids in Biomass Conversion and Friedel-Crafts Acylation Reactions. Ph.D. Thesis, Dalian Institute of Chemical Physics, CAS.

Li, C. Z., Wang, Q. and Zhao, Z. K. (2008). Acid in Ionic Liquid: An Efficient System for Hydrolysis of Lignocellulose. Green Chemistry 10(2): 177-182.

Li, C. Z., Zhang, Z. H. and Zhao, Z. K. (2009). Direct Conversion of Glucose and Cellulose to 5-Hydroxymethylfurfural in Ionic Liquid under Microwave Irradiation. Tetrahedron Letters 50(38): 5403-5405.

Li, C. Z. and Zhao, Z. K. (2007). Efficient Acid-Catalyzed Hydrolysis of Cellulose in Ionic Liquid. Advanced Synthesis & Catalysis 349: 1847-1850.

Li, Q., He, Y. C., Xian, M., Jun, G., Xu, X., Yang, J. M. and Li, L. Z. (2009). Improving Enzymatic Hydrolysis of Wheat Straw Using Ionic Liquid 1-Ethyl-3-Methyl Imidazolium Diethyl Phosphate Pretreatment. Bioresource Technology 100(14): 3570-3575.

Liu, H. B., Sale, K. L., Holmes, B. M., Simmons, B. A. and Singh, S. (2010). Understanding the Interactions of Cellulose with Ionic Liquids: A Molecular Dynamics Study. Journal of Physical Chemistry B 114(12): 4293-4301.

Liu, L. Y. and Chen, H. Z. (2006). Enzymatic Hydrolysis of Cellulose Materials Treated with Ionic Liquid Bmim Cl. Chinese Science Bulletin 51(20): 2432-2436.

MacFarlane, D. R., Pringle, J. M., Johansson, K. M., Forsyth, S. A. and Forsyth, M. (2006). Lewis Base Ionic Liquids. Chemical Communications (18): 1905-1917.

Moniruzzaman, M., Nakashima, K., Kamiya, N. and Goto, M. (2010). Recent Advances of Enzymatic Reactions in Ionic Liquids. Biochemical Engineering Journal 48(3): 295- 314.

Ong, Y. K. and Bhatia, S. (2010). The Current Status and Perspectives of Biofuel Production *Via* Catalytic Cracking of Edible and Non-Edible Oils. Energy 35(1): 111-119.

Pidko, E. A., Degirmenci, V., van Santen, R. A. and Hensen, E. J. M. (2010). Glucose Activation by Transient Cr^{2+} Dimers. Angewandte Chemie-International Edition 49(14): 2530-2534.

Pinkert, A., Marsh, K. N., Pang, S. S. and Staiger, M. P. (2009). Ionic Liquids and Their Interaction with Cellulose. Chemical Reviews 109(12): 6712-6728.

Ragauskas, A. J., Williams, C. K., Davison, B. H., Britovsek, G., Cairney, J., Eckert, C. A., Frederick, W. J., Hallett, J. P., Leak, D. J., Liotta, C. L., Mielenz, J. R., Murphy, R., Templer, R. and Tschaplinski, T. (2006). The Path Forward for Biofuels and Biomaterials. Science 311(5760): 484-489.

Remsing, R. C., Liu, Z. W., Sergeyev, I. and Moyna, G. (2008). Solvation and Aggregation of N,N'-Dialkylimidazolium Ionic Liquids: A Multinuclear NMR Spectroscopy and Molecular Dynamics Simulation Study. Journal of Physical Chemistry B 112(25): 7363-7369.

Remsing, R. C., Swatloski, R. P., Rogers, R. D. and Moyna, G. (2006). Mechanism of Cellulose Dissolution in the Ionic Liquid 1-N-Butyl-3-Methylimidazolium Chloride: A C-13 and Cl-35/37 NMR Relaxation Study on Model Systems. Chemical Communications (12): 1271-1273.

Remsing, R. C., Wildin, J. L., Rapp, A. L. and Moyna, G. (2007). Hydrogen Bonds in Ionic Liquids Revisited: Cl-35/37 NMR Studies of Deuterium Isotope Effects in 1-N- Butyl-3-Methylimidazolium Chloride. Journal of Physical Chemistry B 111(40): 11619-11621.

Rinaldi, R., Palkovits, R. and Schuth, F. (2008). Depolymerization of Cellulose Using Solid Catalysts in Ionic Liquids. Angewandte Chemie-International Edition 47(42): 8047- 8050.

Rinaldi, R., Meine, N., vom Stein, J., Palkovits, R. and Schuth, F. (2010). Which Controls the Depolymerization of Cellulose in Ionic Liquids: The Solid Acid Catalyst or Cellulose? ChemSusChem 3(2): 266-276.

Rogers, R. D. and Seddon, K. R. (2003). Ionic Liquids - Solvents of the Future? Science 302(5646): 792-793.

Stahlberg, T., Sorensen, M. G. and Riisager, A. (2010). Direct Conversion of Glucose to 5- (Hydroxymethyl) furfural in Ionic Liquids with Lanthanide Catalysts. Green Chemistry 12(2): 321-325.

Stephanopoulos, G. (2007). Challenges in Engineering Microbes for Biofuels Production. Science 315(5813): 801-804.

Su, Y., Brown, H. M., Huang, X. W., Zhou, X. D., Amonette, J. E. and Zhang, Z. C. (2009). Single-Step Conversion of Cellulose to 5-Hydroxymethylfurfural (HMF), a Versatile Platform Chemical. Applied Catalysis A-General 361(1-2): 117-122.

Swatloski, R. P., Spear, S. K., Holbrey, J. D. and Rogers, R. D. (2002). Dissolution of Cellose with Ionic Liquids. Journal of the American Chemical Society 124(18): 4974-4975.

Thi, P. T. P., Cho, C. W. and Yun, Y. S. (2010). Environmental Fate and Toxicity of Ionic Liquids: A Review. Water Research 44(2): 352-372.

Vanoye, L., Fanselow, M., Holbrey, J. D., Atkins, M. P. and Seddon, K. R. (2009). Kinetic Model for the Hydrolysis of Lignocellulosic Biomass in the Ionic Liquid, 1-Ethyl-3- Methyl-Imidazolium Chloride. Green Chemistry 11(3): 390-396.

Villandier, N. and Corma, A. (2010). One Pot Catalytic Conversion of Cellulose into Biodegradable Surfactants. Chemical Communications 46(24): 4408-4410.

Werpy, T., Petersen, G. Eds. (2004). Top Value Added Chemicals from Biomass, Report: Doe/Go-102004-1992;. U.S. Department of Energy.

Xie, H., Kilpeläinen, I., King, A., W., T., Järvi, P. and Argyropoulos, D. S., (2009). Opportunities with Dissolved Wood. Cellulose Solvents: For Analysis, Shaping and Chemical Modification. Liebert, T., et al., ACS Symposium Series, 2009, Chapter 19, 343-363.

Xie, H., King, A., Kilpelainen, I., Granstrom, M. and Argyropoulos, D. S. (2007). Thorough Chemical Modification of Wood-Based Lignocellulosic Materials in Ionic Liquids. Biomacromolecules 8(12): 3740-3748.

Yang, F., Li, L. Z., Li, Q., Tan, W. G., Liu, W. and Xian, M. (2010). Enhancement of Enzymatic *in Situ* Saccharification of Cellulose in Aqueous-Ionic Liquid Media by Ultrasonic Intensification. Carbohydrate Polymers 81(2): 311-316.

Youngs, T. G. A., Hardacre, C. and Holbrey, J. D. (2007). Glucose Solvation by the Ionic Liquid 1,3-Dimethylimidazolium Chloride: A Simulation Study. Journal of Physical Chemistry B 111(49): 13765-13774.

Zakrzewska, M. E., Bogel-Lukasik, E. and Bogel-Lukasik, R. (2010). Solubility of Carbohydrates in Ionic Liquids. Energy & Fuels 24: 737-745.

Zavrel, M., Bross, D., Funke, M., Buchs, J. and Spiess, A. C. (2009). High-Throughput Screening for Ionic Liquids Dissolving (Ligno-)Cellulose. Bioresource Technology 100(9): 2580-2587.

Zhang, Y. T., Du, H. B., Qian, X. H. and Chen, E. Y. X. (2010). Ionic Liquid-Water Mixtures: Enhanced Kw for Efficient Cellulosic Biomass Conversion. Energy & Fuels 24: 2410- 2417.

Zhang, Z. H., Wang, Q., Xie, H. B., Dong, K., Liu, W. J. and Zhao, Z. K. (2010). Catalytic Conversion of Carbohydrates into 5-Hydroxymethylfurfural by Germanium (IV) Chloride in Ionic Liquids. ChemSusChem, DOI: 10.1002/cssc.201000279.

Zhang, Z. H. and Zhao, Z. K. (2010). Microwave-Assisted Conversion of Lignocellulosic Biomass into Furans in Ionic Liquid. Bioresource Technology 101(3): 1111-1114.

Zhao, H. (2010). Methods for Stabilizing and Activating Enzymes in Ionic Liquids - a Review. Journal of Chemical Technology and Biotechnology 85(7): 891-907.

Zhao, H. B., Holladay, J. E., Brown, H. and Zhang, Z. C. (2007). Metal Chlorides in Ionic Liquid Solvents Convert Sugars to 5-Hydroxymethylfurfural. Science 316(5831): 1597-1600.

Zhu, S. D., Wu, Y. X., Chen, Q. M., Yu, Z. N., Wang, C. W., Jin, S. W., Ding, Y. G. and Wu, G. (2006). Dissolution of Cellulose with Ionic Liquids and Its Application: A Mini- Review. Green Chemistry 8(4): 325-327.

Zhu, Y. H., Kong, Z. N., Stubbs, L. P., Lin, H., Shen, S. C., Anslyn, E. V. and Maguire, J. A. (2010). Conversion of Cellulose to Hexitols Catalyzed by Ionic Liquid-Stabilized Ruthenium Nanoparticles and a Reversible Binding Agent. ChemSusChem 3(1): 67- 70.

5 Chapter

CHEMICAL MODIFICATION OF CELLULOSE USING SUCCINIC ANHYDRIDE IN IONIC LIQUID, WITH OR WITHOUT CATALYSTS

1. INTRODUCTION: AN OVERVIEW

Presently, environmental challenges such as global warming and depleting energy resources are escalating into urgent concerns. The use of biomass, particularly inedible lignocellulosic materials, to produce fuels, energy, chemicals, composites, and several other goods is essential for the development of a sustainable society. (Zhang et al., 2007; Ogaki et al., 2009; Zhang et al., 2010). U.S. Department of Energy has targeted to achieve 10% of basic chemical building blocks arising from lignocellulosic biomass by 2020, and achieve a further increase to 50% by 2050 (Mohanty et al., 2002). The European Union has formulated a plan to meet 25% of its gasoline demand using biofuels by 2030. (Himmel et al., 2007).

The integrated use and multi-product biorefinery of lignocellulosic biomass is emerging as a critical concern and developmental trend. This potential approach involves the first fractionation of lignocellulosic biomass into three primary components: lignin, hemicelluloses, and cellulose. Subsequently, these isolated components are exploited separately to generate various products based on their distinct qualities. Cellulose, the predominant natural polymer, is a homopolymer consisting of D-glucopyranose units interconnected by β (1→4) glycosidic linkages. (Pandey et al., 2000). It is non-toxic, renewable, biodegradable, and adaptable, with significant potential as an exceptional industrial material (Richardson & Gorton, 2003). Cellulose has promising applications in biofibers, biopolymers, biofuels, and biocomposites. (Reddy & Yang, 2005; Clark et al., 2009). Due to three hydroxyl groups available within one anhydroglucose units (AGU), a great variety of chemical modifications of cellulose are possible (Potthast et al., 2006). Chemical modifications of cellulose can incorporate functional groups into the macromolecules in either heterogeneous or homogeneous phases to enhance the overall usage of cellulosic polymers. Homogeneous functionalization has long been a focal point of cellulose research, as it yields more homogenous and stable compounds in a homogeneous phase compared to a heterogeneous phase. (Regiani et al., 1999; El Seoud et al., 2000; Satge et al., 2004). The inherent insolubility of natural cellulose in water and most organic solvents is a significant barrier to its usage. The effective disintegration of cellulose remains a longstanding objective in cellulose research and development, retaining significant importance. (Heinze et al., 2005).

Recently, due to their unique physico-chemical properties, such as chemical and thermal stability, non-flammability and immeasurably low vapor pressure, ionic liquids (ILs) have been considered as the most potential green solvents in the future, which have attracted a great deal of scientific attention in many fields (Forsyth et al., 2002; Turner et al., 2003). Many kinds of ILs with a variety of structures have been reported as novel solvents for cellulose dissolution, including a series of alkylimidazolium salts containing chloride (Swatloski et al., 2002; Luo et al., 2005; Heinze et al., 2005; Zhang et al., 2005), formate (Fukaya et al., 2006), acetate (Hermanutz et al., 2007; Kosan et al., 2008; de Maria & Martinsson, 2009), and alkylphosphate (Fukaya et al., 2006; Fukaya et al., 2008; Kamiya et al., 2008; Mazza et al., 2009).

Based on the dissolution of cellulose in ILs, the investigation into the use of ILs as reaction media for cellulose functionalization has also been reported. Modification of cellulose represents one of the most versatile transformations as it provides access to a variety of biobased materials with valuable properties. Acylation of cellulose with linear chain acylation reagents such as anhydride or chloride is the most common method to produce cellulosic bioproducts. Because of the wide application of cellulose acetate, cellulose acetylation with acetic anhydride or acetyl chloride in ILs has been extensively studied, and cellulose acetates with high degree of substitution (DS) were easily prepared (Wu et al., 2004; Abbott et al., 2005; Heinze et al., 2005; Barthel & Heinze, 2006; Granstrom et al., 2008). Furthermore, ILs were also be reported as reaction media for cellulose modification with other liquid reagents, such as carbanilation with phenyl isocyanate (Barthel & Heinze, 2006; Schlufter et al., 2006), acylation with lauroyl chloride (Barthel & Heinze, 2006), and perpropionylation with propionic anhydride (Schlufter et al., 2006). Nonetheless, an unwanted byproduct, specifically the associated carboxylic acid, is invariably generated. The acid must be eliminated from the process following modification.

Conversely, cellulose derivatives synthesized using solid reagents like cyclic anhydrides have been extensively employed in diverse applications, including water absorbents for agricultural soil, natural absorbents for heavy metal ion removal in wastewater treatment, medicinal drug delivery systems, and thermoplastic materials. (Hadano et al., 2003; Yoshimura et al., 2006). Significantly, treatment using cyclic anhydrides such as SA does not produce such a result. Furthermore, the reaction yields a pendant carboxylic group linked to the cellulose through a covalent ester connection, creating a locus for further reactive reactions. This study examined the potential for cellulose succinoylation in ionic liquids without the use of catalysts. Three catalysts were investigated to enhance the efficiency of cellulose modification. The cellulose derivatives were characterized by Fourier transform infrared (FT-IR) spectroscopy, solid-state CP/MAS 13C nuclear magnetic resonance (NMR) spectroscopy, and thermal analysis.

2. EXPERIMENTAL

2.1 Materials

Rice straw (RS) was obtained from Guangzhou, China. It was dried in sunlight and then cut into small pieces. The cut RS was ground and screened to prepare 20-40 mesh size particles (450-900 μm). Ionic liquid [C_4mim]Cl was obtained from the Chemer Chemical Co., Ltd., Hangzhou, China, and used as received. All of other chemicals used were of analytical grade and purchased from Sigma-Aldrich, Guangzhou, China.

2.2 Isolation of cellulose from RS

The dried RS powder was first dewaxed with toluene-ethanol (2:1, v/v) in a Soxhlet apparatus for 6 h. The dewaxed RS was oven-dried and then soaked in distilled water (800 mL) at 80 °C for 2 h. The insoluble residue was then dewaxed and water-soluble-free sample was delignified with sodium chlorite at pH 3.8-4.0, adjusted by 10% acetic acid, at 75 °C for 2 h. The residue was collected by filtration, washed with distilled water and ethanol, and then dried in a cabinet oven for 16 h at 50 °C. The holocellulose obtained was extracted with 10% KOH (600 mL) at 20 °C for 10 h. After filtration, the residue was washed thoroughly with distilled water until the filtrate was neutral, and then suspended in 95% ethanol (liquor to solid ratio 25:1) for 10 h. After filtration, the solid residue was dried in an oven at 50 °C for 16 h.

2.3 Succinoylation of cellulose in ionic liquid

The succinylated cellulosic derivatives containing free carboxylic groups were prepared by homogeneous reaction of RS cellulose without any catalysts or with three different catalysts including iodine (I_2), *N*-bromosuccimide (NBS), and 4-dimethylaminopyridine (DMAP) in the solvent system containing [C_4mim]Cl and DMSO. Dried cellulose was added to [C_4mim]Cl in three-necked flask continuously purged with gaseous N_2, and the mixture of cellulose/[C_4mim]Cl was stirred at 80 °C up to 3 h to guarantee the complete dissolution of cellulose. To reduce viscosity and achieve suitable mixing, 5 mL DMSO was added to this mixture. The succinoylation reaction was carried out according to the following procedures.

2.3.1 Succinoylation without any catalysts

To the cellulose solution in [C_4mim]Cl/DMSO system, SA, previously dissolved in 5 mL DMSO, was added at the corresponding reaction temperature with the 4:1 molar ratio of SA/AGU in cellulose in Table 1. The mixture was heated in oil bath under N_2 atmosphere with magnetic stirring for the desired time. After the required time, the resulting mixture was slowly poured into 150 mL of isopropanol with stirring, washed thoroughly with isopropanol to eliminate ILs, unreacted anhydride, and byproducts, and then dried in a vacuum at 50 °C for 16 h.

2.3.2 Succinoylation catalyzed with iodine

To the cellulose solution in [C_4mim]Cl/DMSO system, SA and I_2, previously dissolved in 5 mL DMSO, were added at the corresponding reaction temperature with the 4:1 molar ratio of SA/AGU in cellulose and the corresponding weight ratio of I_2/SA shown in Table 2. The mixture was heated in oil bath under N_2 atmosphere with magnetic stirring for the desired time. Then the reaction mixture was treated with a saturated solution of sodium thiosulfate (2 mL) with stirring, and then vigorously shaken for 2 min to guarantee complete transformation of iodine to iodide. The resulting mixture was slowly poured into 150 mL of isopropanol with continuous stirring. The solid was filtered, washed thoroughly with isopropanol to eliminate ILs, un-reacted anhydride, and byproducts, and then dried in vacuum at 50 °C for 16 h.

2.3.3 Succinoylation catalyzed with N-bromosuccimide

To the cellulose solution in [C_4mim]Cl/DMSO system, SA and NBS, previously dissolved in 5 mL DMSO, were added at the corresponding reaction temperature with the 4:1 molar ratio of SA/AGU in cellulose and the corresponding weight ratio of NBS/SA shown in Table 2. The mixture was heated in oil bath under N_2 atmosphere with magnetic stirring for the desired time. After the required time, the resulting mixture was slowly poured into 150 mL of isopropanol with stirring, washed thoroughly with isopropanol to eliminate ILs, unreacted anhydride, NBS, and byproducts, and then dried in a vacuum at 50 °C for 16 h.

2.3.4 Succinoylation catalyzed with 4-dimethylaminopyridine

To the cellulose solution in [C_4mim]Cl/DMSO system, SA and DMAP, previously dissolved in 5 mL DMSO, were added at the corresponding reaction temperature with the 4:1 molar ratio of SA/AGU in cellulose and the corresponding weight ratio of DMAP/SA shown in Table 2. The mixture was heated in oil bath under N_2 atmosphere with magnetic stirring for the desired time. After the required time, the resulting mixture was slowly poured into 150 mL of isopropanol with stirring, washed thoroughly with isopropanol to eliminate ILs, unreacted anhydride, DMAP, and byproducts, and then dried in a vacuum at 50 °C for 16 h.

2.4 Determination of degree of substitution

The DS of cellulosic preparations was determined by direct titration method (Jeon et al., 1999; Stojanovic et al., 2005). A known weight of the sample was dissolved in 10 mL of DMSO by stirring at 75 oC for 30 min. After cooling, 5-6 drops of phenolphthalein indicator were added. This solution was titrated against 0.01 mol/L standard NaOH solution until a permanent pale pink color was seen. The DS was calculated by using the following equation:

$$DS = \frac{162 \times (V_{NaOH} \times c_{NaOH})}{m - 100 \times (V_{NaOH} \times c_{NaOH})}$$

where: 162 g/mol is the molar mass of an AGU, 100 g/mol is the net increase in the mass of an AGU for each succinoyl substituted, m is the weight of sample analyzed, V_{NaOH} is the volume of standard NaOH solution consumed in the titration, and c_{NaOH} is the molarity of standard NaOH solution.

2.5 Characterization of the native and succinylated cellulose

Viscosity of the cellulose was measured by British Standard Methods for determination of limiting viscosity number of cellulose in dilute solutions, Part 1, cupriethylenediamine (CED) method (BS 6306, Part 1, 1982). The viscosity average DP (degree of polymerization) of cellulose was estimated from their intrinsic viscosity [η] in CED hydroxide solution, $P^{0.90}=1.65[\eta]$, where P is an indeterminate average DP (Evans & Wallis, 1989). Molecular weight (Mw) of cellulose was then calculated from P by multiplied by 162, the Mw of an AGU. The FT-IR spectra of the cellulose and succinylated cellulosic derivatives were recorded on an FT-IR spectrophotometer (Nicolet 510) from finely ground samples (1%) in KBr pellets in the range 4000-400 cm^{-1}. Thirty-two scans were taken for each sample with a resolution of 2 cm^{-1} in the transmission mode. The solid-state CP/MAS ^{13}C NMR spectra were obtained on a Bruker DRX-400 spectrometer with 5 mm MAS BBO probe employing both Cross Polarization and Magic Angle Spinning and each experiment was recorded at ambient temperature. The spectrometer operated at 100 MHz. Acquisition time was 0.034 s, the delay time 2 s, and the proton 90° pulse time 4.85 μs. Each spectrum was obtained with an accumulation of 5000 scans.

Thermal analysis of the cellulose and succinylated cellulosic samples was performed using thermogravimetric analysis (TGA) and differential thermal analysis (DTA) on a simultaneous thermal analyzer (SDT Q600, TA Instrument). The apparatus was continually flushed with nitrogen. The sample weighed between 8 and 12 mg and the scans were run from room temperature to 550°C at a rate of 10°C per minute.

3. RESULTS AND DISCUSSION

3.1 Modification without any catalysts

Delignification of dewaxed and water-soluble-free RS followed by alkaline extraction yielded 44.6% cellulose (based on the dry weight of RS). The intrinsic viscosity, degree of polymerization, and molecular weight of the native cellulosic preparation were determined to be 348 mL· g^{-1}, 1165, and 188730 g· mol^{-1}, respectively.

Due to three hydroxyl groups available within one anhydroglucose units (AGU), a great variety of chemical modifications of cellulose are possible. In the present study, succinoylation of the cellulose isolated from RS in [C_4mim]Cl/DMSO system was investigated with or without catalysts. Usually, SA reacts with cellulose O-H groups to form the monoester as shown in Fig. 1. After cellulose was dissolved in [C_4mim]Cl, the viscous solution obtained was diluted by DMSO to achieve a suitable and clear mixiture. Solid SA and catalyst previously dissolved in DMSO were added to the solution to achieve homogeneous reaction. Table 1 shows the effects of reaction conditions including the molar ratio of SA/anhydroglucose units (AGU) in cellulose, reaction time, and reaction temperature on DS of the cellulose succinylated without any catalysts.

As shown in Table 1, an increase of molar ratio of SA/AGU in cellulose from 1:1 to 2:1, 3:1, 4:1, 6:1, 8:1, and 10:1 resulted in the improvement of the DS of succinylated cellulose from 0.02 to 0.04, 0.07, 0.11, 0.15, 0.22, and 0.31, respectively, which was due to the greater availability of succinic anhydride molecules in the proximity of the cellulosic molecules at higher concentration of succinic anhydride. The DS increased from 0.08 to 0.12, 0.15, 0.16, and 0.20 with the enhancement of reaction temperature from 90 to 100, 110, 120, and 130°C, respectively. The reason for this enhancement of succinoylation was presumed to be due to the favourable effect of temperature on succinoylation reaction. A higher reaction temperature would enhance the mobility of the molecules, the diffusion of succinic anhydride into the cellulosic molecules, and the swell ability of cellulose, therefore, increasing succinoylation rates. The rate and time of collisions of succinic anhydride with cellulosic molecules increased with the improvement of reaction time. The DS of succinylated cellulose was 0.11 within 30 min, 0.15 within 60 min, 0.17 in 90 min and 0.18 in 120 min. The low succinoylation was propably due to the increased chain of succinic anhydride compared with acetylation. The similar reduced carboxymethylation efficiency of cellulose was also reported (Heinze et al., 2005).

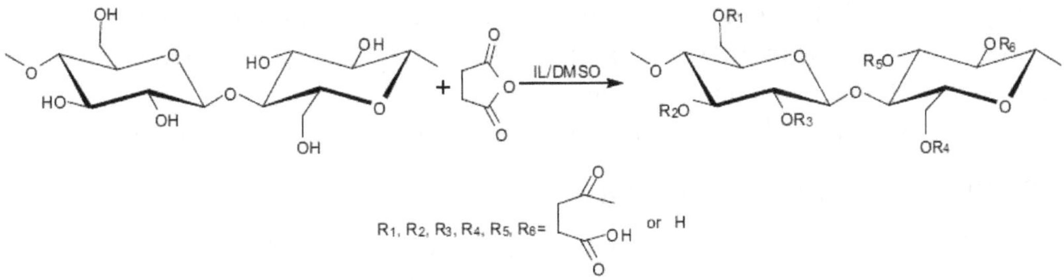

Fig. 1. Reaction of cellulose with succinic anhydride in ionic liquid.

Table 1. The DS of succinylated cellulose without any catalysts in [C_4mim]Cl/ DMSO system.

Succinoylation conditions				Succinylated cellulose	
Cellulose/[C_4mim]Cl(%)	Molar ratio (SA/AGU)	T(°C)	t(min)	Sample	DS
2	1:1	110	60	1	0.02
2	2:1	110	60	2	0.04
2	3:1	110	60	3	0.07
2	4:1	110	60	4	0.11
2	6:1	110	60	5	0.15
2	8:1	110	60	6	0.22
2	10:1	110	60	7	0.31
2	6:1	90	60	8	0.08
2	6:1	100	60	9	0.12
2	6:1	120	60	10	0.16
2	6:1	130	60	11	0.20
2	6:1	110	15	12	0.04
2	6:1	110	30	13	0.11
2	6:1	110	90	14	0.17
2	6:1	110	120	15	0.18

3.2 Iodine-catalysed succinoylation

Recently, it has been demonstrated that iodine (I_2) is an effective catalyst for acetylation of alcohols and polysaccharides without solvents (Biswas et al., 2005; Ren et al., 2007). In the present study, we investigated the possibility of cellulose succinoylation catalyzed with I_2 in [C_4mim]Cl/DMSO system to increase cellulose modification efficiency. As shown in Table 2, the increase of the dosage of I_2 from

2% to 4%, 6%, 8%, 10%, and 15% resulted the improved DS from 0.74 to 0.92, 1.13, 1.25, 1.29, and 1.30, respectively. In comparison to the sample 5 obtained without any catalyst under the same conditions (DS=0.15), the succinoylation efficiency clearly increased. The possible mechanism of succinoylation and the actual role of iodine are not clear. However, a plausible explanation is that iodine might be ionized into I^+ and I^- in ILs. I^+ in turn activates the carbonyl groups of succinic anhydride to form as the acylation reagent for further reaction, as shown in Fig. 2.

3.3 NBS-catalysed succinoylation

It has been demonstrated that NBS is a novel, highly effective catalyst for acetylation of alcohols under mild reaction conditions. It is an inexpensive and commercially available reagent that is traditionally used as an oxidizing agent or brominating agent (Karimi & Seradj, 2001). In the present study, the possibility of cellulose succinoylation catalyzed with NBS was explored in [C4mim]Cl/DMSO system to increase cellulose modification efficiency. As shown in Table 2, the DS was enhanced from 1.14 to 1.76, 1.95, 2.11, 2.17, and 2.18 with the improvement of NBS dosage from 1% to 3%, 5%, 7%, 10% and 15%, respectively, indicating the highly efficient succionylation catalysed with NBS compared with that without any catalyst. The possible mechanism of NBS-catalysed succinoylation is shown in Fig. 3. NBS might act as a source for Br^+, which in turn activates the carbonyl groups of SA to produce the highly reactive acylating agent. This acylating agent reacts with hydroxyl groups of cellulose, which upon elimination of NBS, produce succinoylated cellulose.

Table 2. The DS of succinylated cellulose with different catalysts in [C_4mim]Cl/DMSO system.

Succinoylation conditions					Succinylated cellulose	
SA/AGU	Catalyst	Catalyst/SA(%)	T/oC	t/min	Sample	DS
6:1	I2	2	110	60	16	0.74
6:1	I2	4	110	60	17	0.92
6:1	I2	6	110	60	18	1.13
6:1	I2	8	110	60	19	1.25
6:1	I2	10	110	60	20	1.29
6:1	I2	15	110	60	21	1.30
6:1	NBS	1	110	60	22	1.14
6:1	NBS	3	110	60	23	1.76
6:1	NBS	5	110	60	24	1.95
6:1	NBS	7	110	60	25	2.11
6:1	NBS	10	110	60	26	2.17

6:1	NBS	15	110	60	27	2.18
6:1	DMAP	1	110	60	28	1.26
6:1	DMAP	2	110	60	29	1.54
6:1	DMAP	3	110	60	30	1.85
6:1	DMAP	5	110	60	31	1.92
6:1	DMAP	8	110	60	32	1.99
6:1	DMAP	15	110	60	33	2.27

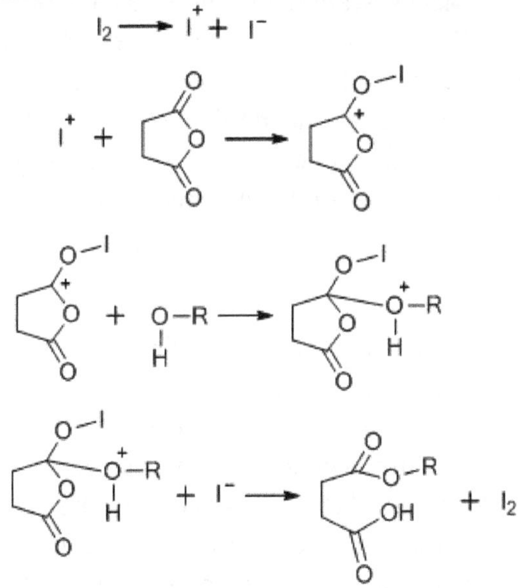

Fig. 2. Possible mechanism of succinoylation of cellulose using I2 as a catalyst.

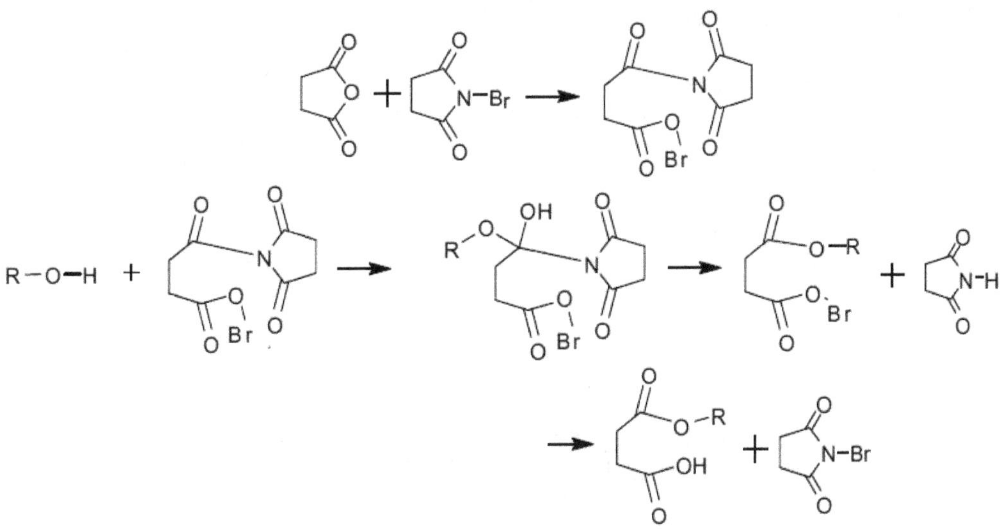

Fig. 3. Possible mechanism of succinoylation of cellulose using NBS as a catalyst.

3.4 DMAP-catalysed succinoylation

Pyridine has been found to be effective in the modification of wood with various long chain anhydrides, because it serves not only to swell the wood structure, thereby permiting effective ingress of reagent, but also catalyses the reaction via nucleopilic mediated catalysis (Sun et al., 2001). In comparison to pyridine, DMAP was found to be approximately 10^4 times more active when used as acylation catalyst (Hill et al., 2000). In the present study, we studied the succinoylation of cellulose in $[C_4mim]Cl$/DMSO system using DMAP as a catalyst. As shown in Table 2, the improvement of DMAP dosage from 1% to 2%, 3%, 5%, 8%, and 15% led to the increment in DS of succinylated cellulose from 1.26 to 1.54, 1.85, 1.92, 1.99, and 2.27, respectively. The possible mechanism of DMAP-catalysed succinoylation is shown in Fig. 3. The nucleofilic attack of DMAP on a carbonyl group of succinic anhydride leads to intermediate, which could react with cellulose hydroxyl groups and produce cellulose ester.

3.5 FT-IR

Fig. 4 illustrates the FT-IR spectra of native cellulose (spectrum 1) and succinylated samples 9 (spectrum 2, without any catalysts) and 24 (spectrum 3, 5% NBS). In spectrum 1, the absorbances at 3431, 2923, 1629, 1381, 1165, and 1048 cm^{-1} are associated with native cellulose. The strong adsorption at 3431 cm^{-1} is due to the stretching of hydroxyl groups and that at 2923 cm^{-1} to the C-H stretching. The band at 1629 cm^{-1} relates to the bending mode of the absorbed water. The peak at 1381 cm^{-1} is attributed to the O-H bending. The absorption band at 1165 cm^{-1} corresponds to C-O antisymmetric bridge stretching. A strong peak at 1048 cm^{-1} arises from C-O-C pyranose ring skeletal vibration.

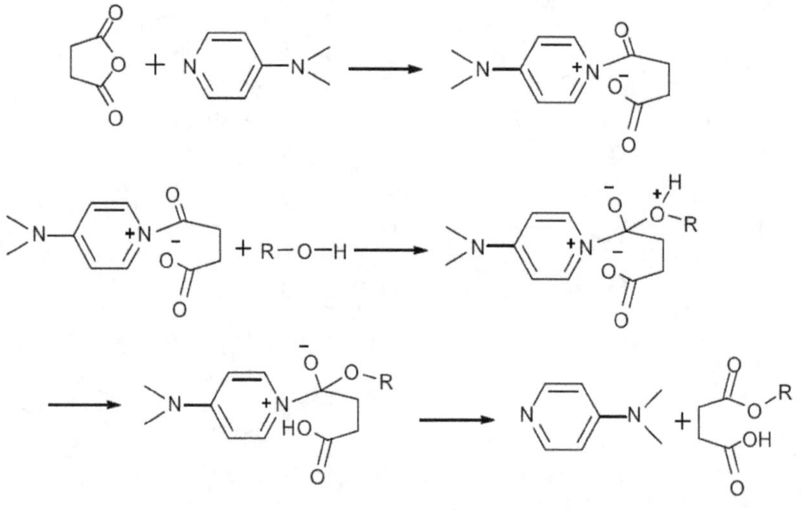

Fig. 4. Possible mechanism of succinoylation of cellulose using DMAP as a catalyst.

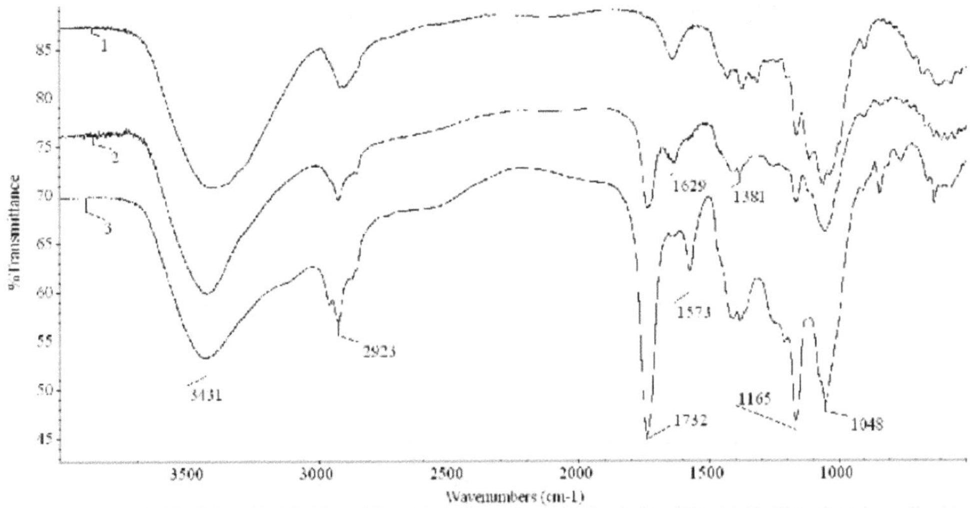

Fig. 5. FT-IR spectra of native cellulose (spectrum 1) and succinylated cellulose samples 9 (spectrum 2) and 24 (spectrum 3)

Compared with spectrum 1, the spectra 2 and 3 of succinylated cellulose provide evidence of succinoylation by the occurrence of the absorbance at 1732 and 1573 cm^{-1}. In general, the absorption by carbonyl bonds in esters gives peaks at 1750 cm^{-1}, and one in carboxylic acids exhibits a band at 1712 cm^{-1} (Jayakumar et al., 2000). The overlapping band at 1732 cm^{-1} is indicative of absorption by carbonyl group in carboxyl and esters. The absorption band at 1573 cm^{-1} is associated with the antisymmetric stretching of carboxylic anions (Yoshimura et al., 2006), indicating the formation of monoester. In addition, the intensity of the absorption band at 1165 cm^{-1} for C-O antisymmetric stretching increased after succinoylation, suggesting that the esterification reaction does occur. These significant changes of the absorption bands indicated that the monoester of succinylated cellulose was formed in $[C_4mim]Cl$/DMSO system under the conditions given. It should be noted that the intensities of the bands at 1732, 1573, and 1165 cm^{-1} are significantly improved from spectrum2 to spectrum 3, indicating the enhanced succinoylation of cellulose. It suggested that NBS is a high efficient catalyst of cellulose succinoylation in $[C_4mim]Cl$/DMSO. As expected, the absence of peaks at 1850 and 1780 cm^{-1} in spectra 2 and 3 for succinylated cellulose confirmed that the products are free of the unreacted SA.

3.6 Solid-state CP/MAS^{13}C-NMR

CP/MAS ^{13}C-NMR spectroscopy is the most effective of the few methods for the characterization of the lignocellulosic materials directly and is capable of providing detailed information directly on solid samples (Maunu, 2002). In the present study, the succinoylation reaction of cellulose with or without catalysts was also studied by solid-state CP/MAS ^{13}C-NMR spectroscopy, and the correspongding spectra are illustrated in Fig. 6.

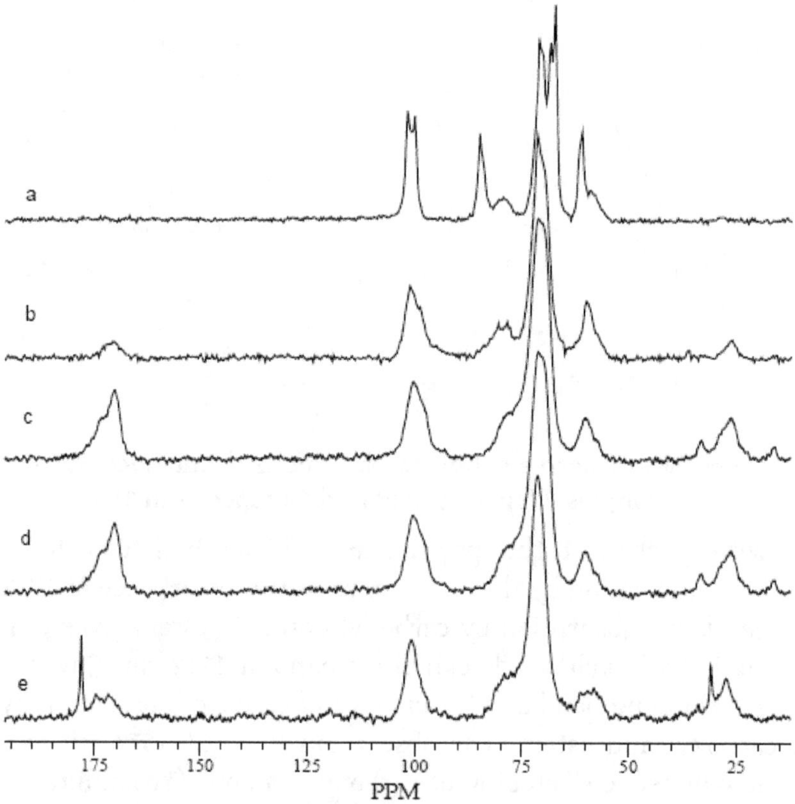

Fig. 6. Solid-state CP/MAS 13C-NMR spectra of native cellulose (spectrum a) and succinylated cellulose without catalysts (sample 5, spectrum b) and with catalyst NBS (sample 24, spectrum c), DMAP (sample 31, spectrum d), and iodine (sample 17, spectrum e).

In spectrum a, all noticeable signals are distributed in the region between 50 and 110 ppm for the carbon atoms of the carbohydrate moiety. The typical signals at 101.4 (C-1), 85.2 (C-4 of crystalline cellulose), 79.7 (C-4 of amorphous cellulose), 71.2 (C-5), 67.8 (C-2 and C-3), 61.1 (C-6 of crystalline cellulose), and 59.8 ppm (C-6 of amorphous cellulose) have all been reported before (Chang & Chang, 2001; Maunu, 2002). Evidently, the presence of the signals of the carboxylic group at 171.0 ppm and the methylene group at 26.4 ppm provided evidence of succinoylation, which indicated that the reaction shown in Fig. 1 does occur. The signals at 85.2 ppm for C-4 of crystalline cellulose and 61.1 ppm for C-6 of crystalline cellulose disappeared in spectra b to e after succinoylation, suggesting the complete disruption of cellulose crystalline structure during the dissolution and functionalization, which indicated that succinoylation reaction occurs in homogeneous phase. The intensity of the signal for C-6 significantly decreased after succinoylation, and that at 67.8 ppm for C-2 and C-3 also decreased, which indicated the succinoylation reaction at C-6, C-2, and C-3 positions does occur. Obviously, the succinoylation reaction is preferred at C-6. The significant decrease of the signal for C-6 indicated that the reactivity of C-6 is higher than that of C-2 and C-3.

3.7 Thermal analysis

Fig. 7 illustrates TGA curves of native cellulose (curve a) and succinylated cellulose without catalyst (sample 6, curve b) and with NBS catalyst (sample 22, curve 3). The native cellulose starts to decompose at 226°C, whereas the regenerated cellulose begins to decompose at 205°C and 202°C for samples 6 and 10, respectively. At 50% weight loss, the decomposition temperature occurs at 338°C for native cellulose, 330°C for sample 6, and 285°C for sample 22. These decreasing trends of decomposition temperature suggested that thermal stability of cellulose decreased after succinoylation, especially true for high DS samples.

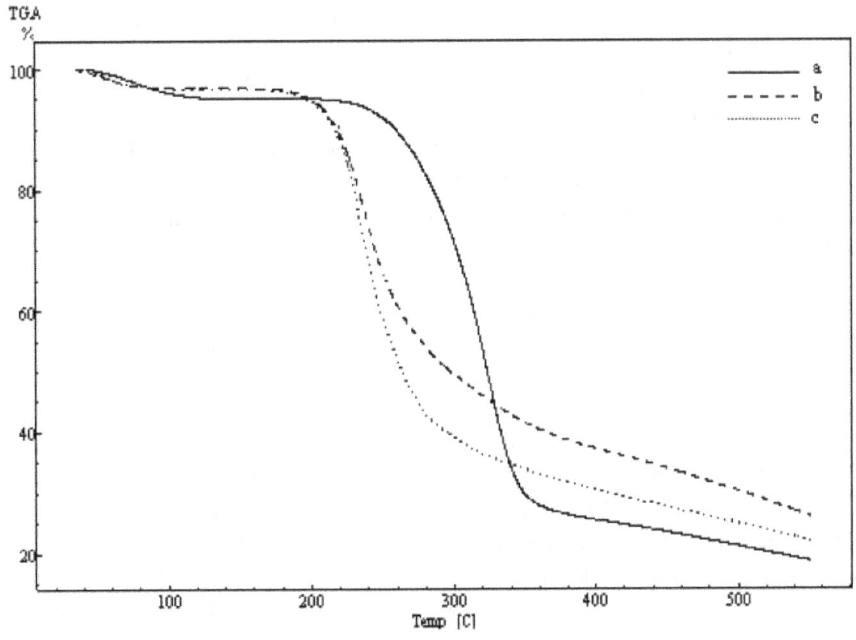

Fig. 7. The thermograms of native cellulose (a) and succinylated cellulose samples 6 (b) and 22 (c)

4. CONCISE SUMMARY

The succinylated cellulose derivatives with a low degree of substitution were synthesized without the use of catalysts. An elevation in reaction temperature, molar ratio of SA/AGU in cellulose, and reaction duration resulted in an augmentation in the degree of substitution (DS) in cellulose samples. The degree of substitution of cellulosic derivatives produced with iodine, NBS, and DMAP catalysts was markedly enhanced compared to products synthesized without catalysts, demonstrating the efficacy of these three catalysts in cellulose succinoylation within ionic liquids. The potential mechanism of homogenous succinoylation facilitated by these catalysts and their real function were also suggested. The Fourier transform infrared and solid-state cross-polarization/magic angle spinning 13C-NMR spectroscopies also demonstrated evidence of the catalyzed homogenous succinoylation process. The results demonstrated that the hydroxyl groups at the C-6, C-2, and C-3 locations in cellulose all reacted. The thermal stability of succinylated cellulose was seen to diminish following chemical treatment.

REFERENCES

Abbott, A.P., Bell, T.J., Handa, S. & Stoddart, B. (2005). O-acetylation of cellulose and monosaccharides using a zinc based ionic liquid, *Green Chemistry* 7(10): 705-707.

Barthel, S. & Heinze, T. (2006). Acylation and carbanilation of cellulose in ionic liquids, *Green Chemistry* 8(3): 301-306.

Biswas, A., Shogren, R.L. & Willett, J.L. (2005). Solvent-free process to esterify polysaccharides, *Biomacromolecules* 6(4): 1843-1845.

Chang, S.T. & Chang, H.T. (2001). Comparisons of the photostability of esterified wood, *Polymer Degradation and Stability* 71(2): 261-266.

Clark, J.H., Deswarte, F.E.I. & Farmer, T.J. (2009). The integration of green chemistry into future biorefineries, *Biofuels Bioproducts & Biorefining-Biofpr* 3(1): 72-90.

de Maria, D. & Martinsson, A. (2009). Ionic-liquid-based method to determine the degree of esterification in cellulose fibers, *Analyst* 134(3): 493-496.

Evans, R. & Wallis, A.F.A. (1989). Cellulose molecular-weights determined by viscometry, *Journal of Applied Polymer Science* 37: 2331-2340.

Forsyth, S.A., MacFarlane, D.R., Thomson, R.J. & von Itzstein, M. (2002). Rapid, clean, and mild o-acetylation of alcohols and carbohydrates in an ionic liquid, *Chemical Communications* 7(7): 714-715.

Fukaya, Y., Sugimoto, A. & Ohno, H. (2006). Superior solubility of polysaccharides in low viscosity, polar, and halogen-free 1,3-dialkylimidazolium formates, *Biomacromolecules* 7(12): 3295-3297.

Fukaya, Y., Hayashi, K., Wada, M. & Ohno, H. (2008). Cellulose dissolution with polar ionic liquids under mild conditions: Required factors for anions, *Green Chemistry* 10(1): 44-46.

Granstrom, M., Kavakka, J., King, A., Majoinen, J., Makela, V., Helaja, J., Hietala, S., Virtanen, T., Maunu, S.L., Argyropoulos, D.S. & Kilpelainen, I. (2008). Tosylation and acylation of cellulose in 1-allyl-3-methylimidazolium chloride, *Cellulose* 15(3): 481-488.

Hadano, S., Onimura, K., Tsutsumi, H., Yamasaki, H. & Oishi, T. (2003). Syntheses of chemical-modified cellulose obtained from waste pulp, *Journal of Applied Polymer Science* 90(8): 2059-2065.

Heinze, T., Schwikal, K. & Barthel, S. (2005). Ionic liquids as reaction medium in cellulose functionalization, *Macromolecular Bioscience* 5(6): 520-525.

Hermanutz, F., Gaehr, F., Uerdingen, E., Meister, F. & Kosan, B. (2007). New developments in dissolving and processing of cellulose in ionic liquids, *Macromolecular Symposia* 262(1): 23-27.

Hill, C.A.S., Cetin, N.S. & Ozmen, N. (2000). Potential catalysts for the acetylation of wood, *Holzforschung* 54(3): 269-272.

Himmel, M.E., Ding, S.Y., Johnson, D.K., Adney, W.S., Nimlos, M.R., Brady, J.W. & Foust, T.D. (2007). Biomass recalcitrance: Engineering plants and enzymes for biofuels production, *Science* 315(5813): 804-807.

Jayakumar, R., Balaji, R. & Nanjundan, S. (2000). Studies on copolymers of 2-(n-phthalimido)ethyl methacrylate with methyl methacrylate, *European Polymer Journal* 36(8): 1659-1666.

Jeon, Y.S., Viswanathan, A. & Gross, R.A. (1999). Studies of starch esterification: Reactions with alkenyl-succinates in aqueous slurry systems, *Starch-starke* 51(2-3): 90-93.

Kamiya, N., Matsushita, Y., Hanaki, M., Nakashima, K., Narita, M., Goto, M. & Takahashi, H. (2008). Enzymatic in situ saccharification of cellulose in aqueous-ionic liquid media, *Biotechnology Letters* 30(6): 1037-1040.

Karimi, B. & Seradj, H. (2001). N-bromosuccinimide (nbs), a novel and highly effective catalyst for acetylation of alcohols under mild reaction conditions, *Synlett* (4): 519- 520.

Kosan, B., Michels, C. & Meister, F. (2008). Dissolution and forming of cellulose with ionic liquids, *Cellulose* 15(1): 59-66.

Luo, H.M., Li, Y.Q. & Zhou, C.R. (2005). Study on the dissolubility of the cellulose in the functionalized ionic liquid, *Polymeric Materials Science & Engineering* 21(2): 233-235.

Maunu, S.L. (2002). NMR studies of wood and wood products, *Progress in Nuclear Magnetic Resonance Spectroscopy* 40(2): 151-174.

Mazza, M., Catana, D.A., Vaca-Garcia, C. & Cecutti, C. (2009). Influence of water on the dissolution of cellulose in selected ionic liquids, *Cellulose* 16(2): 207-215.

Mohanty, A.K., Misra, M. & Drzal, L.T. (2002). Sustainable bio-composites from renewable resources: Opportunities and challenges in the green materials world, *Journal of Polymers and the Environment* 10(1-2): 19-26.

Ogaki, Y., Shinozuka, Y., Hatakeyama, M., Hara, T., Ichikuni, N. & Shimazu, S. (2009). Selective production of xylose and xylo-oligosaccharides from bamboo biomass by sulfonated allophane solid acid catalyst, *Chemistry Letters* 38(12): 1176-1177.

Pandey, A., Soccol, C.R., Nigam, P. & Soccol, V.T. (2000). Biotechnological potential of agro- industrial residues. I: Sugarcane bagasse, *Bioresource Technology* 74(1): 69-80.

Reddy, N. & Yang, Y. (2005). Biofibers from agricultural byproducts for industrial applications, *Trends in Biotechnology* 23(1): 22-27.

Ren, J.L., Sun, R.C., Liu, C.F., Cao, Z.N. & Luo, W. (2007). Acetylation of wheat straw hemicelluloses in ionic liquid using iodine as a catalyst, *Carbohydrate Polymers* 70(4): 406-414.

Richardson, S. & Gorton, L. (2003). Characterisation of the substituent distribution in starch and cellulose derivatives, *Analytica Chimica Acta* 497(1-2): 27-65.

Schlufter, K., Schmauder, H.P., Dorn, S. & Heinze, T. (2006). Efficient homogeneous chemical modification of bacterial cellulose in the ionic liquid 1-n-butyl-3- methylimidazolium chloride, *Macromolecular Rapid Communications* 27(19): 1670-1676.

Stojanovic, Z., Jeremic, K., Jovanovic, S. & Lechner, M.D. (2005). A comparison of some methods for the determination of the degree of substitution of carboxymethyl starch, *Starch-starke* 57(2): 79-83.

Sun, R.C., Sun, X.F. & Zhang, F.Y. (2001). Succinoylation of wheat straw hemicelluloses in n,n-dimethylformamide/lithium chloride systems, *Polymer International* 50(7): 803- 811.

Swatloski, R.P., Spear, S.K., Holbrey, J.D. & Rogers, R.D. (2002). Dissolution of cellose with ionic liquids, *Journal of the American Chemical Society* 124(18): 4974-4975.

Turner, M.B., Spear, S.K., Huddleston, J.G., Holbrey, J.D. & Rogers, R.D. (2003). Ionic liquid salt-induced inactivation and unfolding of cellulase from trichoderma reesei, *Green Chemistry* 5(4): 443-447.

Wu, J., Zhang, J., Zhang, H., He, J.S., Ren, Q. & Guo, M. (2004). Homogeneous acetylation of cellulose in a new ionic liquid, *Biomacromolecules* 5(2): 266-268.

Yoshimura, T., Matsuo, K. & Fujioka, R. (2006). Novel biodegradable superabsorbent hydrogels derived from cotton cellulose and succinic anhydride: Synthesis and characterization, *Journal of Applied Polymer Science* 99(6): 3251-3256.

Zhang, H., Wu, J., Zhang, J. & He, J.S. (2005). 1-allyl-3-methylimidazolium chloride room temperature ionic liquid: A new and powerful nonderivatizing solvent for cellulose, *Macromolecules* 38(20): 8272-8277.

Zhang, M.J., Qi, W., Liu, R., Su, R.X., Wu, S.M. & He, Z.M. (2010). Fractionating lignocellulose by formic acid: Characterization of major components, *Biomass & Bioenergy* 34(4): 525-532.

Zhang, Y.H.P., Ding, S.Y., Mielenz, J.R., Cui, J.B., Elander, R.T., Laser, M., Himmel, M.E., McMillan, J.R. & Lynd, L.R. (2007). Fractionating recalcitrant lignocellulose at modest reaction conditions, *Biotechnology and Bioengineering* 97(2): 214-223.

6 Chapter

FABRICATION OF POLYSACCHARIDE-BASED MATERIALS COMPATIBILIZED WITH IONIC LIQUIDS

1. INTRODUCTION: AN OVERVIEW

Ionic liquids (ILs) are low-melting-point molten salts that exist as liquids at room temperature or at temperatures below the boiling point of water. The feature arises because the liquid state is thermodynamically advantageous because to the substantial size and conformational flexibility of the ions, resulting in low lattice enthalpies and significant entropy shifts that favor the liquid state. (Welton, 1999). Over the past decade, ionic liquids (ILs) have garnered significant attention because to their distinctive attributes, including minimal vapor pressure, exceptional thermal stability, and adjustable physical and chemical properties (Plechkova & Seddon, 2008). In addition to their conventional characteristics, interest and uses of ionic liquids (ILs) have lately expanded to studies involving biomolecules, particularly naturally occurring polysaccharides, due to the peculiar affinities ILs exhibit towards them. (Seoud et al., 2007).

Polysaccharides are prevalent in nature and are considered structural components as well as sources of water and energy (Schuerch, 1986). Natural polysaccharides, including cellulose, starch, and chitin, have garnered significant attention as sources for innovative green and sustainable materials due to their environmentally beneficial characteristics. Nonetheless, natural polysaccharides frequently have issues with solubility, processability, and feasibility due to the abundance of hydrogen bonds within their polymeric chains, hindering their application across diverse materials.

In 1934, Graenacher discovered that molten N-ethylpyridinium chloride, when combined with nitrogenous bases, decomposed cellulose (Graenacher, 1934). This was likely the inaugural instance of cellulose dissolving utilizing IL-type solvents. At that time, this was considered to have no practical significance as the concept of ILs had not yet been introduced. In 2002, Rogers et al. extensively documented that the ionic liquid 1-butyl-3-methylimidazolium chloride (BMIMCl) effectively dissolved cellulose at elevated concentrations, thereby paving the path for the advancement of a new category of cellulose solvent systems (Swatloski et al., 2002). Since this publication, ionic liquids have been employed in the processing of cellulose and other polysaccharides, primarily focusing on dissolution, homogenous derivatization and modification, and regeneration. (Zhu et al., 2006, Liebert & Heinze, 2008, Feng & Cheng, 2008, Pinkert et al., 2009).

In recent years, ionic liquids have been extensively utilized as functional components in polysaccharide-based materials, in addition to their usual roles as solvents and media. The premise of these studies is founded on the characteristics of ionic liquids that have strong affinities for polysaccharides, hence facilitating effective compatibilization. This chapter outlines the synthesis of polysaccharide-based materials compatibilized with ionic liquids (ILs), organized into two sections based on the aforementioned backgrounds and perspectives. The initial issue concerns the synthesis of polysaccharide-polymeric ionic liquid composite materials using in-situ polymerization of polymerizable ionic liquids. The synthesis of IL-gel materials from several natural polysaccharides is presented as a secondary topic.

2. PREPARATION OF POLYSACCHARIDE-POLYMERIC IONIC LIQUID COMPOSITES BY IN-SITU POLYMERIZATION OF POLYMERIZABLE IONIC LIQUIDS

2.1 Polymeric ionic liquids

Polymeric ILs are defined as the polymers obtained by polymerization of ILs having polymerizable groups (polymerizable ILs) (Green et al., 2009). Thus, 'polymeric ILs' are termed just the polymeric forms of ILs, but they are not necessary to show liquid form at room temperature or even at some ambient temperatures. The polymeric ILs, therefore, are often called 'polymerized ILs' too. The major advantages for providing the polymeric ILs are to be enhanced stability, and improved processability and feasibility in application as practical materials. Polymerizable ILs as a source of the polymeric ILs can be available by incorporating the polymerizable groups at both anionic and cationic sites in the IL structures (Fig. 1). In the former case, polymerizable anions are ionically exchanged to some anions of general ILs (Fig. 1(a)), giving the polymerizable ILs. In the latter case, vinyl, meth(acryloyl), and vinyl benzyl groups have typically been appeared as the polymerizable group (Fig. 1(b)). 1-Vinylimidazole is a commercially available reagent. Therefore, the vinylimidazolium-type polymerizable ILs are prepared by quaternization of 1- vinylimidazole with a variety of alkyl halides. The reaction of haloalkyl (meth)acrylates or vinylbenzyl halides with 1-alkylimidazoles gives the corresponding imidazolium-type polymerizable ILs having the (meth)acryloyl or vinyl benzyl polymerizable groups (Fig. 2). Furthermore, when haloalkyl (meth)acrylates or vinylbenzyl halides are reacted with 1- vinylimidazole, the polymerizable ILs having two polymerizable groups are produced. Because these ILs can be converted into insoluble and stable polymeric ILs with the cross- linked structure by the radical polymerization (Fig. 3), they have a highly potential as the source of the components in the practical materials. On the basis of the above viewpoints, the polymerizable ILs having the acrylate or vinylbenzyl group have been employed for the preparation of polysaccharide-polymeric IL composite materials.

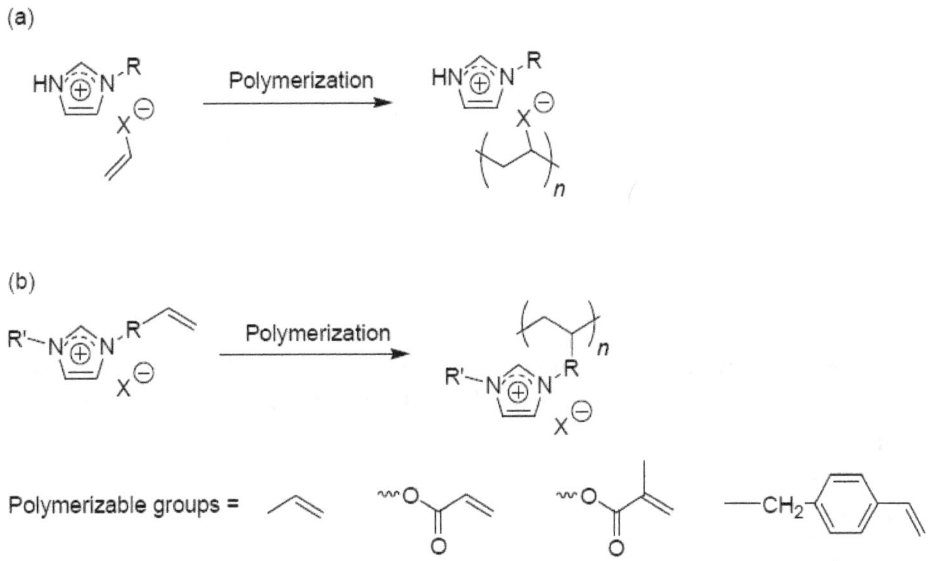

Fig. 1. Polymerization of polymerizable ILs having a polymerizable group at anionic site (a) and cationic site (b).

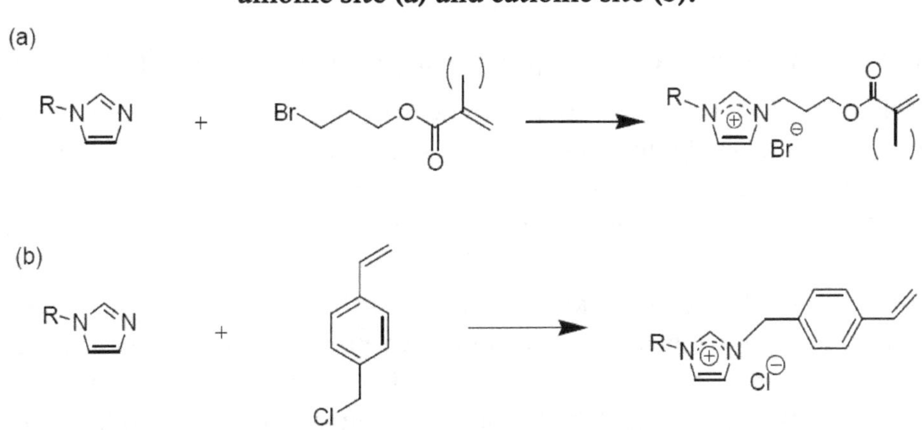

Fig. 2. Typical synthetic schemes for polymerizable ILs having (meth)acrylate (a) and vinylbenzyl (b) groups.

2.2 Preparation of cellulose-polymeric IL composites

Cellulose is the most abundant organic substance on the earth, which consists of ®-(1→4)- linked glucose repeating units. Cellulose has limited applications because of the strong inter- and intramolecular hydrogen bonds between the hydroxy groups of the glucose residues. Therefore, considerable efforts have been still devoted to the compatibilization of cellulose with synthetic polymers for improvement of the processability of cellulose. In-situ polymerization method, for example, those giving interpenetrating polymer network (IPN) system has been useful for formation of composites from cellulose (Fig. 4).

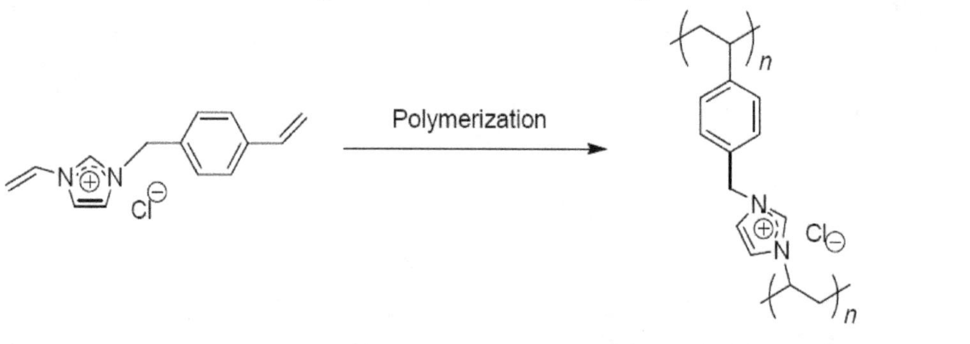

Fig. 3. Polymerization of a polymerizable IL having two polymerizable groups to produce a cross-linked insoluble polymeric IL.

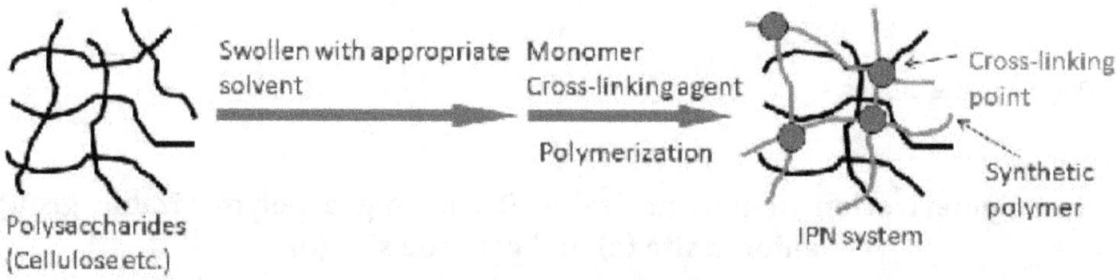

Fig. 4. Schematic image of interpenetrating polymer network (IPN) system.

The IPNs are a special class of polymer blends or composites based on two or more polymers, with each polymer chemically cross-linked or at least one network being prepared in the presence of the others. If all polymers are cross-linked, the blend is called a full-IPN. Semi-IPN differs from the full-IPN, in which it is composed of a non-cross-linked polymer entrapped in another polymer networks. Since the imidazolium-type ILs such as BMIMCl are good solvents for cellulose, it has been interesting to employing the imidazolium-type polymeric ILs as one component for good compatibilization with cellulose. To obtain the IPN- composite of cellulose with the polymeric IL by the in-situ polymerization method, in the first study, a polymerizable IL, 1-(4-acryloyloxybutyl)-3-methylimidazolium bromide (AcMIMBr) was used (Fig. 5) (Murakami et al., 2007).

First, cellulose was dissolved in BMIMCl (9.1 wt%) by heating at 100 °C for 24 h. Then, AcMIMBr (an equimolar amount of the glucose units in cellulose) and a radical initiator, AIBN, were added to the solution and the radical polymerization of the mixture was carried out at 80 °C for 5 h. After the polymerization mixture was washed with acetone and treated further with refluxed acetone, the residue was subjected to Soxhlet extraction with methanol to give the composite. The unit ratio of cellulose to AcMIMBr was calculated by the elemental analysis to be 1.0 : 0.63 and the ^{1}H NMR spectrum of the Soxhlet extract indicated that a part of poly(AcMIMBr) was washed out by the extraction. The IR spectrum of the resulting material suggested the structure composed of cellulose and poly(AcMIMBr). Furthermore, the thermal gravimetric analysis (TGA) and the X-ray diffraction (XRD) data supported the efficient compatibilization of cellulose with poly(AcMIMBr) in the composite. When cellulose and poly(AcMIMBr) were simply mixed in BMIMCl and the obtained mixture was subjected to the same isolation procedure as that for the aforementioned composite, almost of poly(AcMIMBr) was removed out. This result indicated that the simple mixing method of cellulose with poly(AcMIMBr) in BMIMCl was not useful for the efficient compatibilization.

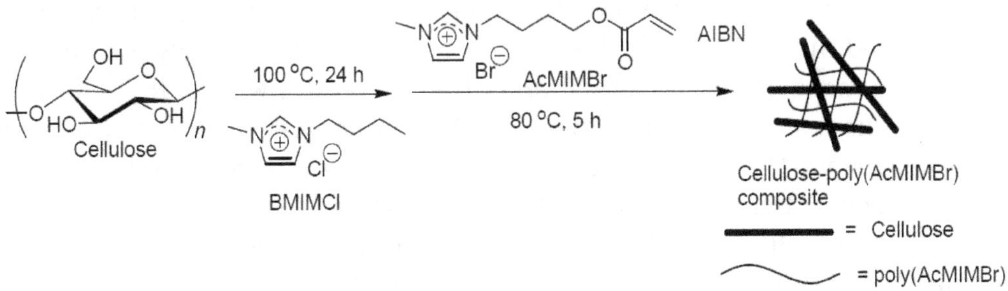

Fig. 5. Preparation of cellulose-polymeric IL composite using two ILs, BMIMCl and AcMIMBr.

Although the aforementioned approach was the efficient for the preparation of the cellulose- polymeric IL composites, the procedure should be improved due to the following disadvantageous points; (1) the necessity of two kinds of ILs and (2) the instability of the obtained composites toward solvents. To solve the second problem, the polymerizable IL having two polymerizable groups was employed, which was polymerized by a radical initiator to give the insoluble and stable cross-linked polymeric IL. Thus, in the improved procedure for the preparation of cellulose-polymeric IL composites in the next paper, a combined use of a polymerizable IL having acryl and vinyl groups (1-(3-acryloyloxypropyl)-3-vinylimidazolium bromide, AcVIMBr) with AcMIMBr (an alkyl group = propyl) was conceived (Fig. 6) (Takegawa et al., 2009). The former acted as a cross-linker in the produced polymeric ILs by the radical copolymerization of these two ILs. Therefore, the property of the polymeric IL can be changed by the feed ratio of the two ILs in the polymerization. Furthermore, it was found that a mixture of these polymerizable ILs mostly disrupted the crystalline structure of cellulose by pre-treatment, indicating the preparation of the composites to exclusion of use of the solvent IL such as BMIMCl unlike the case in the first study.

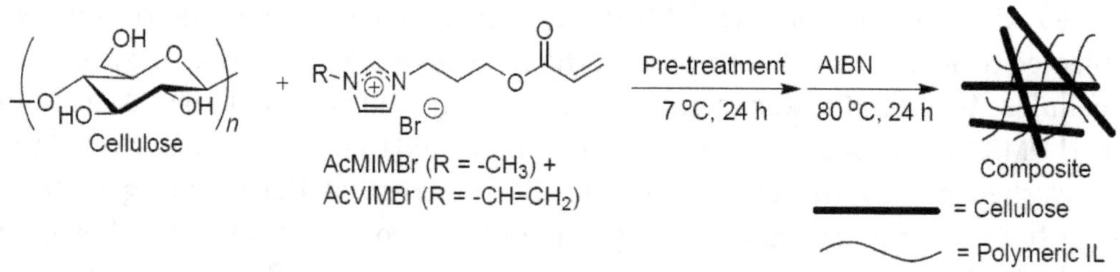

Fig. 6. Preparation of cellulose-polymeric IL composite using acrylate-type polymerizable ILs.

Therefore, for the preparation of the composites, first, cellulose (30 wt% for AcMIMBr + AcVIMBr) was pre-treated in mixtures of AcMIMBr and AcVIMBr with various weight ratios (100 : 0 – 50 : 50) at 7°C for 24 h. The ability of the ILs that disrupted the crystalline structure of cellulose was evaluated by the XRD and TGA measurements of the pre-treated mixtures. Consequently, the analytical data indicated that the crystalline structure of cellulose was largely disrupted by the pre-treatment. Then, the formation of the composites was performed by the in-situ polymerization of the polymerizable ILs in the pre-treated mixtures. After AIBN was added to the pre-treated mixtures, the mixtures were heated at 80 °C for 24 h, resulting in the composites without any further isolation and purification procedures. The XRD and TGA results of the resulting composites indicated good compatibility of cellulose and the polymeric ILs in the composites in all cases. The compatibility between cellulose and the polymeric ILs in the composites was also evaluated by the SEM observation. The SEM images of the surface and inside area in the composite showed the completely different morphology from those of the standard cellulose, supporting the good compatibility of two components in the composite (Fig. 7). The stability of the composites was examined by washing with methanol by the Soxhlet extraction manner. After washing the composites obtained by using the mixtures of AcMIMBr and AcVIMBr, the unit ratios of cellulose to the polymeric ILs were hardly changed from those before washing. On the other hand, the content of the polymeric IL in the composite resulted by using AcMIMBr alone decreased by washing. The results indicated that the composites from the mixtures of AcMIMBr and AcVIMBr were very stable because of consisting of the cross-linked polymeric IL components formed by the copolymerization of the two polymerizable ILs. In contrast, the component of the polymeric IL in the composite from AcMIMBr alone was washed out, due to its linear structure, which was soluble in methanol. The mechanical property of the composites obtained from the mixtures of AcMIMBr and AcVIMBr with the various weight ratios was evaluated by the stress-strain curves under tensile mode. The composite from AcMIMBr alone showed the more elastic property compared with other composites containing cross-linked poly(AcMIMBr/AcVIMBr). The higher contents of the units from AcVIMBr in the composites affected to sustain the higher stresses until 30 wt% of AcVIMBr in the mixture of the two polymerizable ILs was used.

Fig. 7. SEM images of cellulose and composite.

The above in-situ polymerization method was extended to use another polymerizable IL, which was an IL having vinylbenzyl and vinyl polymerizable groups (1-vinyl-3-(4-vinylbenzyl)imidazolium chloride, VVBnIMCl) (Fig. 8) (Kadokawa et al., 2008a). This IL gives the polymeric IL composed of the polystyrene main-chain by radical polymerization. The preparation of the composites composed of cellulose and poly(VVBnIMCl) was carried out by the same procedure as that using AcMIMBr/AcVIMBr. Thus, cellulose (9.1 – 50.0 wt%) was first pre-treated with VVBnIMCl at 7 °C for 24 h. The XRD profile of the pre- treated mixture (33.3 wt%) showed only the slight diminution of the crystalline peaks. This result in comparison with that using AcMIMBr/AcVIMBr indicated that the ability of AcMIMBr/AcVIMBr disrupting the crystalline structure of cellulose was much higher than that of VVBnIMCl. However, the TGA result of the pre-treated mixture (33.3 wt%) using VVBnIMCl exhibited the similar curve as that using AcMIMBr/AcVIMBr, where an onset weight loss at around 250 °C due to the thermal degradation of cellulose was observed, which was ca. 50 °C lower than that of a standard cellulose. The TGA data indicated that the cellulose chains were partially swollen in the pre-treated mixture using VVBnIMCl.

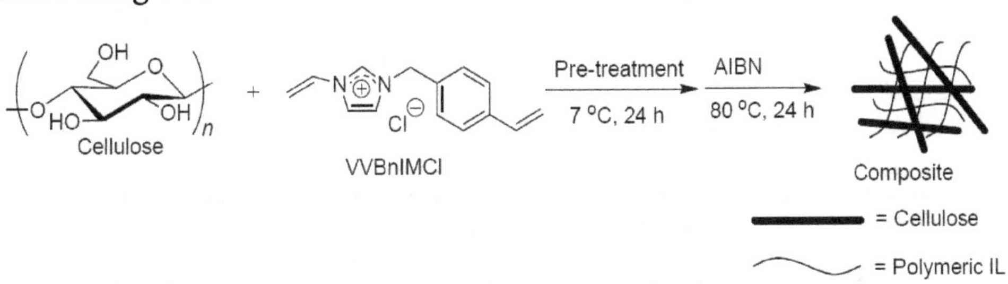

Fig. 8. Preparation of cellulose-polymeric IL composite using vinylbenzyl-type polymerizable IL.

After AIBN was added to the pre-treated mixtures, the mixtures were heated at 80 °C for 24 h to occur the radical polymerization. The composites were thus obtained without any isolation and purification procedures. The XRD and TGA results of all the composites indicated the partial disruption of the crystalline structure of cellulose. The SEM images of the composite supported good compatibility between cellulose and poly(VVBnIMCl) in the composite. The mechanical properties of the composites with 9.1, 33.3, and 50.0 wt% cellulose contents were evaluated by the tensile testing. The composites sustained the stresses in the range 2.5 – 6.5 MPa and the strains at break were 1.3 – 2.0 %. The higher content of cellulose in the composite (50 wt%) caused a brittle property. However, the appropriate content of cellulose affected to sustain higher stress. Consequently, the composite containing cellulose of 33.3 wt% indicated good mechanical property. Because of the presence of poly(VVBnIMCl) with the cross-linked structure in the composite, it showed the good stability toward washing with methanol.

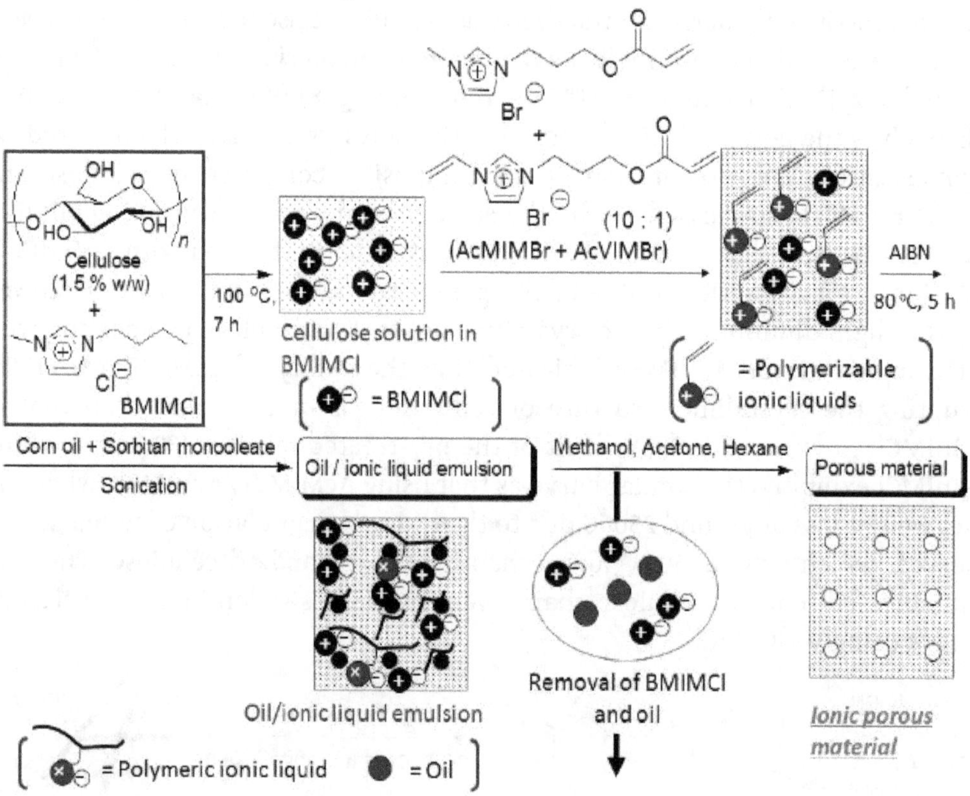

Fig. 9. Schematic image for formation of cellulose-based ionic porous material compatibilized with polymeric IL.

This in-situ polymerization method was applied to the preparation of cellulose-based ionic porous material compatibilized with the polymeric IL (Fig. 9) (Prasad et al., 2010). This was achieved by combining the in-situ polymerization method with the templating technique using the oil/IL emulsion system. The in-situ polymerization of a mixture of AcMIMBr and AcVIMBr was first performed in a solution of cellulose in a solvent of BMIMCl. Then, the sonication of the mixture coexisting with corn oil and sorbitan monooleate, followed by the treatment with the appropriate solvents gave the porous material. The pore sizes in the materials ascertained from the SEM images were found to be around 0.15 – 1.3 m accompanied with the smaller sizes of 30 – 70 nm (Fig. 10).

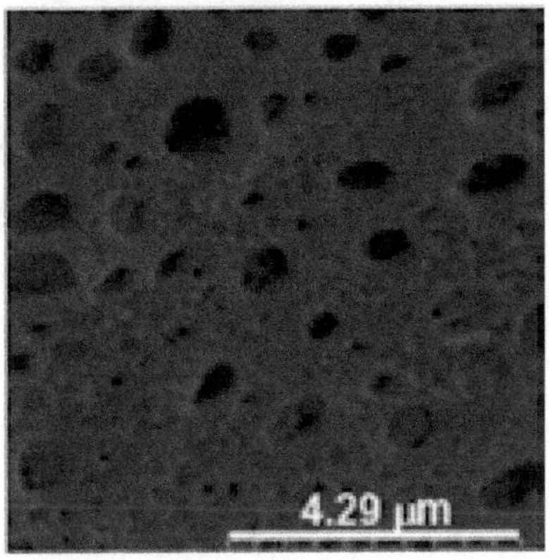

Fig. 10. SEM image of porous material.

2.3 Preparation of carrageenan-polymeric IL composites

Seaweed-originated polysaccharides like agarose and carrageenan have been used to prepare some functional materials such as electrical conductive materials by incorporating salts into their gel matrices as well as by doping with conducting polymers such as polyaniline and polypyrrole. Following the preparation of the aforementioned cellulose composites with the polymeric ILs, the new composite materials using ι-carrageenan of a seaweed polysaccharide was also prepared by compatibilizing with the polymeric ILs by means of in-situ polymerization technique (Fig. 11) (Prasad & Kadokawa, 2010). ι-Carrageenan is a water- soluble phycocolloid extracted from red algae consisting of alternating 1,3-linked 〈-D- galactopyranose unit and 1,4-linked ®-3,6-anhydro-D-galactopyranose unit with sulfates on C- 4 of the galactopyranose unit and C-2 of the anhydro-D-galactopyranose unit (Section 3.2). Due to the presence of the charges, it is expected to form well-miscible composites with the polymeric ILs through cross-linking via ionic exchange, which possibly contributes to improving the mechanical properties of the materials. It has also been considered that the material may exhibit the electrical conductivity due to the ionic nature. Therefore, the preparation of the composite materials composed of ι-carrageenan and the polymeric ILs by the in-situ polymerization of the polymerizable ILs was performed. The polymerizable ILs used were AcVIMBr and VVBnIMCl. For the preparation of the composite using AcVIMBr, BMIMCl was used for the dissolution of ι-carrageenan because its direct soak in AcVIMBr was not possible due to the high viscosity of AcVIMBr. Thus, ι-carrageenan was first dissolved in BMIMCl by heating at 100 oC for 7 h. Then, the radical polymerization of AcVIMBr with AIBN in the solution was performed, followed by the work-up procedure, to give the composite. Because unlike AcVIMBr, VVBnIMCl was no viscous, on the other hand, the composite using VVBnIMCl was prepared without use of any solvent for dissolution of ι-carrageenan. The CP- MS ^{13}C NMR and IR spectra of the resulting composites fully supported which were composed of ι-carrageenan and the corresponding polymeric ILs.

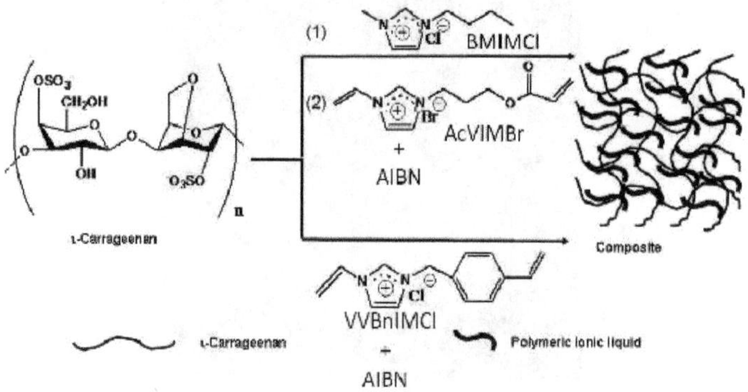

Fig. 11. Preparation of ι-carrageenan-polyIL composites.

Immiscible components were not seen in the charge coupled device (CCD) camera views (mm scale) of the surfaces of the composites. However, the SEM images with the nm scale of the two composites exhibited the different phase structures. The SEM image of the composite with poly(AcVIMBr) showed loose structures with nano scale distributions, whereas closer packed nano structures were observed in the SEM image of the composite with poly(VVBnIMCl). The morphology for the former material was probably caused by the removal of the solvent, BMIMCl, during the work-up procedure.

The mechanical properties of the composites of ι-carrageenan with poly(AcVIMBr) and poly(VVBnIMCl), which were evaluated under compressive mode, showed the fracture stresses of 108.07 and 102.0 MPa with the fracture strains of 25.0 and 10.7 %, respectively. These values were larger than those of the respective cellulose composite with each polymeric IL. The good mechanical properties of the present composites were reasonably explained by the formation of the good miscibility as observed in the SEM images, which were owing to the cross-linking through the ionic exchange between ι-carrageenan and the polymeric ILs. The electrical conductivities of the composites with poly(AcVIMBr) and poly(VVBnIMCl) were measured to be 6.2 x 10-4 and 1.6 x 10-4 S cm-1, respectively, which were comparable to the conductance of semiconductor. On the other hand, cellulose composites with poly(AcVIMBr) and poly(VVBnIMCl) showed the electrical conductivity values of ca. 1.3 x 10-5 and 2.8 x 10-5 S cm-1, respectively. The higher electrical conductivity of the present composites indicated the role of the effective ionic exchange during the formation of the composites, due to the presence of charges in ι-carrageenan.

3. PREPARATION OF IONIC LIQUID-GELS OF POLYSACCHARIDES

3.1 Preparation of ionic liquid-gels of cellulose

When a solution of cellulose in BMIMCl (13.0 wt%) was left standing at room temperature for 7 days, it was found that the formation of a gel gradually took place with exclusion of excess BMIMCl from the gel matrix (Fig. 12) (Kadokawa et al., 2008b). Thus, the gel was purified by washing with ethanol to remove the excluded BMIMCl and dried under reduced pressure. The resulting gel had flexible and manipulatable natures and was characterized by the elemental analysis, XRD, and TGA. The XRD result of the gel indicated that the crystalline structure of cellulose was mostly disrupted in the gel. The elemental analysis and TGA data suggested that gel was composed not only of cellulose and BMIMCl, but also of water. On the basis of the above analytical results, it was considered that the gel was produced by the formation of non-crystalline aggregates of cellulose in the solution during gradually absorbing water owing to the strong hygroscopic nature of BMIMCl (Fig. 13). The aggregates probably acted as cross-linking points for the gelation.

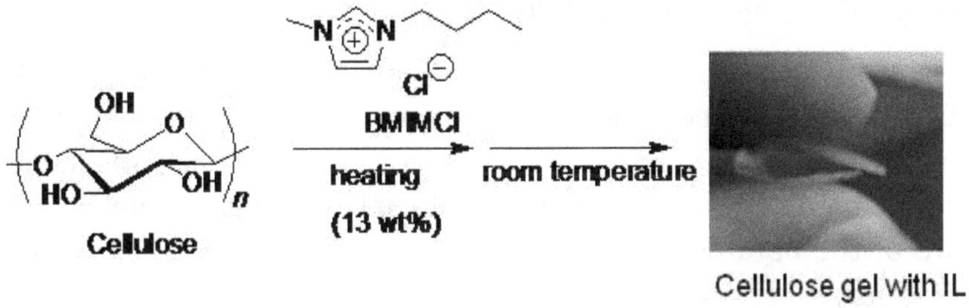

Fig. 12. Procedure for preparation of cellulose gel with IL.

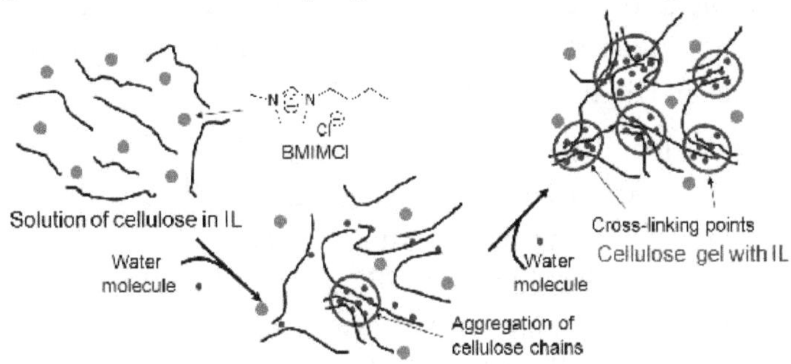

Fig. 13. Schematic image for procedure of formation of cellulose gel with IL.

When the gel was heated at 120 oC, it was gradually softened, and converted into a fluid at 150 oC. The soft material, which was obtained by heating the gel at 120 oC for

2 h, could be transferred to the gel form again by standing it at room temperature for 2 days. The regenerated gel was still flexible and was more transparent than the primary formed gel. Because the aggregates of cellulose as the cross-linking points in the regenerated gel were probably dispersed well by the heating-cooling process compared with those in the primary gel, the gel after regeneration was more transparent than before.

This gelation technique was extended to the preparation of cellulose/starch composite gel with BMIMCl (Kadokawa et al., 2009). When the mixture of cellulose (9.1 wt%) and starch (4.8 wt%) in BMIMCl was heated at 100 °C for 24 h, a homogeneous liquid was formed, which was left standing at room temperature for 4 – 6 days. Consequently, the excess BMIMCl was excluded from the liquid to take place the gelation. Thus, the resulting gel was purified by washing with methanol to remove the excluded excess BMIMCl and dried under reduced pressure. In the XRD profile of the gel, the diffraction peaks due to the crystalline structures of the standard cellulose and starch were not obviously exhibited, indicating that the crystalline structures of the polysaccharides were largely disrupted in the gel. The results of the elemental analysis and TGA indicated that the gel was composed of the polysaccharides, BMIMCl, and water, as same as the aforementioned gel from cellulose and BMIMCl. From the data, the similar gelation process was conceived as that for the gel of cellulose with BMIMCl.

The homogeneous mixture of cellulose and starch (9.1 wt% each) in BMIMCl, obtained by heating as same as above, had a very viscous nature, and thus, could be raised from the surface. The resulting fine linear material was subjected to the formation of the composite fibrous material by soaking into acetone. The SEM picture of the obtained material showed the fiber form with ca. 100 – 200 m diameters, but the crystalline and particle images, which were observed in the SEM pictures of the standard cellulose and starch, were not seen. The XRD result of the material indicated that the crystalline structures of cellulose and starch chains were not regenerated well, suggesting the good compatibilization of the two polysaccharides in the fibrous material.

It was found that a clear liquid of chitin (4.8 wt%) with ionic liquid, 1-allyl-3-methylimidazolium bromide (AMIMBr) was obtained by heating a mixture of chitin with AMIMBr at 100 oC for 48 h (Prasad, Murakami et al., 2009). The disappearance of the chitin powder in the liquid was confirmed by the CCD camera view and the SEM observation. The IR spectrum, TGA curve, and XRD profile of the regenerated chitin from the liquid as well as the analytical data of its hexanoyl derivative supported that the degradation and decreasing the molecular weight of chitin did not frequently occur during the dissolution process. On the other hand, when a mixture of 6.5 wt% chitin was soaked in AMIMBr at room temperature for 24 h, heated at 100 oC, and then cooled to room temperature, it turned into the gel-like form. The dynamic rheological measurements showed that both 4.8 wt% and 6.5 wt% chitins with AMIMBr behaved as the weak gels. However, it was seen that the obtained 6.5 wt% chitin with AMIMBr did not flow upon leaning a test tube, while the 4.8 wt% chitin with AMIMBr started to flow upon leaning (Fig. 14). The much lower yield stress of 4.8 wt% chitin with AMIMBr than that of 6.5 wt% chitin with AMIMBr supported that 4.8 wt% chitin with AMIMBr can flow under gravitation.

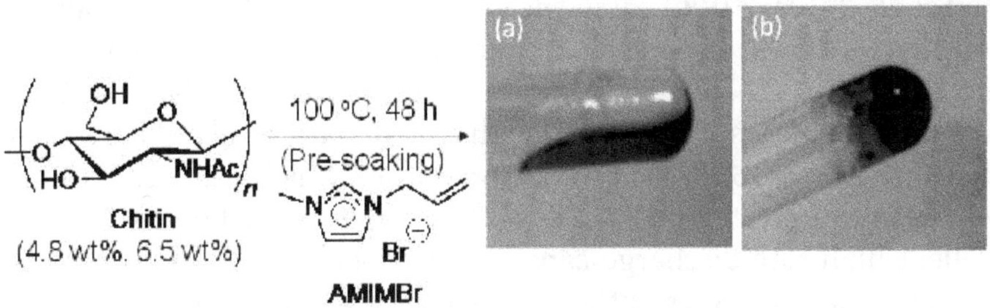

Fig. 14. Dissolution and gelation of chitin with IL; 4.8 wt% (a) and 6.5 wt% (b).

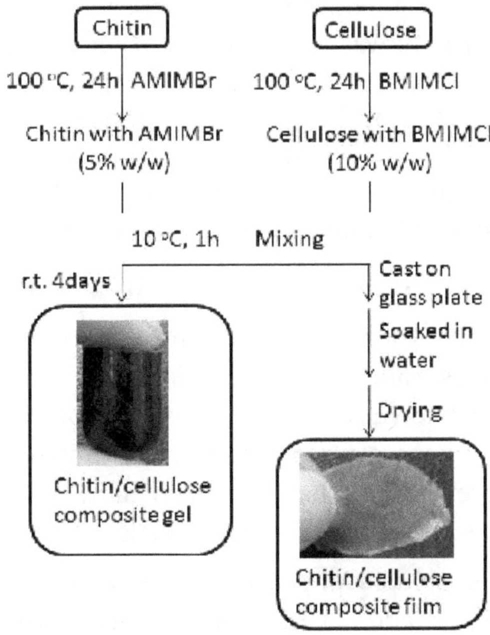

Fig. 15. Preparation procedure for chitin / cellulose composite gel and film with ILs.

Thus, the preparation of the chitin/cellulose composite gel and film using the two ionic liquids, AMIMBr and BMIMCl, was performed (Fig. 15) (Takegawa et al., 2010). First, chitin and cellulose were dissolved in each appropriate ionic liquid. Then, the two liquids were mixed in the desired ratios at 100 °C to give the homogeneous mixtures. The gels were obtained by standing the mixtures at room temperature for 4 days and the films were obtained by casting the mixtures on a glass plate, followed by soaking in water and drying. The resulting gels and films were characterized by the XRD and TGA measurements, which showed relatively good miscibility among the polysaccharides and the ionic liquids in the materials. The mechanical properties of the gels and films were changed depending on the ratios of chitin to cellulose in the materials.

The chitin/cellulose composite gel with ionic liquids was employed as the novel electrolyte for an electric double layer capacitor (EDLC) (Yamazaki et al., 2009, Yamazaki et al., 2010a, 2010b). First, the gel was treated with an aqueous 2.0 mol/L H_2SO_4 solution for 3 h. Electrochemical characteristics of the obtained acidic chitin/cellulose composite gel electrolyte were investigated by galvanostatic charge-discharge measurements. The test cell with the composite gel electrolyte showed a specific capacitance of 162 F/g at room temperature, which was higher than that for a cell with an H_2SO_4 electrolyte (155 F/g). The composite gel electrolyte exhibited the excellent high-rate discharge capability in a wide range of current densities as well as an aqueous H_2SO_4 solution. Moreover, the discharge capacitance of the test cell retained over 80 % of its initial value in 10^5 cycles even at a high current density of 5000 mA/g. The temperature dependence of the ionic conductivity for the composite gel electrolyte was the Arrhenius-type and similar to an aqueous H_2SO_4 solution. Furthermore, the self-discharge of the EDLC cell with the composite gel electrolyte was advantageously slow in comparison with the cell containing only the aqueous H_2SO_4 solution. The above results indicated that the acidic chitin/cellulose composite gel electrolyte had a practical applicability to an advanced EDLC with the excellent stability and working performance.

3.2 Preparation of ionic liquid-gels of food hydrocolloids

Besides the abundant polysaccharides like cellulose, starch, and chitin, many kinds of natural polysaccharides from various sources such as plants, animals, seaweeds, and bacteria have been known. For example, some polysaccharides are used as hydrocolloids for a stabilizer, a viscous agent, and a structure provider in food industries (Fig. 16). However, there have not been many reports on the use of these food hydrocolloids as a source of practical materials.

Carrageenan is a water soluble phycocolloid extracted from red algae and is used mostly as a stabilizer and a structure provider in the food and ice cream industries. Three major varieties of carrageenans are known, which are κ-, ι-, and λ-carrageenans. They differ in the number and position of sulfate groups on the repeating garactopyranose units. To efficiently employ the food hydrocolloids as a source of the materials, the gelling system of κ-, ι-, and λ-carrageenans with BMIMCl was reported (Prasad, Kaneko et al., 2009). First, 13.0 wt% carrageenan was dissolved in BMIMCl by heating at 100 °C for 10 h with stirring. After the solution was cooled to room temperature, a gel was formed, which could suitably be handled after 12 h of standing upon addition of acetone and excess ionic liquid leaching out from the gel was washed out with ethanol. κ-Carrageenan gave the formation of hard gel while other two carrageenans gave the formation of softer gels with BMIMCl (Fig. 17). A schematic representation for the formation of the gels is shown in Fig. 18. At elevated temperatures, carrageenans are soluble in BMIMCl, and thus exist as random coils in the solutions. Upon cooling, three dimensional networks develop, leading to the formation of junction zones in the polysaccharide chains (Phase 1). The subsequent treatment of the mixture with acetone causes further aggregation and stabilization of the junction zones, particularly for the parts close to the surface, resulting in a stronger gelling system with the exclusion of BMIMCl (Phase 2). The excluded excess BMIMCl was washed out with ethanol to give the formation of the compact gels. The composite gels of carrageenans and cellulose with BMIMCl could be obtained by the same procedure as that described above. The mechanical properties of the carrageenan gels, as well as the composite gels with cellulose and BMIMCl, under the compressive mode were better than those of the hydrogels of κ- and ι-carrageenans and the cellulose gel with BMIMCl.

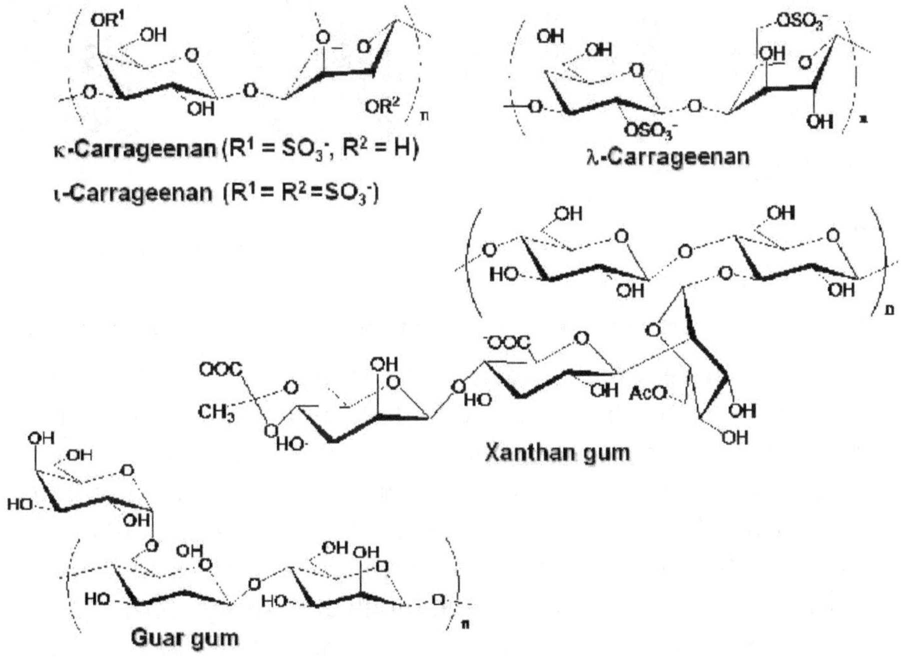

Fig. 16. Structures of typical food hydrocolloids, carrageenans, xanthan gum, and guar gum.

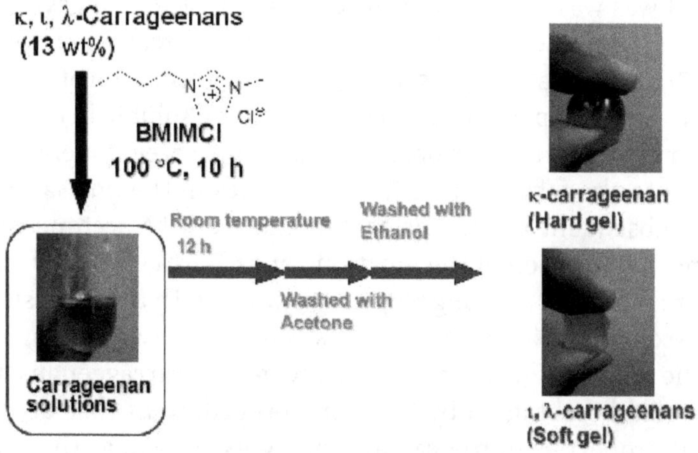

Fig. 17. Preparation of carrageenan gels with IL.

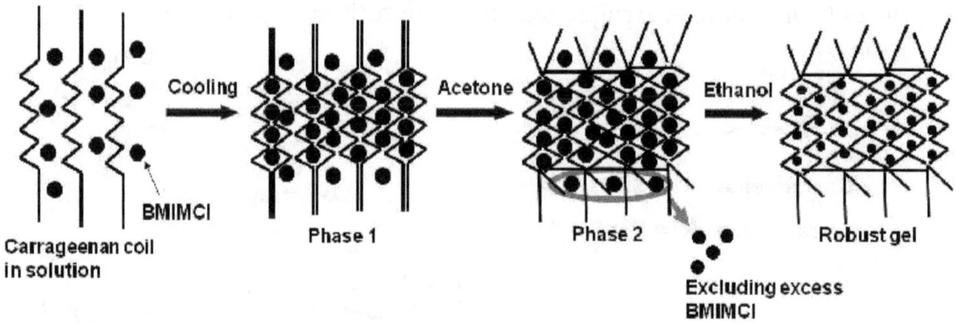

Fig. 18. Schematic representation for formation of carrageenan gels with IL.

Xanthan gum, which is a water soluble polysaccharide produced by *Xanthomonas campestris*, is also a useful food hydrocolloid. It has a cellulose main-chain with anionic trisaccharide side chains attached to alternating glucose units in the main-chain. It was found that xanthan gum/BMIMCl gels in various contents were obtained from the 9.1 – 50 wt% solutions of xanthan gum in BMIMCl (Izawa et al, 2009, Izawa & Kadokawa, 2010). When the solutions were left standing at room temperature for 30 min, they turned totally into the gel form (Fig. 19). The process of gelation was completely different from that for the aforementioned gels of cellulose and carrageenan with BMIMCl because long duration or treatment with appropriate organic solutions was required to generate the stable gels in their cases. It was considered that the ion-exchange of carboxylate metal cations in the original xanthan gum with BMIMCl took place during the gelation to form imidazolium- carboxylate pair in xanthan gum (Fig. 20). All the gels exhibited good mechanical properties under compressive mode. The 9.1 wt% gel showed elastic nature and the properties gradually changed to harder with increasing the contents of xanthan gum in the gels. It should be additionally noted that these gels did not have apparent fracture points. Interestingly, the gels had the sufficient strength to apply to the tensile test. The 9.1 and 16.7 wt% gels showed good elasticity, in which they were elongated to strain values near by 400%. In contrast, the 50 wt% gel had a hard nature and a fracture stress was 2.1 MPa. Furthermore, the xanthan gum/ BMIMCl gels exhibited thermally induced shape-memory effect. First, a desired permanent shape of the gel was prepared by heating-cooling process using an appropriate mold. Subsequently, the gel was softened and deformed by heating at around 50 oC, and then the intended temporary shape was fixed by cooling the gel at room

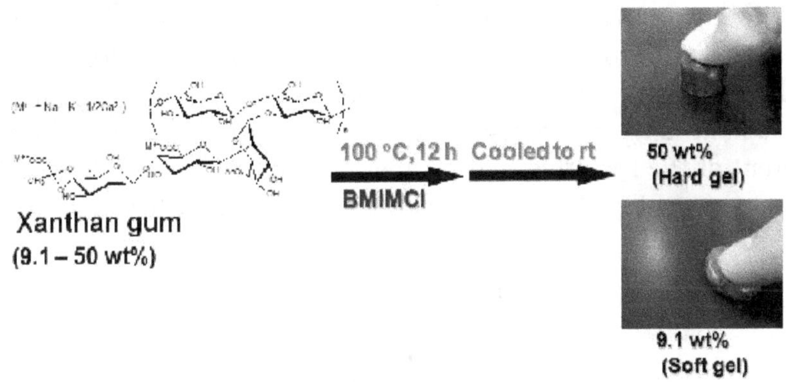

Fig. 19. Preparation of xanthan gum gels with IL.

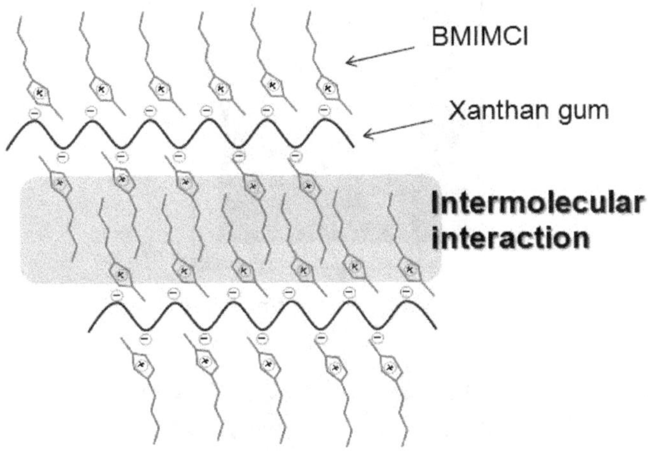

Fig. 20. Ionic exchange between xanthan gum and BMIMCl.

temperature. The temporary shape of the gel gradually returned back to the permanent shape by heating at around 50 oC. Consequently, the gel was almost recovered from temporary shape to permanent shape. It was supposed that the shape memory behavior of the gel was probably generated by the effect of ion-exchange during the gel formation and the rigid nature of the xanthan gum main-chain (Fig. 20).

The 9.1 wt% xanthan gum/BMIMCl gel was converted into a hydrogel of xanthan gum by soaking it (1.15 g) in water (100 mL) (Fig. 21). A water content of the obtained hydrogel (primary form) was estimated to be 94.1 wt%, indicating the formation of the xanthan gum hydrogel with high water content. This material was completely different from the aqueous xanthan gum dispersion because that was thixotropic gel-like paste. The elemental analysis data showed that the gel still contained 2.4 equivalents of the imidazolium groups to a repeating unit of xanthan gum. This is probably due to the presence of ion pair between the imidazolium and carboxylate formed by ion-exchange of carboxylate metal salts in the original xanthan gum with BMIMCl during the formation of the xanthan gum/BMIMCl gel. Further ion-exchange of the imidazolium groups in the primary formed hydrogel with Ca^{2+} was attempted by soaking it in 0.2 mol/L $CaCl_2$ aqueous solution to obtain an ionically cross-linked xanthan gum hydrogel with divalent cations (Ca^{2+} form) (Fig. 21). The water content of the resulting hydrogel was estimated to be 85.2 wt%, indicating shrinkage of this material compared with the primary hydrogel. In addition, the ionically cross-linked

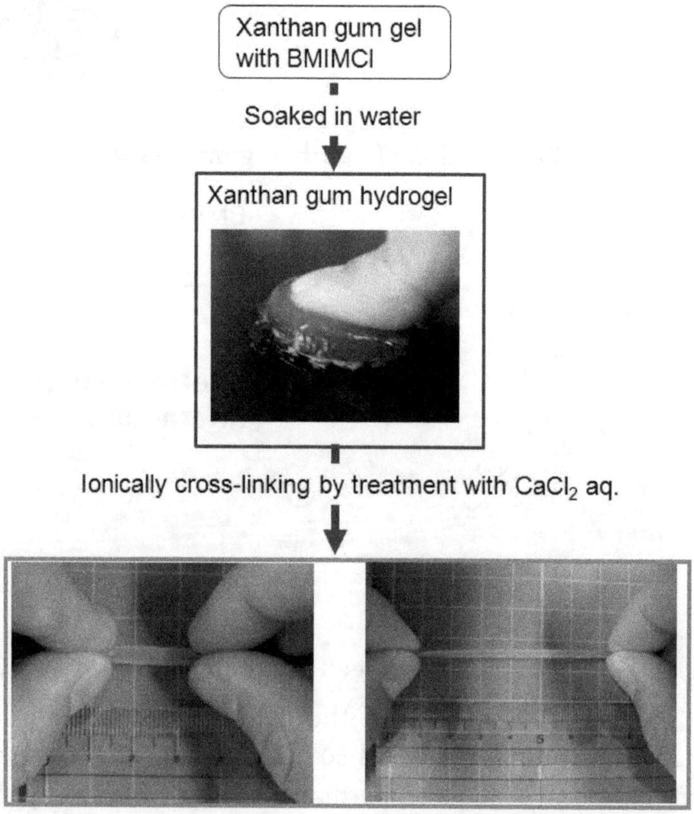

Fig. 21. Preparation of xanthan gum hydrogels from xanthan gum gel with IL.

hydrogel showed a reversible swelling-shrinking behavior. When the ionically cross-linked hydrogel was soaked in water for 1 day, it was gradually swollen to be 134.2% weight, which contained water of 24.4 times weights of xanthan gum. Soaking the swollen hydrogel in 0.2 mol/L $CaCl_2$ aqueous solution for 1 day caused rapid shrinkage to be 7.7% weight. The swelling-shrinking cycle of the hydrogel was repeated further twice. These results indicated that the present hydrogel exhibited salt concentration-induced responsive property. This behavior was probably caused by the conformation change of xanthan gum, which was induced by ionic strength of the aqueous solution. Under compressive mode, the hydrogel in the Ca^{2+} form showed much better mechanical property than the primary formed hydrogel. The primary hydrogel was too weak to be subjected to a tensile test, whereas the hydrogel in the Ca^{2+} form had sufficient strength for tensile testing, which exhibited elongation up to 550%. These results suggested that cross-linking with Ca^{2+} in the hydrogel strongly contributed to enhancement of mechanical properties. The ion-exchanged hydrogels with Na^+ and Fe^{3+} were also prepared by soaking the primary hydrogel in NaCl and $FeCl_3$ aqueous solutions, respectively.

The gel formation of food hydrocolloids with BMIMCl was not limited to the anionic polysaccharides, but extended to a neutral polysaccharide, which was guar gum (GG) (Prasad, Izawa et al., 2009). Guar gum (GG) is a galactomannan extracted from the seed of the leguminous shrub *Cyamopsis tetragonoloba* and consists of a (1→4)-linked ®-D- mannopyranose main-chain with a branched ⟨-D-garactopyranose unit at 6 position. As a typical procedure for the preparation of the gel of GG with BMIMCl (Fig. 22), first, GG (13.0 wt%) was dissolved in BMIMCl by heating at 100 °C for 5 h. After the solution was cooled to room temperature, it formed a gel, which was suitably handled after 12 h of standing upon addition of acetone and the excess BMIMCl leaching out from the gel was washed out with ethanol. However, the obtained gel started to leach out BMIMCl again after 12 h of preparation unlike other polysaccharide gels with BMIMCl described above. When the gel was further soaked in ethanol for 12 h, a stable hard material of GG with BMIMCl was obtained. The elemental analysis data of the material indicated that during the soaking process, most of BMIMCl was leached out from the gel matrices and subsequent aggregation of the GG chains existing on the surface area of the gel probably took place, causing stabilization of the material, because BMIMCl was soluble, but GG was insoluble in ethanol. The resulting hard material was further applied to the film formation. The hard material with an appropriate size was put on a compression machine and compressed applying gradually variable forces up to a maximum of ca. 10 MPa. Since expulsion of some BMIMCl from the material due to the compression was observed, the film thus obtained was further soaked in ethanol for 2 h to remove the excluded BMIMCl and dried at room temperature (Fig. 20). The film was stable and did not leach out BMIMCl even after long storage under ambient conditions. The tensile testing of the film showed a fracture stress of 17 MPa with a fracture strain of 18.2%. It was observed that the film became hard upon heating and hence it could be arranged to give a shape at higher temperature (Fig. 23). For example, the film was twisted to a desired shape and was heated to a temperature of ca. 70 – 80 °C. The film became hard immediately and retained the shape for a few minutes after the heating was stopped. When the film was then left standing at room temperature, it gradually became soft and returned to the original shape after ca. 15 minutes.

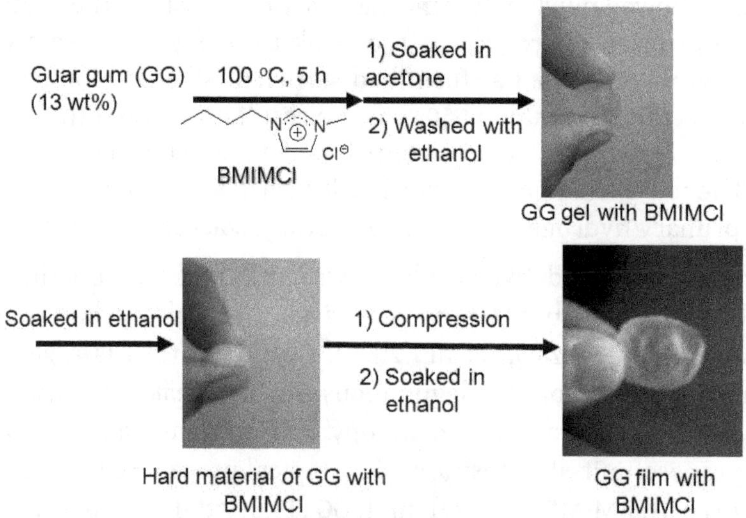

Fig. 22. Preparation of guar gum gel and film with IL.

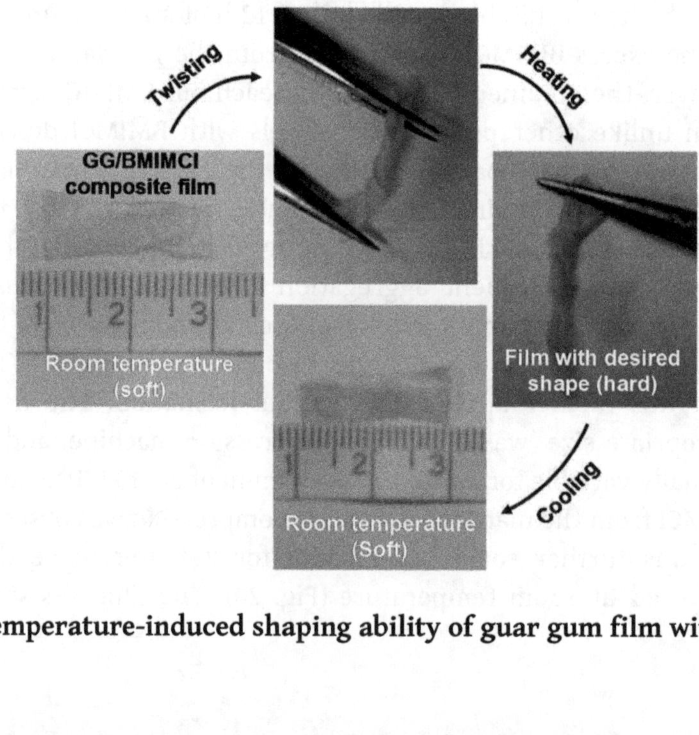

Fig. 23. Temperature-induced shaping ability of guar gum film with BMIMCl.

4. CONCISE SUMMARY

This chapter examined the synthesis of polysaccharide-based compounds compatibilized with ionic liquids. The solubility, processability, and practicality issues of polysaccharides, attributed to their many hydrogen bonds, have frequently hindered their application across a diverse range of materials. Based on the perspective that imidazolium-type ionic liquids exhibit strong affinity for polysaccharides, suitable materials have been developed from both abundant polysaccharides and food hydrocolloids. This research delineates novel polysaccharide-based functional materials, facilitating the efficient utilization of polysaccharides as an organic resource in practical applications. Consequently, research in this sector will progressively advance, garnering significant attention, thereby contributing to the establishment of a green and sustainable society in the future.

REFERENCES

Feng, L. & Chen. Z. –I. (2008). Research progress on dissolution and functional modification of cellulose in ionic liquids. *Journal of Molecular Liquids*, 142, 1-5, 0167-7322

Graenacher, C. (1934). Cellulose solution. *US patent*, 1943176

Green, O.; Grubjesic, S.; Lee, S. & Firestone, M. A. (2009). The design of polymeric ionic liquids for the preparation of functional materials. *Journal of Macromolecular Science, Part C: Polymer Reviews*, 49, 339-360, 1558-3724

Izawa, H.; Kaneko, Y. & Kadokawa, J. (2009). Unique gel of xanthan gum with ionic liquid and its conversion into high performance hydrogel. *Journal of Materials Chemistry*, 19, 6969-6972, 0959-9428

Izawa, H. & Kadokawa, J. (2010). Preparation and characterizations of functional ionic liquid-gel and hydrogel of xanthan gum. *Journal of Materials Chemistry*, 20, 5235- 5241, 0959-9428

Kadokawa, J.; Murakami, M. & Kaneko, Y. (2008a). A facile method for preparation of composites composed of cellulose and a polystyrene-type polymeric ionic liquid using a polymerizable ionic liquid. *Composite Science and Technology*, 68, 493-498, 0266-3538

Kadokawa, J.; Murakami, M. & Kaneko, Y. (2008b). A facile preparation of gel materials from a solution of cellulose in ionic liquid. *Carbohydrate Research*, 343, 769-772, 0008- 6215

Kadokawa, J.; Murakami, M.; Takegawa, A. & Kaneko, Y. (2009). Preparation of cellulose- starch composite gel and fibrous material from a mixture of the polysaccharides in ionic liquid. *Carbohydrate Polymers*, 75, 180-183, 0144-8617

Liebert, T. & Heinze, T. (2008). Interaction of ionic liquids with polysaccharides 5. Solvents and reaction media for the modification of cellulose. *BioResources*, 3, 576-601, 1930- 2126

Murakami, M.; Kaneko, Y. & Kadokawa, J. (2007). Preparation of cellulose-polymerized ionic liquid composite by in-situ polymerization of polymerizable ionic liquid in cellulose-dissolving solution. *Carbohydrate Polymers*, 69, 378-381, 0144-8617

Pinkert, A.; Marsh, K. N.; Pang, S. & Staiger, M. P. (2009). Ionic liquids and their interaction with cellulose. *Chemical Reviews*, 109, 6712-6728, 0009-2665

Plechkova, N. V. & Seddon, K. R. (2008). Applications of ionic liquids in the chemical industry. *Chemical Society Reviews*, 37, 123-150, 0306-0012

Prasad, K.; Izawa, H.; Kaneko, Y. & Kadokawa, J. (2009). Preparation of temperature- induced shapeable film material from guar gum-based gel with an ionic liquid. *Journal of Materials Chemistry*, 19, 4088-4090, 0959-9428

Prasad, K.; Kaneko, Y. & Kadokawa, J. (2009). Novel gelling systems of κ-, ι- and λ- carrageenans and their composite gels with cellulose using ionic liquid. *Macromolecular Bioscience*, 9, 376-382, 1616-5187

Prasad, K.; Murakami, M.; Kaneko, Y.; Takada, A.; Nakamura, Y. & Kadokawa, J. (2009). Weak gel of chitin with ionic liquid, 1-allyl-3-methylimidazolium bromide. *International Journal of Biological Macromolecules*, 45, 221-225, 0141-8130

Prasad, K. & Kadokawa, J. (2010). Preparation of composite materials composed of ι- carrageenan and polymeric ionic liquids. *Polymer Composites*, 31, 799-806, 0272-8397

Prasad, K.; Mine, S.; Kaneko, Y. & Kadokawa, J. (2010). Preparation of cellulose-based ionic porous material compatibilized with polymeric ionic liquid. *Polymer Bulletin*, 64, 341-349, 0170-0839

Schuerch, C. (1986). Polysaccharides. *Encyclopedia of polymer science and engineering, 2nd edition*, Vol. 13, pp. 87-162, John Wiley & Sons, New York

Seoud, O. A. E.; Koschella, A.; Fidale, L. C.; Dorn, S. & Heinze, T. (2007). Applications of ionic liquids in carbohydrate chemistry: A windows of opportunities. *Biomacromolecules*, 8, 2629-2647, 1525-7797

Swatloski, R. P.; Spear, S. K.; Holbrey, J. D. & Rogers, R. D. (2002). Dissolution of cellulose with ionic liquids. *Journal of the American Chemical Society*, 124, 4974-4975, 0002-7863

Takegawa, A.; Murakami, M.; Kaneko, Y. & Kadokawa, J. (2009). A facile preparation of composites composed of cellulose and polymeric ionic liquids by in situ polymerization of ionic liquids having acrylate groups. *Polymer Composites*, 30, 1837-1841, 0272-8397

Takegawa, A.; Murakami, M.; Kaneko, Y. & Kadokawa, J. (2010). Preparation of chitin/cellulose composite gels and films with ionic liquids. *Carbohydrate Polymers*, 79, 85-90, 0144-8617

Welton, T. (1999). Room-temperature ionic liquids. Solvents for synthesis and catalysis. *Chemical Reviews*, 99, 2071-2083, 0009-2665

Yamazaki, S.; Takegawa, A.; Kaneko, Y.; Kadokawa, J.; Yamagata, M. & Ishikawa, M. (2009). An acidic cellulose-chitin hybrid gel as novel electrolyte for an electric double layer capacitor. *Electrochemistry Communications*, 11, 68-70, 1388-2481

Yamazaki, S.; Takegawa, A.; Kaneko, Y.; Kadokawa, J.; Yamagata, M. & Ishikawa, M. (2010a). Performance of electric double-layer capacitor with acidic cellulose- chitin hybrid gel electrolyte. *Journal of Electrochemical Society*, 157, A203-A208, 0013- 4651

Yamazaki, S.; Takegawa, A.; Kaneko, Y.; Kadokawa, J.; Yamagata, M. & Ishikawa, M. (2010b). High/low temperature operation of electric double layer capacitor utilizing acidic cellulose-chitin hybrid gel electrolyte. *Journal of Power Sources*, 195, 6245-6249, 0378-7753

Zhu, S.; Wu, Y.; Chen, Q.; Yu, Z.; Wang, C.; Jin, S.; Ding, Y. & Wu, G. (2006). Dissolution of cellulose with ionic liquids and its application: a mini-review. *Green Chemistry*, 8, 325-327, 1463-9262

7
Chapter

POLYMERIZATION OF CYCLODEXTRIN-IONIC LIQUID COMPLEXES FOR THE ELIMINATION OF ORGANIC AND INORGANIC CONTAMINANTS FROM WATER

1. INTRODUCTION: AN OVERVIEW

The elimination of hazardous pollutants from water continues to provide a significant issue for water delivery corporations and municipalities. Water distribution networks frequently contain both organic and inorganic contaminants, predominantly heavy metals, due to agricultural, household, and industrial activities. Effective techniques must be developed to eliminate these contaminants from water delivery networks. Our laboratories have successfully proved that polymers of cone-shaped cyclodextrins (CDs) coupled by linkers can effectively remove organic contaminants from water. We have initiated a program to synthesize polymers with moieties that can eliminate both organic and inorganic pollutants from water, referred to as multifunctional complexes. To access these multifunctional complexes, cyclodextrins were conjugated to salts of organic cations (i.e., imidazolium and pyridinium) in a manner analogous to the creation of ionic liquids. The cyclodextrin-ionic liquid (CD-IL) derivatives were further polymerized with bifunctional linkers, anticipating that the resultant polymers would effectively eliminate a diverse array of both organic and inorganic contaminants from water.

Pollutants from many sectors frequently contaminate water systems, while additional pollutants arise from home and agricultural runoff. Organic pollutants, including solvents, pesticides, and fungicides, together with inorganic pollutants such as nitrates, cyanides, and dissolved gases, are prevalent in numerous water systems (Schwarzenbach et al., 2003). Upon introduction into the environment, these pollutants undergo physical, chemical, and biological processes, resulting in the formation of derivatives with varying environmental behaviors and impacts (Schwarzenbach et al., 2003). These pollutants may permeate the soil and adulterate groundwater. They also have a tendency to bioaccumulate within the creatures' bodies. The diminutive size and chemical inertness of these contaminants complicate their removal from water using existing treatment methods (Li and Ma, 1999). Additionally, they induce physiological consequences including renal and hepatic impairment, damage to the nervous and circulatory systems, and other psychological effects harmful to human well-being. Communities require ultra-pure water, necessitating the removal of pollutants even at minimal quantities to safeguard the health and safety of water users.

Recent laboratory investigations have focused on the synthesis and application of standard (i.e., unfunctionalized) and functionalized cyclodextrin polymers for the removal of organic pollutants from water at concentration levels as low as parts per billion (μgL-1). CDs can create inclusion complexes with organic molecules, so aiding in their removal. To utilize CDs for the elimination of organic contaminants from water, they must be transformed into water-insoluble polymers. Ionic liquids (ILs) have exhibited the capacity to remove heavy metal ions from water. While technologies for the removal of inorganic contaminants from water are established, to our knowledge, no comprehensive method has been created that can simultaneously eliminate both organic and inorganic species from water. Consequently, the application of CDs associated with ionic liquid appendages is anticipated to yield a dual system proficient in simultaneously eliminating both organic and inorganic contaminants from water. The CD moiety in the polymer will continue to absorb organic pollutants, while the IL component will extract heavy metals from water.

This study aims to broaden the utilization of cyclodextrin polymers for the extraction of inorganic contaminants from water. The aims of this investigation were as follows:

i. To synthesize and characterize 6-deoxy-6-(p-tosyl)-β-cyclodextrin (CDOTs) and 6-deoxy-6-iodo-β-cyclodextrin (CDI) precursors. The objective is to create and optimize a dependable chemical pathway for the synthesis of CD-IL precursors. The characterization of the CDOTs and CDI precursors was mostly conducted using IR and NMR spectroscopy.

ii. To synthesize and characterize CD-IL complexes by reacting CDOTs and CDI precursors synthesized in (i) with 3-alkylimidazole and pyridine. The CD-IL complexes were characterized using IR and NMR spectroscopy.

iii. To polymerize the CD-IL complexes produced in (ii) using appropriate bifunctional linkers, like hexamethylene diisocyanate and toluene-2,4-diisocyanate. Infrared spectroscopy was employed for the characterization of the polymers.

iv. To investigate the thermal stability of the CD-IL polymers via thermogravimetric analysis (TGA) and differential scanning calorimetry (DSC).

v. To examine the morphology of the polymers utilizing scanning electron microscopy (SEM) and Brunauer-Emmet-Teller (BET) analysis.

vi. To evaluate the efficacy of CD-IL polymers in eliminating organic and inorganic contaminants from water through model contamination tests. The studies entail intentionally introducing the polymers into an aqueous solution with established concentrations of the pollutants. GC-MS and UV-Visible spectroscopy were employed to measure the extent of absorption of the organic pollutants. The quantification of inorganic pollutants was conducted using atomic absorption (AA) spectroscopy.

A schematic representation for the synthesis of CD-IL complexes and CD-IL polymers are shown in Scheme 1.1.

2. LITERATURE REVIEW

2.1 Organic pollutants

The problems associated with the use of the conventional water treatment techniques for the removal of these pollutants from the water systems and the approaches undertaken in our laboratories to address these problems are also described. The chapter then concludes with an overview of the applications of cyclodextrins and ionic liquids in the attempt to remove organic and inorganic contaminants from drinking water systems. Organic pollutants in water pose a major threat to human health even when present at low concentrations.

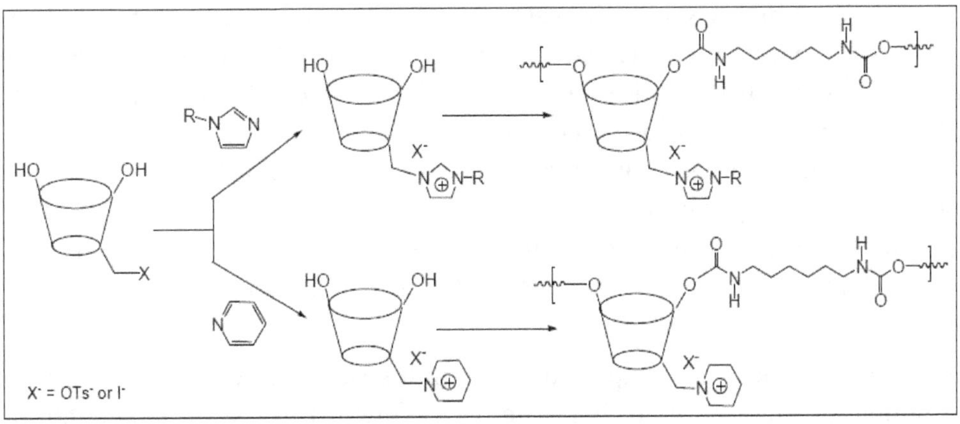

Scheme 1.1. Synthetic pathway for the preparation of CD-IL polymers

Some organic pollutants are referred to as persistent organic pollutants (POPs) because they remain in the environment for long periods of time. POPs can be absorbed by plants or animals thereby finding their way to poisoning food chains of living organisms (Oleszczuk and Baran, 2004). They have also been linked to adverse human effects such as cancer, nervous system damage, reproductive disorders, as well as the disruption of the immune system (EPA, 2006). POPs released in one part of the world can be transported globally *via* oceans and atmospherically, and their effects can be felt in regions which are distant from where they originate. POPs are mostly from human activity but there are other organics that are also of major concern.

Also, drinking water disinfection by-products pose a health risk to humans. Chlorine is still widely used for water disinfection. A residual amount of free chlorine (0.5-1.0 ppm) can keep levels of micro organisms such as bacteria to their lowest (Halmowitz, 1991). However, during chlorination, chlorine combines with water to form hypochlorous acid. Hypochlorous acid in turn reacts with NOM to form DBPs. The low volatility of DBPs, combined with their extreme stability, has led to their high persistence both in the environment and in the bodies of organisms after exposure (Gardio et al., 2006). They are mainly found in fatty tissues and bioaccumulate in food chains. DBPs are known to induce or aggravate adverse health problems in humans such as cancer, liver damage, failure of immune systems and the disruption of hormonal functions (Gardio et al., 2006). Due to the problems associated with DBPs, it is imperative to find better techniques to remove both NOM and its by-products from drinking water systems.

Phenolic compounds are common constituents of many industrial and commercial wastes (Rogers et al., 1999; Wikstrom and Marklund 2000). They may be introduced into the environment by accidental spills, illegal release of industrial and municipal wastewater, and excessive use of pesticides (Park et al., 1999). Phenolic compounds can be easily incorporated into the soil organic matter or dissolved humic materials because they are highly soluble in water. The presence of minute amounts of some phenolic compounds in water causes bad tastes and odours, which are greatly aggravated by the usual addition of chlorine for germicidal purposes (Niegowski, 1952; Kawaguchi 2004). Chlorinated and nitrated phenols pose a health hazard to both humans and animals. For instance, *para*- nitrophenol (PNP), though slightly soluble in water, is regarded as a possible mutagen to humans. 2,4,6-Trichlorophenol (TCP) is also known to affect the respiratory system and is a recognised carcinogen (NPI, 2007). Examples of important nitrated and chlorinated phenols are shown in Figure 2.1.

The removal of phenolic compounds from industrial wastewater is currently an environmental and economic challenge. This effluent is resistant to conventional water treatment techniques due to the poor biodegradability and toxicity of phenolic compounds (Santos et al., 2007; Sabio et al., 2006). Moreover, it is also expensive to use these water treatment technologies to remove these organic pollutants at ppb levels ($\mu g.L^{-1}$).

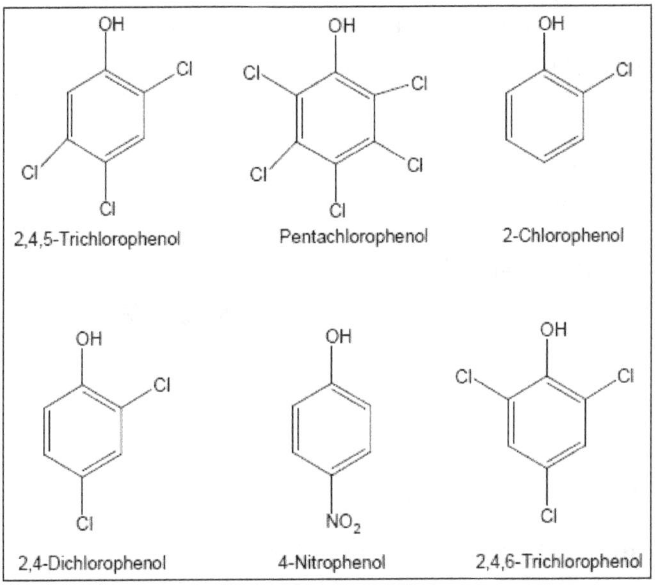

Fig. 2.1. Structures of chlorinated and nitrated phenols

2.2 Current water treatment methods for organic pollutants

Water treatment technologies used for the removal of organic species around the world include filtration using activated carbon, reverse osmosis, zeolites, and the use of dendrimers (Li and Ma, 1999). However, these techniques often do not remove organic contaminants to accepted levels. Due to the failure of most of the aforementioned water treatment techniques to remove organic pollutants from water, the world is facing challenges in meeting the rising demand for clean water. This is because the sources of available fresh water supplies are decreasing due to extended droughts, population growth and more stringent health-based regulations (Savage and Diallo, 2005). Also, water scientists are questioning the viability of current water treatment practices in meeting the water demands of all users according to regulations by the World Health Organisation (WHO). Since the current water treatment techniques fail to remove pollutants to the required levels due to their diversity and varying chemical properties, better technologies need to be developed. In our laboratories the use of cyclodextrin polyurethanes has been studied. These CD polymers have been found to have a high efficiency for the removal of organic pollutants from water, even at ppb levels (Mhlanga et al., 2007). Cyclodextrin polyurethanes are discussed in the following sections.

2.3 Cyclodextrins

Cyclodextrins (CDs) were first discovered by Villiers in 1891 (Szeijtli, 1998). CDs are cyclic oligomers formed by the enzymatic hydrolysis of starch by *Bacillus macerans.* The glucose units in a CD molecule are linked by α-(1,4) linkages (Bender and Komiyana, 1978; Croft and Bartsch 1983). Figure 2.2 illustrates the *a*-(1, 4) linkages and the numbering system in a typical CD moiety.

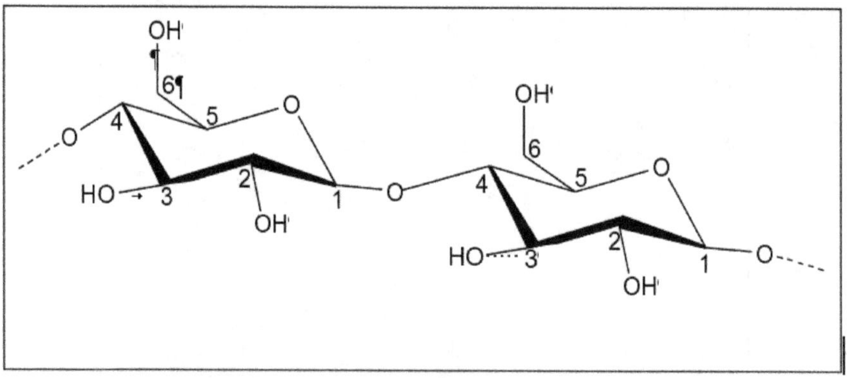

Fig. 2.2. ***a*-(1, 4) linkages and the numbering system in a CD molecule**

2.3.1 Physical properties of cyclodextrins

The most common CDs are made of six, seven or eight glucose units and are called α-, β- and γ-CDs, (Szeijtli, 1998) respectively. The β-CD has been extensively studied by our research group and other researchers and was as a result used in this study. Higher molecular weight CDs are known, however their use is not common even in industry (Martin, 2004). Figure 2.3 shows the three common types of CDs.

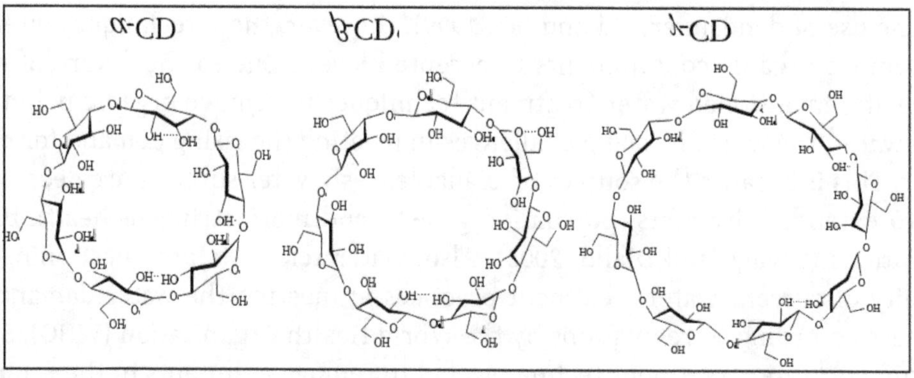

Fig. 2.3. Structures of α-, β- and γ-CDs, respectively showing the glucose units

CDs have a characteristic toroidal shape that form well defined cavities (Bender and Komiyana, 1978). These cavities are about 8Å deep. Depending on the number of the glucose units, the diameter of these cavities range between 5 and 10 nm. They have a hydrophobic interior because of the presence of carbon and hydrogen atoms and this feature allows them to host several compounds in their cavities. Their exterior cavities are hydrophilic because of the presence of hydroxyl groups and this makes them soluble in water. Figure 2.4 shows a representation of CD moiety with a hydrophilic exterior and a hydrophobic cavity.

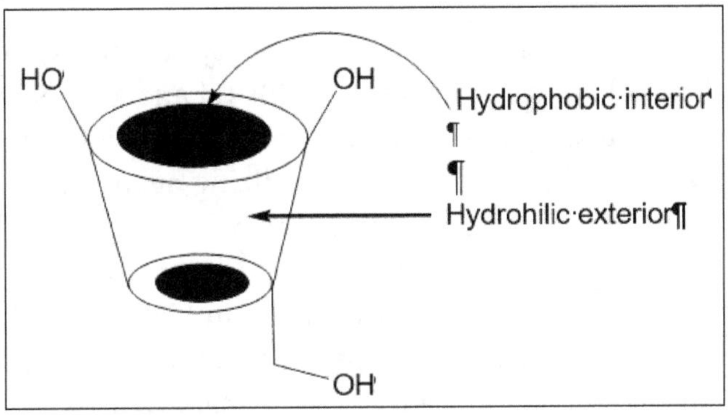

Fig. 2.4. Hydrophobic interior and hydrophilic exterior of a CD

2.3.2 Formation of inclusion complexes

Cyclodextrins can form inclusion complexes with various compounds (guests) of low molecular weight (Figure 2.5) (Li and Tan, 2001; Phan et al., 2002). Examples of the guest molecules include acids, apolar aliphatic and aromatic hydrocarbons and amines. The CD moiety harbours these small, suitably shaped organic compounds in its tubular cavities by shielding the bound species from the surrounding aqueous environment (Weinz et al., 1997). This phenomenon is a result of the hydrophobic-hydrophobic interaction between the host CD and the organic species (Liu et al., 2003; Weinz et al., 1997). The formation of inclusion complexes does not involve the formation of bonds but is an attraction between the host and guest as a result of their polarities (Szeijtli, 1998).

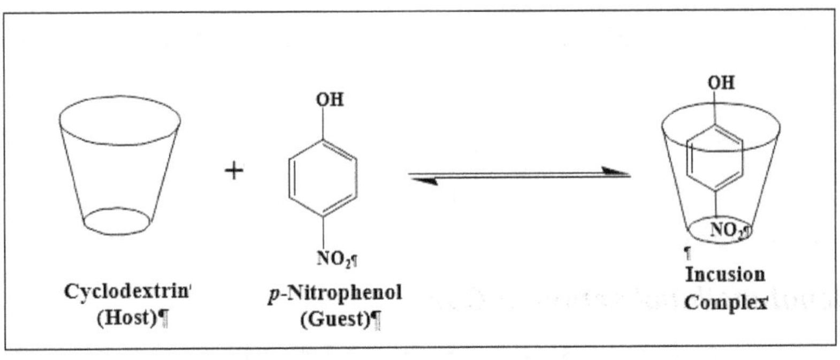

Fig. 2.5. Schematic representation of an inclusion complex

Because the lengths of the diameter of the CD vary, the organic species must be able to fit (at least partly) into the CD cavity (compatible geometry). The size of the organic compound and the type of the CD used is important for the formation of inclusion complexes. Direct exploitation of this phenomenon for water treatment is however limited by the solubility of CDs in water. Hence, making insoluble CD derivatives and polymers is a prerequisite if CDs are to be used in water treatment.

2.3.3 Cyclodextrin derivatives

The presence of hydroxyl groups in the outer cavity can be exploited in order to alter the physical and chemical properties of CDs (Li and Ma, 1999; Szeijtli, 1998). For example, functionalising CDs can decrease their solubility, increase their stability in the presence of light, heat and oxidizing agents and also make them more volatile (Harada, 1997; Liu et al., 2003). The primary hydroxyl groups of a CD molecule (C-6 OHs) are more reactive compared to their secondary hydroxyl groups (C-2, C-3 OHs). These hydroxyl groups are shown in Figure 2.6. The C-2 hydroxyl groups are acidic. The C-3 hydroxyl groups are less reactive because they are not easily accessible due to sterric hindrance. The primary hydroxyl groups at C-6 are basic hence an acid is used to remove the H in the OH, i.e. and are the most easily accessible thus often more reactive during nucleophilic substitution reactions.

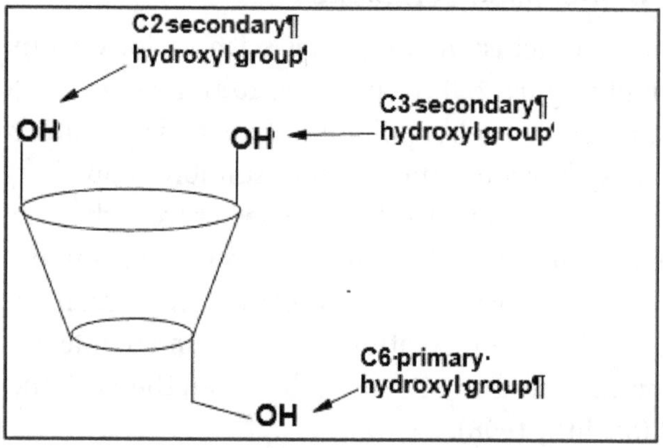

Fig. 2.6. The structure of a CD showing the three different hydroxyl groups.

$$R{-}OH + OH^- \rightarrow R{-}O^- + H_2O \quad (1)$$

and are the most easily accessible thus often more reactive during nucleophilic substitution reactions.

2.3.3.1 Monofunctionalisation of CDs

The purpose of functionalising CDs is to modify their physico-chemical properties and also to introduce groups with specific activity. The two most common types of CD substitution reactions are mono- and per-functionalisation reactions. Monofunctionalisation refers to the reaction of only one hydroxyl group while per-functionalisation involves the whole set of hydroxyl groups. In monofunctionalisation, regioselective reactions should be the target (Sutyagin et al., 2002). Regioselective reactions usually involve the protection of other OH groups to prevent them from participating in the reaction while exposing the target group to reactions. Monofunctionalisation at C-2 and C-3 is well documented (Rong and D'Souza, 1990; Fugita et al., 1984). In this study monofunctionalisation at C-6 was mainly exploited. Monofunctionalisation of C-6 can be achieved because in the presence of a base (e.g. NaOH), the proton at C-6 is removed to make O- which is a nucleophile that can be easily substituted. Thus when a CD is reacted with a tosyl group in a basic medium, a monofunctionalised CD derivative (CDOTs) is afforded (Byun et al., 1998; Muderawan et al., 2005; Zhong et al., 1998). The CDOTs is an important precursor for all C-6 CD derivatives because the tosylate is a good leaving group and can be easily replaced by other nucleophiles (e. g. halogens, azides and acetates). In this study, the CDOTs was also used as a precursor for the CD derivatives that were synthesised.

2.3.4 Cyclodextrin polymers

The solubility of CDs and their derivatives limits their application in water treatment. Converting them to water insoluble entities is therefore crucial if CDs are to be utilised in the removal of contaminants from water. To attain this property, CD derivatives have been polymerised with suitable bifunctional linkers in our laboratories to produce water insoluble polymers (Mhlanga et al., 2007; Mamba et al., 2007; Salipira et al., 2007). The presence of the OH groups in the CDs or CD-derivatives allows a reaction in which the monomers react with linkers to produce insoluble polymers. A reaction scheme for the polymerisation of CDs with diisocyanate linkers (hexamethylene diisocyanate (HDI)) is shown in Figure 2.7 (Li and Ma, 1999).

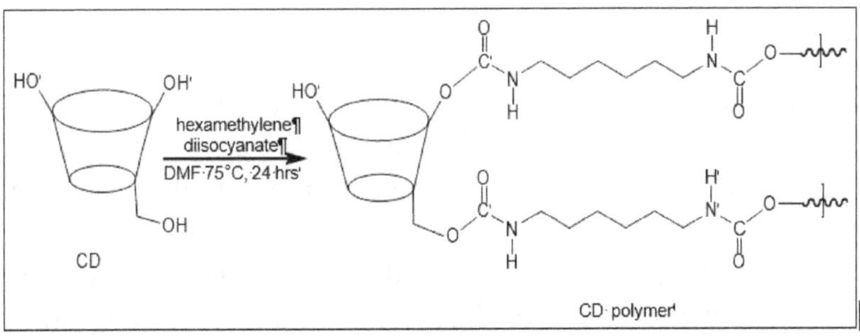

Fig. 2.7. Polymerisation of CDs with an isocyanate linker (HDI)

The CD polyurethanes have demonstrated a high efficiency for the absorption of organic pollutants in water (Mhlanga et al., 2007; Mamba et al., 2007; Salipira et al., 2007). However, in real water systems organic and inorganic contaminants co-exist; the challenge is to craft a water treatment system that is capable of simultaneously removing these contaminants from water. In an attempt to overcome this challenge, we functionalised CDs with ionic liquids (ILs). Since ILs are known to chelate heavy metals from aqueous media and CDs are able to remove organic pollutants from water, a system which combines CDs and ionic liquids is expected to absorb both organic and inorganic pollutants from water.

The CD polyurethanes have demonstrated a high efficiency for the absorption of organic pollutants in water (Mhlanga et al., 2007; Mamba et al., 2007; Salipira et al., 2007). However, in real water systems organic and inorganic contaminants co-exist; the challenge is to craft a water treatment system that is capable of simultaneously removing these contaminants from water. In an attempt to overcome this challenge, we functionalised CDs with ionic liquids (ILs). Since ILs are known to chelate heavy metals from aqueous media and CDs are able to remove organic pollutants from water, a system which combines CDs and ionic liquids is expected to absorb both organic and inorganic pollutants from water.

2.4 Inorganic pollutants in water

Inorganic contaminants are compounds that typically do not contain the element carbon in their structure (waterborne, 2006). Inorganic pollutants are introduced into the environment naturally, for example, from weathering of parent rocks, industrial wastes and agricultural run-offs (waterborne, 2006; Petanen and Romantschuk, 2002). Inorganic pollutants include dissolved gases (e. g. oxygen, carbon dioxide, nitrogen and radon), heavy metals (e. g. Hg^{2+}, Cd^{2+}, Cr^{6+}) and anions such as fluorides (F^-), nitrates (NO_3^-) and cyanides (CN^-). These contaminants also pose health risks to living organisms because they bioaccumulate in the body resulting in chronic health effects (Morales et al., 1999). Inorganic contaminants are not biodegradable; hence they can typically poison the food chains of living organisms (waterborne, 2006).

2.5 Ionic liquids

Ionic liquids (ILs) are salts of organic cations which have aroused an increasing interest for their promising role as alternative media for volatile solvents (Wei et al., 2003). Ionic Liquids (ILs) are low melting point molten salts composed of organic cations (e.g. pyridinium and imidazolium ions) and mostly inorganic anions like Cl^-, Br^-, $[PF_6]^-$ and $[BF_4]^-$. Compared to conventional salts, ILs are liquid at room temperature (Wei et al., 2003). The most common types of organic cations are the imidazolium and pyridinium ions. Examples of the structures of ionic liquids are depicted below in Figure 2.8.

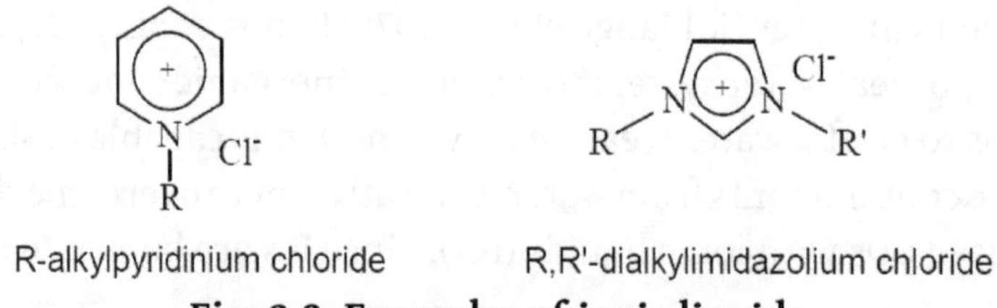

Fig. 2.8. Examples of ionic liquids

The organic cationic component and the alkyl substituents are known to be responsible for the low melting points associated with these compounds. The ability to dissolve organic solvents has also been linked to the cationic component and the alkyl substituents since both of them are organic in character (Welton, 1999). The alkyl group is usually an alkane chain whereby increasing the length of the alkane chain one can change the resulting properties (e.g., viscosity, hydrophobicity, and melting point) (Ferreira et al., 2000). Besides pyridinium and imidazolium cations, other less common organic cations have been used for the preparation of ionic liquids. These include ammonium, pyrrolidium, sulfonium and phosphonium ions (Figure 2.9) (Pereiro et al., 2007).

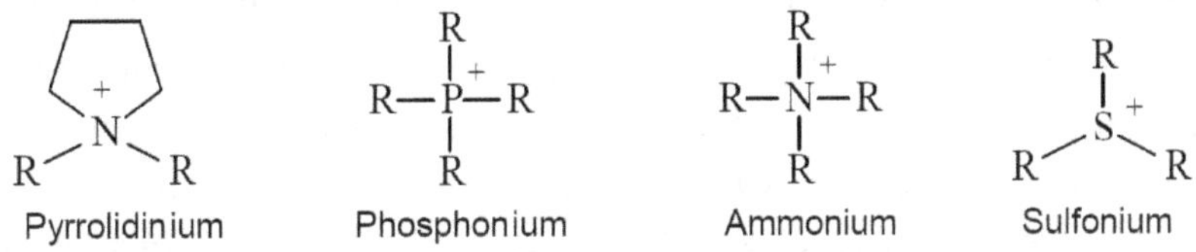

Fig. 2.9. Less common organic cations used for the synthesis of ILs.

2.5.1 Synthesis of ILs

Ionic liquids are made up of both a cation and an anion, whose forces of attraction are not sufficiently strong to hold them together as solids at ambient temperatures. It is therefore possible, by proper choice of starting material, to synthesize ionic liquids that are liquid at or below room temperature (Ferreira et al., 2000). Volatile organic chemicals (VOC's) are used to prepare non-volatile ionic liquids which have negligible vapour pressure; therefore new and better methods need to be researched since many processes used for cleaning ionic liquids involve washing with water and VOC's. This will also help in terms of eliminating or decreasing the hazards in the system and eliminate the waste before it is produced. The two most common methods of synthesising ionic liquids are metathesis reaction involving a halide salt and acid-base neutralisation reactions (Welton, 1999). For example, the metathesis reaction of 1-ethyl-3-methylimidazolium iodide with ammonium tetrafluoroborate is shown in Scheme 2.1.

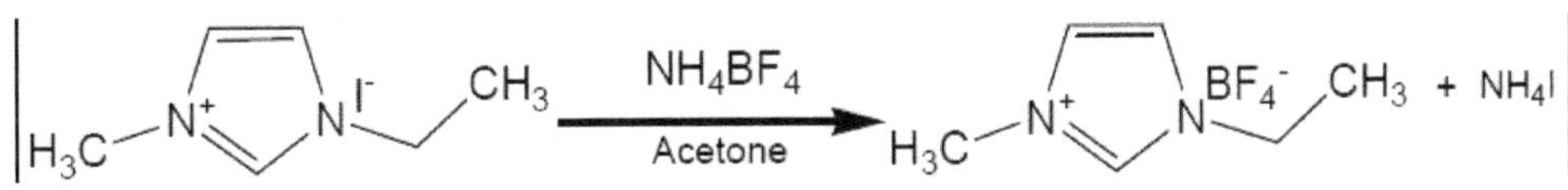

Scheme 2.1. Synthetic pathway for 1-ethyl-3-methylimidazolium tetrafluoroborate

Also, a reaction an alkyl halide with pyridinium and imidazolium cations yields ionic liquids, as shown in Scheme 2.2.

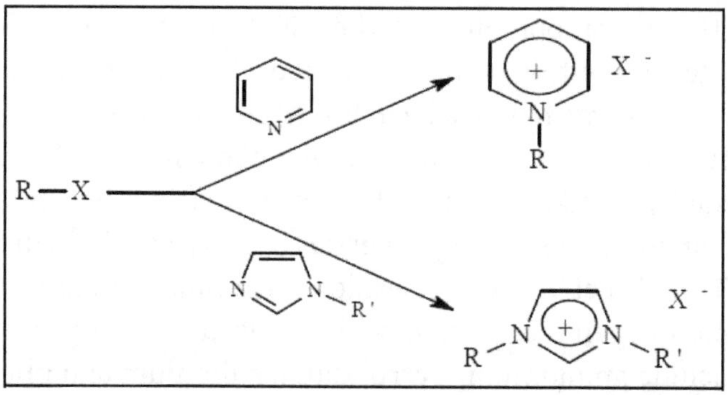

Scheme 2.2. Pathway for the synthesis of pyridinium and imidazolium based ILs

The anions, to a great extent influence both the chemical and physical properties of the ionic liquids. For example, although both anions are air- and moisture-stable, $[BF_4]^-$ and $[PF_6]^-$ anions are used for the preparation of water-miscible and water-immiscible ionic liquids, respectively (Ferreira et al., 2000). Besides anions such as hexafluorophosphate, tetrafluoroborate and halide ions, other common inorganic anions that are used for the preparation of ILs include $[SbF_6]$ and $[(CF_3SO_2)_2N]^-$. The anion need not necessarily have to be inorganic; ILs possessing organic anions such as alkylsulfate, tosylate and methanesulfonate are known (Figure 2.10) (Sigma-aldrich, 2007).

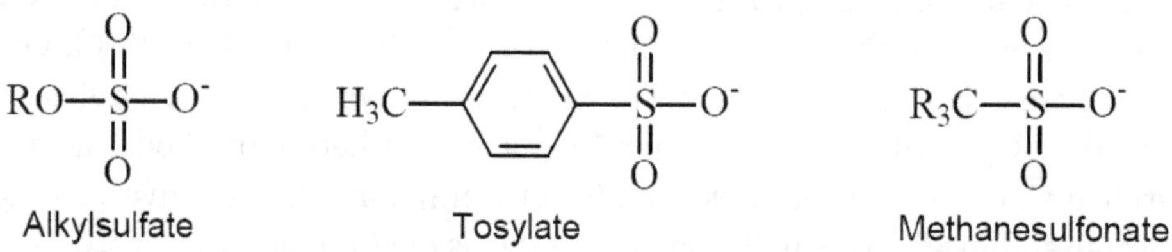

Fig. 2.10. Examples of organic anions used for the synthesis of ILs.

An example of such an IL (ethylpyridinium tosylate) is shown in Figure 2.11.

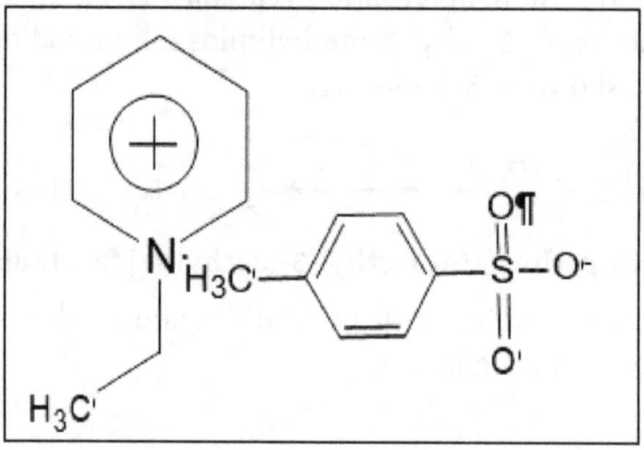

Fig. 2.11. Structure of ethylpyridinium tosylate

2.5.2 Physical properties of ILs

The physical properties of ionic liquids depend on the nature and the size of their cationic and anionic species (Visser and Rogers, 2004). These properties relate to the choice of the two ionic components (Ferreira et al., 2003; Visser and Rogers, 2004). Hence, ILs are sometimes referred to as "designer solvents". ILs are gaining a wide recognition in analytical chemistry because they possess several unique properties (Liu et al., 2004). These physical properties include negligible vapour pressure (Liu et al., 2003), an ability to dissolve a wide range of organic and inorganic materials (Welton, 1999), high thermal stability (they have low melting points and high boiling points, for example, some ILs are liquid at 400°C, while others are liquid at -96°C) (Liu et al 2005) as well as variable viscosity and miscibility with water and other organic solvents (Liu et al., 2003). These physical properties have triggered an interest in applying ionic liquids in separation chemistry and synthesis. In addition, these properties form the basis for the utilisation of ILs in replacing traditional toxic, flammable and volatile organic compounds (VOCs), which are not environmentally friendly. To this end, ILs have the potential to be used as alternative reaction media for "Green Chemistry" (Wei et al., 2003).

2.5.2.1 Hydrophobic and hydrophilic ILs

The hydrophilic and hydrophobic nature of ILs depends on the type of anion used in the synthesis of the target ionic liquid (Welton, 1999; Cocalia et al., 2006). This characteristic feature determines the extent of miscibility of ILs with water and other organic solvents. It also provides an alternative for the use of ILs in nonaqueous, polar phase separation systems, i.e. liquid-liquid extraction (Welton, 1999). For instance, hydrophobic ILs can be used as immiscible polar phases with water. Examples of common hydrophobic and hydrophilic anions are shown in Figure 2.12. In this study, one hydrophobic (tosylate) and one hydrophilic (iodide) anion were utilised in the synthesis of ionic liquids.

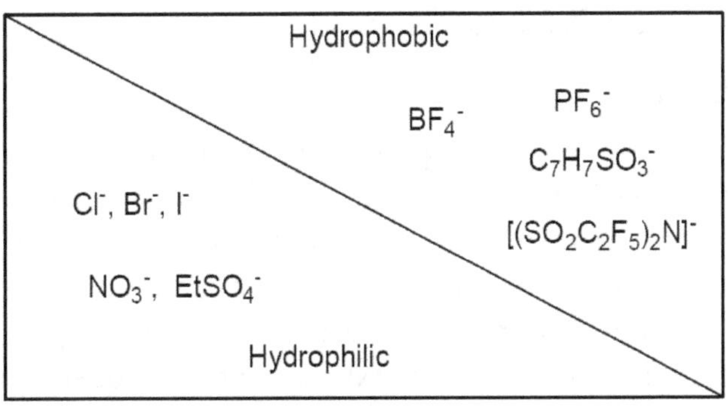

Key: $[PF_6]^-$ = hexafluorophosphate, $[BF_4]^-$ = tetrafluoroborate, $[C_7H_7SO_3]^-$ = tosylate $[(SO_2C_2F_5)_2N]^-$ = bis (trifyl) imide and $[EtSO_4]^-$ = ethylsulfonate.

Fig. 2.12. Examples of hydrophobic and hydrophilic IL anions

2.5.3 Chemical properties of ILs

Ionic liquids have been widely considered as greener solvents which are suitable for application in a wide range of organic chemical reactions (Chowdhury et al., 2007). Applications of ILs include control of product distribution, enhanced reactivity rates, recyclability, catalysis and ease of recovering products from a reaction mixture (Chowhury et al., 2007). Further, they have shown to possess much needed properties such as inflammability and less volatility. Thus they are emerging as novel replacements for the widely used VOC's which are detrimental to the environment (Liu et al., 2003; Cruz and Moens, 2000). ILs are generally referred to as polar solvents because they are able to dissolve and stabilise dipolar and charged solutes (Welton, 1999). The localisation of the anions and the delocalisation of the cationic charge also influence the polarity of ILs. Ionic liquids can be applied in battery industries since they are good electrolytes and good solvents for Friedel-Crafts alkylations, hydrogenations and Diels-Alder reactions, to mention just a few (Welton, 1999; Cruz and Moens, 2000). However, the reactivity of ILs is also affected by the acidity and basicity of the IL under investigation.

2.5.3.1 Acidic and basic ILs

The acidic and basic nature of ionic liquids is largely determined by the ability of the salt to act as a hydrogen bond donor (acid) or acceptor (base) (Welton, 1999; Chowdhury et al., 2007). Incorporation of an aliphatic carboxylic group to the cationic component of an IL can result in the formation of an acidic ionic liquid. Also, reacting the cationic component with an acidic anion, e. g. [HSO_4^-], results in an acidic IL (Chowdhury et al., 2007). Increasing the acidity of the anion has been found to improve the catalytic activity of the IL. On the other hand, basic ILs are a result of combining the cationic component of the IL with a basic anion, e. g. [OH^-]. Knowing the acidic and basic nature of an IL is important because it allows for the proper choice of an IL for a specific application. This has led to a tailored synthesis of task specific ionic liquids (TSILs).

2.5.4 Applications and uses of ILs

The application of any ionic liquid depends on the purpose for which it is designed. ILs that have been specifically designed for use in catalysis, chromatography, spectrometry, electrochemical applications, as chemo sensors and extraction solvents have been reported in a number of publications (Visser and Rogers, 2004; Liu et al., 2004; Cruz and Moens, 2000). Task specific ionic liquids that have been synthesised for the purposes of metal extraction from aqueous media have also been reported (Pereiro et al., 2007; Liu et al., 2005; Visser et al., 2002). Metal ions such as Hg^{2+}, Cd^{2+}, Sr^{2+} and Cs^{+} have been removed from aqueous media using ILs (Visser and Rogers, 2004). The ability of ILs to form complexes with most metal-halides, especially transition metal-halides, is the basis for their application in metal extraction from aqueous media (Welton, 1999). Metal ions form complexes with both the anion and cation of the IL with a specific coordination. The coordination mechanism depends on the ionisation state of the metal ion. For instance, metals with ionisation state of two, [M (II)], form complexes with bidentate coordination (Welton, 1999). To enhance the metal ion extraction capability of ionic liquids, metal ligating groups such as urea and sulphur, are used (Visser et al., 2002). Affiliating such ligands on the IL cations improve the extraction efficiency of the targeted metal ions.

2.6 Problems associated with current water treatment technologies

Current water treatment techniques can either absorb organic or inorganic pollutants from an aqueous medium, but not both. In addition, these technologies fail to remove pollutants to the required levels and this is attributed to the diversity and varying chemical properties of organic pollutants. Furthermore, in real water systems organic and inorganic pollutants co-exist hence techniques that can remove both contaminants simultaneously would be ideal. Cyclodextrin polyurethanes have been applied in the removal of organic pollutants from water (Mhlanga, 2006). The new challenge is to functionalise these polymers such that they are capable of removing both organic and inorganic pollutants present in water. One possibility is to functionalise the CD moieties with salts of organic cations known as ionic liquids. On their own, ionic liquids have demonstrated the ability to remove inorganic contaminants (heavy metal ions such as Hg^{2+}, As^{3+} and Cr^{3+}) from an aqueous solution (Visser et al., 2002; Tran and De Paoli, 2002). Since cyclodextrin polyurethanes and ionic liquids have independently shown the capability to remove organic and inorganic contaminants, respectively, it is expected that cyclodextrin polymers functionalised with ionic liquids can simultaneously remove organic and inorganic contaminants from water.

2.7 Summary

From the literature overview in this section, the health risk associated with the presence of organic and inorganic pollutants in water has been highlighted. The failure of conventional water treatment technologies to remove organic and inorganic pollutants to ppb levels remains an ongoing research concern. Water insoluble CD polymers have demonstrated the ability to remove organic pollutants at ppb levels. On the other hand ILs are able to remove metal ions from water. To the best of our knowledge, there is no known technique that simultaneously removes organic and inorganic contaminants from water. Water insoluble polymers of CDs functionalised with ILs should provide a dual system that is capable of removing both organic and inorganic pollutants from water simultaneously.

3. EXPERIMENTAL METHODOLOGY

3.1 Reagents and solvents

Unless otherwise specified, all chemicals and reagents were obtained from suppliers and used without further purification. All reactions were carried out under inert conditions of argon or nitrogen. All solvents used in this study were dried over calcium hydride for 2 days and distilled before being used. The distilled solvents were stored under molecular sieves.

3.2 Synthetic procedures

The synthesis of monosubstituted cyclodextrins mainly takes place at the C-6 hydroxyl group. A strong nucleophilic reagent (i.e. sulfonic anhydride (Ts_2O)), was used for the displacement of the hydroxyl group at the C-6 position. Monotosylated cyclodextrins are generally good precursors for derivatives of CDs due to the ease of formation of the mono- tosylate and ease with which the tosyl leaving group can be replaced. Therefore, in order to access a CD monotosylate, Ts_2O had to be first synthesised.

3.2.1 Synthesis and characterisation of p-toluene sulfonic anhydride (Ts_2O)

p-Toluene sulfonyl chloride (2.00g, 10.4mmol) was dissolved in dichloromethane (12.5mL) and *p*-toluene sulfonic acid (TsOH) (0.52g, 2.63mmol) was added gradually with vigorous stirring under nitrogen atmosphere. The resulting mixture was stirred overnight at room temperature. The milky reaction mixture was then filtered through silica gel to remove the unreacted *p*-toluene sulfonic acid. Hexane (50mL) was added to the clear filtrate and a precipitate was afforded. The crystals of Ts_2O (1) were collected as white needle-shaped granular/solids after drying overnight under vacuum.

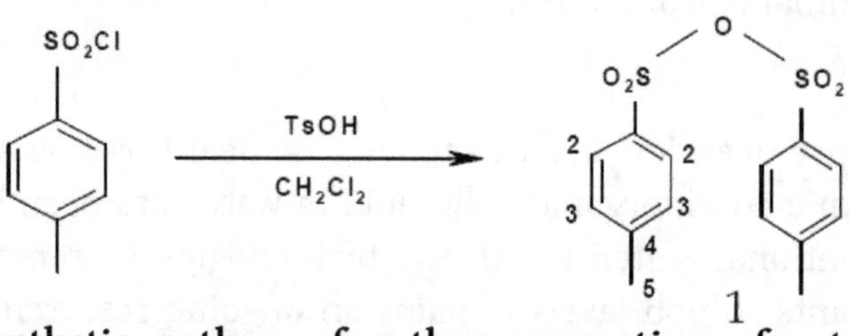

1

Scheme 3.1. Synthetic pathway for the preparation of *p*-toluene sulfonic anhydride.

IR/KBr, cm^{-1}: 3095 (CH, Ar), 3058 (CH, Ar), 2925 (C-H), 1928, 1809, 1652, 1592, 1488 (SO_2), 1373, 1305, 1925, 1175 (SO_2), 1121, 1079, 1015. ^{1}H NMR, DMSO-$_{d6}$: 7.40 (d, 4H, H-2), 7.0 (d, 4H, H-3), 2.0 (s, 6H, CH_3). ^{13}C NMR, DMSO-$_{d6}$: 145 (C-1), 139 (C-2), 129.0 (C-3), 126.0 (C-4), 21.5 (C-5). Yield: 5.89g, 18.1mmol, 73%. M.pt = 124-126°C. TLC:Eluent 2:1 hexane/ethylacetate, R_f: 0.80.

3.2.2 Synthesis of monofunctionalised CD precursors

3.2.2.1 thesis and characterisation of mono-6-deoxy-6-(p-tosyl)-β-cyclodextrin (2) (β- CDOTs)

A solution of *β*-CD (2.86g, 2.53mmol), *p*-toluene sulfonic anhydride (1.24g, 3.80mmol) and distilled water (60 mL) was stirred at room temperature under an inert atmosphere. A solution of NaOH (1.25g, 31.3mmol) in water (13 mL) was then added after 2 hours. After 10 minutes, unreacted Ts_2O was removed by filtration through silica gel. The pH of the filtrate was adjusted to ~ 8 by the addition of ammonium chloride (3.36g). Overnight refrigeration of the resultant precipitate followed by drying under reduced pressure led to the isolation of 2 as a white powdery solid (2.25g, 1.75mmol, 68%).

3.2.2.2 ynthesis and characterisation of mono-6-iodo-β-cyclodextrin (β-CDI)

Dry β-CDOTs (1.00g, 0.88mmol) was dissolved in anhydrous DMF (40 mL). To this solution, sodium iodide (1.20g, 7.76mmol, 10 equiv), was added under nitrogen with vigorous stirring which was continued for 5 hours at 90°C in an inert atmosphere. Addition of acetone (25 mL) led to the formation of a precipitate. The mixture was filtered and thereafter

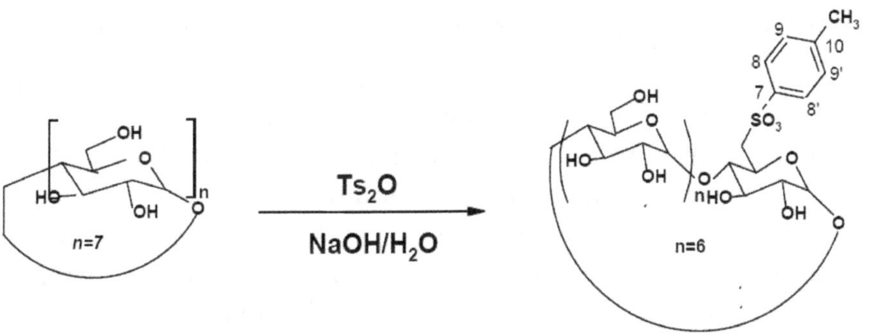

2

Scheme 3.2. A synthetic pathway for the synthesis of β-CDOTs

IR/KBr, cm^{-1}: 3377 (O-H), 2934 (C-H), 1635 (C=C), 1599 (C-C), 1357 (SO_2, Assy.), 1156 (SO_2, Sym), 1037 (C-O). ^{1}H NMR/ppm, DMSO-$_{d6}$: 7.53(d, H_{Ar}, 2H), 7.21 (d, H_{Ar}, 2H), 4.55 (s, OH- 6), 5.40-5.80 (m, H-6, 2H), 4.0 (m, H-6), 3.20-3.55 (m, H-3, H-5, H-6), 5.40-5.80 (br, OH-2, OH-3) , 2.90-3.20 (m, H-2, H-4) 4.63 (d, H-1, 7H), 2.21, (s, $-CH_3$, 3H). ^{13}C NMR/ppm, DMSO-$_{d6}$: 145.0 (C_{Ar}), 132.8 (C_{Ar}), 129.9 (C_{Ar}), 127.6 (C_{Ar}), 101.9 (C-1), 81.5 (C-4), 72.4 (C-2), 73.1 (C-3), 72.2 (C-5), 59.9 (C-6), 21.2 ($-CH_3$). Yield = 2.25g, 1.75mmol, 69%.M.pt = 171-174°C. TLC: R_f= 0.49.

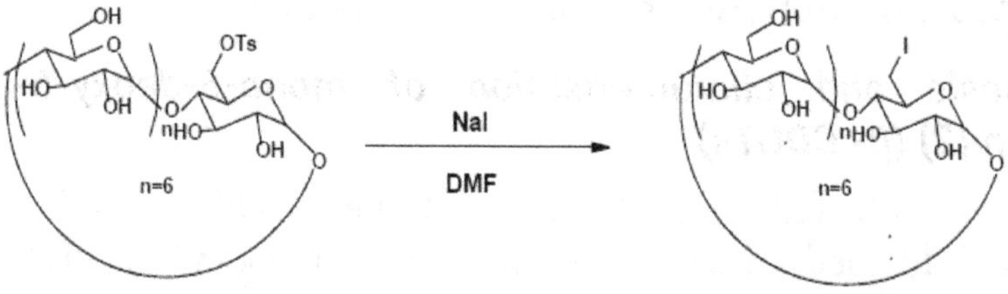

2 3

Scheme 3.3. Synthetic pathway for the preparation of β-CDI.

FTIR/KBr, cm^{-1}: 3346 (O-H), 2924 (C-H), 1027 (C-O), 586 (C-I). ^{1}H NMR/ppm, DMSO-$_{d6}$: 5.40-5.63 (m, OH-2, OH-3), 4.62 (d, H-1), 4.30 (br, OH-6), 4.30(m, H-6), 3.20-3.55 (m, H-3, H-5, H-6), 3.00-3.20 (m, H-2, H-4), 1.90 (CH_2-I). ^{13}C NMR/ppm, DMSO-$_{d6}$: 102.0 (C-1), 81.6 (C-4), 73.1 (C-3), 72.4 (C-2), 72.1 (C-5), 60.0 (C-6), 30.9 (C-6') M.pt = 206-209°C. TLC: R_f = 0.43.

washed with acetone (100 mL). The acetone was then removed under reduced pressure (rotary evaporator). To ensure complete removal of unreacted NaI, this procedure (washing and evaporation of acetone) was repeated three times. The resulting solid was dissolved in deionised water (50 mL) and precipitated by the addition of copious amounts of acetone. The precipitate was filtered off and dried in a desiccator to give 3 as white powdery solid (0.79g, 0.63 mmol, 82%).

3.2.3 General procedure for the synthesis of CD-IL complexes

Typically an alkyl imidazole or pyridine (10 equiv.) was added dropwise to a stirred solution of dry β-CDI or β-CDOTs (1.00g, 0.88mmol) dissolved in anhydrous DMF (40 mL). Stirring was continued at elevated temperatures (80°C and 90°C for β-CDI an β-CDOTs, respectively) under nitrogen for a further 24 hours. After cooling to room temperature, acetone (25 mL) was added to precipitate the product. The reaction mixture was then stirred for 30 minutes. Evaporation of organic solvents using a rotary evaporator afforded a white solid. The solid was finally dissolved in deionized water (50 mL) and precipitated by the addition of acetone (200 mL). The precipitate was filtered off and dried under vacuum to yield a white powdery solid.

3.2.3.1 thesis and characterisation of mono-6-deoxy-6-(3-methylimidazolium)-β- cyclodextrin tosylate (4) (β-CD-MIMOTs)

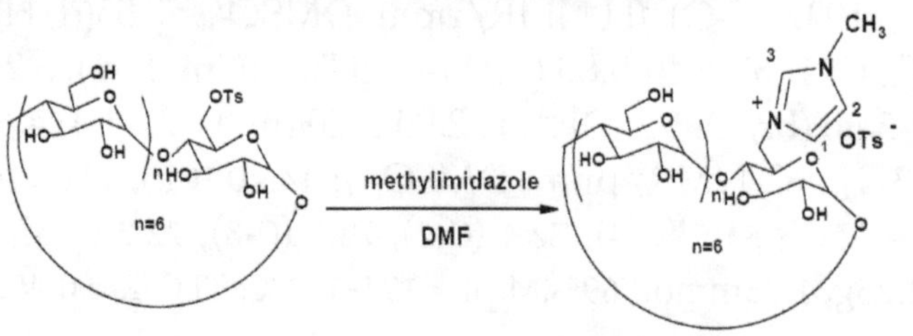

2 4

Scheme 3.4. Synthetic pathway for the preparation of β-CD-MIMOTs

IR/KBr, cm^{-1}: 3388 (O-H), 2929 (C-H), 1027 (C-O), 1641 (C=C), 1156 (C-N). ^{1}H NMR/ ppm, $DMSO_{d6}$: 8.99 (s, $=CH_{im}$), 7.92 (s, $=CH_{imid}$), 7.65 (s, $=CH_{imid}$), 7.45 (d, $H_{Ar(ortho)}$), 7.09 (d, $H_{Ar(meta)}$), 5.20-6.21 (OH-2, OH-4), 4.76 (s, H-1), 4.27 (t, OH-6) 3.87 (s, H-6), 3.40-3.66 (m, H-3, H-5, H-6), 3.20-3.41 (m, H-2, H-4), 2.25 (s, $-CH_{3imid}$), 2.04 (s, $-CH_{3OTs}$). ^{13}CNMR/ ppm, $DMSO_{d6}$: 145.4 ($C_{Ar(meta)}$), 137.9 ($C_{Ar(ortho)}$), 128.3 ($C_{Ar(ipso)}$), 125.6 ($C_{Ar(para)}$), 137.2 (C_{im}) 123.17 (C_{im}), 123.0 (C_{im}), 102.0 (C-1), 81.6 (C-4), 73.1 (C-3) 72.5 (C-2), 72.1 (C-5), 60.1 (C-6), 48.76 (C- 6'). Yield = 0.91g, 0.66 mmol, 97%. M.pt = 226-229°C, TLC: = 0.58.

3.2.3.2 thesis and characterisation of mono-6-deoxy-6-(3-butylimidazolium)-β- cyclodextrin tosylate(5) (β-CD-BUTIMOTs)

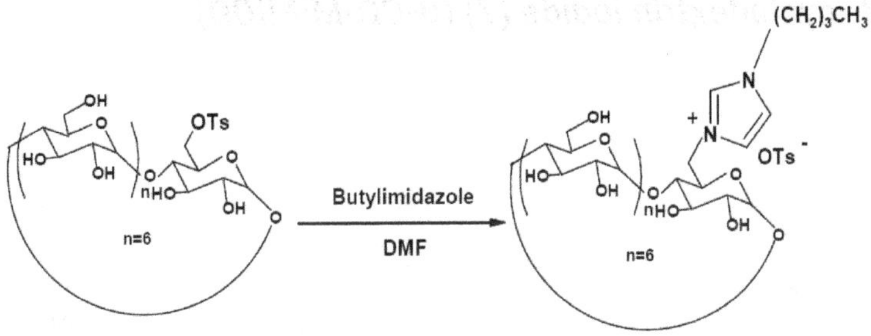

2 5

Scheme 3.5. Synthetic pathway for the preparation of the *β*-CD-BUTIMOTs

IR/KBr, cm^{-1}: 3346 (O-H), 2924 (C-H), 1035 (C-O), 1633 (C=C), 1156 (C-N). ^{1}HNMR/ ppm DMSO-$_{d6}$: 9.06 (s, CH-$_{im}$), 7.74 (s, CH-$_{im}$), 7.68 (CH-$_{im}$), 7.57 (d, $CH_{Ar(ortho,OTs)}$), 7.08 (d, $CH_{Ar(meta,OTs)}$), 5.52-5.82 (OH-2, OH-3), 4.49 (d, H-1), 4.10 (t, OH-6), 3.41-3.72 (m, H-3, H-5, H- 6), 3.10-3.41 (H-2, H-4), 2.08 (t, N-CH_2-), 1.76-2.00 (m, -CH_2-), 1.20-1.40 (m, -CH_2-CH_3) 0.84 (CH_{3-OTs}), 0.005 (s, CH_{3-im}). ^{13}CNMR/ppm DMSO-$_{d6}$: 145.4 (C-$2'_{OTs}$), 137.1 (C-2_{OTs}), 128.2 (C- 3_{OTs}), 125.6 (C-$3'_{OTs}$), 138.0 (C-$_{8but}$), 123.5 (C-$_{10but}$), 123.1 (C-$_{9but}$), 102.0 (C-1), 81.6 (C-4), 73.1 (C-3), 72.49 (C-2), 72.09 (C-5), 60.00 (C-6), 48.71 (C-6') ,31.55 (N-CH_2), 31.2 (-CH_2-), 21.29 - CH_2-CH_3, 19.3 (CH_{3-OTs}), 13.75 (CH_{3-im}). Yield: 0.96g, 0.67 mmol, 98%, M.pt: 240-243°C. TLC: R_f= 0.61.

3.2.3.3 thesis of mono-6-deoxy-6-pyridinium-β-cyclodextrin tosylate (6) (β-CD-PYROTs)

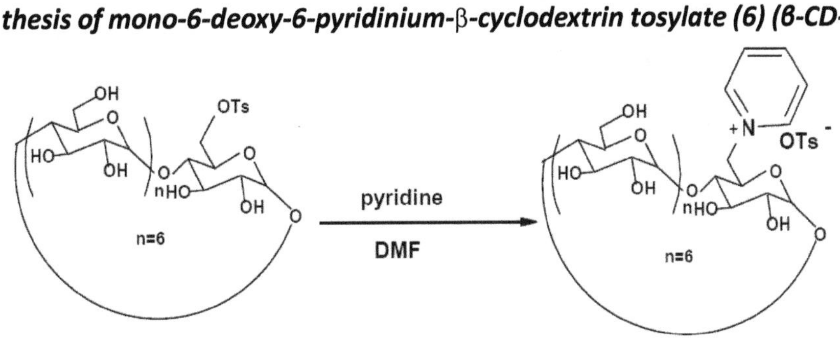

2 6

Scheme 3.6. A synthetic pathway for the preparation of *β*-CD-PYROTs

IR/KBr, cm^{-1}: 3395 (O-H), 2933 (C-H), 1033 (C-O), 1636 (C=C), 1159 (C-N). ^{1}H NMR/ppm, DMSO-$_{d6}$: 9.08 (d, =CH_{pyr}), 8.20 (t, =CH_{pyr}), 7.80 (dd, =CH_{pyr}), 7.57 (br d, $H_{Ar(ortho)}$), 7.21 (d, $H_{Ar(meta)}$), 5.01 (s, H-1), 4.40-4.82 (m,OH-2, OH-3), 3.98 (t, OH-6), 3.52-3.88 (m, H-3, H-5, H-6), 3.21-3.52 (m, H-2, H-4), 2.15 (d,-CH_{3OTs}). ^{13}CNMR/ppm, DMSO-$_{d6}$: 147.69 ($C_{py(C-8)}$), 145.16($C_{py(C-9)}$), 125.66 ($C_{py(C-10)}$), 143.00 ($C_{OTs(C-1)}$), 138.25 ($C_{OTs(C-2)}$), 137.92 ($C_{OTs(C-3)}$), 128.38 ($C_{OTs(C-4)}$), 102.08 (C-1), 81.68 (C-4), 73.22 (C-3), 72.53 (C-2), 72.17 (C-5), 60.09 (C-6), 45.42 (C- 6'). Yield = 2.49g, 0.70 mmol, 90.7%. M.pt = 242-245°C. TLC: R_f = 0.55.

3.2.3.4 thesis and characterisation of mono-6-deoxy-6-(3-methylimidazolium)-β- cyclodextrin iodide (7) (β-CD-MIMIOD)

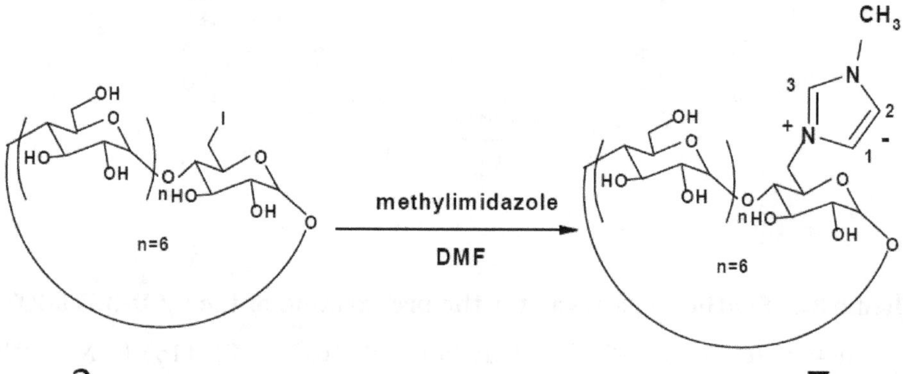

3 7

Scheme 3.7. A synthetic scheme for the preparation of β-CD-MIMIOD

FTIR/KBr, cm^{-1}: 3393 (O-H), 2919 (C-H), 1027 (C-O), 1645 (C=C), 1156 (C-N). ^{1}H NMR/ppm, DMSO-$_{d6}$: 5.60-5.80 (br m, OH-2, OH-3), 4.81 (s, H-1), 4.54 (br, OH-6), 3.41-3.80 (m, H-3, H-5, H-6), 3.22-3.41 (m, H-2, H-4), 7.48 (d, H_{Ar}-1), 7.10 (d, H_{Ar}-2), 7.92 (s, H_{Ar}-3), 2.06 (s, -CH_3). ^{13}C NMR/ppm, DMSO-$_{d6}$: 128.4($C_{Ar-imid}$), 125.7($C_{Ar-imid}$), 102.2 (C-1), 81.8 (C-4), 73.2 (C-3), 72.6 (C-2), 72.2 (C-5), 60.1 (C-6), 30.9 (-CH_{3-imid}). Yield = 0.86g, 0.63 mmol, 92%, M.pt: 197- 202°C. TLC: R_f= 0.54.

3.2.3.5 thesis and characterisation of mono-6-deoxy-6-(3-butylimidazolium)-β- cyclodextrin iodide (8) (β-CD-BUTIMIOD)

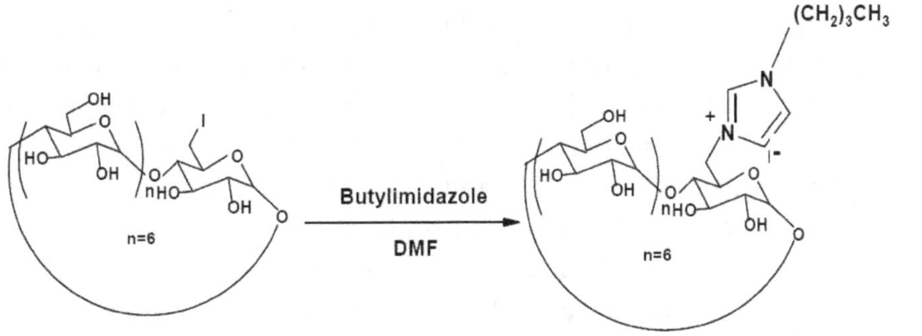

3 8

Scheme 3.8. Synthetic pathway for the preparation of β-CD-BUTIMIOD

FTIR/KBr, cm^{-1}: 3378 (O-H), 2924 (C-H), 1032 (C-O), 1655 (C=C), 1151 (C-N). ^1H NMR/ppm, DMSO-$_{d6}$: 7.10 (d, H_{Ar}), 7.46 (d, H_{Ar}), 7.92 (s, H_{Ar}), 5.60-5.81 (m, OH-2, OH-3), 4.81 (s, H-1), 4.43-4.62 (br, OH-6), 3.29-3.65 1.73 (s, $CH_{2\text{-im}}$), 1.21 (s, $CH_{2\text{-im}}$), 0.85 (s, $CH_{2\text{-im}}$), (m, H-3, H-5, H-6), 3.15-3.39 (m, H-2, H-4), 2.27 (m), 2.06 (s), 1.93 (s). ^{13}C NMR/ppm, DMSO-$_{d6}$: 128.3($C_{Ar\text{- imid}}$), 125.6($C_{Ar\text{-imid}}$), 102.1 (C-1), 81.7 (C-4), 73.2 (C-3), 72.5 (C-2), 72.2 (C-5), 60.1 (C-6), 30.9(- $CH_{3\text{-imid}}$), 36.0(-$CH_{2\text{-imid}}$), 20.9. Yield = 0.82g, 0.58 mmol, 85%. M.pt: 219-221°C. TLC: R_f= 0.57.

3.2.3.6 thesis and characterisation of mono-6-deoxy-6-pyridinium β-cyclodextrin iodide (9) (β-CD-PYRIOD)

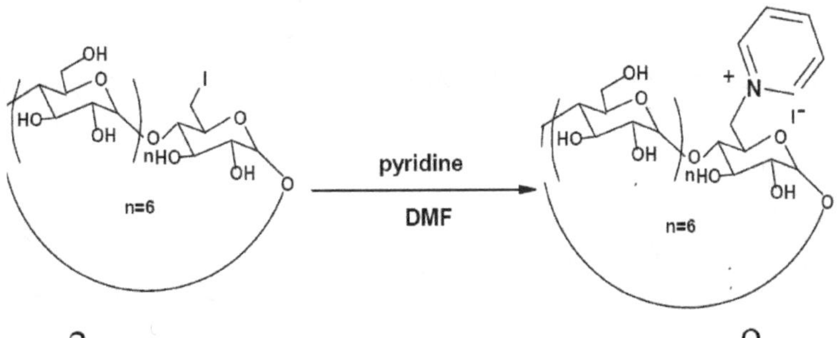

3 9

Scheme 3.9. A schematic pathway for the preparation of β-CD-PYRIOD

FTIR/KBr, cm^{-1}: 3388 (O-H), 2929 (C-H), 1027 (C-O), 1656 (C=C), 1151 (C-O). ^1H NMR/ppm, DMSO-$_{d6}$: 9.01 (d, H_{Ar}), 7.48 (d, H_{Ar}), 7.12 (t, H_{Ar}), 5.50-5.84 (m, OH-2, OH-3), 4.81 (s, H-1), 4.51 (t, OH-6), 3.42-3.78 (m, H-3, H-5, H-6), 2.98-3.42 (m, H-2, H-4). ^{13}C NMR/ppm, DMSO-$_{d6}$: 128.2 ($C_{Ar\ (ortho)}$), 125.6 ($C_{Ar\ (meta)}$), 120.3 ($C_{Ar\ (para)}$), 102.4 (C-1), 81.9 (C-4), 73.6 (C-3), 72.8 (C-2), 72.4 (C- 5), 60.4 (C-6) 48.3 (C-6'). Yield = 0.855g, 0.63 mmol, 88%, M.pt: 216-218°C. TLC: R_f= 0.51

3.3 Synthesis of CD-IL polymers

The CD-IL complexes (both tosylate and iodide derivatives) were reacted with bifunctional diisocyanate linkers to produce CD-IL polymers. Typically, the β-CD-IL complex (1.00g, 0.88 mmol) was dissolved in DMF (20 mL), and reacted with the bifunctional linker (10 equiv.) in DMF at 75°C for 18-24 hours. Two bifunctional linkers were used for the polymerisation of the CD-IL complexes. These are hexamethylene diisocyanate (HDI) and toluene-2,4- diisocyanate (TDI). The polymerisation reaction was monitored by IR spectroscopy. The completion of the polymerisation was confirmed by the total disappearance of the isocyanate peak at 2270 cm^{-1} after 18-24 hours (Li and Ma, 1999). The reaction mixture was then precipitated by the addition of acetone (100 mL). The solid formed was then left to settle in acetone for 10 minutes to allow for the removal of residual DMF from the polymers. To remove traces of DMF which may have been present, the polymers were filtered and washed with copious amounts of acetone (100 mL). The polymers were then dried overnight under reduced pressure. The examples of the synthesised polymers are shown in Figures 3.10 and 3.11.

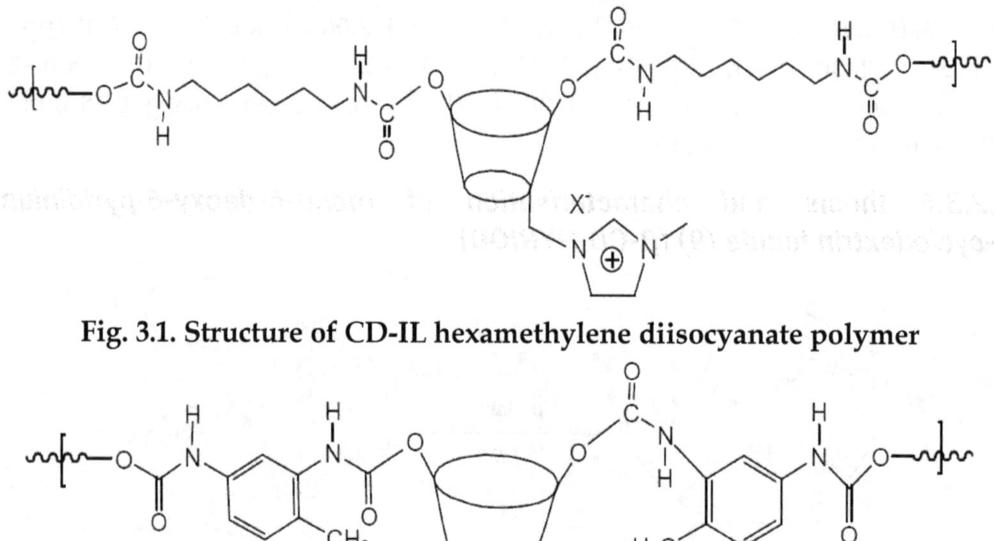

Fig. 3.1. Structure of CD-IL hexamethylene diisocyanate polymer

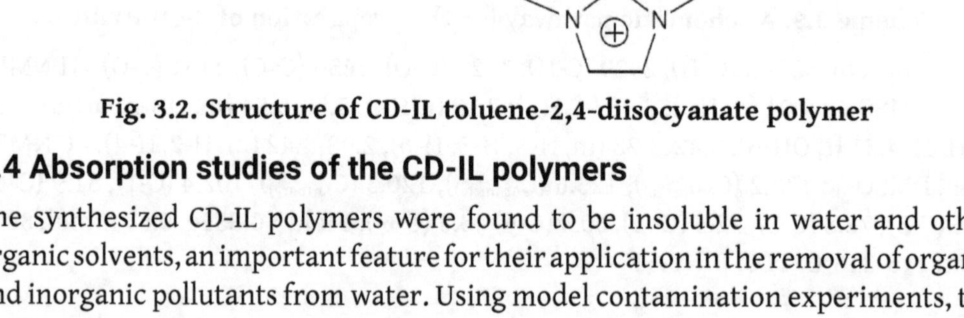

Fig. 3.2. Structure of CD-IL toluene-2,4-diisocyanate polymer

3.4 Absorption studies of the CD-IL polymers

The synthesized CD-IL polymers were found to be insoluble in water and other organic solvents, an important feature for their application in the removal of organic and inorganic pollutants from water. Using model contamination experiments, the ability of the polymers to absorb organic and inorganic pollutants from an aqueous medium was studied. In these model experiments, a known concentration of the pollutant in water was passed through the polymers. Quantification of organic pollutants was achieved by using UV-visible spectroscopy and GC-MS analysis. Inorganic pollutants (heavy metals) werequantified using AAS.

3.4.1 UV-Visible spectroscopy absorption experiments

To measure the UV absorbance of *para*-nitrophenol (PNP), a Cary 50 UV-visible spectrophotometer was used. PNP was chosen because phenolic compounds are prevalent as pollutants in most aquatic systems. Compounds that can be analysed by this technique must have chromophores (i.e. covalently bonded but unsaturated groups such as NO_2, C=C and C=O) that absorb electromagnetic radiation in the ultraviolet and visible regions of the spectrum (Field et al., 1995). An added advantage of using PNP is that it can be easily monitored by UV-visible spectroscopy since it strongly absorbs at 318 nm (visible region of the spectrum). A 10 mm quartz cuvet cell was used as a sample holder. The instrument was calibrated using PNP standards (2, 5, 10, 15 and 20 ppm) and a calibration curve was plotted and used to determine the amount of the PNP absorbed by the polymers. For the actual analysis of the pollutants, the pollutant under investigation, (30 mL, 10 mg.L^{-1}) was typically passed through a CD-IL polymer (300 mg) (i.e. HDI and TDI polymers) packed in a C-18 SPE chromatography column. PNP was allowed to pass through the polymers at a flow rate of 5 mL per minute. It was observed that after the extraction process that the polymers assumed the colour of the pollutant (i.e. turned yellow) while the filtrate became clearer. The filtrate was then analysed for the amount of PNP present after it had been passed through the polymer. The residual concentration of PNP absorbed by the polymers was determined from a calibration curve drawn from standards.

3.4.2 GC-MS analysis

The GC-MS uses gas chromatography and mass spectroscopy as a combined analytical technique for analysing mixtures of compounds (Frederic, 2004). GC analysis separates the components in a sample and provides a representative spectral output while MS identifies the substances by electrically charging the specimen molecules; this entails accelerating them through a magnetic field thus breaking them into charged fragments and detecting the mass ions. By combining these techniques, a solution containing a number of compounds can be analysed qualitatively and quantitatively. The sample is injected into the injector port and passed through the column by a mobile phase. Helium was used as a carrier gas in all experiments in this study. The gas sweeps the analyte into a column packed with a 0.25 μm coating of a polymeric stationary phase in order to promote maximum separation. As the compounds emerge from the column, a signal is created whenever a compound is detected. The time duration that a compound is retained in the GC column is referred to as the retention time (T_R). Though the retention time can assist in differentiating between compounds, it is sometimes not a reliable determiner since similar compounds can have similar retention times. Therefore, more information other than retention times is required to identify a particular compound. As the compounds elute, the MS identifies the substances by electrically charging the specimen molecules, accelerating them through a magnetic field, breaking them into charged fragments and thus detecting the different charges. These fragments are charged ions with a specific mass. The mass of the fragments divided by the charge is called mass to charge ratio (m/z). The mass spectrum can thus be used to confirm the identity of a given compound using m/z values. Also, an internal library system can be used to relate an unknown compound by comparing the mass spectrum of a compound with mass spectra of compounds having similar fragmentation patterns and m/z values.

3.4.2.1 Preconcentration of water samples

The concentration of water samples when performing analyses at low concentrations is crucial because it ensures that samples are within the detection limit of the instrument (Valerie, 2003). The method used for this procedure was EPA Method 525.2, which is an EPA method for the analysis of semi-volatile organic compounds in drinking water samples (Zymark Coporation, 2001). It utilises solid phase extraction (SPE) for isolation and concentration of the analyte and GC-MS for the determination of the preconcentrated analyte. For the purposes of this study, this method was used with some modifications. For instance, a 30 mL water sample, instead of a 1000 mL, was loaded onto a conditioned C-18 SPE cartridge. The cartridge was then dried and eluted with ethyl acetate (EtOAc) and dichloromethane (CH_2Cl_2). After this the sample was concentrated under nitrogen to about 1 mL before analysis. The first step involved in the SPE method is the conditioning of the column with methanol. The second step involves loading of the sample in the solid sorbent. Thirdly, the sorbent is washed with a solvent that has low elution strength and lastly, the desired analyte is eluted with a suitable solvent

3.4.2.1.1 2,4,6-Trichlorophenol absorption tests

2,4,6-Trichlorophenol (TCP) was selected as an additional organic pollutants to be studied because it is known to affect the respiratory system and is a recognised carcinogen. For the preparation of 50 ppm TCP standard solution (1000 mL), TCP (0.500 g) was dissolved in acetone (2 mL) and then made up to the mark by adding CH_2Cl_2. Serial dilutions were then carried out to prepare 1, 2, 5, 10, 15 and 20 ppm standard solutions. Residual TCP was determined using a calibration curve obtained from these standards. An empty C-18 SPE cartridge was filled with the polymer under investigation (0.300 g). The cartridge was then connected to a filtering unit comprising of separating funnel and a receiving Erlenmeyer flask. The Erlenmeyer flask was in turn connected to a suction pump (see Figure 3.3 for the manual set up of the SPE) to enhance a smooth and controlled flow of the water samples through the polymer. The polymer was conditioned by first passing through CH_2Cl_2 (5 mL) after which a spiked water sample (30 mL, 10 ppm) was passed through the polymer at a flow rate that was maintained between 3 and 5 mL per minute. Residual TCP was extracted from the eluent with CH_2Cl_2.

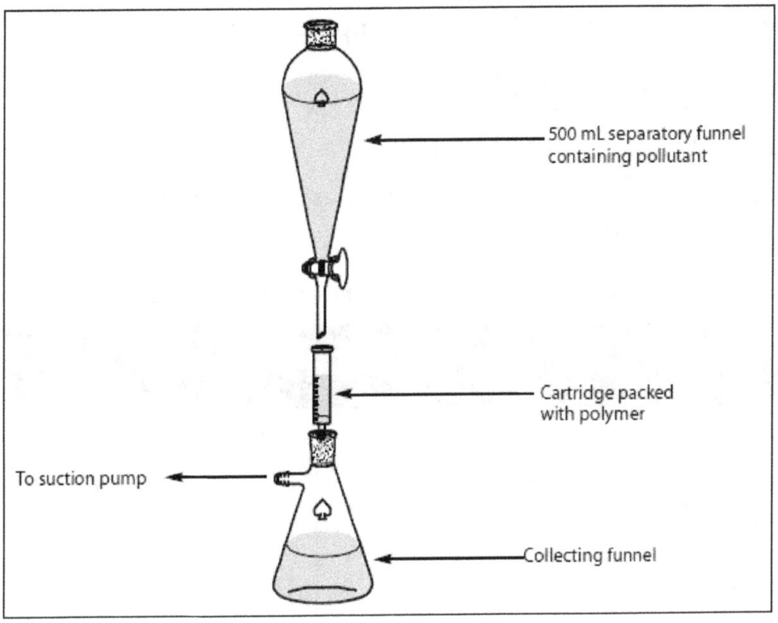

Fig. 3.3. Diagram showing the manual SPE used in this project

3.5 Extraction of heavy metals

Chromium and cadmium were used as model pollutants for the extraction study of heavy metals from water by the polymers. Besides being available in our laboratory, these pollutants were selected because they are toxic and lethal to the human body, particularly the central nervous system. Apparently these metals bind to sulphur bonds of enzymes resulting in the malfunctioning of the enzymes. A chromium standard (1000 mg.L^{-1}, 1000 mL) was prepared by dissolving potassium dichromate ($K_2Cr_2O_7$) (2.84 g), which was first pre-dried in an oven at 110°C for 2 hours, in distilled water (200 mL). HNO_3 (1.5 mL) was then added. To prepare cadmium standard solution (1000 mg.L-1, 1000 mL), cadmium metal (1.00 g) was dissolved in a 1:1 solution of distilled water/HCl. $La(NO_3)_2$ (1.00g) was added to stabilise the cadmium standards.

3.5.1 AAS analysis

For AAS analysis, serial dilutions were made from the 1000 mg.L-1 standard chromium and cadmium solutions. This led to the preparation of 2, 4, 5, 6 and 8 ppm for chromium standards and 0.4, 0.6. 0.8 1, 1.5 and 2 ppm for cadmium standards. These concentrations were chosen based on the detection limits for chromium and cadmium. These are 0.2-10 ppm and 0.05-2 ppm, respectively. Absorption of chromium was determined at 5 ppm while cadmium was measured at 1 ppm. Distilled water was used to zero the instrument as a reference sample. All measurements were performed at room temperature.

4. RESULTS AND DISCUSSIONS

4.1 CD-IL complexes/derivatives

The CD-IL derivatives were successfully synthesized and their physical parameters are summarised in Table 4.1.

Table 4.1. Yields and physical properties of the β-CD-IL complexes

CD and CD-IL derivatives	Melting point (°C)	Colour and form	Yield (%)
β-CDOTs (2)	171-174	White fine powder	68
β-CDI (3)	206-209(dec)	White fine powder	82
β-CDMIMOTs (4)	229-229(dec)	White solid	98
β-CDBUTIMOTs (5)	240-242	White powder	97
β-CDPYROTs (6)	242-245(dec)	White solid	91
β-CDMIMIOD (7)	197-202(dec)	White solid	92
β-CDBUTIMIOD (8)	219-221(dec)	White powder	85
β-CDPYRIOD (9)	216-218(dec)	White powder	88

β-CDOTs: mono-6-tosyl, .. β-CDMIMOTs: mono-6-methyl imidazolium tosylate
β-CDBUTIMOTs: mono-6-butyl imidazolium tosylate,
β-CDPYROTs: mono-6-pyridinium tosylate, β-CDI: mono-6-CD iodide,
β-CDMIMIOD: mono-6-methyl imidazolium iodide
β-CDBUTIMIOD: mono-6-butyl imidazolium iodide,
β-CDPYRIOD: mono-6-pyridinium iodide. Dec = decomposes at its melting point.

The melting points of the synthesised CD-IL derivatives were found to range between 190 and 245°C. With the exception of the β-CDBUTIMOTs derivative, all the CD-IL complexes melted with decomposition. These compounds were found to be soluble in waterand several organic solvents (e.g. DMF, ethanol, DMSO and ethyl acetate). For application in water treatment, it was therefore necessary to polymerise the CDI-IL complexes into water- insoluble polymers. The synthetic reactions and characterisations of these CD-IL polymers are discussed in the section that follows.

4.2 Synthesis and characterisation of the CD-IL polymers

Polymerisation of the CD-IL complexes was performed as per the procedure described in section 3. Each CD-IL complex was polymerised with bifunctionalised linkers; hexamethylene diisocyanate (HDI) and toluene 2,4-diisocyanate (TDI). The polymerisation reactions were monitored using infrared spectroscopy. Figures 4.1a and 4.1b show the gradual disappearance of the isocyanate peak (N-C=O) at 2270 cm^{-1} as the polymerisation reaction proceeded. The complete disappearance of the isocyanate peak at 2270cm^{-1} (after 24 hrs) marked the end of the polymerisation reaction.

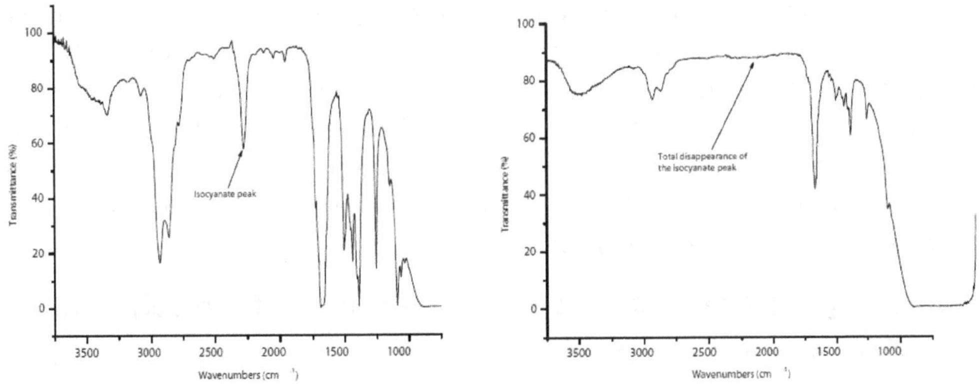

Fig. 4.1. IR spectrum showing isocyanate peak at 00 mins andafter 24 hrs

4.3 Physical properties of the β-CD-IL polymers

The physical properties (colour, texture and the yields) of the synthesised CD-IL polymers are shown in Table 4.2.

Table 4.2. Physical properties of the β-CD-IL polymers

CD-IL polymer	Texture	Colour	Yield (%)
Β-CD/HDI	Powder	White	100
Β-CD/TDI	Powder	White	98
β-CDMIMOTs/HDI	Powder	White	97
β-CDMIMOTs/TDI	Powder	Light brown	99
β-CDBUTIMOTs/HDI	Powder with granules	Crème white	95
β-CDBUTIMOTs/TDI	Powder	Light brown	96
β-CDPYROTs/HDI	Granulated powder	White	93

β-CDPYROTs/TDI	Powder	Yellowish brown	96
β-CDMIMIOD/HDI	Granulated powder	Crème white	100
β-CDMIMIOD/TDI	Granulated powder	Light brown	97
β-CDBUTIMIOD/HDI	Powder	White	100
β-CDBUTIMIOD/TDI	Powder with granules	Light brown	95
β-CDPYRIOD/HDI	Crystalline with powder	Beige	100
β-CDPYRIOD/TDI	Crystalline	Yellow	94

β-CDMIMOTs: mono-6-methyl imidazolium tosylate

β-CDBUTIMOTs: mono-6-butyl imidazolium tosylate

β-CDPYROTs: mono-6-pyridinium tosylate β-CDI: mono-6-CD iodide

β-CDMIMIOD: mono-6-methyl imidazolium iodide β-CDBUTIMIOD: mono-6-butyl imidazolium iodide β-CDPYRIOD: mono-6-pyridinium iodide.

While these polymers were produced in high yields, it should be noted that they are insoluble in water, an important feature for their application in water treatment. Unfortunately, NMR studies could not be performed due to the unavailability of solid-state NMR equipment. However, to study the morphology of these polymers SEM and BET surface area analyses were carried out. SEM is important because it gives us an idea of the physical morphology of the polymers while BET analysis provides an insight into the effect of the pore volume and surface area on the absorption efficiency of the polymers.

4.4 Scanning electron microscopy (SEM) analysis of the polymers

SEM was utilised in order to determine and compare the surface features of both the native β-CD and the CD-IL polymers. The SEM micrographs of both the β-CD/HDI and the CD- IL/HDI polymers are shown from Figure 4.13a to Figure 4.13e. SEM images of the TDI linked polymers were also investigated and the SEM micrographs are presented in Figures 4.2 h-n.

From these SEM micrographs it can be observed that the SEM micrographs of the native β- CD/HDI or TDI polymers and the CD-IL polymers are different. Some of the polymers exhibit a spongy surface (e.g. β-CDPYROTs/HDI and β-CDMIMOTs/TDI) while others are crystalline (e.g. β-CDBUTIOD/HDI and β-CDMIMIOD/TDI). From these results it can be concluded that the type of cation (pyridine or alkyl imidazole) or anion (iodide or tosylate) have an effect on the physical surface of the polymers. However, it is not clear at this stage whether these differences are as a result of some discrepancies during the processing of the polymers, thus BET analysis was carried out to further elucidate their physical properties.

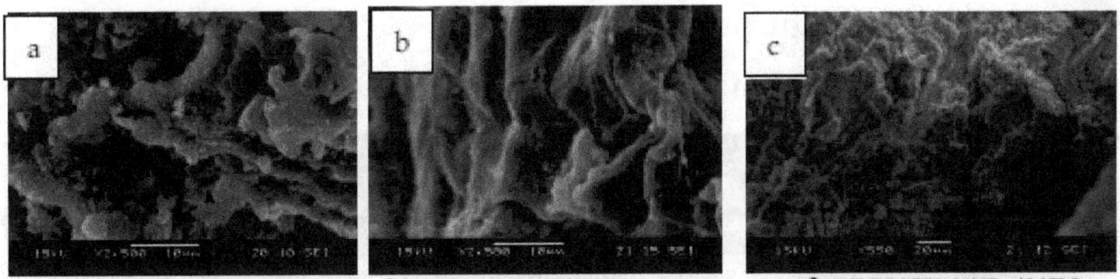

Figure 4.13a: β-CD/HDI, b: β-CDMIMOTs/HDI, c: β-CDBUTIMOTs/HDI

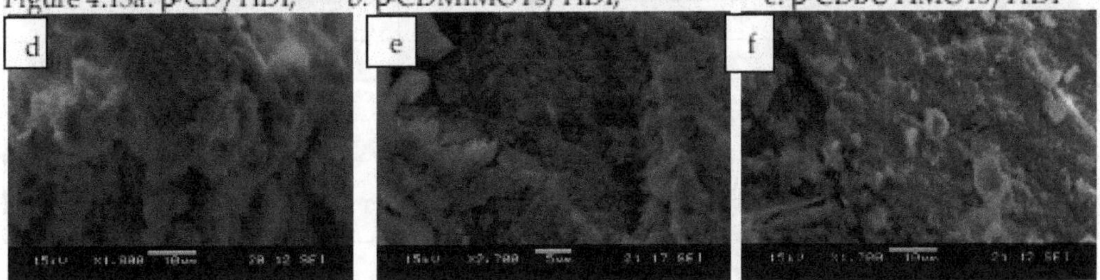

d: β-CDPYROTs/HDI, e: β-CDMIMIOD/HDI, f: β-CDBUTIMIOD/HDI

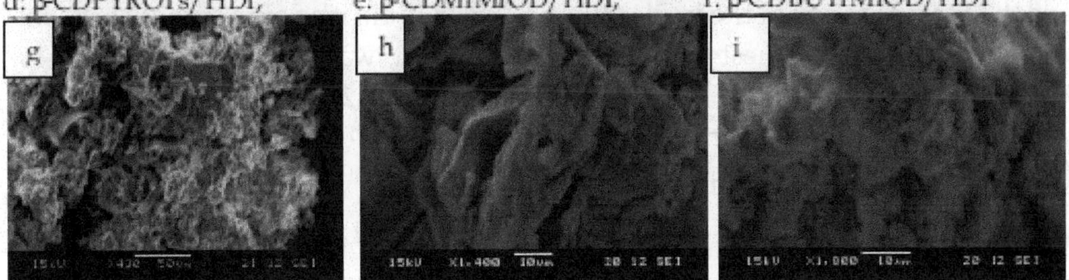

g: β-CDPYRIOD/HDI, h: β-CD/TDI, i: β-CDMIMOTs/TDI

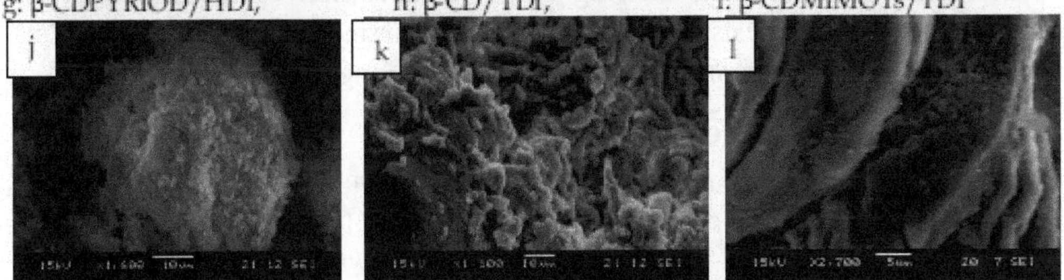

j: β-CDBUTIMOTs/TDI, k: β-CDPYROTs/TDI, l: β-CDMIMIOD/TDI

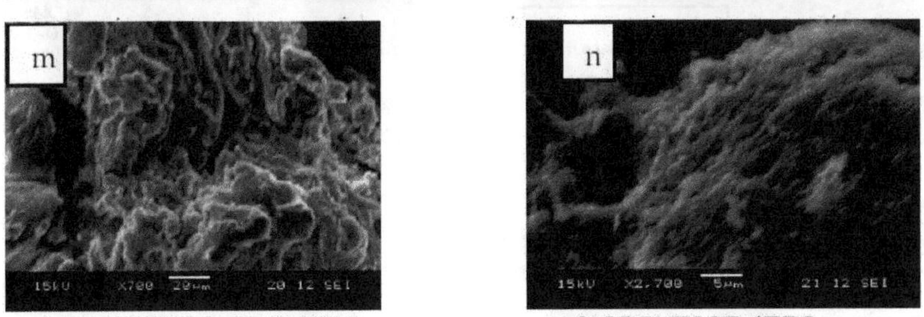

m: β-CDBUTIMIOD/TDI, n: β-CDPYRIOD/TDI,

Fig. 4.2.

4.5 BET analysis of the polymers

Table 4.3. BET results of the polymers

Polymers	HDI		TDI	
	Surface area (m^2/g)	Pore volume (cm^3/g)	Surface area (m^2/g)	Pore volume (cm^3/g)
B-CD	2.53	0.009	1.706	0.009
β-CDMIMOTs	23.26	0.084	22.71	0.11
β-CDBUTIMOTs	2.99	0.014	10.51	0.023
β-CDPYROTs	2.89	0.063	1.64	0.026
β-CDMIMIOD	2.51	0.012	1.81	0.080
β-CDBUTIMIOD	10.88	0.058	1.78	0.0066
β-CDPYRIOD	0.4427	0.0036	7.00	0.030

These BET results show that there is no pattern observed between the surface area and the incorporation of the imidazolium or pyridinium ring onto the CD backbone. For example, the highest surface area recorded was for the *β*-CDMIMOTs/HDI polymer (23.26 m^2/g) while the lowest was 0.4427 m^2/g (*β*-CDPYRIOD/HDI polymer). However, the surface area significantly increased for the *β*-CDMIMOTs (both HDI and TDI polymers.

4.6 Thermal analysis of the polymers

To determine the stability, melting and decomposition temperatures of the polymers, thermal analysis of the polymers was carried out using TGA and DSC. The findings of this study are presented in the following sections.

4.6.1 TGA analysis of the polymers

The polymers generally had similar heating profiles. Representative TGA graphs of the native *β*-CD/HDI, *β*-CDMIMIOD/HDI and *β*-CDPYROTS/TDI polymers are shown in Figure 4.3. Other TGA thermographs are shown in Appendix B.

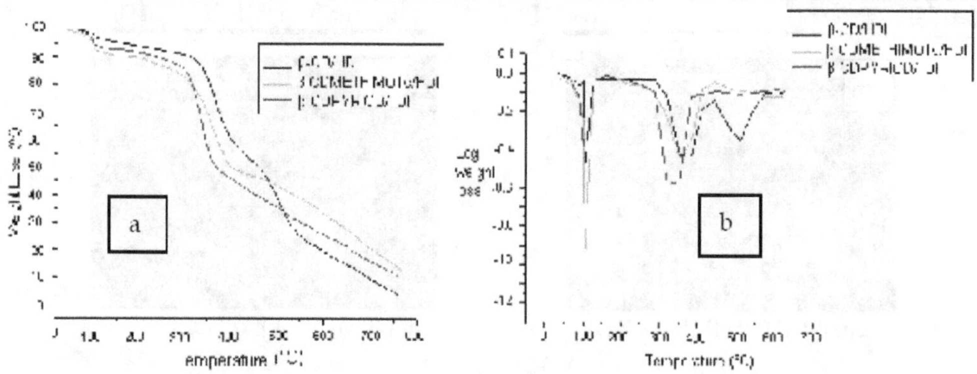

Fig. 4.3. (a) TGA thermographs and (b) derivative curves of the native *β*-CD and CD-IL polymers

The TGA thermographs (Figure 4.14(a) showed some differences between the native β-CD polymers and the CD-IL polymers. The polymers showed a three step degradation process; the first step at about 100°C (loss of water), the second step at about 350°C (melting point of the polymer) and lastly the decomposition of the polymers at about 520°C. However, to fully study the thermal behaviour of these polymers derivative curves of both the β-CD polymers and the CD-IL polymers were done. These derivative curves are presented in Figure 4.14(b). Analysis of the derivative curves confirmed that the native β-CD polymers showed three different regions which were characterised by the loss of water at 100°C, the melting point at about 350°Cand the decomposition of the polymers at about 520°C. The imidazolium CD-IL polymers on the other hand showed only two regions; loss of moisture at 100°C and the melting point was accompanied by degradation at about 350°C.However, the pyridinium based CD-IL polymers had lower melting points at 320°C.

4.6.2 DSC analysis of the polymers

DSC analysis was also performed in order to compare the thermal properties of the polymers with or without the attachment of the imidazolium or pyridinium moieties. Noteworthy is that the DSC thermograms are generally similar thus Figure 4.4 shows representative DSC thermographs of the polymers.

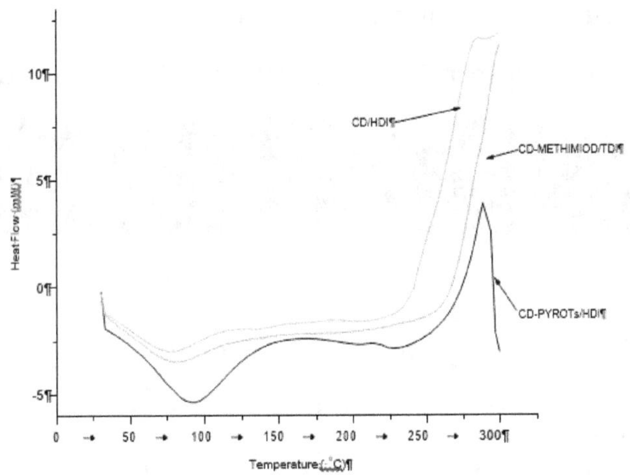

Fig. 4.4. DSC chromatograms of the native β-CD and CD-IL polymers

The DSC thermograms all showed an endothermic peak associated with the evaporation of water and other organic solvents at about 95°C. The thermograms also showed an extra endothermic peak at 250°C which corresponds to the melting points of these polymers. For all three polymers degradation occurred at around 300°C.

4.7 Absorption studies of the polymers

Four pollutants (two organic and two inorganic) were used as model pollutants for the study of the absorption capabilities of the polymers. These pollutants are para-nitrophenol (PNP), 2,4,6-trichlorophenol (TCP), chromium VI and cadmium II and were readily available in our laboratory.

4.7.1 Absorption of para-nitrophenol (PNP) and 2,4,6-trichlorophenol (TCP)

p-Nitrophenol (PNP) was chosen as a model pollutant because phenolic compounds are prevalent as pollutants in most aquatic systems. Also, PNP can be easily monitored by UV- Visible spectroscopy because of its yellow colour. Polymers formed by cross-linking the synthesised CD-IL complexes with HDI and TDI linkers were investigated for their ability to absorb PNP. *p*-Nitrophenol (30 mL, 10mg.L^{-1}) was passed through the polymers (300 mg) as described in section 3. The absorbance values were recorded in triplicates at room temperature. TCP, which is known to affect the respiratory system and is carcinogenic, was also used as a model pollutant and was studied using GC-MS. TCP standard solution (30 mL, 10 mg.L^{-1}) was used to determine the absorption efficiencies of the polymers. The residual amount of the TCP was extracted by liquid-liquid extraction with CH_2Cl_2 as outlined in section 3 after it was passed through the polymers. Results obtained from these analyses are summarised in Tables 4.3. The native β-CD polymer was included in the study for comparison purposes.

Table 4.3. Percentage absorption efficiency of PNP by HDI and TDI polymers

Polymers	Amount of PNP and TCP removed (%)			
	HDI linker		TDI linker	
	PNP	TCP	PNP	TCP
β-CD	64	54	68	58
β-CDMIMOTs	80	69	77	69
β-CDBUTIMOTs	68	69	64	77
β-CDPYROTs	15	63	40	60
β-CDMIMIOD	65	75	55	74
β-CDBUTIMIOD	63	67	71	71
β-CDPYRIOD	24	69	48	61

As mentioned in section 2, the amount of the pollutant absorbed can be affected by the type of cation or anion, or the length of the alkyl chain attached to the imidazolium or pyridinium ring. It was, therefore, necessary to study the effect of these properties on the absorption efficiency of PNP and TCP by the polymers. Also, the effect of the linker on the overall performance of the polymer was investigated. These variations are discussed in the section that follows.

4.7.1.1 of the type of the cation on the extraction of the organic pollutants

It is very clear from Table 4.3 that the incorporation of the imidazolium ring onto the CD backbone generally enhances the absorption efficiencies of the polymers (i.e. with respect to PNP). In fact, the highest absorption (80%) was recorded for β-CDMIMOTs/HDI. However, the pyridinium-based showed less affinity for PNP. For example, the absorption efficiencies of β-CDPYROTs/HDI and β-CDPYRIOD/HDI were a mere 15% and 24%, respectively. The corresponding TDI polymers followed a similar pattern. In contrast, both the imidazolium- and pyridinium-based polymers are better at absorbing TCP when compared to the native cyclodextrin polymers. It is not clear at this stage why the extraction of PNP is influenced by the type of cation.

4.7.1.2 of the imidazolium alkyl chain on the extraction of the organic pollutants

The effect of the length of the alkyl chain attached to the imidazolium ring on the absorption efficiency of the polymers was investigated. According to previous studies (Visser et al, 2001; Pandey, 2006), the longer the alkyl chain the less effective is the ionic liquid for metal ion extraction. In our study, two substituents (methyl and butyl) have been incorporated onto the ionic liquid component (imidazolium) of the polymer and this allowed us to test this concept. Results presented in Table 4.3 indeed confirmed that the alkyl chain had an effect on the amount of pollutant absorbed by the polymers. As expected, the methylimidazolium polymers were generally found to have a higher absorption efficiencies compared to their butylimidazolium counterparts.

4.7.1.3 of the type of anion on the extraction of the organic pollutants

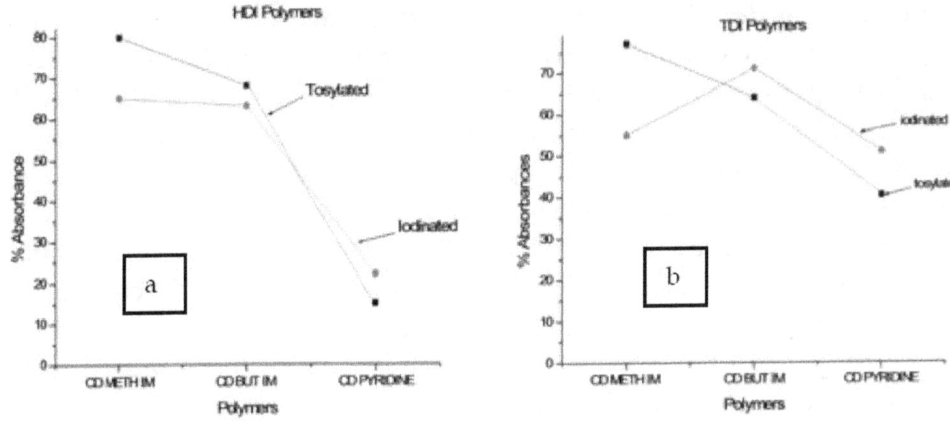

Fig. 4.5. Comparison of PNP absorbed by tosylated versus iodinated HDI-linked (a) and (b) TDI-linked polymers.

To further study the absorption efficiencies of the polymers, the effect of the counter-anion (tosylate and iodide) on the absorption capacity of the polymers was investigated. It was observed that the amount of pollutant absorbed and the type of anion present inthe polymer were not well correlated. As summarised in Figure 4.5a and Figure 4.5b, the type of the anion does not seem to significantly affect the absorption of both PNP and TCP.

4.7.1.4 of the linker on the extraction of the organic pollutants

Figures 4.17a and 4.17b were also used to determine how the linkers affected the absorption efficiency of the polymers. According to these results, the tosylated polymers showed better absorption efficiencies than the TDI linked polymers (especially the imidazolium polymers) for the HDI-linked polymers. However, this was not the case with the pyridinium polymers where higher absorption efficiency was observed for the iodinated polymers. For the TDI- linked polymers, it was observed that the iodinated polymers had better absorption efficiencies compared to the tosylated polymers with the exception of β-CD- methylimidazolium tosylate polymer.

4.7.2 Absorption of chromium (Cr^{6+}) and cadmium (Cd^{2+}) by the polymers

Chromium is a toxic inorganic pollutant (heavy metal), especially when present in water as Cr^{6+}. This metal ion can penetrate the skin causing irritation, liver and kidney damage as well as adecrease in male sperm counts (Li and Ma, 1999; Morales et al., 1999). Chromium standards were prepared as outlined in Chapter 3 and a Cr^{6+} concentration of 5 $mg.L^{-1}$ (30 mL) was passed through the polymers (300 mg). The detection limit of Cr^{6+} by AAS is 0.002 $mg.L^{-1}$. A working range of 0.2-10 $mg.L^{-1}$ was recommended thus 5 $mg.L^{-1}$ was used for its analysis. Cadmium ion (Cd^{2+}) was also used as a model inorganic pollutant to study the absorption efficiencies of the CD-IL polymers. Cadmium was studied because it is toxic and even lethal to the human body, particularly the central nervous system (Morales et al., 1999). This metal ion has a great affinity for sulphur hence it attacks the sulphur bonds of enzymes making them to function improperly (Lewis and Evans, 2001). Cadmium standards were prepared as outlined in section 3 and a concentration of Cd^{2+} (1 $mg.L^{-1}$, 30 mL) was passed through the polymers (300 mg). The detection limit of Cd^{2+} by AAS is 0.002 $mg.L^{-1}$. Since a working range of 0.005-2 $mg.L^{-1}$ is recommended, a 1 $mg.L^{-1}$ was used for Cd^{2+} analysis. After passing the Cr^{6+} and the Cd^{2+} spiked water sample through the CD-IL polymers, AAS analysis of the eluent was performed. Results obtained are shown in Table 4.4. Native β-CD polymers were also used to see if they can absorb inorganic pollutants from an aqueous medium.

Table 4.4. Comparison of absorption efficiency of Cr^{6+} by HDI and TDI polymers

Polymers	Amount of Cr^{6+} and Cd^{2+} removed (%)			
	HDI linker		TDI linker	
	Cr^{6+}	Cd^{2+}	Cr^{6+}	Cd^{2+}
β-CD	34	18	60	24
β-CDMIMOTs	100	33	100	28
β-CDBUTIMOTs	98	32	97	23
β-CDPYROTs	100	24	95	20
β-CDMIMIOD	69	15	43	18
β-CDBUTIMIOD	51	21	95	17
β-CDPYRIOD	34	10	60	13

It is clear from Table 4.4 that the incorporation of the ionic liquid component onto the CD backbone resulted in high absorption of Cr^{6+}. Absorption of Cd^{2+} was however significantly reduced. For example, β-CDMIMOTs/HDI polymer absorbed 100% of the Cr^{6+} but only 33% of the Cd^{2+} was absorbed. This can be the attributed to the fact that ionic liquids are selective in the absorption of heavy metal ions (Welton, 1999; Chowbury et al., 2007). For example, the incorporation of sulphur or urea in the alkyl chain of the cation has been reported to greatly enhance the absorption of Cd^{2+} and Hg^{2+} from an aqueous medium (Visser et al., 2002). This observation suggests that, to improve the extraction of Cd^{2+}, further studies that incorporate sulphur or urea on the cations need to be carried out. A discussion on how variations of cation, anion, alkyl chain as well as the linker impact on the extraction efficiency of the polymers follows in the next section.

4.7.2.1 of the type of the cation on the extraction of the inorganic pollutants

The incorporation of the imidazolium or pyridinium rings onto the CD backbone was found to have no bearing on the amount of the pollutant absorbed by the polymers. Some imidazolium-based polymers (e.g. β-CDMIMOTs) and some pyridinium-based polymers (e.g. β-CDPYROTs) had 100% absorptionefficiencies forCr^{6+} while some imidazolium- based polymers (e.g. β-CDBUTIMIOD/HDI) and some pyridinium-based polymers (e.g. β- CDPYRIOD/HDI) had low absorption efficiencies of 51 and 34%, respectively. The same polymers, however, showed less efficiency in the absorption of Cd^{2+}.

4.7.2.2 of the imidazolium alkyl chain on the extraction of the inorganic pollutants

The length of the alkyl chain of the imidazolium ring was also investigated to determine its effect on the absorption efficiency of the polymers. From Table 4.4 it was observed that increasing the alkyl chain of the imidazolium ring did not have much bearingonthe amount of either Cr^{6+} or Cd^{2+} absorbed by the polymers.

No pattern could be established between the amount of pollutant (either Cr^{6+} or Cd^{2+}) absorbed and the length of the alky chain attached to the imidazolium ring.

4.7.2.3 of the type of anion on the extraction of the inorganic pollutants

The type of anion of the ionic liquid seems to have an effect on the amount of Cr^{6+} and Cd^{2+} absorbed by the polymers (Table 4.4). The tosylated polymers (both HDI and TDI) showed high absorption efficiencies of Cr^{6+} and Cd^{2+} than the iodinated polymers (Figure 4.18). The tosylate anion enhances the chelation (hexavalent dentition) of the chromium with the cationic component of the CD-IL polymers because it is a hydrophobic anion (Welton, 1999), while the iodide anion is a hydrophilic anion; hence the tosylate anion enhances the chelation of the heavy metal by repelling the water molecules from the cationic component (Cocalia et al., 2006).

Fig. 4.18. Comparison of Cr^{6+} absorbed by average tosylated versus average iodinated polymers (both (TDI and HDI)

4.7.2.4 of the linker on the extraction of the inorganic pollutants

For the absorption of Cd^{2+} and the Cr^{6+}, there appears to be no correlation between the amount of pollutant absorbed and the type of linker used. For instance, the *β*-CDMIMOTs polymer absorbed 100% of Cr^{6+} irrespective of the linker used. However, it is noteworthy that the native *β*-CD polymers were able to absorb some chromium and cadmium from water. The HDI and TDI polymers showed 34% and 60% of Cr^{6+}, respectively, while 18 and 24% absorption efficiencies were respectively observed for the Cd^{2+}polymers. This observation is attributed to the ability of cyclodextrins to form complexes with chromium (Cardas et al., 2005). Similar findings (Brusseau et al., 1997), confirm that not only can modified cyclodextrins form complexes with organic compounds, but they can also complex with cationic heavy metals.

4.8 Comparison of absorption and surface area

After attaining the absorption efficiencies of the polymers, it was necessary to see if there was any correlation between the amount of pollutant absorbed and the surface area of the polymers. The results obtained are summarised in Table 4.5.

Table 4.5. Comparison of surface area versus percentage absorption

Sample	HDI					TDI				
	Surface area (m^2/g)	% PNP absorbed	% TCP absorbed	% Cr^{6+} absorbed	%Cd^{2+} absorbed	Surface area (m^2/g)	% PNP absorbed	% TCP absorbed	% Cr^{6+} absorbed	% Cd^{2+} absorbed
β-CD	2.53	64	54	34	18	1.706	68	58	60	24
β-CDMIMOTs	23.26	80	69	100	33	22.71	77	69	100	28

β-CDBUTI-MOTs	2.99	68	69	98	32	10.51	64	77	97	23
β-CDPY-ROTs	2.89	15	63	100	24	1.64	40	60	95	20
β-CDMIM-IOD	2.51	65	75	69	15	1.81	55	74	43	18
β-CDBUTI-MIOD	10.88	63	67	51	21	1.78	71	71	95	17
β-CDPY-RIOD	0.4427	24	69	34	10	7.00	48	61	60	13

From these results, it was observed that there is no direct relationship between the amount of pollutant absorbed and the surface areas of the polymers. For example, both β- CDMIMOTs/HDI and β-CDPYROTs/HDI polymers showed 100% absorption efficiency for PNP yet their surface areas are significantly different, 23.26 m^2/g and 2.89 m^2/g, respectively.

4.9 Absorption efficiencies of the polymers on a mixture of organic and inorganic pollutants

In real water samples, organic compounds and heavy metals co-exist (Brusseau et al., 1997). It was, therefore, necessary to study the absorption efficiency of the CD-IL polymers in a synthetic mixture containing both organic and inorganic pollutants. These experiments were carried out to determine if the organic and inorganic pollutants would compete for the absorption sites of the CD-IL.

4.9.1 PNP and chromium

From the absorption studies of PNP and Cr^{6+}, it was evident that the CD-IL polymers exhibited enhanced absorption efficiencies on exclusive absorption tests; hence studies to investigate whether they would still exhibit such enhanced efficiencies were carried out. PNP and Cr^{6+} were thus chosen to study the effect of the presence of both organic and inorganic pollutants on the absorption efficiency of the polymers (Table 4.6).

Table 4.6. Comparison of HDI and TDI linked CD-IL polymers on the absorption efficiencies of a mixture of PNP (10 ppm) and Cr^{6+} (5 ppm)

Polymers	HDI linked		TDI linked	
	PNP absorbed	Cr^{6+} absorbed	PNP absorbed	Cr^{6+} absorbed
β-CD	63 (64)	14 (34)	60 (68)	40 (60)
β-CDMIMOTs	62 (80)	72 (100)	34 (77)	36 (100)
β-CDBUTIMOTs	13 (68)	21 (98)	34 (64)	50 (97)
β-CDPYROTs	27 (15)	23 (100)	40 (40)	29 (95)
β-CDMIMIOD	16 (65)	36 (69)	20 (55)	14 (43)

β-CDBUTIMIOD	69 (63)	25 (51)	47 (71)	27 (95)
β-CDPYRIOD	14 (24)	10 (34)	43 (48)	50 (60)
Values in parenthesis refer to individual absorption of the pollutant by the polymer				

From the results in Table 4.6, it was observed that the presence of both organic and inorganic pollutants generally resulted in reduction in the absorption efficiency of the pollutants. This is an indication that there is competition between the organic and inorganic pollutantsforcomplexingsitesonthepolymers. However, for β-CDBUTIMIOD/HDI and β-CDPYROTs/HDI polymers, the presence of Cr^{6+} enhanced the absorption efficiency of PNP.

4.9.2 PNP and cadmium

Since the CD-IL polymers showed low absorption efficiency for the mixture of PNP and chromium spiked water samples (though separate analysis showed high absorption efficiencies), a mixture of PNP and cadmium spiked water samples was passed through the polymers in order to determine if the same observations would be seen. The results for this analysis are shown in Table 4.7.

Table 4.7. Comparison of HDI and TDI linked CD-IL polymers on the absorption efficiencies of PNP (10 ppm) and Cd^{2+} (1 ppm)

Polymers	HDI linked		Amount removed Cd^{2+} (%)	
	% PNP absorbed	% Cd^{2+} absorbed	% PNP absorbed	% Cd^{2+} absorbed
β-CD	24 (64)	12 (18)	42 (68)	17 (24)
β-CDMIMOTs	48 (80)	33 (33)	23 (77)	28 (28)
β-CDBUTIMOTs	15 (68)	32 (32)	25 (64)	23 (23)
β-CDPYROTs	18 (15)	24 (24)	51 (40)	20 (20)
β-CDMIMIOD	16 (65)	15 (15)	23 (55)	18 (18)
β-CDBUTIMIOD	66 (63)	21 (21)	43 (71)	17 (17)
β-CDPYRIOD	13 (24)	10 (10)	37 (48)	13 (13)
Values in parenthesis refer to individual absorption of the pollutant by the polymer				

From Table 4.7 it was observed that the absorbances of the CD-IL polymers remain the same, (low regardless of the type of heavy metal ion used (i.e. whether Cd^{2+} or Cr^{6+})). It was observed that the absorption efficiency of the polymers generally decreased in the presence of both organic and inorganic pollutants an indication that Cd^{2+} retards the absorption of both organic and inorganic pollutants from water. However, a thorough investigation needs to be carried out to ascertain these observations.

5. CONCISE SUMMARY

In accordance with the purposes and objectives stated earlier in this chapter, it may be determined that this inquiry was successful, and the following conclusions may be formed.:

Alkyl imidazolium and pyridinium-based CD-IL complexes were synthesized from monotosylated and monoiodinated CDs in high yields and characterized using IR, 1H NMR, and 13C NMR spectroscopic methods.The CD-IL complexes were effectively polymerized utilizing hexamethylene diisocyanate (HDI) and toluene-2,4-diisocyanate to produce CD-IL polymers.In contrast to the water-soluble CD-IL complexes, the synthesized CD-IL polymers were insoluble in water and prevalent organic solvents, including DMF, CH2Cl2, and CH3OH.This feature is significant as it is commonly utilized, facilitating its application in water treatment.Subsequent examination of the polymers using SEM indicated that certain polymers (e.g., β-CDPYROTs/HDI and β-CDMIMOTs/TDI) exhibited a spongy surface, whereas others displayed a crystalline structure (e.g., β-CDMIMIOD/TDI and β-CDBUTIOD/HDI).Nonetheless, it remains ambiguous whether this occurrence resulted from the inclusion of the pyridinium or the imidazolium rings.The BET results were inconclusive since no correlation was found between pore volume, surface area, and the quantity of pollutant absorbed by the polymers.

Thermal studies of the polymers demonstrated that CD-IL polymers exhibited greater stability than native CD polymers, with the thermal stability of CD-IL polymers influenced by the kind of cation.Research indicated that polymers incorporating pyridinium are less stable than those containing imidazolium.The CD-IL polymers were proficient at eliminating organic and inorganic contaminants from water.The β-CDMIMOTs/HDI polymer shown significant absorption efficiency for both organic (80% relative to PNP) and inorganic (100% relative to Cr6+) contaminants.The CD-IL polymers exhibited minimal affinity for cadmium.This indicates that the synthesized CD-IL polymers are unsuitable for the absorption of Cd2+ from aqueous environments.The study's results indicated that the absorption efficiency of the CD-IL polymers was generally diminished when a synthetic mixture of organic and inorganic contaminants was analyzed.This indicated that organics and inorganics fight for the complexing sites of the CD-IL polymers.Also, as part of future work on this subject, the following is under study for future workThis study only explored the use of imidazolium and pyridinium cations in the synthesis of CD-IL complexes. Synthesis of CD-IL complexes using other cations like ammonium and phosphonium cations need to be explored. Other anions such as hexafluorophosphate (PF_6), tetrafluoroborate (BF_4) and trioctylmethylammonium thiosalicylate (TOMATS) have also not been tried. Conversion of the tosylate or iodide anions (of the CD-IL complexes we synthesised) into hydrophobic anions (e.g. the less toxic TOMATS) would generate new and insoluble CD-IL complexes. This would thus eliminate the need for the preparation of the CD-ILpolymers. Attachmentofthio,thioether,thioureaandureaonthecationic components of the ionic liquid has been reported to have high complexation capabilities for cadmium and mercury (Visser et al., 2002). Such attachments, especially on the imidazolium ring, need to be explored.

The absorption efficiencies of the polymers were evaluated just at elevated concentrations (ppm levels) and with only four contaminants.Testing these polymers at lower concentrations (ppb levels) and across a diverse array of organic and inorganic contaminants is essential.Initial investigations in our laboratory indicate competition for complexation sites between the anion and the cation; therefore, conditions must be identified to mitigate this competition in order to fully utilize the absorption potential of the polymers.The efficiency of the polymers was examined for only a single cycle; therefore, more cycles must be investigated to assess the polymers' durability and their capacity to continue absorbing pollutant species after several cycles.

REFERENCES

Bender M. L. and Komiyana M. (1978) *Cyclodextrin Chemistry*p2.

Brusseau M. L., Wang X. & Wang W. (1997) Enhanced Solubilization of a Metal-Organic Contaminant Mixture (Pb, Sr, Zn, and Perchloroethylene) by Cyclodextrin. *Environ. Sci. Technol.* 31 1087-1092.

Byun H. S., Zhong N. and Bittman R. (2000) *Organic Syntheses* 77 225-228.

Cardas S., Foray M-F., Gadelle A., Gerbaud G. & Bardet M. (2005) High-resolution solid state ^{13}C NMR study of per(3,6-anhydro)-*a*-cyclodextrin based polymers and of their chromium complexes *Carbohydrate Polymers* 61 88-94.

Chowdhury S., Mohan R. S. & Scott J. L. (2007) Reactivity of ionic liquids *Tetrahedron* 63 2363-2389.

Cocalia V. A., Holbrey J. D., Gutowski K. E., Bridges N. J. & Rogers R. D. (2006) Use of ionic liquids in the study of fruit ripening by high-resolution ^{13}C NMR spectroscopy: 'green' solvents meet green bananas *Tshinghua Science and Technology* 11183-193.

Croft A. P. & Bartsch R. (1983) Synthesis of chemically modified cyclodextrins *Tetrahedron* 39 1417-1474.

Cruz D. M. P. & Moens L. (2004) Thermal Analysis and Oxidation Reactions with Room- Temperature Ionic Liquids. DOE ERULF Program, University of Puerto Rico, National Renewable Energy Laboratory, Golden, Colorado

Ferreira P., Goncalves I. S., Pillinger M., Rocha J., Santos P. & Teixeira-Dias J. J. C. (2000) Modifications of β-cyclodextrin with ferrocenyl groups by ring opening of an encapsulated [1] ferrocenophane *Organometallics* 19 1455-1457.

Field L. D., Sternhell S. & Kalman J. R. 1995 *Organic Structures from Spectra,* John Wiley and Sons, New York, p6

Frederic D. GC/MS Analysis. *http://www.scientific.org/tutorials/articles/gcms.html* Accessed 19.05.2004.

Fujita K., Nagamura S. & Imoto T. (1984) Convenient preparation and effective separation of the C-2 and C-3 tosylates of α-cyclodextrin *Tetrahedron Letters* 255673-5676.

Garrido Frenich A., Martinez Vidal J. L., Cruz Sicilia A. D. Gonzalez-Rodriguez M. J., & Plaza Bolanos P. (2006) Multiresidue analysis of organochlorine and organophosphorus pesticides in muscle of chicken, pork and lamb by gas chromatography–triple quadrupole mass spectrometry *Analytica Chimica* 558 45-52.

Halmowitz S. (1991) *Chemical Engineering*75-80.

Harada A. (1997) Construction of supramolecular structures from cyclodextrins and polymers. *Carbohydrate Polymers* 34183-192.

http://www.epa.gov/pesticides Accessed 01-05-06.

http://www.npi.gov.au/database/substance-info/profiles/90.htm Accessed 28-05-07.

http://www.sigma-aldrich.com/chemicalsynthesis. ChemFiles 5 (6) 1-24.

http://www.waterbornne.org.doc;factsheets.doc;inorganic.htm Accessed 26-04-06.

Kawaguchi M., Ishii I., Okanouchi N., Sakni N., Ito R. & Inowe K. (2004) *Organochlorine Compounds* 66 181-185.

Lewis R. & Evans W. *Chemistry* 2nd Ed. Palgrave Macmillan, Hampshire United Kingdom 2001, pp 415.

Li D. Q. & Ma M. (1999) New organic nanoporous polymers and their inclusion complexes *Chem. Mater.* 11872-874.

Li D. Q. & Ma M. (1999) Nanosponges: From inclusion chemistry to water purifying technology *Chemical Science and Technology* 26-28.

LiD.Q.&MaM.(1999)Frominclusionchemistrytowaterpurificationtechnology *Chemtech.* 3531-37.

Li J. & Tan D. (2001) Inclusion complexes formation between cyclodextrins and poly(1,3- dioxolane) *Macromolecules* 34 1542-1544.

Liu J., Chi Y., Jiang J., Tai C., Peng J. & Hu J. (2004) Ionic liquid-based liquid-phase microextraction, a new sample enrichment procedure for liquid chromatography *Journal of Chromatography A.* 1026 143-147.

Liu J., Jiang G., Chi Y., Cai Y., Zhou Q. & Hu J. (2003) Use of ionic liquids for liquid-phase microextraction of polycyclic aromatic hudrocarbons *Journal of Anal. Chem.* 75 5870- 5876.

Liu J., Johnsson J. A. & Jiang G. (2005) Application of ionic liquids in analytical chemistry. *Trends in Analytical Chemistry* 2420-27.

Mamba B. B., Krause R. W., Malefetse T. J. & E. N. Nxumalo (2007) Monofuctionalized cyclodextrin polymers for the removal of organic pollutants from water. *Environ. Chem. Lett.* 5 79-84.

Martin Del Ville E. M. (2004) Cyclodextrins and their uses: a review *Process Biochemistry* 39 1033-1046.

Mhlanga S. D., Mamba B. B., Krause R. W. & Malefetse T. J. (2007) Removal of organic contaminants from water using nanosponge cyclodextrin polymers *J. Chem. Technol. Biotechnol.* 82 382-388.

Mhlanga S. D. MTech Dissertation (2006) University of Johannesburg.

Morales M. M., Marti P., Llopis A., Campos L. & Sagrado S. (1999) An environmental study by factor analysis of surface seawaters in the Gulf of Valencia (Western Mediterranean) *Analytica Chimica Acta* 394 109-117.

Muderawan W., Ong T., Tang W., Yong D., Ching C. & Ng S. (2005) Synthesis of ammonium substituted β-cyclodextrins for enantioseparation of anionic analytes. *Tetrahedron Letters* 46 1747-1749.

Niegowski S. J. (1952) Destruction of phenols by oxidation with ozone. *Industrial and Engineering Chemistry* 45 632-634.

Oleszczuk P. & Baran S. (2004) Application of solid-phase extraction to determination of polycyclic aromatic hydrocarbons in sewage sludge extracts. *Journal of Hazardous Materials* 113 237-245.

Pandey S. (2006) Analytical applications of room-temperature ionic liquids: A review of recent efforts *Analytica Chimica Acta* 556 38-45.

Park J., Dec J., Kim J. & Bollag J. (1999) Effect of Humic Constituents on the Transformation of Chlorinated Phenols and Anilines in the Presence of Oxidoreductive Enzymes or Birnessite *Environ. Sci. Technol.* 33 2028-2039.

Pereiro A. B., Legido J. L. & Rodriguez A. (2007) Physical Properties of Ionic Liquids Based on 1-Alkyl-3-methylimidazolium Cation and Hexafluorophosphate as Anion and Temperature Dependence *J. Chem. Thermodynamics* 391219-1230.

Petanen T & Romantschuk M. (2002) Use of bioluminescent bacterial sensors as an alternative method for measuring heavy metals in soil extracts *Analytica Chimica Acta* 456 55-56.

Phan T. N. T., Bacquet M. & Morcellet M. The removal of organic pollutants from water using new silica-supported β-cyclodextrin derivatives *Reactive and Functional Polymers* 52 (2002) 117-125.

Rogers J. D., Jedral W. & Bunce N. J. (1999) Electrochemical Oxidation ofChlorinated Phenols *Environ. Sci. Technol.* 33 1453-1457.

Rong D. & D'Souza V. T. A convenient method for functionalization of the 2-position of cyclodextrins *Tetrahedron Letters* 31 (1990) 4275-4287.

Sabio E., Zamora F., Gonzalez J. F., Gonzalez-Garcia C. M., Roman S. & Al-Kassir (2006) Characterisation under static and dynamic conditions of commercial activated carbons for their use in wastewater plants *Applied Surface Science*6058-6063.

Salipira K. L., Mamba B. B., Krause R. W., Malefetse T. J. & Durbach S. H. (2007) Carbon nanotubes and cyclodextrin polymers for the removal of organic pollutants from drinking water. *Environ. Chem. Lett.* 55 13-17.

Santos A., Yustos P., Rodriguez S., Simon E. & Garcia-Ochoa F. (2007) Abatement of phenolic mixtures by catalytic wet oxidation enhanced by Fenton's pretreatment: Effect of H_2O_2 dosage and temperature *Journal of Hazardous Materials* 146 595-601.

Savage N. & Diallo M. S. (2005) Nanomaterials and water purification: Opportunities and challenges *Journal of Nanoparticle Research* 7 476-486.

Schwarzenbach R. P., Gschweld P. M. & Imboden D. M. *Environmental Organic Chemistry* 2nd Ed 2003, 4-7 .

Sutyagin A. A., Glazyrin A. E., Kurochkina G. I., Gracher K. K. & Nifant'ev E. E. (2002) Regioselective Acetylation of β-Cyclodextrin *Russian Journal of Chemistry* 72 147- 150.

Szeijtli J. (1998) Introduction and General Overview of Cyclodextrin Chemistry *Chem. Rev.* 98 1743-1753.

Tran C. D. & De Paoli Lacendra S. H. (2002) Determination of Binding Constants of Cyclodextrins in Room-Temperature Ionic Liquids by Near-Infrared Spectrometry. *Anal. Chem.* 74 5337-5341.

US Environmental Protection Agency (EPA), (2000) *Chemical Precipitation*, US EPA, Washington, DC, (EPA832-f-00-018).

Valerie C. (2003) Solid phase extraction of trace elements. *Spectrochimica Acta Part B:* 58 1177- 1233.

Visser A. E. & Rogers R. D. (2004) Influence of structural variation in room-temperature ionic liquids on the selectivity and efficiency of competitive alkali metal salt extraction by a crown ether. *Ind. Eng. Chem. Res.* 39 3596-3604.

Visser A. E., Swatloski R. P., Griffin S.T., Hartmen D. H. & Rogers R. D. (2001) Liquid/Liquid Extraction of Metal Ions in Room Temperature Ionic Liquids *Sep. Sci. Technol.* 36 785-804.

Visser A. E., Swatloski R. P., Reichert W. M., Mayton M., Sheff M., Wierzbicki A., Davis Jr. J.

H. & Rogers R. D. (2002) Task-Specific Ionic Liquids Incorporating Novel Cations for the Coordination and Extraction of Hg^{2+} and Cd2+: Synthesis, characterization and Extraction Studies. *Environ. Sci. Technol.* 36 2523-2529.

Wei G.T., Yang Z. & Chen C. J. (2003) Room temperature ionic liquid as a novel medium for liquid/liquid extraction of metal ions. *Analytica Chimica Acta* 488 183-192.

Weinz G., Bons Steinbrum M. & Landfester K. (1997) Solid state polycondensation within cyclodextrin channels leading to watersoluble polyamide rotaxanes. *Tetrahedron* 5315575-15592.

Welton T. (1999) Room-Temperature Ionic Liquids: Solvents for Synthesis and Catalysis. *Chem. Rev.* 99 2071-2083.

Wikstrom E. & Marklund S. (2000) Secondary Formation of Chlorinated Dibenzo-*p*-dioxins, Dibenzofurans, Biphenyls, Benzenes, and Phenols during MSW Combustion. *Environ. Sci. Technol.* 34 604-609.

Zhong N., Byun H. S. & Bittman R. (1998) An improved synthe- sis of 6-O-monotosyl-6- deoxy-β-cyclodextrin. *Tetrahedron Letters* 39 2919-2920.

Zymark Corporation. Application Brief (2001) Method 252.2, *Semi-Volatile Organics in Waterby Solid Phase Extraction and GC/MS Detection, Zymark Corporation.*

8 Chapter

IONIC LIQUID AS AN INNOVATIVE SOLVENT FOR EXTRACTION AND SEPARATION IN ANALYTICAL CHEMISTRY

1. INTRODUCTION: AN OVERVIEW

Direct examination of samples, irrespective of their origin, is preferable; nevertheless, it is often impractical due to the complexity of the sample matrix, insufficient concentration of target analytes, or incompatibility with the detector.A sample pretreatment step is necessary for the elimination of interferences and the separation/preconcentration of analytes.Notwithstanding significant progress in separation science, conventional solvent extraction is extensively employed for sample preparation.The primary disadvantage of solvent extraction is the necessity for substantial quantities of high-purity solvents, which are costly, poisonous, and generate hazardous waste.Consequently, the pursuit of novel solvents is a critical theme in the evolution of solvent extraction.Ionic liquids, composed of organic cations and diverse anions, have garnered significant interest due to their distinctive properties, including low vapor pressure, high viscosity, dual polarity, excellent thermal stability, and extensive miscibility with water and various organic solvents, thereby mitigating numerous environmental and safety concerns linked to organic solvents.The distinctive chemical and physical features of ionic liquids have generated growing interest in their potential as alternate mediums for traditional solvent extraction and organic synthesis.Recently, ionic liquids have been swiftly advanced as environmentally benign acceptor phases for diverse microextraction techniques in analytical chemistry, including liquid phase microextraction, single drop liquid phase microextraction, solid phase microextraction, dispersive liquid phase microextraction, and cold induced aggregation microextraction.The microextraction utilizing ionic liquid frequently demonstrated superior extraction efficiency and enrichment factor compared to traditional solvent extractions. Dispersive liquid phase microextraction is exceptionally quick and efficient among microextraction techniques.The complete dispersion of the ionic liquid in the aqueous phase facilitates the migration of analytes into the ionic liquid phase due to the significantly larger contact surface, allowing the dispersive liquid phase microextraction method to be completed within a few minutes.In recent years, ionic liquid extraction, in conjunction with various analytical technologies, has been extensively utilized for the determination of ultra-trace analytes in water and other samples. Techniques such as high-performance liquid chromatography (HPLC) and gas chromatography-mass spectrometry (GC-MS) are employed for organic compounds, while flame atomic absorption spectrometry (AAS) and electrothermal atomic absorption spectrometry (ETAAS) are used for metal ions. Overall, extractions utilizing ionic liquids have rapidly emerged as the predominant sample preparation techniques in analytical chemistry..

2. APPLICATION OF IONIC LIQUIDS AS MEDIUM FOR CLASSICAL SOLVENT EXTRACTION

2.1 Metal ion extraction and its application

For metal ion extraction, one needs to find the right combination of ionic liquid and ligand to obtain similar behavior as in organic solvents or one needs to use the differences from organic solvents for achieving more advantageous separations. Although the metal ion extraction in the ionic liquid/aqueous two phase system often indicated high efficiency and selectivity that excelled most of organic solvents, the metal ion partitioning always rely on the species of the ionic liquid, metal ion and ligand seriously [1-7]. For examples, Bartsch group reported on the influence of structural variations in room temperature ionic liquid on the selectivity and efficiency of competitive alkali metal salt extraction by crown ethers [11]. They showed that extraction efficiency generally diminished as the length of the 1alkyl group on the ionic liquid was increased. It was demonstrated that metal ion affinity for the hydrophobic phase necessitates the presence of an extractant [12]. Professor Rogers has investigated Cyaex-923 as an ligand for actinide ions in liquid/liquid separation using the hydrophobic ionic liquid [C_{10}mim][Tf_2N] as the extracting phase, a mixture of tri-alkyl-phosphine oxides of varying mixture of tri-alkyl-phosphine oxides of varying chain lengths (hexyl- and octyl-) and has been used for actinide extraction into molecular solvents. Distribution ratio measurements for Pu^{4+} and Am^{3+} showed little acid dependency for Pu^{4+}, but strong acid dependency studies seem to support 1:1 metaltoligand binding for Pu^{4+}, but significantly the notion that ionic liquids can provide different coordination environments [13]. In such application, imidazolium cations were also derived to include the task specific functionalities, such as metal ligating groups that when used as part of the solvent or doped in the less expensive ionic liquids could dramatically enhance the extraction of targeted metal ions [14]. Moreover, the effect of pH on the extraction efficiencies showed a big difference due to its differences in solvent behavior of ionic liquid and conventional organic solvents. After metal ion was extracted to ionic liquid, a strong solution such as nitric acid was applied to stripping metal ion from the ionic liquid. We have investigated the conditions of stripping metal ions from ionic liquid too, the experiments indicated metal ion is easy to be stripped from ionic liquid phase and enter aqueous phase iftheneutral complex was formed in a base medium. Otherwise, metal ion stripping required a relatively strong base solution. However, the stripping process is slow very much at room temperature. In order to overcome the problem, a new stripping procedure was developed for stripping metal ion from ionic liquid, in which the equal volume of acetone and the ionic liquid containing metal ion complex mixed before a strong solution was added, whole process can complete within several minutes. At the same, above stripping method may be used to renew of the ionic liquids.

Recently, we attempt that metal ion extraction in the ionic liquid/aqueous two phase system combined with spectrometry technology including AAS and spectrophotometry to determine ultra trace amounts of metal elements in environmental samples. It was found that the extraction procedure can decrease the detect limit and improve the selectivity of analytical method remarkably. For atomic absorption spectrometry, the metal ion extraction must be back-extracted into aqueous solution, then was introduced into atomic absorption spectrometer for determination. For example, a new room temperature ionic liquid [C_4tmsim][PF_6] was developed as a novel medium for liquid/liquid extraction of inorganic mercury with CDAA, the extraction and back-extraction efficiencies were 99.9 and 100.1 % for 5.0 $\mu g.L^{-1}$ standard mercury in 1000 mL of water solution, respectively, the detection limit is 0.01 $\mu g.L^{-1}$ mercury in real water sample, proposed method has been used to the determination of trace inorganic mercury in natural water with satisfactory results [2]. If the ionic liquid phase containing metal complex is transparent, it can be directly used to spectrophotometric determination. So that, metal ion extraction in the ionic liquid/aqueous two phase system in which some excellent chromogenic reagent was selected as ligand to form metal complex is very selective, this may overcome the drawback in sensitivity of conventional spectrophotometric methods. For example, we reported a highly sensitive and selective extraction spectrophotometric method for determination of trace germanium in natural water with a new chromogenic reagent methy-benzeneazosalicyl fluorone abbreviated as MBASF, in which ionic liquid [C_4mim][PF_6] was used as novel medium for liquid/liquid extraction of germanium (IV). In the presence of TritonX-100, MBASF reacted with germanium (IV) to form a red complex rapidly, the complex was then extracted into the $[C_4mim][PF_6]$ phase (shown in Fig.1), absorbance of the complex in ionic liquid at 496 nm was recorded and used to determine trace Ge^{4+}. The apparent molar absorptivity of the complex and the detection limit for the real sample were found to be 3.12×10^6 $L.moL^{-1}.cm^{-1}$ and 0.2 $ng.mL^{-1}$ respectively. The absorbance of the complex at 496 nm increases linearly with the concentration up to 4mg of Ge^{4+} in 250 mL of aqueous solution. The interference study show the determination of germanium is free from the interference of almost all positive and negative ions found in the natural water samples. The determination of Ge^{4+} in natural water was carried out by the present method and ETAAS. The results were satisfactorily comparable so that the applicability of the proposed method was confirmed using the real samples [40]. Metal ion extraction in ionic liquid/aqueous two phase system and their application were listed in Table 1.

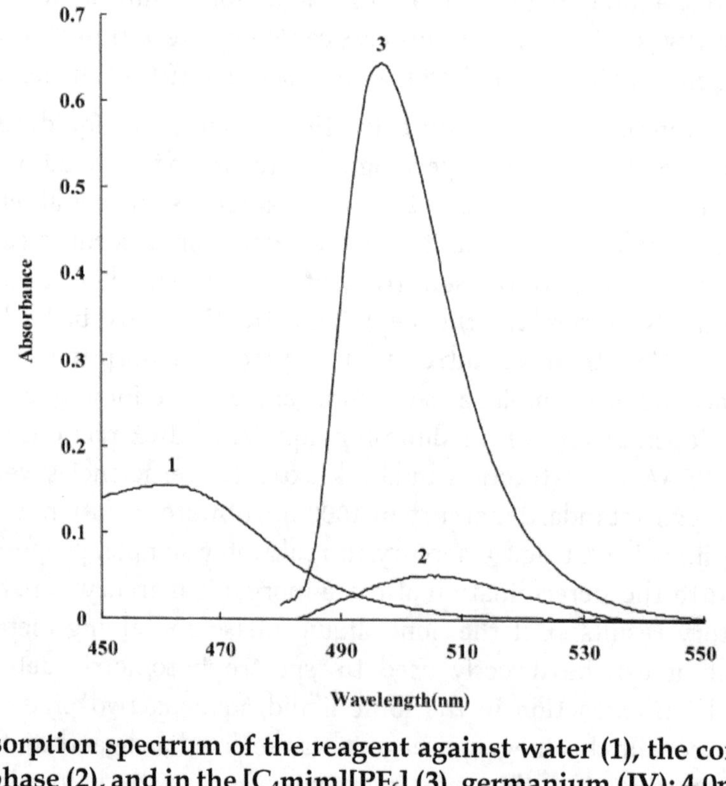

Fig. 1. Absorption spectrum of the reagent against water (1), the complex in the aqueous phase (2), and in the $[C_4mim][PF_6]$ (3), germanium (IV): 4.0mg in 250 mL of aqueous solution, $[C_4mim][PF_6]$

Table 1. Reviews in extraction of metal ions with ionic liquid as medium

Metal ion	Ionic liquid	Ligand	Detection technology /detection limit	Ref.
Ni^{2+}	[C4mim][PF_6],	PAN	FI-FAAS/12.5 µg.L^{-1}	15
Fe^{3+}, Cu^{2+}, Ni^{2+}, Mn^{2+}	trioctyl methylammonium salicylate ionic liquid functionalized 2-mercaptobenzothiazol	ammonium pyrrolidine dithiocarbamate and diethydithiocarbamate	AAS	16 17
Pb2+, Cu2+, Hg^{2+}, Cd^{2+}	[C_4mim][PF_6]	2-aminothiophenol		18
Co^{2+}	1-hexylpyridinium hexafluorophosphate	1-phenyl-3-methyl-4- benzoyl-5-pyrazolon	AAS/0.70 mug.L^{-1}	19
Cd^{2+}	[C4mim][PF_6]	2-mercaptobenzothiazole	AAS/4.6 ng.L^{-1}.)	20
La^{3+}	[C4mim][Tf2N]	18-Crown-6 and 2 thenoyltrifluoroacetone-		21
Sr^{2+}	[C4mim][PF_6]	DCH-18-C-6		22
Am^{3+}	[C4mim][Tf2N]	tri-n-butylphosphate and n-octyl(phenyl)-N,N-diisobutylcarbamoylmethylphosphine oxide		23
$UO2^{+}$	[C4mim][PF_6]	tri-n-butyl phosphate		24
Mn^{2+}, Co^{2+},Ni^{2+}, Zn^{2+}, Cd^{2+}	[Cnmim][PF_6], [Cn-mim][Tf_2N]			25
Pb^{2+}, Cd^{2+}	[C4mim][PF6]	tetra-n-propylammonium iodide	differential-pulse anodic stripping voltammetry/0.001 mug.mL^{-1}	26
Th^{4+}	[C_8mim][PF_6]	primary amine N-1923		27
$UO2^{2+}$	[C_4mim][Tf_2N]	ri-n-butylphosphate		28
Y^{3+}	[C8mim][PF_6]	Cyanex-923 and EDTA		29
Pu^{4+}	[C_8mim][PF_6]	carbamoyl methyl phosphine oxide		30
La^{3+}	[C4mim][Tf_2N]	2-thenoyltrifluoroacetone and 18-crown-6		31
Ni^{2+}	[C4mim][PF_6]	2-pyridinealdoxime		32

Sr^{2+}, Cs^{+}	tri- butylmethylammo-niu m bis((trifluoromethyl)su lfonyl)imide	ionophores dicyclohex-ano-18-crown- 6 and calix[4]arenebis- (tert-oc-tylbenzo-crown-6)		33
Pb^{2+}	$[C_4mim][PF_6]$	dithizone	ETAAS/1.0 ng.L^{-1}	5
Al^{3+}	$[C_4mim][PF_6]$	3,5-ditertbutylsalicylfluorone	ETAAS/0.06 mug.L^{-1}	6
Zn^{2+}	Butyl-methyl-trifluo-romethylsulfonate	2,4-dinitrophenolate		34
Pb^{2+}	$[C_8mim][PF_6]$	dithizone	ETAAS/1.0 ng.L^{-1}	7
Sc^{3+}	$[C_8mim][PF_6]$	Cyanex -925		35
M^{2+}	$[C_4mim][PF_6]$	benzoyltrifluoroacetone, 2- naphtoyltrifluoroace-tone and trifluoroacety-lacetone		36
Pd^{2+}	tri-n- octylmethylam-monium chloride and tri-n- octylmethylam-moniu	Aliquat-336		37
Cu^{2+}	$[C_4mim][PF_6]$	1-methylimidazole		38
La^{3+}	$[C_4mim][PF_6]$	N,N,N ',N '-tetrakis(2-pyridylmethyl)ethylenedi amine		39
Ge^{4+}	$[C_4mim][PF_6]$	methybenzeneazosalicyl-fluorone	Spectrophotometry/0.2 ng.mL^{-1}	40
UO_2^{2+},	$[C_4mim][PF_6]$			41
Am^{3+}, Nd^{3+}, Eu^{3+}		dialkylphosphoric or dialkylphosphinic acids		
UO_2^{2+}	$[C_4mim][PF_6]$	tri-n-butylphosphate		42
Cu^{2+}	$[C_4mim][Tf_2N]$	beta-diketonate		43
Cu^{2+}	[C4mim]$[PF_6]$	dithizone, 8-hydroxyquin-oline, and 1- (2-pyridyla-zo)-2-naphthol		44

2.2. Organic compound/gas extraction and separation

Ionic liquids have been widely applied to organic compound/gas extraction and separation too. Basing on the ionic liquid various unique solvent characteristics and extraction performance, room temperature ionic liquids may be used as liquid membrane materials to separation organic compound and gas. For example,

$[C_4mim][PF_6]$ as membrane solution can be utilized for hydrocarbon separation by using a supported liquid membrane. Although the permeation rates through the membrane based on the ionic liquids were less than those of water, the selectivity of aromatic hydrocarbons was greatly improved [45]. Ionic liquid as special solution combined with classical separation technologies such as liquid/solid phase microextraction was used to beforehand separation and enrichment of the sample for high performance liquid chromatography, this may improve the separation effect of analytical technology remarkably. Ionic liquid as electrolyte can enhance electric capability to capillary zone electrophoresis. For example, professor Qi reported the ionic liquid used as running electrolyte in capillary zone electrophoresis with cyclodextrin (β-CD) as modifier for the separation of anthraquinones extract of Chinese herb Paedicalyx attopevensis Pierre ex Pitard [46]. Ionic liquid as ion pair reagent was applied to reversedphase high performance liquid chromatography. Moreover, utilization of supercritical carbon dioxide (scCO2) in concert with various popular ionic liquids is shown to have modified and enhanced the overall extraction capabilities and applications of these neoteric solvents [47-50]. Brennecke group have investigated the recovery of organic products from ionic liquids using scCO2. The results indicated the recovery rates of various aromatic and aliphatic solutes from [C4mim][PF6] in the presence of environmentally benign scCO2 are presented. In a subsequent study, interesting outcomes showing utility of scCO2 as a separation switch for ionic liquid/organic mixtures were published by the same group [47, 48]. In general, we conjecture ionic liquid will rapidly be applied to many new fields of the extraction and separation with understanding on the extraction performance of room temperature ionic liquid.

3. APPLICATION OF IONIC LIQUIDS IN MICROEXTRACTIONS

3.1 Liquid phase microextraction

Ionic liquid has been widely applied to various liquid phase microextractions for sample preparation. Commonly, volume of the ionic liquid for liquid phase microextraction is often lower than 25 μL in order to obtain high enrichment factor. To fine mix the ionic liquid with aqueous phase, a relatively long extraction time and strong stir process are strongly required. However, so small volume of the ionic liquid is difficult to be precisely collected from aqueous phase due to its dissolution in water and adhere to wall of the vial (Table 2). To overcome the problems, some new types of liquid phase microextraction have been reported in literatures. For examples, single drop microextraction (SDME) (Table 3) and dispersive liquid microextraction (DLME) (Table 4).

Table 2. Reviews for common liquid phase microextraction using ionic liquid as acceptor phase

Analyte	Ionic liquid	Sample	Remark	Ref.
Steroidal antiinflammatory drugs	$[C_4mim][PF_6]$	Human urine	Enrichment factors varied between 10.69 for naproxen and 13.93for fenbufen. HPLC system wasemployedforanalysis. A dynamic liquid-phase microextraction using an ionic liquid as acceptor phase is reported.	51
Aromatic and aliphatic hydrocarbons	$[C_4mim][PF_6]$	Water	GC-MS was coupled with determination of anlyte; for the first time hollow fiber-protected ionic liquid supported threephase (liquid–liquid–liquid) microextraction; enrichment factor and extraction time are 210 and 40min.	52
4-Nonylphenol, 4-tert-octylphenol	$[C_4mim][PF_6]$	Water	HPLC was employed for analysis; enrichment factors are between 19.5 and 33.8, and extraction time was about at 30 min.	53
Chlorpromazine, promethazine, levomepromazine, prochlorperazine, trifluoperazine, fluphenazine and thioridazine	$[C_4mim][PF_6]$	Human urine	HPLC was employed for analysis.	54
Phorate,parathion and poxim	$[C_4mim][PF_6]$	Water	HPLC was employed for analysis; enrichment factors are in the range from 553 to 665.	55
Benzene, toluene, ethylbenzene, and xylene), polycyclic aromatic hydrocarbons, phthalates, phenols, aromatic amines, herbicides, organotin, and organomecury	$[C_4mim][PF_6]$, $[C_8mim][PF_6]$	Water	HPLC, AAS and or cold-vapor atomic fluorescence spectrometry determination was employed for analysis, respectively; the enrichment factors determined were in the range of 5-168 for 15 min extraction by [C4mim][PF6] and 4-178 for 30 min extraction by 4-178 for 30 min extraction by $[C_8mim][PF_6]$.	56

Chlorpromazine, promethazine, levomepromazine, prochlorperazine, trifluoperazine, fluphenazine and thioridazine	[C_4mim][PF_6]	Human urine	The proposed method is a valuable alternative for the analysis of these drugs in urine within the concentration range 0.07-10μg.mL^{-1}. Limits of detection were in the range from 21 ng.mL^{-1} for thioridazine to 60ng.mL^{-1} for levomepromazine.	102
Cobalt	[C_6mim][PF_6]	Environmental and biological samples	ETAAS was employed for detection of cobalt.	111
Triazines	[C_6mim][PF_6]	water	HPLC was employed for detection of triazines, the detection limits are for the triazine range from 0.36 to 1.41mug.L^{-1}.	116
Palladium	[C_6mim][PF_6]	food additive	The limit of detection was 0.2ng.mL^{-1}.	117
Cadmium	[C_6mim][PF_6]	saline	o,o-Diethyldithiophosphate was used as a ligand. FAAS was employed for detection of cadmium, the limit of detection (3s) and the enhancement factor were 0.07 mug.L^{-1} and 78, respectively.	127

SDME includes direct-immersion liquid phase microextraction and headspace liquid phase microextraction. Because during headspace liquid microextraction the ionic liquid has no loss because it lack effective vapor pressure even at high temperature, headspace liquid phase microextraction often provide a higher enrichment factor for most of analytes than direct-immersion liquid phase microextraction. A typical experimental set-up of headspace liquid phase microextractionisillustrated inFig.2 [59].Before each extraction, the microsyringewas rinsedwithmethanoltoavoidanal ytecarry-overandair bubble formation. For the extraction an aliquot of the sample solution was placed in a vial immersed in the recirculating cell. After the uptake of the 10-25 μL of the ionic liquid, the microsyringe needle was inserted through the septum and then tightly fitted with a long polytetrafluoroethylene tube, and the syring to begin with the extraction process, the syringe plunger was depressed completely to expose an ionic liquid drop and the magnetic stirrer was turned. After extracting, the magnetic stirrer was switched off, and then the microdrop was retracted back into the microsyringe and injected directly into detector for analysis. SDME is based on the equilibrium of the three phases: the condensed aqueous phase, the headspace phase and the microdrop phase. Mass transfer in the headspace is assumed to be a faster process since diffusion coefficients in the gas phase are typically 10^4 times greater than corresponding diffusion coefficients in the condensed phase. Thus, both slow mass transfers in the aqueous phase and diffusion of analyte into the microdrop are limiting steps in the overall extraction process was clamped so that the tip of the needle was located in a constant position in the headspace of the vial. The penetration theory indicates the aqueous phase mass transfer coefficient of solute increases with increasing stirring rate, and heating improves the mass transfers in the aqueous phase and induces convection in the headspace. Therefore, consequently the equilibrium between the aqueous and vapor phase can be attained more rapidly. Moreover, an addition of salt (such as NaCl and Na_2SO_4 etc) to the sample can decrease the solubility of analytes in the aqueous phase and enhance their partitioning into the organic phase, the extraction is often enhanced with increasing salt concentration and polarity of compound. Thus, a salt concentration of about 30 % was used in most SDME.

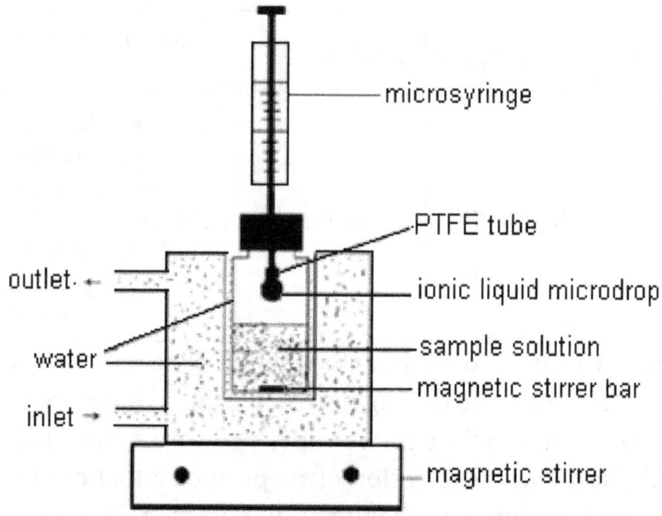

Fig. 2. Schematic diagram of SDME

Table 3. Reviews for single drop microextraction using ionic liquid as acceptor phase

Analyte	Ionic liquid	Sample	Remark	Ref.
Chlorobenzenes	$[C_4mim] [PF_6]$	Water	HPLC was coupled with analysis; extraction time is 37min; an obvious salting-out effect was observed.	57
Formaldehyde	$[C_8mim]\_[PF_6]$	Shiitake mush-room	2,4-Dinitrophenylhydrazine was used as derivative agent; HPLC was coupled with analysis; extraction time is 30 min	58

Dichlorodiphenyl-trichl oroethane and its metabolites	[C$_4$mim]_[PF$_6$]	Water	HPLC was coupled with analysis; an obvious salting-out effect was observed; extraction time is 30min	59
Phenols	[C$_8$mim] [PF$_6$]	Water	GC was used for detection of phenols; effect of salting-out is obvious.	60
Phenols	[C$_4$mim] [PF$_6$]	Water	HPLC was employed for analysis; enrichment factors were in the range from 17.2–160.7.	61
Chloroform, bro-modichloromethane, dibromochlorometh-ane and bromoform	[C$_4$mim] [PF$_6$]	Water	GC-MS was employed for analysis; extraction time was about 30min; an obvious salting-out effect was observed well.	62
Chlorinated anilines	[C$_4$mim][PF$_6$]	Water	HPLC was employed for analysis; extraction temperature and salt-ing-out have seriously influenced on the extraction efficiencies.	63
Benzophenone-3	[C$_6$mim][PF$_6$]	Urine	HPLC was employed for analysis.	64
Dichloromethane, xylene, and undecane	*p*- [C$_4$mim][PF$_6$] *n*-	Water	GC-MS was employed for the analysis; a new removable interface that enables the introduction of the extracted analytes into the GC system, while preventing the ionic liquid from entering the column, has been developed	65
Benzene, toluene, ethylbenzene xylene isomers	[C$_4$mim][PF$_6$]	Waters	GC-MS was employed for the anal-ysis; extraction temperature and salting-out effect on the extraction performance.	66
Cobalt, mercury and lead	[C$_4$mim][PF$_6$]	Biological and	Electrothermal vaporization inductively coupled plasma mass spectrometry was employed for determination; after 10 min of extrac-tion, the enrichment factors were 350 (cobalt), 50 (mercury) and 60 (lead); 1-(2-pyridylazo)- 2-naphthol as both extractant and chemical modifier.	67
Chlorobenzenes	[C$_4$mim] [PF$_6$]	Water	A 20 min of extraction time was 68 required; a microwave was used for assist extraction.	68
Mmanganese	[C$_4$mim] [PF$_6$]	Water	ETAAS was employed for analysis; 1-(2-69 thiazolylazo)-2-naphthol was used for chelator of manganese; the drop of the ionic liquid was directly. injected into the graphite furnace.	69

1-Naphthylamine, N,N-dimethylaniline and diphenylamin	$[C_4mim] [PF_6]$	Water	Enrichment factors are in the range from 93 13.7 to 116.3.	93
4-Dichlorophenol, 2- naphthol, 2-nitrophenol and 4- chlorophenol	$[C_4mim] [PF_6]$	Water	The effect of complex matrices natural water samples could be resolved with addition of sodium ethylene diamine tetracetate into the samples.	71
4-Nonylphenol, 4-tert-octylphenol	$[C_6mim] [PF_6]$	Water	HPLC was employed for analysis; the proposed procedure has a detection limit and enrichment factor of 0.3mug.L-1 and 163 for 4-nonylphenol, and 0.7mug.L-1 and 130 for 4-tert-octylphenol, respectively.	72
chlorobenzenes	$[C_6mim] [PF_6]$	Water	Limits of detection ranging between 1 and 4ng.L^{-1}.	103
mercury species	$[C_6mim] [PF_6]$	Water	Limits of detection were between 1.0 and 22.8 mug.L^{-1} for the four species of mercury.	104
lead	$[C_4mim] [PF_6]$	certified reference materials	Lead was complexed with ammonium pyrroldinedithiocarbamate; The limit of detection and the enhancement factor were 0.015 and 76μg.L^{-1}.	105
Trihalomethanes	$[C_4mim] [PF_6]$	water	The limits of the detection range between 0.1 ng/mL (bromoform) and 0.9 ng.mL^{-1} (chloroform).	106
Tributyltin and Triphenyltin	$[C_4mim] [PF_6]$ $[C_4mim] [PF_6]$	water Chinese Medicines	Detection limits are between 1 and 100 mug.mL^{-1}. The limit of detection was 0.39mug.L^{-1}.	108 115
phenols	$[C_4mim] [PF_6]$		capillary electrophoresis was employed for analysis, detection limits to three phenols were less than 0.05μg.L^{-1}.	120
carbonyl compounds	$[C_6mim] [PF_6]$	water	Enrichment factor is up to 150. HPLC was employed for detection of carbonyl compounds, the limit of detection was in the range of 0.04-2.03ng.mL^{-1}.	124
cobalt	$[C_6mim] [PF_6]$	water	ETAAS was employed for detection of cobalt, the detection limit was 0.04mug.mL^{-1}, and the enrichment factor was 120.	131

2-hydroxy-4- methoxybenzophenone, isoamyl-4- methoxycinnamate, 3-(4'-methylbenzylidene)camphor, 2-ethylhexyl 2- cyano-3,3-diphenylacrylate, 2-ethylhexyl 4- dimethylaminobenzoate and 2-ethylhexyl 4- methoxycinnamate	$[C_6mim][PF_6]$	water	HPLC was employed for detection, limits of detection were found in the low mug. L^{-1} range.	128
lead	Tetradecyl (trihexyl) phosphonium chloride	water	2-(5-bromo-2-pyridylazo)-5-diethylaminophenol was used as ligand. ETAAS was employed for detection of lead, its detection limit is 4.5 mug.L-1.	129

Recently, a novel liquid phase microextraction technology termed dispersive liquid phase micro-extraction (DLME) was developed and validated (Table 4). From the name, it is obviously that it was based on the temperature change or using organic disperser making ionic liquid completely disperse in the aqueous phase and increase the chance of mass transfer into ionic liquid phase, in which the analytes will more easily migrate into the ionic liquid phase because of the much larger contact area than that of conventional single drop liquid micro-extraction, and the ionic liquid was concentrated into one drop by cooling and centrifugation. Experimental indicated that this system could be tuned to a great extent because ionic liquids can be designable and the method does not suffer from the limitations of that in conventional solvent micro-extraction. DLME has been applied to detect trace organophosphorus pesticides [73], fipronil, chlorfenapyr, buprofezin, hexythiazo [74] and mercury [75] with high extraction efficiencies. In general, DLME is more simple and rapid than SDME.

Table 4. Reviews for dispersive liquid phase microextraction using ionic liquid as acceptor phase

Analyte	Ionic liquid	Sample	Remark	Ref.
Organophosphorus pesticides	$[C_6mim][PF_6]$	water	HPLC was coupled with analysis; a 70℃ of heating and 30min of the centrifugation process were required strongly in order to form a fine dispersive phase and extraction efficiency; serious salting-out	73

Fipronil, chlorfenapyr, buprofezin, and hexythiazo	$[C_4mim][PF_6]$	Water	HPLC was employed for analysis; enrichment factors are in the range from 209 to 276.	74
Hg^{2+}	$[C_4mim][PF_6]$	Water	Spectrophotometry was employed for analysis; 4,4'-bis(dimethylamino)- thiobenzophenone was used as chelator of mercury; enrichment factor is about to 18.8.	75
Sm^{3+}, Eu^{3+}, Gd^{3+}, Dy^{3+}	$[C_4mim][PF_6]$	uranium dioxide powder	1-hydroxy-2,5-pyrrolidinedione was used as agent.	90
polycyclic aromatic hydrocarbons	$[C_8mim][PF_6]$	tap, bottled, fountain, well, river, rainwater, treated and raw wastewater	Analysis of extracts was performed by HPLC coupled with fluorescence detection; Quantification limits obtained for all of these considered compounds are between 0.1 and 7 $ng.L^{-1}$.	91
Zn^{2+}	1-hexylpyridinium hexafluorophosph ate	water and milk	Zinc was complexed with 8-hydroxyquinoline; The ionic liquid was proposed for the preconcentration of trace levels of zinc as a prior step to determination by flame atomic absorption spectrometry; the limit of detection (3s) and the enhancement factor were 0.22 mug. L^{-1} and 71.	92
non-steroidal antiinflammatory drugs	$[C_6mim][PF_6]$	urine	The suitability of the proposal is evaluated by means of the determination of non-steroidal antiinflammatory drugs in urine by liquid chromatography ultraviolet detection; Limits of detection were in the range from 8.3 $ng.mL^{-1}$ for indomethacin to 32 $ng.mL^{-1}$ for ketoprofen.	93
Cd^{2+}	$[C_6mim][PF_6]$	Water	Dispersive liquid-liquid microextraction followed by electrothermal atomic absorption spectrometry; The enrichment factor of the method is 67. The detection limit was 7.4 $ng.L^{-1}$.	94

Dichlorodiphen-yltrichloroethane and its main metabolites	$[C_6mim][PF_6]$		The experimental results indicated that the detection limits obtained for p,p'-DDD, p,p'-DDT, o,p'-DDT and p,p'-DDE were 0.24, 0.24, 0.45, 0.24 ng.mL-1, respectively	95
Thiophanate-methyl, carbofuran, carbaryl, tebuconazole, iprodione, oxyfluorfen, hexythiazox and fenazaquin	$[C_6mim][PF_6]$	Water	This paper describes a dispersive liquid-liquid microextraction procedure using room temperature ionic liquids coupled to high-performance liquid chromatography with diode array detection capable of quantifying trace amounts of eight pesticides; Limits of detection are between 0.320-4.66μg.kg^{-1}).	96
phenols	$[C_4mim][PF_6]$	Water	The settled phase was injected into HPLC for separation and detection of phenols; The main advantages of the proposed method are high speed, high recovery, good repeatability and environmental friendliness.	97
aromatic compounds	$[C_4mim][PF_6]$	Water	The limits of detection for seven polycyclic aromatic hydrocarbons varied from 0.02 to 0.3 mug.L^{-1}.	98
dicofol and DDT	1-isooctyl-3-methylimidazolium hexafluorophosphate	Water	The enrichment factors were 532 for dicofol and 540 for DDT; Proposed extraction combined with gas chromatography mass spectrometry was a valuable alternative for analysis of dicofol and DDT in environmental water samples within the concentration range 0.005-3.0 mug.L^{-1}. The detection limits were 1.3 ng.L-1 for DDT and 3.2 ng.L-1 for dicofol.	99
parathion, phoxim, phorate and chlorpyifos	$[C_8mim][PF_6]$	Water	Enrichment factor of analytes is 200 and the limits of detection calculated at a signal-to-noise ratio of 3 were in the range of 0.1-5.0 mug.L^{-1}.	100

thiophanate-methyl, carbofuran, carbaryl, tebuconazole, iprodione, oxyfluorfen, hexythiazox, and fenazaquin	$[C_8mim][PF_6]$	fruit	Limits of detection were in the range 0.651-5.44 mug/kg for table grapes and 0.902-6.33 A mug.kg^{-1} for plums.	101
three triazine and two phenylurea herbicides	$[C_6mim][PF_6]$	Water	HPLC was used for detection of five herbicides, their limit of detections were between 0.46 and 0.89µg.L^{-1}.	109
triclosan and triclocarban	$[C_6mim][PF_6]$	Water	HPLC-MS was used for detection of triclosan and triclocarban, limits of detection in the range from 0.040 to 0.58 µg.L^{-1}	110
Pb^{2+}, Cd^{2+}	$[C_6mim][PF_6]$	saline	FAAS was employed for detection of lead and cadmium, detection limits are 0.6µg/L for lead and 0.03µg.L^{-1} for cadmium	112
triclosan and triclocarban	$[C_6mim][PF_6]$	Water	HPLC-ESI-MS-MS was employed fortriclosan and triclocarb, detection limits are in the range from 0.040 to 0.58 mµg.L^{-1}.	113
organophosphorus pesticides	1,3-dibutylimidazolium hexafluor phosphate	water and fruit	Enrichment factors are more than 300 times. HPLC was employed for detection of pesticides, Limits of detection were 0.01-0.05 µg.L^{-1} for four analytes.	121
V^{5+}	$[C_4mim][PF_6]$	environmental and biological samples	2-(5-bromo-2-pyridylazo)-5-diethylaminophenol was used as ligand. ETAAS was employed for detection of vanadium. The detection limit achieved after preconcentration of 5 mL of sample solution, was 4.8 ng.L^{-1}	122
triclosan and triclocarban	$[C_6mim][PF_6]$	Water	HPLC-ESI-MS/MS was employed for detection of triclosan and triclocarban, detection limit is 0.04 and 0.3 mug.L^{-1}	125

anthraquinones	$[C_6mim] [PF_6]$	Radix	HPLC was employed for detection of anthraquinones, the limits of detection for all target analytes were ranged from 0.50 to 2.02 μg.L^{-1}.	126
Cr^{6+}	$[C_6mim] [PF_6]$	lake and tap water	ETAAS was employed for detection of chromium, the limit of detection of the proposed method was 0.07ng. mL^{-1}.	130

3.2 Solid phase microextraction

Solid-phase microextraction (SPME) has developed rapidly and it has been widely applied in sample analysis. This is because this simple and solvent-free sample preparation technique, which combines extraction, concentration and sample introduction in one step, is portable, sensitive and convenient to couple with various analytical instruments, especially GC. Almost a dozen different fiber coatings have been commercialized. Among them, the polydimethylsiloxane shows excellent selectivity for non-polar compounds while the polyacrylate coating facilitates the extraction of relatively polar analytes. To date, most of these fiber coatings suffered from limited life span, relatively low operating temperatures in GC applications, and incompatibility with the organic solvents in HPLC mobile phases. In recent years, ionic liquid was widely applied to coating for SPME (Table 4). A typical experimental set-up of headspace SPME mode was illustrated in Fig.3 [79]. The SPME fiber was coated with IL prior to every extraction, the analytes were then extracted and desorbed on the injection port of gas chromatography, and finally the ionic liquid coating on the fiber was washed out with solvents. The coating and washing out of the ionic liquid from the fiber can be finished in a few minutes. Experimental demonstrated that the relatively large

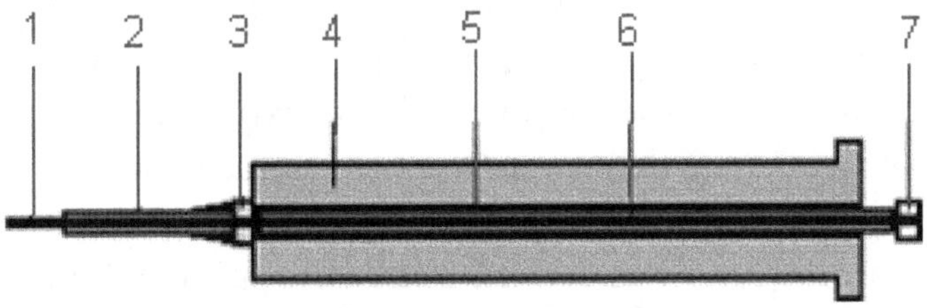

Fig. 3. Schematic diagram of the home-made SPME device: (1) ionic liquid-coated fiber tip; stainless steel needle; (3) silicone rubber O-ring; (4) glass syringe body; (5) moveable stainless steel tubing; (6) stainless steel fiber or polyimide coated fused-silica fiber; (7) localizer.

viscosity of ionic liquid aided its coating on the SPME fiber and permitted easy washing out after sample injection. On the other hand, the involatility and good thermal stability of the ionic liquid will keep it staying on the fiber when the analytes were thermally desorbed in the injection port of GC, though loss of some of the ionic liquid film occurred due to the simple coating procedure. This proposed disposable ionic liquid-coated fibers show reproducibility comparable with the widely used commercially available SPME fibers, but its sensitivity is lower due to the relatively thin coating. The major advantage of the disposable ionic liquid-coated fiber is the very low cost because only a very low amount of ionic liquid is consumed for each determination. Another advantage of this disposable coating is the carryover was avoided as new fiber coating is used for every determination. However, the detection limits were higher than those obtained using commercially available SPME fibers due to the relative thin film of ionic liquid that can be coated on the fiber. Therefore, this proposed method is only applicable to samples with relatively high levels of analytes. Considering that ionic liquid possess good extractability for various organic compounds and metals ions, and task-specific ionic liquid can be designed and synthesized for selective extraction of target analytes, disposable SPME/LPME coating ionic liquid might has great potential in sample preparation.

Baghdadi group reported a new microextraction with ionic liquid termed cold-induced aggregation microextraction procedure (CIAME) [85]. In this method, very small amounts of 1-hexyl-3-methylimidazolium hexafluorophosphate and 1-hexyl-3-methylimidazolium bis (trifluoromethylsulfonyl)imide (as extractant solvents) were dissolved in a sample solution containing Triton X-114 (as an anti-sticking agent). Afterwards, the solution was cooled in the ice bath and a cloudy solution was formed. After centrifuging, the fine droplets of extractant phase were settled to the bottom of the conical-bottom glass centrifuge tube. CIAME is a simple and rapid method for extraction and preconcentration of metal ions from water samples and can be applied for the sample solutions containing high concentration of salt and water miscible organic solvents. Furthermore, this technique is much safer in comparison with the organic solvent extraction. Performance of the technique was evaluated by determination of the trace amounts of mercury as a test analyte in several real water samples. Michler thioketone was chosen as a complexing agent. Analysis was carried out using spectrophotometric detection method. Type and amount of IL and the surfactant, temperature and the other parameters were optimized. Under the optimum conditions, the limit of detection of the method was 0.3 ng mL^{-1} and the relative standard deviation was 1.32% for 30 ng.mL^{-1} mercury.

Table 5. Reviews for solid phase microextraction using ionic liquid as acceptor phase

Analyte	Ionic liquid	Sample	Remark	Ref.
	trihexyltetrad cylphos-phoniu m, N-butyl-4-methylphyridinium tetrafluoroborate		Ionic liquid-mediated sol-gel hybrid organic-inorganic materials present enormous potential for effective use in analytical microextraction; Detection limits ranging from 4.9 to 487ng.L^{-1}.	76
Polycyc lic aromatic hydroc arbons	[C_6mim] Br, 1-hexade-cyl-3- butylimidazolium bromide and 1,3-didode-cylimidazolium bromide	sediment BCR-535	GC-MS was employed for the analysis; an aqueous ionic liquid aggregate was used as acceptor; the extraction tube was introduced into the microwave cavity.	77
Esters	polymeric ionic liquids	Water	For the fist time polymeric ionic liquids was used as fiber coatings for extraction; GC was employed for analysis; long lifetime and high thermal stability of the polymeric ionic liquid solid phase microextraction coatings may provide them particular advantages in GC-MS applications involving highly selective ester and FAME extractions from complex	78
benzene, toluene, ethylbe nzene, and xylenes	[C_8mim] [PF_6]	Paint	GC was employed for analysis; the SPME fiber was coated with ionic liquid prior to every extraction; an extraction time of 30min was selected; an obvious salting-out effect was observed well.	79

ethanol	[C_4mim]Cl and [C_2mim] acetate		The application of ionic liquids as nonderivatizing solvents for the pretreatment and regeneration of cellulose is a growing area of research; GC was employed for analysis.	80
nonane, methyl caproate, ocresol, secphenet hylalcohol	six dicationic ionic liquids	water	GC was employed for analysis; the partitioning behavior of six imidazolium-based ionic liquid aggregates in water has been studied for the first time. Monocationic and dicationic ionic liquids-aggregates behave as traditional surfactant systems; the obtained partition coefficients show that hydrophobic analytes are preferably extracted with ionic liquids-aggregates compared to more polar analytes.	81
4-Chloro phenol, 3-chorop henol, 2,4-dichlor opheno l and 2,4,6- trichlor opheno l	[C_8mim] [PF_6]	water	HPLC was employed for analysis; a 60 min of extraction time was required; an obvious salting-out effect was observed.	82
Aromatic compo-unds	[C_{16}mim]Br, 1-hexade-cyl-3-butylimidazolium bromide, 1,3-didodecylim-idazolium bromide	water	HPLC was employed for analysis.	83
Polycyclic aromatic hydroc arbons	[C_8mim] [PF_6] 1-benzyl-3- methylim-idazolium trifluo-romethanesulfonate:and 1-methyl-3-phenylpro-pylimidazoliumtrifluo-romethanesulfonate	water	nafion membrane was used to support the ionic liquid; GC-MS was employed for analysis.	84

shortchain alcohols, polar and basic amines	[C_4mim] [PF_6]	water	two ionic liquids contained styrene units that allowed for polymerization and higher carbon content of the bonded silica particles; The sensitivities of the two ionic liquid fibers differing by the anion were similar. Their efficacy compares favorably to that of commercial fibers for polar analytes. The mechanical strength and durability of the polymeric ionic liquid fibers were excellent.	87
Metha mpheta mine and amphet amine	1-Ethoxyethyl-3-methylimidazloium bis(trifluoromethane sulfonyli mide	human urine	combined with gas chromatography/mass spectrometry working in selected ion monitoring mode, the new method showed good linearity in the range of 20-1500µg/L and low detection limits (0.1µg.L^{-1} for MAP and 0.5 µg.L^{-1} for AP.	88
polycyclic aromatic hydroc arbons	[C_4mim] [PF_6]	mosquito coil incense	A simple and cost effective solid-phase microextraction device has been developed	89
Sulfadi azine, sulfame razine sulfame thazine, sulfadi methoxine and sulfame thoxazole	[C_4mim] [PF_6]	water	By using ionic liquid as membrane liquid and tri-n-octylphosphine oxide as additive, hollow fiber supported liquid phase microextraction was developed for the determination of five sulfon-amides in environmental water samples by high performance liquid chromatography with ultraviolet detection; Detection limits are between 0.1 and 0.4 mug.L^{-1}.	107

Acetoni trile	Bonded ionic liquid polymeric material		GC was employed for detection.	114
CO_2	poly(1-vinyl-3-hexylimidazolium)-bis[(trifluoromethyl) sulfonyl]i mide and poly(1-vinyl-3-hexylimidazolium) taurate		GC was employed for detection of CO_2.	118
polycyclic aromatic hydroc arbons and substit uted phenols	poly(1-vinyl-3-hexadecylimidazolium) bis((trifluoromethyl sulfonyl)i mide		GC-MS was employed for analysis, detection limits oscillate from 0.005 ng.mL^{-1} for fluoranthene to 4.4 ng/mL for 4-chloro-3-methylphenol.	119
lead and nickel	$[C_8mim][PF_6]$		Ammonium pyrroldin-edithiocarbamate was used as ligand. ETAAS was employed for detection lead and nickel, the detection limits of the method were 0.03 and 0.02 mug.mL^{-1}, for Ni and Pb, respectively	123

4. THE COMBINATION OF ANALYTICAL TECHNOLOGIES WITH THE EXTRACTION

Many modern analytical technologies have been applied to couple with microextraction using ionic liquid as extracting media for different analytical purpose such as AAS for metal ions, HPLC, GC and GC-MS for organic compounds. When ionic liquids are employed as medium in various LPME and SPME, HPLC is preferred to GC as separation technique since this ionic liquid is incompatible with GC due to the non-volatility of the ionic liquid. Consequently, the utilizations of ionic liquid in GC are limited to their use as stationary phases, as coaters of the surface of fibers in solid-phase microextraction or as solvents in various liquid phase microextraction including SDME and DLME etc. In order to overcome the problem, some effective attempts have been reported recently in the literatures. For example, Miguel Valcarcel group has successfully developed a new interface for the direct coupling between ionic liquid-SDME and GC-MS [65] (shown in Fig.4). Using this interface, no ionic liquid can reach the GC system dirtying the chromatographic column and a proper volatilization and subsequent transference of the substances to the GC is achieved. Therefore, the advantages of the use of ionic liquids with SDME, avoiding the irreproducibility associated with the use of organic solvents to form the drop, are added to the analytical possibilities of GC-MS. Furthermore, using the interface analytes of an extensive range of polarity/volatility can be determined on account of the fact that no solvent delay is necessary in the detection step. Moreover, Zeng Baizhao group also provided anther scheme to avoid ionic liquid leaking into the chromatographic column (shown in Fig.5). In the scheme through placing a small glassy tube in the sample injection part, ionic liquid no long enters into the chromatographic column, thus ionic liquid based microextraction technique can be coupled with GC. To evaluate the proposed method, five chlorobenzene derivatives are used as model. Results show that the chlorobenzenes can be extracted effectively by the ionic liquid and the extractant exhibits good chromatographic behavior. The linear ranges are up to 1.5mg.L^{-1} and the detection limits range from 0.1 to 0.5 μg.L^{-1}. As ionic liquids have many unique properties, ionic liquid based microextraction technique coupled with GC has potential application in the detection of pollutants etc [86]. In our laboratory, above two schemes have been investigated too, the results exhibited above two schemes can not avoid the ionic liquid entering into chromatographic column.

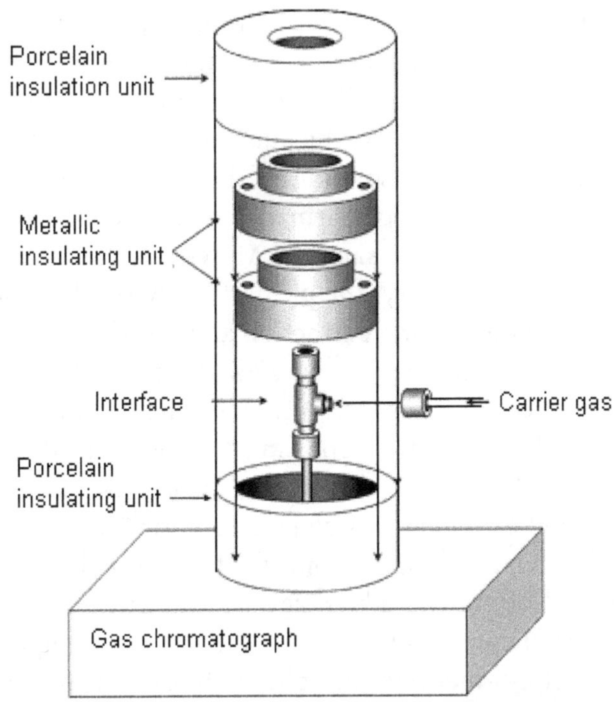

Fig. 4. Scheme of the household unit of the interface for the direct introduction of analytes extracted in a drop of ionic liquid into a gas chromatograph

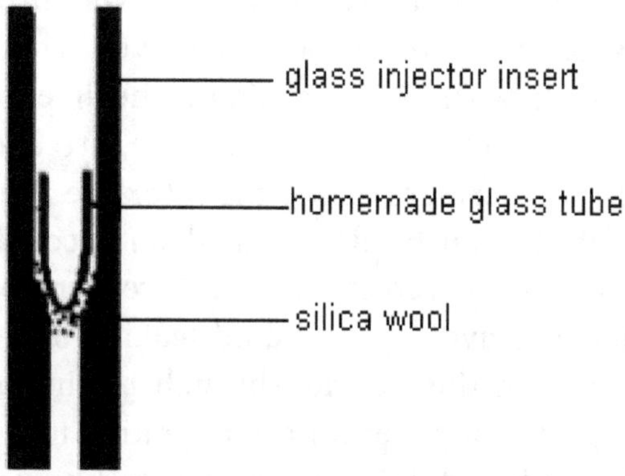

Fig. 5. The modified sample injection part of GC instrument. The dimension of the homemade tube: outer diameter, 2.0 mm; length, about 15 mm

5. EFFECT OF SOLVENT PROPERTIES ON THE EXTRACTION PERFORMANCES

Although several 1-alkyl-3-methylimidazolium hexafluorophosphate room temperature ionic liquids, abbreviated as [Cnmim][PF6], have been widely developed as friendly acceptor phase for various extractions in literatures, their application also suffers from many drawbacks. Firstly, these ionic liquids are difficult to be dispersed into aqueous phase and tend to tightly adhere on vessel wall due to high viscosity. The special property results that a long heating process (or mechanical agitation) and extract operation were required strongly in order to obtain good extraction efficiency. Secondly, most of the ionic liquids have relatively high water-solubility, the aqueous phase in the extraction system contains large amounts of imidazolium and hexafluorophosphate. The salting-out will effect on the extraction performance and following analysis in present of high ionic strength for complex samples. Moreover, the imidazolium dissolved in water is of good surface activity, the property may result in producing serious emulsification phenomenon during extraction and prolong phase separation time. In order to resolve above problems, some new ionic liquids have been designed, synthesized and applied to microextraction in the laboratory. For examples, 1,3- diisooctylimidazolium hexafluorophosphate (shown in Fig.6) was developed as a novel solvent for DLPME. In this study, dicofol and its three degradation products were selected as model compounds to optimize the factors relevant to the extraction and evaluate the applicability of the method. Experimental showed the proposed method offers the advantage of simplicity of operation, rapidity, high extraction efficiency and enrichment factor. The temperature-controlled ionic liquid 1-isooctyl-3-methylimidazolium hexafluorophosphate has been developed as novel medium for DLPME. This new mode of LPME with temperature-controlled ionic liquid increases the rapidity due to good hydrophobic property and relatively high melting point of the ionic liquid, the stability of extraction because of the larger volume of extraction solvent that may be used, enhances the sensitivity because of the much larger contact area than that of conventional LPME and SPME. Comparing with room temperature ionic liquid, proposed method has all performance during liquid/liquid and co-precipitation extraction, it can extract various polar or non-polar analyte from aqueous phase into the ionic liquid with high extraction efficiency. Moreover, new extraction can easily avoid the degradation of the pesticide during enrichment step due to no use a long heating time. Thirdly, the incompatibility between the extraction with ionic liquid and GC or GC-MS has become a problem in analytical chemistry due to ionic liquid's decomposition and leakage. Recently, Zhao group carried out a beneficial attempt to resolve above difficulty [78]. In this study, a new class of stationary phase coatings for SPME based on polymeric ionic liquids is presented. Proposed SPME coatings exhibit exceptional film stability, high thermal stability, reproducible extraction efficiencies, and long lifetimes. Moreover, the fiber coatings were used to extract esters and fatty acid methyl esters from aqueous solution followed by gas chromatographic separation and flame ionization detection. To examine the effect of the matrix on the coatings, extractions were carried out in a synthetic wine solution followed by recovery experiments in two real wine samples.

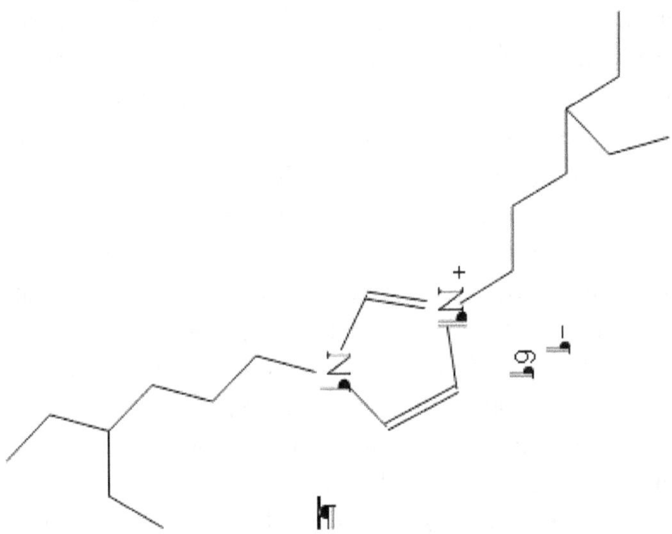

Fig. 6. Molecular structure of 1,3-diisooctylimidazolium hexafluorophosphate

When carrying out the extractions in aqueous solution, the detection limits for most analytes ranged from 2.5 to 50mug.L^{-1} whereas lower detection limits were obtained for larger fatty acid methyl esters. Recovery experiments carried out in red and white wines ranged from 70.2 to 115.1% using the poly(ionic liquid) fibers compared to 61.9 to 102.9% using a commercial polydimethylsiloxane fiber of similar film thickness. The structural tuning capability of these new coating materials makes them widely amendable to performing task- specific microextractions.

In selection of solvents for extraction and extractive distillation processes knowledge of the activity coefficients at infinite dilution (γ^∞) is very important and useful. A large number of works concerns measurements of activity coefficients at infinite dilution of organic solvents in ionic liquids. In separation processes both properties of extractant are important, namely selectivity S and capacity k which can be directly calculated from activity coefficients at infinite dilution for different separation problems. Activity coefficients at infinite dilution are also very helpful for characterizing the behaviour of liquid mixtures, estimation of mutual solubilities, fitting the excess molar energy (GE) model parameters, predicting the existence of an azeotrope, analytical chromatography, calculation of Henry constant and partition coefficients, development of thermodynamic models based on the group contribution methods such as mod. Typical solvents used in separation of aromatic hydrocarbons from aliphatic hydrocarbons are polar components such as sulfolane and N- methylpyrrolidone (NMP). Replacement of these volatile solvents in the separation processes by the nonvolatile ionic liquids can offer advantages such as less complex processes, more simple regeneration of the extractant using techniques as simple as flash distillation or stripping and the environmental protection. Extraction of aromatics from aromatic/aliphatic mixture with ionic liquids is expected to require less process steps and less energy consumption than extraction with conventional solvents due to negligible vapor pressure of ionic liquids. Andrzej Marciniak has made the analysis of cation and anion structure of the ionic liquid and effect of the temperature on the selectivity and the capacity for aliphatics/aromatics and n-hexane/hex-1-ene separation problems [132]. It was found the highest values of selectivity show ionic liquids with less aliphatic character of the anion and the cation, e.g. based on following cations 1,3-dimethylimidazolium, 1-ethyl-3- methylimidazolium, 1-ethylpyridinium, triethylsulfonium and with –C≡N group in the structure, like [C3mim]CN. Unfortunately always when the ionic liquid reveals high values of the selectivity, the capacity takes low values. Therefore a selection of proper anion and cation is required. The highest values of capacity have Tf_2N^- and $FeCl_4$ anions. As was shown most of ionic liquids with high values of both selectivity and capacity is based on Tf_2N^- anion. 1-Butyl-4-methyl-pyridinium bis(trifluoromethylsulfonyl)amide have higher values of selectivity than sulfolane and NMP and much higher value of capacity than these two mentioned entrainers for the n-hexane/benzene separation. The highest selectivity for the n-hexane/ benzene separation has [C_4mim][SCN] above 4 times higher than for sulfolane with a slightly better capacity. There are a large number of ionic liquids with selectivity and capacity better than sulfolane. A large number of ionic liquids have comparable or better both of selectivity and capacity values than NMP, e.g.: [1-ethyl-3-methyl-imidazolium bis(trifluoromethylsulfonyl)amide, triethyl-sulfonium bis(trifluoromethylsulfonyl)-amide, 1- ethyl-2,3-dimethylimidazolium bis(trifluoromethylsulfonyl)amide, hexyl-trimethyl- ammonium bis(trifluoromethylsulfonyl)amide, tributylmethylphosphonium methylsulfate, 1-hexyl-oxymethyl-3-methylimidazolium bis(trifluoromethylsulfonyl)amide and [C_6mim][Tf_2N]. For n-hexane/hex-1-ene the 3-(3-(Acryloyloxy)propyl)-1-methyl- imidazolium bromide has the highest selectivity of 15.2. For this separation process majority of ionic liquids have selectivity on the level of 2. To improve selectivity and capacity lower temperature is favourable.

From the values of selectivities and capacities, obtained for the large number of ionic liquids, it can be concluded that majority of ionic liquids may replace conventional entrainers applied for the separation processes of aliphatic/aromatic hydrocarbons and alkanes/alkenes.

6. EXTRACTION MECHANISM

Numerous investigations pertain to extraction mechanisms, primarily including solvent ion pair extraction, ion exchange, and their simultaneous combination. The ion exchange mode was extensively utilized to analyze separation performance. Generally, in the ion exchange mode, metal ions are anticipated to displace the imidazolium cation or the anion associated with the ionic liquid in the ionic liquid phase. Dietz's group has proposed that elevated distributions in the ionic liquid/aqueous system resulted from a modification in extraction equilibria relative to the octanol/aqueous system. In octanol, Sr2+ is extracted as a strontium nitratocrown ether complex, while in ionic liquids, the two axially coordinated nitrates are replaced by two water molecules associated with the metal ion. The charged species Sr(CE)(H2O)2, where CE denotes dicyclohexano-18-crown-6-ether, is transferred to the ionic liquid phase, with system neutrality preserved by the migration of two imidazolium cations to the aqueous phase for each extracted metal ion. The loss of ionic liquid components in the aqueous phase is a significant restriction for the practical application of ionic liquids in separations and enrichments. [143]. However, the nature of the ionic liquid is not the only factor which determines the metal extraction pathway, ligand type also has a very important contribution. Professor Jensen has suggested La^{3+} was shown to follow an anion exchange mechanism when extracted in an ionic liquid/aqueous system, in contrast to the cation exchange observed in molecular solvents. Here, changing the lipophilicity of the ionic liquid did not change the partitioning mechanism [144]. Recently, the ion exchange theory was also applied to explain the extraction performance of amino acids using a room temperature ionic liquid 1butyl3methylimidazolium hexafluorophosphate with dicyclohexano-18-crown-6 system[145]. The indication of an ion exchange mechanism is the significant leakage of the dialkylimidazolim cation into the aqueous phase on extraction of amino acids-confirmed by monitoring the corresponding UV band. After extraction of amino acids the clear absorption peak of the dialkylimidazolium cation appeared in the spectrum of the aqueous phase at ca. 300 nm. Increasing the initial amino acid concentration in the aqueous phase from 1×10^{-3} to 1×10^{-2} mol L^{-1} results in an increase of the dialkylimidazolium concentration in water after extraction. Clearly this is caused by ion exchange of the dialkylimidazolium cation with the amino acid cation. Thus, the absence of any appreciable anion effect and distribution of dialkylimidazolium cation into water, which accompanies extraction, strongly suggest that extraction of amino acids with $DC_{18}C_6$ into the ionic liquid is an ion exchange process. However, above all extraction mechanism mode did not consider the extraction functions of the surface charge on the ionic liquid phase and the ionic liquid dissolved in water. In our previous work, the surface tension and Zeta potential of the ionic liquid solution were investigated in detail in order to understand the extraction performance in the room temperature ionic liquid $[C_4tmsim][PF_6]$ with CDAA system. It was found the surface tension of the solution decreased with increase of the concentration of $[C_4tmsim][PF_6]$ solution, its critical micelle concentration and surface tension of saturated $[C4tmsim][PF_6]$ solution are 1.59×10^{-3} mol L^{-1} and 65.95 Nm/m, respectively. The data showed ionic liquid has surface activity, this property allow the ionic liquid to fully interact with the complex or CDAA in aqueous phase and easy to extract the complex into the ionic liquid phase due to micelle extraction function, which is unique property of surfactant. On the other hand, the Zeta potential value of saturated $[C_4tmsim][PF_6]$ solution was found to be +2.1mV. The result showed the double electron layer at the $[C_4tmsim][PF_6]$ water interface existed the positive charge. The positive charge will produce strong interaction between Hg-CDAA complex or CDAA species and ionic liquid surface. Above two factors allow ionic liquid has a higher the extraction efficiency than ordinary organic solvent. It should be stated that Zeta potential may be positive or negative value for different ionic liquid, that decide various ionic liquid has different extraction performance. In order to test suggested extraction mechanism, the extraction efficiencies of CDAA in different acidity was researched in detail. The result is very interesting that the order of the extraction efficiency of CDAA is in base > neutral > acid medium. Because CDAA is a weak organic acid, its dissociation product three reagent species in water, these are RH_2, RH^- and R^{2-} respectively, in which CDAA was replaced by RH_2 in order to simple. In strong base solution, reagent specie is R_2 mainly, it is easy to extract into ionic liquid due to its strong negative electric charge, the order accord with suggested extraction mechanism [2]. Recently, professor Yan

Yongsheng suggested a new extraction mechanism by use of ionic liquid-salt aqueous two phase system (shown in Fig.7). Aqueous two-phase system consisting of a hydrophilic ionic liquid (1-butyl-3- methylimidazolium tetrafluoraborate, [C_4mim][BF_4]) and Na_2CO_3, which is a novel, simple, non-toxic and effective sample pretreatment technique coupled with molecular fluorescence spectrophotometry, was developed for the simultaneous separation, enrichment and rapid analysis of roxithromycin. The extraction yield of roxithromycin (ROX) in [C_4mim][BF_4]-Na_2CO_3 is influenced by the types of salts, concentrations of Na_2CO_3 and [C_4mim][BF_4], as well as the extracting temperature. The creation mechanism of the ionic liquid-salt aqueous two-phase system was examined by hydration theory, while the extraction mechanism of [C4mim][BF4]-salt was analyzed using FT-IR and UV-vis spectroscopy. The findings indicate that no chemical bonding interactions occur between the ionic liquid and roxithromycin (ROX), and the intrinsic characteristics of ROX remain unchanged [146]. The extraction performance of ionic liquids is intricate due to their complex solvent properties; thus, the theoretical models regarding the extraction mechanism require enhancement and validation through dependable experimental techniques in the future.

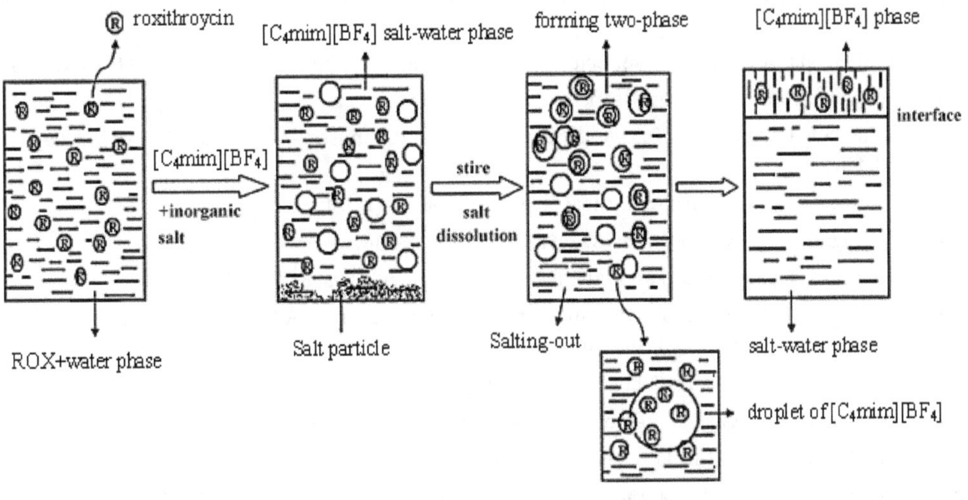

Fig. 7. The schematic diagram of separating ROX in ionic liquid-salt aqueous two phase system

REFERENCES

H.X. Shan, Z.J. Li, Spectroscopy and Spectral Analysis, 2008, 28, 214-217.

Z.J. Li, Q. Wei, R. Yuan, X. Zhou, H.Z. Liu, H.X. Shan, Q.J. Song, Talanta, 2007, 71, 68-72.

Z.J. Li, J. Chang, H.X. Shan, J.M. Pan, Reviews in Analytical Chemistry, 2007, 26, 109-153.

Z.J. Li, Q.P. Peng, H.X. Shan, Journal of AOAC International, 2007, 90, 1191-1196.

Z.J. Li, N.P. Lu, X. Zhou, Q.J. Song, Journal of Pharmaceutical and Biomedical Analysis, 2007, 43, 1609-1614.

H.X. Shan, Z.J. Li, M. Li, Microchimica Acta, 2007, 159, 95-100.

K. Sekiguchi, M. Atobe, T. Fuchigami, Electrochemistry communications, 2002, 4, 881-885.

I.A. Ansari, R. Gree, Organic letters, 2002, 4, 1507-1509.

H.X. Shan, Z.J. Li, M. Li, G.X. Ren, Y.J. Fang, Journal of Chemical Technology and Biotechnology, 2008, 83, 886-891.

S.K.Chun, S.V.Dzyuba and R.A.Bartsch, Anal. Chem., 2001, 73(15), 3737.

A.E.Visser, W.M.Swatloski, S.T.Griffin,D.H.Hartman and R.D.Rogers, Sep. Sci. Technol., 2001, 36, 785.

V.A.Cocalia, J.D.Holbrey, K.E.Gutowski, N.J.Bridges and R.D.Rogers, Tsinghua Science and Technology,2006, 11(2), 188.

A.Visser, R. Swatloski, W. Matthewreichert, Rebeccamayton, Seansheff, Andrzejwierzbicki, J.R. Jamesh Davis and R.D.Rogers, Environmental Science and Technology, 2002, 36 (11), 2523.

D. Shayesteh, S. Ali Mohammad Haji, B. Mahboubeh Shirani, J. Abbas Ali, Journal of Hazardous Materials, 2010, 173 (1-3), 534-538.

V.M. Egorov, D.I. Djigailo, D.S. Momotenko, D.V. Chernyshov, I.I. Torocheshnikova, S.V. Smirnova, I.V. Pletnev, Talanta, 2010, 80(3), 1177-1182.

N. Li, G.Z. Fang, B. Liu, J. Zhang, L.J. Zhao, S. Wang, Analytical Sciences, 2010, 26(4), 455-459.

R. Lertlapwasin, N. Bhawawet, A. Imyim, S. Fuangswasdi, Separation and Purification Technology, 2010, 72(1), 70-76.

H. Abdolmohammad-Zadeh, E. Ebrahimzadeh, Central Europen Journal of Chemistry, 2010, 8(3), 617-625.

N. Pourreza, K. Ghanemi, Journal of Hazardous Materials, 2010, 178 (1-3), 566-571.

H. Okamura, N. Hirayama, K. Morita, K. Shimojo, H. Naganawa, H. Imura, Analytical Sciences, 2010, 26 (5), 607-611.

C. Xu, X.H. Shen, Q.D. Chen, H.C. Gao, Science in China Series B-Chemistry, 2009, 52 (11), 1858-1864.

A. Rout, K.A. Venkatesan, T.G. Srinivasan, P.R.V. Rao, Radiochimica Acta, 2009, 97(12), 719-725.

J.S. Wang, C.N. Sheaff, B. Yoon, R.S. Addleman, C.M. Wai, Chemistry-A Europan Journal, 2009, 15(17), 4458-4463.

K. Kidani, N. Hirayama, H. Imura, Analytical Sciences, 2008, 24 (10), 1251-1254.

A. Kamio, Y. Nagaosa, Analytical Sciences, 2008, 24 (10), 1363-1367.

Y. Zuo, J. Chen, D.Q. Li, Separation and purification Technology, 2008, 63 (3), 684-690.

P. Giridhar, K.A. Venkatesan, S. Subramaniam, T.G. Srinivasan, P.R.V. Rao, Journal of Alloys and Compounds, 2008, 448 (1-2), 104-108.

X.Q. Sun, B. Peng, J. Chen, D.Q. Li, F. Luo, Talanta, 2008, 74 (4), 1071-1074.

K.V. Lohithakshan, S.K. Aggarwal, Radiochimica Acta, 2008, 96 (2), 93-97.

N. Hirayama, H. Okamura, K. Kidani, H. Imura, Analytical Sciences, 2008, 24 (6), 697-699.

Kozaka, K. Fulinaga, J. Noro, Y. Watanabe, Y. Komatsu, Bunseki Kagaku, 2008, 57(8), 593-597.

P.Y. Chen, Electrochimica Acta, 2007, 52 (17), 5484-5492.

N.V. Prondzinski, A. Babai, A.V. Mudring, K. Merz, Zettschrift Fur Anorganische Und Allgemeine Chemie, 2007, 633 (9), 1490-1492.

Q. Sun, D.B. Wu, J. Chen, D.Q. Li, Journal of Chemical Technology and Biotechnology, 2007, 82 (3), 267-272.

N. Hirayama, M. Deguchi, T. Honjo, Solvent Extraction Research and Devlopment- Japan, 2006, 13, 83-88.

P. Giridhar, K.A. Venkatesan, T.G. Srinivasan, P.R.V. Rao, Hydrometallurgy, 2006, 81(1), 30-39.

H.L. Huang, H.P. Wang, G.T. Wei, T.W. Sun, J.F. Huang, Y.W. Yang, Environmental Science and Technology, 2006, 40 (15), 4761-4764.

K. Shimojo, H. Naganawa, F. Kubota, M. Goto, Chemicstry Letters, 2006, 35(5), 484-485.

X. Zhou, Z.J. Li, R. Yuan, H.Z. Liu, Analytical Letters, 2006, 39(5), 863-877.

V.A. Cocalia, M.P. Jensen, J.D. Holbrey, S.K. Spear, D.C. Stepinski, R.D. Rogers, Dalton Transactions, 2005, (11), 1966-1971.

P. Giridhar, K.A. Venkatesan, T.G. Srinivasan, P.R.V. Rao, Journal of Radioanalytical and Nuclear Chemistry, 2005, 265 (1), 31-38.

S. Mekki, C.M. Wai, I. Billard, G. Moutiers, C.H. Yen, J.S. Wang, A. Ouadi, C. Gaillard, P. Hesemann, Green Chemistry, 2005, 7 (6), 421-423.

G.T. Wei, J.C. Chen, Z.S. Yang, Journal of the Chinese Chemical Society, 2003, 50(6),1123-1130.

M. Matsumoto, Y. Inomoto and K. Kondo, J. Membrane Sci., 2005, 246, 77.

S.D.Qi, S.Y.Cui, X.G.Chen and Z.D.Hu, J. Chromatogr. A, 2004, 1059, 191.

L.A. Blanchard and J.F. Brennecke, Ind. Eng. Chem. Res., 2001, 40, 287.

A.M. Scurto, S.N.V.K. Aki and J.F. Brennecke, J. Am. Chem. Soc.,2002, 124, 10276.

J.F. Brennecke, L.A. Blanchard, J.L. Anthony, Z.Y. Gu, I. Zarraga and D.T. Leighton, Clean solvents, ACS Symp. Ser.,2002, 819, 82.

M. Solinas, A. Pfaltz, P.G. Cozzi and W. Leitner, J. Am. Chem. Soc.,2004, 126, 16142.

M. Cruz-Vera, R. Lucena, S. Cardenas, M. Valcárcel, Journal of Chromatography A, 2008, 1202, 1-7.

C. Basheer, A.A. Alnedhary, B.S. Madhava Rao, R. Balasubramanian, H.K. Lee, Journal of Chromatography A, 2008, 1210, 19-24.

X.M. Fu, S.G. Dai, Y. Zhang, International Journal of Environmental Analytical Chemistry, 2006, 86, 985-993.

Cruz-Vera, R. Lucena, S. Cardenas, M.Valcarcel, Journal of Chromatography B, 2009, 877, 37-42.

H.X. Xie, L.J. He, Y. Wu, K. Lu, X.Z. Si, Chinese Journal of Analytical Chemistry, 2007, 35, 187-190.

J.F. Liu, Y.G. Chi, G.B. Jiang, Journal of Separation Science, 2005, 28, 87-91.

L. Vidal, E. Psillakis, C.E. Domini, N. Grane, F. Marken, Analytica Chimica Acta, 2007, 584, 189-195.

J.F. Liu, J.F. Peng, Y.G. Chi, G.B. Jiang, Talanta, 2005, 65, 705-709.

C.L. Ye, Q.X. Zhou, X.M. Wang, Analytica Chimica Acta, 2006, 572, 165-171.

F.Q. Zhao, J. Li, B.Z. Zeng, Journal of Separation Science, 2008, 31, 3045-3049.

C.L. Ye, Q.X. Zhou, X.M. Wang, J.P. Xiao, Journal of Separation Science, 2007, 30, 42-47.

Eva Aguilera-Herrador, Rafael Lucena, Soledad Cardenas, Miguel Valcarcel. Journal of Chromatography A, 2008, 1209, 76-82.

J.F. Peng, J.F. Liu, G.B. Jiang, C. Tai, M.J. Huang, Journal of Chromatography A, 2005, 1072, 3-6.

L. Vidal, A. Chisvert, A. Canals, A. Salvador, Journal of Chromatography A, 2007, 1174, 95-103.

E. Aguilera-Herrador, R. Lucena, S. Cardenas, M. Valcarcel, Analytical Chemistry, 2008, 80, 793-800.

E. Aguilera-Herrador, R. Lucena, S. Cardenas, M.Valcarcel, Journal of Chromatography A, 2008, 1201, 106-111.

L.B. Xia, X. Li, Y.L. Wu, B. Hu, R. Chen, Spectrochimica Acta Part B, 2008, 63, 1290-1296.

L. Vidal, C.E. Domini, N. Grane, E. Psillakis, A. Canals, Analytica Chimica Acta, 2007, 592, 9-15.

J.L. Manzoori, M. Amjadi, J. Abulhassani, Talanta, 2009, 77, 1539-1544.

Q.X. Zhou, C.L. Ye, Microchimica Acta, 2008, 162, 153-159.

Q.X. Zhou, J.P. Xiao, C.L. Ye, X.M. Wang, Chinese Chemical Letters, 2006, 17, 1073-1076.

J.F. Liu, Y.G. Chi, G.B. Jiang, C. Tai, J.F. Peng, J.T. Hu, Journal of Chromatography A, 2004, 1026, 143-147.

Q.X. Zhou, H.H. Bai, G.H. Xie, J.P. Xia, Journal of Chromatography A, 2008, 1188, 148- 153.

Y. Liu, E.C. Zhao, W.T. Zhu, H.X. Gao, Z.Q. Zhou, Journal of Chromatography A, 2009, 1216, 885-891

M. Gharehbaghi, F. Shemirani, M. Baghdadi, International Journal of Environmental Analytical Chemistry, 2009, 89, 21-33.

A.M. Shearrow, G.A. Harris, L. Fang, P.K Sekhar, L.T. Nguyen, E.B. Turner, S. Bhansali, Journal of Chromatography A, 2009, 1216, 5449-5458.

L. Guerra-Abreu, V. Pino, J.L. Anderson, A.M. Afonsoa, Journal of Chromatography A, 2008, 1214, 23–29.

F. Zhao, Y.J. Meng, J.L. Anderson, Journal of Chromatography A, 2008, 1208, 1-9.

J.F. Liu, N. Li, G.B. Jiang, J.M. Liu, J.A. Jonsson, M.J. Wen, Journal of Chromatography A, 2005, 1066, 27-32.

F. Zhao, T.K. Ponnaiyan, C.M. Graham, C.A. Schall, S. Varanasi, J.L. Anderson, Analytical Bioanalytical Chemistry, 2008, 392, 1271-1275.

V. Pino, Q.Q. Baltazar, J.L. Anderson, Journal of Chromatography A, 2007, 1148, 92-99. [82] J.F. Peng, J.F. Liu, X.L. Hu, G.B. Jiang, Journal of Chromatography A, 2007, 1139, 165-170.

C. Yao, V. Pino, J. L. Anderson, Journal of Chromatography A, 2009, 1216, 948-55.

Y.N. Hsieh, P.C. Huang, I.W. Sun, T.J. Whang, C.Y. Hsu, H.H. Huang, C.H. Kuei, Analytica Chimica Acta, 2006, 557, 321-328.

M. Baghdadi, F. Shemirani, Analytica Chimica Acta, 2008, 613, 56-63.

F.Q. Zhao, S. Lu, W. Du, B.Z. Zeng, Microchim Acta, 2009, 165, 29-33.

E. Wanigasekara, S. Perera, J.A. Crank, L. Sidisky, R. Shirey, A. Berthod, D.W. Armstrong, Analytical and Bioanalytical Chemistry, 2010, 396, 511-524.

Y. He, J. Pohl, R. Engel, L. Rothman, M. Thomas, Journal of Chromatography A, 2009, 1216, 4824-4830.

K.P. Huang, G.R. Wang, B.Y. Huang, C.Y. Liu, Analytica Chimica Acta, 2009, 645, 42-47.

M.H. Mallah, F. Shemirani, M.G. Maragheh, Environmental Science and Technology, 2009, 43, 1947-1951.

M.T. Pena, M.C. Casais, M.C. Mejuto, R. Cela, Journal of Chromatography A, 2009, 1216, 6356-6364.

H. Abdolmohammad-Zadeh, G.H. Sadeghi, Analytica Chimica Acta, 2009, 649, 211-217.

M. Cruz-Vera, R. Lucena, S. Cardenas, M. Valcarcel, Journal of Chromatography A, 2009, 1216, 6459-6465.

S.Q. Li, S. Cai, W. Hu, H. Chen, H.L. Liu, Spectrochimica Acta Part B-Atomic Spectroscopy, 2009, 64, 666-671.

H.H. Bai, Q.X. Zhou, G.H. Xie, J.P. Xiao, Analytica Chimica Acta, 2009, 651, 64-68.

L.M. Ravelo-Perez, J. Hernandez-Borges, M. Asensio-Ramos, M. Angel Rodriguez- Delgado, Journal of Chromatography A, 2009, 1216, 7336-7345.

Y.C. Fan, M.L. Chen, C. Shen-Tu, Y. Zhu, Journal of Analytical Chemistry, 2009, 64, 1017-1022.

C. Yao, J.L. Anderson, Analytical and Bioanalytical Chemistry, 2009, 395, 1491-1502.

Z.J. Li, P.P. Chen, C.P. Yu, Y.J. Fang, Z.Y. Wang, M. Li, H.X. Shan, Current Analytical Chemistry, 2009, 5, 324-329.

L.J. He, X.L. Luo, H.X.Xie, C.J. Wang, X.M. Jiang, K. Lu, Analytical Chimica Acta, 2009, 655, 52-59.

L.M. Ravelo-Perez, J. Hernandez-Borges, A.V. Herrera-Herrera, Analytical and Bioanalytical Chemistry, 2009, 395, 2387-2395.

P. Berton, E. M. Martinis, L. D. Martinezc, R. G. Wuilloud, Analytica Chimica Acta, 2009, 640, 40-46

A. Chisvert, I.P. Roman, L. Vidal, Journal of Chromatography A, 2009, 1216, 1290-1295.

F. Pena-Pereira, I. Lavilla, C. Bendicho, L. Vidal, A. Canals, Talanta, 2009, 78, 537-541.

J.L. Manzoori, M. Amjadi, J. Abulhassani, Analytica Chimica Acta, 2009, 644, 48-52.

E. Aguilera-Herrador, R. Lucena, S. Cardenas, M. Valcarcel, Journal of Chromatography A, 2009, 1216, 5580-5587.

Y. Tao, J.F. Liu, X.L. Hu, H.C. Li, T. Wang, G.B. Jiang, Journal of Chromatography A, 2009, 1216, 6259-6266.

H. Sheikhloie, M. Saber-Tehrani, P. Abrumand-Azar, S. Waqif-Husain, Acta Chromatographica, 2009, 21, 577-589.

S.L. Wang, L.P. Ren, C.Y. Liu, J. Ge, F.M. Liu, Analytical and Bioanalytical Chemistry, 2010, 397 (7), 3089-3095.

R.S. Zhao, X. Wang, J. Sun, S.S. Wang, J.P. Yuan, X.K. Wang, Analytical and Bioanalytical Chemistry, 2010, 397(4), 1627-1633.

P.Berton, RG. Wuilloud, Analytica Chimica Acta, 2010, 662 (2), 155-162.

Yousefi, Seyed Reza; Shemirani, Farzaneh, Analytica Chimica Acta, 2010, 669(1-2), 25-31.

RS. Zhao, X. Wang, J. Sun, S.S.Wang, J.P. Yuan, X.K.Wang, Analytical and Bioanalytical Chemistry, 2010, 397(4), 627-1633.

E. Wanigasekara, S. Perera, J.A.Crank, L. Sidisky, R. Shirey, A. Berthod, D.W. Armstrong, Analytical and Bioanalytical Chemistry, 2010, 396 (1), 511-524.

H. Xiao-Wen, J.Ye, L. Yan, Chinese Journal of Analytical Chemistry, 2010, 38 (5), 727-730.

Y Wang, JY. You, CL Bao, H.Q. Zhang, A.M. Yu, Y.Yu, Chinese Journal of Chemistry, 2010, 28 (5), 785-790.

M. Vaezzadeh, F. Shemirani, B. Majidi, Food and Chemical Toxicology, 2010, 48 (6), 1455-1460.

QC. Zhao, J.L.Anderson, Journal of Chromatography A, 2010, 1217 (27), 4517-4522.

J. Lopez-Darias, V. Pino, J.L. Anderson, C.M. Graham, A.M. Afonso, Journal of Chromatography A, 2010, 1217 (8), 1236-1243.

Q. Wang, H.D. Qiu, J. Li, X. Liu, S.X. Jiang, Journal of Chromatography A, 2010, 1217 (33), 5434-5439.

L.J. He, X.L. Luo, X.M. Jiang, L.B. Qu, Journal of Chromatography A, 2010, 1217 (31), 5013-5020.

P. Berton, E.M.Martinis, R.G. Wuilloud, Journal of Hazardous Materials, 2010, 176 (1-3), 721-728.

J. Abulhassani, J.L. Manzoori, M. Amjadi, Journal of Hazardous Materials, 2010, 176 (1- 3), 481-486.

Q.Y. Liu, Y.J. Liu, S.Z. Chen, Q.J. Liu, Journal of Separation Science, 2010, 33 (15), 2376-82.

R.S. Zhao, X. Wang, J. Sun, J.P. Yuan, S.S. Wang, X.K. Wang, Journal of Separation Science, 2010, 33 (12), 1842-1848.

H.F. Zhang, Y.P. Shi, Talanta, 2010, 82 (3), 1010-1016.

S. Mahpishanian, F. Shemirani, Talanta, 2010, 82 (2), 471-476.

L.Vidal, A.Chisvert, A. Canals, A. Salvador, Talanta, 2010, 81 (1-2), 549-555.

E.M. Martinis, P. Berton, J.C.Altamirano, U. Hakala, R.G. Wuilloud, Talanta, 2010, 80 (5), 2034-2040.

H. Chen, P. Du, J. Chen, S.H. Hu, S.Q. Li, H.L. Liu, Talanta, 2010, 81 (1-2), 176-179.

M. Amjadi, J.L. Manzoori, J. Abulhassani, Journal of AOAC International, 2010, 93 (3), 985-991.

Marciniak, Fluid Phase Equilibria, 2010, 294, 213-233.

M.L. Dietz and J.A.Dzielawa, Chem. Commun.,2001, 20, 2124.

P. Jensen, J.Neuefeind, J.V.Beitz, S.Skanthakumar and L.Soderholm, J. Am. Chem.Soc.,2003, 125, 15466.

V.S. Svetlana, I.T.Irina, A.F. Andrey and V.P.Igor, Anal. Bioanal. Chem.,2004, 378, 1369.

C.X. Li, J. Han, Y. Wang, Y.S. Yan, X.H. Xu, J.M. Pan, Analytica Chimica Acta, 2009, 653, 178-183.

Chapter 9

EXEMPLARY THERAPIES UTILIZING IONIC LIQUIDS

1. INTRODUCTION: AN OVERVIEW

The term ionic liquid (IL) currently refers to a diverse category of semi-organic salts or salt combinations that consist solely of ions and remain liquid within the temperature range of 180-600 K. They are hydrogen-bonded compounds with robust interionic interactions that determine their crystallographic structure. The existence of at least one constituent ion possessing a delocalized charge inhibits the establishment of a stable crystal lattice (Palacio & Bhushan, 2008).

Although ionic liquids have been known since 1914, their proliferation began with the advent of air and moisture-stable variants. This second generation of ionic liquids integrates the distinctive characteristics of the previous version with enhanced manageability. Their most pertinent qualities are as follows: minimal vapor pressure, excellent thermal stability, adjustable viscosity and miscibility with water, inorganic and organic compounds, an extensive electrochemical window, elevated conductivity, high heat capacity, and appropriateness for reaction control.

The electronic and steric characteristics of the cation and anion determine their physicochemical properties. Their synthetic flexibility allows for the manipulation or adaption of specific features of the IL for a designated purpose. Density, viscosity, and solubility, which are essential in every extraction process, are among the more adjustable characteristics. Furthermore, the characteristics of cation and anion structures affect the acid-base and solvent properties, highlighting the distinctions from traditional molecular solvents. The prospective incorporation of functional groups that can engage in complimentary interactions with other solvents or molecules introduces new diversity in the characteristics and behavior of ionic liquids.

A vast array of potential cation-anion pairings exists, resulting in an exceptionally large number of distinct molecules. The predominant high molecular weight organic cations include imidazolium, pyridinium, pyrrolidinium, quaternary ammonium, and tetraalkylphosphonium. The predominant compounds are the 1,3-dialkylimidazolium salts, attributed to their readily customizable characteristics. Among the most favored organic anions, perfluorinated compounds (e.g., trifluoromethanesulphonate) and halogen-free alternatives (e.g., n-alkylsulphates) are frequently documented in the literature.

The presence of impurities significantly influences the properties of any ionic liquid, as trace amounts of chemicals from unreacted starting materials or water can substantially alter its characteristics. This phenomenon has resulted in irreproducibility among responses and methods. Surmounting such limitations is essential for the integration of ionic liquids in analytical science and their establishment as a viable alternative to conventional molecular organic solvents.

2. IONIC LIQUIDS IN ANALYTICAL CHEMISTRY

Ionic liquids can be successfully used in different steps of the (bio)chemical measurement process. In addition to selectivity and sensitivity improvement of existing methodologies, ILs can offer to the analytical scientists original alternatives to face up new applications.

Probably, their use in the preliminary steps is of advantage as they can markedly reduce the sample pretreatment step while improving its efficiency. This section presents a current picture of the contribution of ILs to improve the analytical process. It must be taken into consideration that the following sections will be exclusively devoted to the role of ionic liquids in sample preparation.

2.1 Extraction

Ionic liquids have been extensively used for the liquid-liquid extraction of metals (aided by complexation reagent), small organic molecules (hydrocarbons, PAHs) and also large biomolecules such as proteins and DNA thanks to hydrophobic interactions helped by electrostatic and also salting out effects. This wide applicability derives from the possibility of ad hoc design of ionic liquid used as extractant. In this way, the selectivity and efficiency can be modulated from the specific application as well as the solubility and miscibility between the extractant and the sample matrix. Several studies can be found in the literature dealing with this topic. For example, it has been established that the distribution coefficients for different metal cations were maximum for shorter 1-alkyl-3-methyl imidazolium ILs (C4 expected). On the contrary, the extraction of proteins is favored with the use of ILs presenting a long alkyl chain in the cation. Particularly interesting is the applicability of ionic liquids to break a variety of azeotropic systems and chiral separations.

Liquid-liquid extraction has been moved to the miniaturized version of the technique in order to overcome its negative connotations. As expected, ILs have played a crucial role in the development of several modalities of liquid phase microextraction (LPME). 1-octyl-3-methylimidazolium-PF_6 demonstrated excellent performance for the extraction of PAHs from waters in both, direct immersion and headspace LPME. Since then, ILs have been used in dynamic LPME, hollow fiber protected LPME, single drop microextraction and dispersive liquid-liquid microextraction.

The applicability of ILs in solid-phase microextraction (SPME) was established by Liu *et al.* (Liu *et al.*, 2005) for the determination of benzene, toluene, ethylbenzene and xylene in paints. The thermal stability, extraction capacity and the easy immobilization on fused silica capillaries are among the most remarkable advantages of this coating. Their usefulness has been demonstrated in both headspace and direct immersion approach.

All these extraction modalities will be deeply commented in section 4.

2.2 Chromatography

Room temperature ionic liquids have been used as novel stationary phases in a variety of chromatographic separations. The research in gas chromatographic uses of ILs has been leaded by Armstrong´s group, who studied the behavior of $[C_4MIM][PF_6]$ and $[C_4MIM][Cl]$ as coatings in fused silica capillary columns (Armstrong *et al.*, 1999). The most valuable features in this application field are the high viscosity and thermal stability together with the easy immobilization on the inner surface of the fused silica capillary. The versatile combinations of ILs permit the efficient separation of polar and non-polar compounds, although long retention times and tailing peaks can be obtained for alcohols and organic acids and even irreversible interaction of alkylamines has been reported. Notwithstanding this, the design of specific ionic liquids can overcome this limitation (Anderson & Armstrong, 2003).

In liquid chromatography, ILs can be used either as mobile phase modifiers or as stationary phases. The high viscosity and UV absorption of ILs limits their use as mobile phases. However, their presence at low concentrations (as additives) results in a decreased peak tailing, reduced band broadening and thus improved resolution. This effect is justified by a competitive reaction between imidazolium cations and the polar groups of the analytes for the silanol groups of the silica surface. Additional hydrophobic interactions between the alkyl chain of the cation and the C_{18} groups can also be behind this favorable effect of the ionic additive. More recently, ionic liquid-based stationary phases with immobilized imidazolium based ILs have been synthesized. In most cases, they work under strong anion exchange character and reversed phase interactions.

2.3 Electrophoresis

Ionic liquids have been employed to improve the electrophoretic separation in non aqueous capillary electrophoresis as their use in the conventional mode is hindered by their high viscosity and conductivity. ILs have been used to separate water-insoluble dyes due to the modification of the overall electrophoretic mobility of the system and the dissociation of the analytes in presence of the ILs. Further research has demonstrated the capability of ILs to improve the electrophoretic separation of chiral and achiral compounds aided by polymeric surfactants (Mwongela *et al.*, 2003).

Ionic liquids have also been covalently bonded to the inner surface of the fused silica capillary to reduce the analytes adsorption and reverse the electrosmotic flow (Qin & Li, 2002). Biomolecules (DNA and basic proteins), drugs, inorganic cations and alkylphosphonic acids have been successfully separated using capillaries modified with ILs. The reusability of the capillary (up to 96 runs) and the reproducibility improvement have been reported as additional advantages of this modality.

2.4 Sensors

The applicability of ionic liquids in electrochemistry has been boosted by the high ionic conductivity, low volatility and good redox robustness. However, most of the references deal with the electrodes construction and voltammetric data generation, being the analytical applications very limited yet. The high viscosity of ILs has been exploited for the development of membrane-free electrode modified with a thin layer of IL (Buzzeo *et al.*, 2004). They would be especially useful under extreme temperature conditions owing to their thermal stability and low volatility. The combination of hydrophobic ionic liquids, such as 1-ethyl-3-methylimidazolium trifluoromethylsulfonate, with membranes (e.g. Nafion) has been proposed as alternative to water for the construction of electromechanical actuators and sensors (Bennett & Leo, 2004). Although they are more stable operating in air, their response is rather slow in comparison with the use of water as consequence of the higher viscosity of the medium. Moreover, irreversible redox reactions can also occur.

More relevant is the use of ionic liquids in sensing approaches for the determination of organic vapors and gases. In this regard, variation on the viscosity of ionic liquids has been successfully employed as sensing material for the fabrication of quartz crystal microbalances (Liang *et al.*, 2002).

3. IONIC LIQUIDS AS SOLVENTS

The solvent power of ILs, recently reviewed (Poole & Poole, 2010), has attracted much attention not only for extraction but also for the use of ILs as solvents in synthesis. The use of ILs as solvents is based on their peculiar physical and solvation properties. Among physical characteristics, we can highlight their high density, viscosity and conductivity, their low vapor pressure, and their miscibility with other organic solvents and water. ILs involve a great variety of compounds, in the range of 10^4 (Chiappe & Pieraccini, 2005), which clearly indicates their versatility. Moreover, some of their characteristics are tunable and therefore task-specific ILs (Davis, 2004) can be synthesized for a defined application. The environmental friendly nature of ILs, which is under scrutiny (Couling *et al.*, 2006), is also a positive aspect. In fact their recycling is easier compared to conventional organic solvents due to their special physical properties.

ILs have been extensively used in liquid-liquid extraction (LLE) for the isolation from aqueous samples of a great variety of compounds. In this sense, they have been selected for the extraction of metals following three different procedures (Zhao *et al.*, 2005):

- Extraction of metals by using crown ethers. The resulted crown- ether complexes present a high hydrophobicity but they still remain a residual electric charge. The use of ILs for extraction of these complexes presents efficiencies up to 104 better compared to conventional organic solvents (Dai *et al.* 1999).
- Extraction of metals as neutral complexes using anionic ligands (Wei *et al.*, 2003) (Hirayama *et al.* 2005). This approach has been extensively used for the extraction of different metals such as mercury (Li *et al.*, 2007a), aluminium (Li *et al.* 2007b) or nickel (Dadfarnia *et al.*, 2010).
- Extraction of metals using task-specific ILs which present a cationic interchangeable group in their structure (Visser *et al.*, 2001).

The chemical nature of ILs (tunable polarity and miscibility with water) allows also their use for the extraction of organic compounds from water (Simone *et al.*, 2010).

In summary, ILs have become an important tool in sample preparation, especially in microextraction techniques. Their peculiar characteristics have allowed the improvement of the existing techniques and moreover the development of new ones. These aspects will be considered in depth in the following section of the chapter.

4. IONIC LIQUIDS IN MICROEXTRACTION TECHNIQUES

Sample pre-treatment is a key step in the analytical process in order to improve the sensitivity (by the preconcentration of the target analytes) and selectivity (by avoiding the presence of potential interferences) as well as to overcome the incompatibility with the detection system. The ideal sample treatment should be: a) simple, reducing the number of steps; b) automatic, reducing operator manipulation; c) suitable for miniaturization, reducing sample and solvent consumption; d) rapid, to allow a high sample throughput in order to resolve problems in real time; e) inexpensive; and f) safe to the operators and environment (Lucena *et al.* 2009).

Microextraction techniques, including solid phase microextraction (SPME) and liquid phase microextraction (LPME), emerged in the 1990′s as a consequence of analytical chemistry evolution towards automation, miniaturization and simplification. In this context, the unusual physical properties of ionic liquids as well as their excellent solvent properties have been exploited in a large extent.

4.1 Ionic liquid in liquid phase microextraction

Liquid-liquid extraction (LLE), a classical sample treatment technique, is based on the partitioning (distribution) of a chemical substance between two immiscible phases. In the usual approach the aqueous sample containing the analytes is mixed with an appropriate organic solvent which presents a chemical affinity towards the target analyte at the operating conditions (temperature, pH, ionic strength ...etc). After a vigorous shaken both phases are conveniently mixed, the transference of the analytes from the donor to the acceptor phase taking place. Despite the different approaches developed, LLE remains as a manual technique involving different steps (subsequent extractions, back-extractions, evaporation of the extracts ...etc) which are a source of potential contamination or analyte losses, directly affecting the reproducibility of the final measurements. Moreover, the technique is tedious, time consuming and it requires the use of large volumes of pure organic solvents which are expensive and hazardous to operators and environment.

In this scenario, LPME emerged as an alternative to classical LLE. Dasgupta (Liu & Dasgupta, 1996) proposed in 1996 for the very first time the use of solvents in the low microliter range as extractants in LLE. Apart from the obvious economical and safety connotations, the reduction of the volume of extractant has allowed an evident improvement in the enrichment factors since the sample/extractant volumes ratio is dramatically increased.

From this origin, LPME has evolved spectacularly and it has become a reality in routine laboratories. In this sense, a large number of sub-techniques have been developed in order to solve different analytical problems. Taking into account that this chapter is focused on the role of ILs in this context, only those techniques which make use of the peculiar characteristic of ILs will be considered in depth.

4.1.1 Ionic liquids in single drop microextraction

Single drop microextraction technique (SDME) was firstly proposed by Jeannot and Cantwell in 1996 (Jeannot & Cantwell, 1996). Despite the utility of this contribution, the widespread use of SDME is due basically to the works published in 1997, almost simultaneously, by Jeannot and Cantwell (Jeannot & Cantwell, 1997) and He and Lee (He & Lee, 1997). In these articles, the authors proposed the use of a conventional microsyringe as extraction manifold for SDME. Theoretically, SDME is based on the transference by passive diffusion of the analytes from the sample to a micro-drop of extractant which is located in the tip of a microsyringe. The technique involves in the same step the isolation and preconcentration of the target analytes allowing also the direct injection of the extracts in the analytical instrument. Depending on the phases involved, two different extractions schemes are possible. In the direct immersion mode (DI-SDME), the microsyringe tip is inserted into the aqueous phase and the drop is directly exposed to the sample. In the headspace mode (HS-SDME) the extractant is exposed to the vapor phase generated from the sample, the extraction involving three different phases (aqueous, headspace and organic extractant).

In both modes, conventional organic solvents have been successfully applied for the resolution of diverse analytical problems. However, ILs present special characteristics like tunable polarity and miscibility with water as well as exceptional physical (viscosity, surface tension...) and chemical (extraction capabilities) properties which make them so attractive as alternatives to conventional solvents. Moreover, ILs have been considered as green solvents due to their negligible vapor pressure compared with volatile organic solvents, making them an interesting tool with respect to the safety to operators and environment.

Liu *et al.* (Liu *et al.*, 2003) proposed for the first time the use of ILs in SDME (in both extraction modes) for the extraction of polycyclic aromatic hydrocarbons from water samples. The IL employed, 1-octyl-3-methyl imidazolium hexafluorophospate, ([C_8MIM][PF_6]), presented superior capabilities compared to 1-octanol in terms of enrichment factors. On the one hand, the adequate viscosity and surface tension of ILs allow the suspension of larger drops in the tip of the microsyringe. On the other hand, their negligible vapor pressure makes the microdrop more stable (avoiding potential evaporation during the extraction) in the headspace mode.

DI-SDME is the preferred technique for the isolation of non-volatile and/or thermolabile compounds from aqueous samples. The solvents employed in these procedures should fulfil some requirements like immiscibility with water (in order to avoid extractant losses), affinity towards the analytes (to achieve good preconcentration factors) and good stability on the microsyringe tip. 1-alkyl-3-methyl imidazolium hexafluorophospate ([C_nMIM][PF_6] with n=4,6,8) ILs have been extensively used in DI-SDME for the extraction of organic compounds as well as inorganic species. Vidal *et al.* have proposed the use of [C_6MIM][PF_6]

for the extraction of typical UV-filters in water samples (Vidal *et al.*, 2007a) (Vidal *et al.* 2010). The chemical compatibility of these ILs with the chromatographic columns allows the separation and determination of the extracted analytes by liquid chromatography coupled to UV detection. Preconcentration factors up to 98 (for 3-benzophenone) were obtained. In these articles, the authors present an interesting study of the influence of different chemical variables on the extraction of a given analyte.

In the same way, ILs have been employed for the extraction of inorganic species such as mercury, manganese or lead. According to their ionic nature, these analytes cannot be directly isolated in the ILs phase, a previous complexation with an appropriate ligand being necessary. Pena-Pereira *et al.* used $[C_6MIM][PF_6]$ for the extraction of mercury species, including inorganic and organic ones, making speciation possible by HPLC-UV (Pena- Pereira *et al.*, 2009). In other sense, for the total determination of metals electrothermal atomic absorption spectrometry (ETAAS) is preferred due to its selectivity and sensitivity levels. Manzoori *et al.* proposed for the first time the direct coupling of DI-SDME using ILs with ETAAS (Manzoori *et al.*, 2009a). This coupling, which is possible due to an adequate selection of the pyrolysis temperature and time, allows the determination of manganese in the low part per billion range using $[C_4MIM][PF_6]$ as extractant. The same procedure has been proposed for the determination of lead in water and food samples with similar success (Manzoori *et al.*, 2009b). Although the use of $[C_nMIM][PF_6]$ ILs is predominant, Martinis *et al.* have proposed the use of tetradecyl(trihexyl)phosphonium chloride as extractant for the determination of lead in waters (Martinis *et al.*, 2010).

DI-SDME effectiveness is restricted by the stability of the drop in the microsyringe tip, which limits the stirring of the sample solution and involves the use of drops into the 2-10 μL range. Different approaches have been developed in order to overcome these limitations. Cycle flow configuration (Xia *et al.* 2005) allows the free flowing of the sample through the IL microdop avoiding its potential detaching during the extraction. This approach has allowed the determination of cobalt, mercury an lead in biological and environmental samples by electrothermal vaporization inductively coupled plasma mass spectrometry (ETV-ICP-MS) (Xia *et al.*, 2008). In other sense, dynamic liquid-phase microextraction has become an excellent alternative when higher volumes of ILs (in the range of 50 μL) are required for liquid chromatographic analysis (Cruz- Vera *et al.* 2008). This approach makes use of a special manifold, completely mechanized, which performs all the steps involved in the extraction. The obtained enrichment factors, in the range from 10 to 20, are acceptable for the resolution of different analytical problems such as the determination of non steroidal anti-inflammatory drugs or phenotiazines derivatives in biological samples (Cruz-Vera *et al.* 2009a). In samples of biological concern, ILs seem to present an excellent selectivity towards the target analytes, clean extracts being obtained after sample pretreatment.

The non-volatility of ILs, due to their low vapor pressure, makes them superior solvents than the conventional organic ones in HS-SDME. In fact, the evaporation of the solvent during the extraction is a critical limitation since it affects the extraction effectiveness and reproducibility. Therefore, ILs have been extensively used in this context for the determination in environmental samples of several contaminants such as chlorinated anilines (Peng *et al.*, 2005), organochlorine pesticides (Ye *et al.*, 2006), phenols (Ye *et al.* 2007), chlorobenzenes (Vidal *et al.*, 2007b) and aromatic amines (Zhou & Ye, 2008). Although HS- SDME is usually performed in conventional furnaces or heating baths, the use of microwave energy can be employed improving the expeditiousness of the extractions (Vidal *et al.*, 2007c). Moreover, in the latter approach the design allows the heating of the sample while the ILs remains unheated, which clearly improves the efficiency of the extraction.

All the above described analytical methods use liquid chromatography or atomic techniques for the determination of the extracted analytes, even when volatile compounds are under study. This fact is ascribed to the non-volatility of ILs which is a clear limitation in gas chromatography (GC). In fact, the presence of ILs dirties the chromatograph and even blocks the analytical column making the analysis unfeasible. Some research groups have focused their efforts in the development of interfaces and/or strategies that overcomes this shortcoming making possible the use of IL in GC. The first approach, which was proposed in 2008, made use of a special removable interface directly coupled to the chromatograph which allowed analytes desorption from the ILs avoiding the entering of the solvent in the system (Aguilera-Herrador *et al.*, 2008a). The excellent performance of the interface has been exploited in some applications which use ILs in HS-SDME for the determination of monoaromatic hydrocarbons (Aguilera-Herrador *et al.*, 2008b) and halogenated hydrocarbons (Aguilera-Herrador *et al.*, 2008c) in waters. The versatility of the interface makes possible its application with other instrumental techniques such as ion-mobility spectrometry (Aguilera-Herrador *et al.*, 2009).

Zhao and co-workers proposed another alternative for making compatible the use of ILs with GC. In this case, a minimal modification of the injection liner and a special injection procedure are required (Zhao *et al.*, 2008a). The microsyringe with the IL containing the extracted analytes is inserted in the injector of the chromatograph and the microdrop is exposed to the carrier gas in the liner, which is employed as desorption chamber. Once injected the analytes in the chromatographic column, the IL is retracted into the syringe avoiding its introduction in the chromatograph. A similar approach has been recently proposed for the determination of pesticides is soil samples (Zhang & Lee, 2010). Finally, the application of a commercially available thermal desorption unit especially employed for stir bar sorptive extraction (SBSE) has been also proposed for this coupling (Chisvert *et al.* 2009). The commercial desorption unit is modified including an inner tube in order to increase the sample throughput.

In all the described applications, both in DI-SDME and HS-SDME, pure ILs are used as extractants. However, special combinations of IL with chemical modifiers can be also employed. Micellar ionic liquids, which are the result of combining ILs with surfactants, have been recently evaluated for the extraction of aromatic compounds from aqueous samples (Yao *et al.*, 2010). The obtained results indicate the good extractability of micellar ILs as well as a special selectivity towards the evaluated analytes.

4.1.2 Ionic liquids in hollow fiber protected liquid phase microextraction

The instability of the drop in the syringe tip may cause its detachment during the extraction, which is undesirable. Hollow fiber protected liquid phase microextraction (HF-LPME) faces up this shortcoming using a polymeric cylindrical membrane as container for the extraction solvent (Pedersen-Bjergaard & Rasmussen, 1999). The extraction solvent is located in the lumen of a hollow fiber of minimal dimensions (600 μm I.D., 200 μm of wall thickness and variable porosity) usually made in polypropylene. According to the phases involved in the extraction, two different modalities are possible. In the two phases mode (2P-HF-LPME) the extraction solvent is located in the lumen of the fiber as well as in the pores. This type of extraction, which is similar to the classical LLE, is ideal for hydrophobic analytes. In the three phases mode (3P-HF-LPME) the organic solvent is located in the pores of the membrane acting as physical barrier (supported liquid membrane, SLM) between two aqueous phases, namely: the donor (sample) and the acceptor (extractant). This extraction mode, similar to classical LLE with back-extraction, is the alternative of choice for hydrophobic ionizable compounds. In fact, in 3P-HF-LPME the pH gradient established at both sides of the liquid membrane is the driving force of the extraction. Conventional organic solvents (like toluene, undecane, dihexylether and octanol) have been successfully applied in both extraction modes although their tendency to evaporation is a critical aspect being a key source of irreproducibility. Once again, ILs are a good alternative to this solvents due to their peculiar characteristics.

Fortunato *et al.* suggested the potential of [C_nMIM]-based ILs as supported liquid membranes due to their high viscosities and interfacial tensions and their low solubility in water when certain types of anions are employed (Fortunato *et al.*, 2004). This usefulness has been later on demonstrated by different research groups. Peng *et al.* proposed the use of [C_8MIM][PF_6] as supported liquid membrane for the extraction of chlorophenols from water samples (Peng *et al.*, 2007). The effective immobilization of the IL in the pores of the hollow fiber was demonstrated by the authors using scanning electron microscopy (SEM). The SEM pictures obtained before and after the impregnation process are presented in Figure 1.

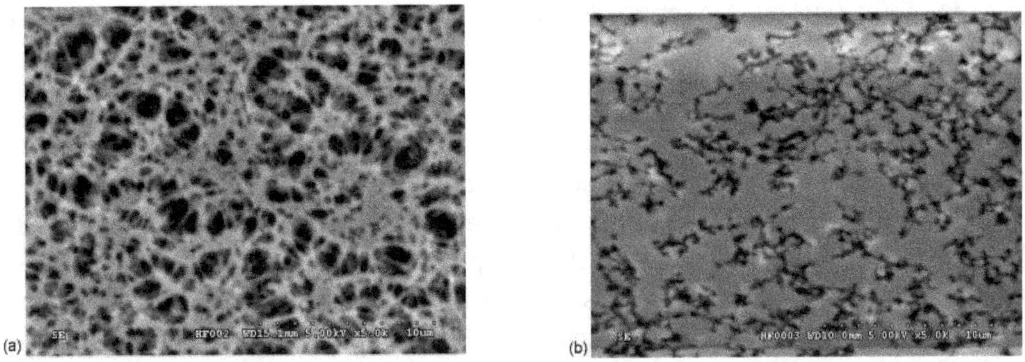

Fig. 1. Scanning electron microscopy pictures of the inner surface of a polypropylene hollow fiber (a) before and (b) after its impregnation with ionic liquid. Picture magnified by 5000 times. Figure from Peng *et al.* (2007) and reproduced with permission of Elsevier.

The potential of ILs as SLM can be also ascribed to their chemical characteristics which make possible the development of special applications. On the one hand, ILs present higher polarities compared to the conventional organic solvents used as SLM. This fact has been exploited for the extraction of sulphonamides in environmental samples which are hard to extract with conventional solvents (Tao *et al.*, 2009). On the other hand, ILs present variable solubility in water and organic solvents. In this sense Basheer *et al.* proposed a special 3P- HF-LPME with the IL acting a SLM between an aqueous phase (the sample) and an organic phase (the acceptor). The method allows the determination of aliphatic and aromatic hydrocarbons in storm water by GC since the organic extracts can be directly injected in the chromatograph (Basheer *et al.*, 2008). Finally, the good performance of ILs in the 2P-HF- LPME has been recently demonstrated by the determination of lead and nickel using ETAAS (Abulhassani *et al.* 2010).

4.1.3 Ionic liquids in dispersive liquid-liquid microextraction

Microextraction techniques are surface dependent processes. Due to the small contact area between sample and extractant, the kinetic of the extraction process is too low and the time necessary to achieve the distribution equilibrium is unpractical for analysis purposes. *Dispersive liquid-liquid microextraction* (DLLME) faces up this shortcoming using an innovative approach (Rezaee *et al.*, 2006). In

DLLME the extractant solvent is dispersed physically, chemically or assisted by an external energy source, in the sample in order to obtain a cloud of fine extractant drops, increasing dramatically the contact surface with the sample. DLLME allows the rapid extraction, almost instantaneously, of the analytes providing therefore excellent enrichment factors since the distribution equilibrium is achieved. In classical DLLME the extraction solvent should fulfil some general requirements such as: (a) immiscibility with water; (b) good chemical affinity towards the target analytes and (c) low vapor pressures. Moreover, taking into account that the extractant is recovered by centrifugation after the extraction, the solvent must be denser than water. In the light of these requirements, ILs are excellent solvents for DLLME.

Fan *et al.* proposed for the first time the use of ILs in DLLME for the determination of aromatic amines in water samples (Fan *et al.*, 2008). The IL, $[C_4MIM][PF_6]$, is directly added to the sample and the mixture is aspirated in a 1mL-microsryringe and later on discharged in a test tube. The process is repeated twice producing the physical dispersion of the IL into the sample and providing a cloudy solution of fine microdrops. Enrichment factors in the range of 31 to 269 are obtained allowing the determination of the analytes in the low µg/L level. Physical dispersion has been also applied for the determination of phenols in waters (Fan *et al.*, 2009).

According to Liu and co-workers (Liu *et al.*, 2009a), chemical dispersion is more effective than physical dispersion when ILs are used for the extraction of heterocyclic insecticides in water samples. In fact, the extraction recoveries were 5 fold better when an organic solvent is employed as disperser. Chemical dispersion is based on the use of an organic solvent (called disperser), which is miscible with the sample (water) and the extractant (IL). A mixture of the disperser and extractant solvents, in the appropriate proportion, is rapidly injected into the sample forming the cloudy solution immediately. After a determined period of time (named extraction time) the resulted mixture is centrifuged in order to recover the IL containing the isolated analytes. This strategy has been employed for the determination of organophosphorus pesticides (He *et al.*, 2009a), polycyclic aromatic hydrocarbons (Pena *et al.*, 2009) and bactericides (Zhao *et al.*, 2010a) in water samples. As well as other microextraction techniques, DLLME can be used for the extraction and determination of inorganic species in liquid samples. In order to favor the extraction, an appropriate ligand has to be added to the sample to form the corresponding chelate which is the extractable substance. Thus, the extraction of zinc (Abdolmohammad-Zadeh & Sadeghi, 2009), cobalt (Berton & Wuilloud, 2010a) and aluminium (Abdolmohammad-Zadeh & Sadeghi, 2010) have been successfully accomplished. DLLME can also operate with solid samples but a previous treatment is always required. A specific amount of sample is located in a container and the analytes are extracted by leaching with an appropriate solvent (usually acetonitrile). The extracts, which also contain matrix compounds, are later on evaporated and reconstituted in water. Once the aqueous matrix is obtained, the DLLME procedure is developed. In this way the determination of pesticides in bananas (Ravelo- Pérez et al., 2009a) and table grapes and plums (Ravelo-Pérez et al. 2009b) by liquid chromatography have been possible. Moreover, the versatility of DLLME allows the treatment of complex samples. In fact, Wang *et al.* have determined triazines in honey. The sample (2 g) has to be conveniently diluted with 20 mL of water before the extraction. The authors proposed the use of a surfactant (Triton X 114) as disperser for the extractant ([C_6MIM][PF_6]). Both, disperser and extractant, are directly added to the diluted sample and the mixture is vigorously shaken for 10 min. After the extraction, the IL is recovered by centrifugation and finally analyzed by HPLC (Wang *et al.*, 2010).

The use of IL-based microextraction techniques for the treatment of samples with a high content of electrolytes is usually problematic since the solubility of a given IL in water dramatically depends on the ionic strength. At high levels of ionic strength, the IL is completely soluble in water and its recovery after the extraction is unaffordable. This phenomenon is caused essentially by an ion exchange process, the so-called metathesis reaction, between the IL and the electrolytes present in the sample. In this way, when a [C_nMIM][PF_6]-based IL is introduced in a sample with a high concentration of chloride, the chloride anion may exchange with the hexafluophosphate. As a consequence of this metathesis reaction a new IL ([C_6MIM][Cl]), completely soluble in water, is produced. Yousefi *et al.* have proposed a strategy in order to minimize this reaction by the external addition of $NaPF_6$ to the sample (Yousefi & Shemirani, 2010). Due to the common ion effect, the solubility of the IL decreases even in the presence of high concentration of electrolytes. This approach has been applied in a DLLME configuration for the determination of lead and cadmium in saline aqueous samples by FAAS. The common ion effect has also supported the development of a novel microextraction technique called *in situ solvent formation microextraction* (ISFME) (Baghdadi & Shemirani, 2009). In ISFME, a water-miscible IL (such as [C_6MIM][BF_4]) is added to the sample, being completely dissolved and therefore maximizing the contact surface with the target analytes. Later on, a common ion reagent (like $NaPF_6$) is added to the extraction vial inducing the metathesis reaction. As a consequence, a water-immiscible IL (in this case [C_6MIM][PF_6]) is formed which is finally recovered by centrifugation. ISFME has been characterized in depth by Yao & Anderson (Yao & Anderson, 2009). The authors have demonstrated the superior features of ISFME compared to traditionally DLLME or DI-SDME, using ([C_4MIM][Cl] and $LiNTf_2$ as IL and common ion reagent, respectively. Enrichment factors in the range of 189 to 753 have been obtained for the determination of 13 aromatic compounds in water samples. ISFME has been also used for the determination of cadmium in saline samples by FAAS providing limits of detection of 0.07 μg/L (Mahpishanian & Shemirani, 2010).

The dispersion of the extraction solvent into the sample is the crucial step in any DLLME and different alternatives, apart from the physical and chemical approaches, have been developed in recent years. In these novel approaches, an external energy source is employed in order to assist the dispersion process. The technique, *temperature controlled ionic liquid dispersive liquid phase microextraction* (TILDLME) was proposed in 2008 by Zhou and co- workers (Zhou *et al.*, 2008a) and it uses an external heat source. TILDLME is based on different well established steps. First of all, a specific amount of IL is added to the sample which is subsequent incubated at a high temperature (in the range of 45-90ºC) for an appropriate time. During the heating of the sample, the IL is completely dispersed, even solubilized, into the sample matrix maximizing the contact area. Later on, the vial is introduced in an ice-bath decreasing the temperature of the mixture. In such conditions, the IL becomes insoluble and a cloudy solution appears in the extraction vial. After a proper extraction time, the mixture is centrifuged and the IL, with the extracted analytes, is recovered for its subsequent analysis. In TILDLME three especial properties of ILs are exploited, namely: (a) their good extractability towards a great variety of compounds; (b) their negligible vapor pressure which avoids the evaporation of the extractant during the heating process, and (c) their variable water-solubility which directly depends on the working temperature. Compared to classical DLLME, TILDLME provides better enrichment factors, especially for polar substances, since no chemical disperser is employed. Moreover, the contact area between sample and extractant is dramatically increased. TILDLME has been successfully applied for the determination of organophosphorous pesticides (Zhou *et al.*, 2008b), herbicides (Zhou et al., 2009a) or bactericides (Zhao *et al.*, 2010b). As other extraction techniques previously described, TILDLME can be also employed for the extraction of inorganic species such as vanadium (Berton *et al.*, 2009) or lead (Bai *et al.*, 2010). In 2008, the *cold induced aggregation microextraction* (CIAM) technique is introduced for the first time (Baghdadi & Sheminari, 2008). CIAM is similar to TILDLME but lower temperatures are used in the incubation step. CIAM has been successfully used for the determination of cobalt in water samples (Gharehbaghi *et al.*, 2009).

Ultrasounds can also be used to assist DLLME originating a new technique which is called *ultrasound-assisted ionic liquid dispersive liquid-liquid phase microextraction* (USA-DLLME) (Zhou *et al.*, 2009b). In USA-DLLME the IL is dispersed by using an ultrasound source and once the IL is dispersed, the extraction vessel is introduced in an ice-bath inducing the aggregation of the IL. The resultant cloudy solution is finally centrifuged and the IL, with the extracted analytes, is recovered for its subsequent analysis. Conventional or high energy ultrasounds (Mao *et al.*, 2009) can be employed. As the previously described techniques, the versatility of USA-DLLME allows its application for the resolution of diverse analytical problems. In this sense, it has been used for the extraction of cadmium from water samples (Li *et al.*, 2009).

In the light of the results, DLLME is a powerful technique for the extraction of different analytes in a great variety of sample matrixes. DLLME allows an efficient miniaturization and simplification of the sample pre-treatment although its automation is clearly difficult. This difficulty arises from the unavoidable centrifugation step which is a typical off-line process. Some efforts have been made to avoid this step in order to open up a new horizon in DLLME automation. Berton *et al.* (Berton *et al.*, 2010b) have recently proposed an automatic TILDLME method, based on a flow injection manifold, for the determination of vanadium species in water samples. With no centrifugation the IL is recovered after the extraction using a florisil column, where the IL remains due to its high viscosity. The centrifugation step is also avoided in the technique *one-step in-syringe ionic liquid-based dispersive liquid-liquid microextraction* (Cruz-Vera *et al.*; 2009b). In this technique the complete DLLME procedure is developed in a conventional plastic syringe. Once the cloudy solution is formed, the IL is recovered by the slight movement of the syringe plunger. This separation is possible due to the tendency of the IL to interact with the plastic walls of the syringe.

4.2 Ionic liquids in solid phase microextraction

Solid phase microextraction (SPME) was introduced by Pawliszyn and co-workers (Berladi & Pawliszyn, 1989) (Arthur & Pawliszyn, 1990) as a simple, time efficient and solvent-free technique. SPME integrates the sampling, isolation and preconcentration of the target analytes in one step as well as it allows the direct coupling with different instrumental techniques. SPME is based on the partition of the analytes of interest between the sample and a small amount of extracting phase dispersed and immobilized in a solid support and the subsequent thermal or solvent-based desorption (Cruz-Vera et al., 2009c). SPME has been in a continuous evolution and different approaches, based on the sorptive phase minimization, have been developed.

Classical SPME used a fiber with an appropriate sorptive coating as extraction unit. The fiber is integrated in a syringe manifold which permits the development of the extraction as well as the subsequent injection of the analytes in the instrument, commonly a gas chromatograph. The fiber coating plays a key role on the extraction of the analytes. The ideal coating must fulfil some general requirements such as: (a) good affinity towards the target analytes in order to allow a good preconcentration; (b) thermal stability, especially when analytes desorption is assisted by an external heat source; (c) chemical stability, in particular when the desorption is performed by means of a chemical elution; (d) long life span; (e) low memory effect to avoid carry-over effects, (f) appropriate batch to batch reproducibility and low price. As it is evident, the ideal coating does not exist and it should be selected according to the analytical problem under study. Moreover, the development of new coatings has turned into the goal of an extensive research in recent years. In this context, ILs can play an important role due to their characteristics that have been widely described in this chapter.

Liu and co-workers proposed for the first time the use of IL in a hybrid LPME-SPME approach (Liu *et al.*, 2005). In fact, $[C_8MIM][PF_6]$ is employed to cover a fused silica-fiber producing a disposable IL-coating. The IL-coating is of physical nature, without a covalent bonding, and it is performed by dipping a pretreated fused silica fiber into an IL solution in dichloromethane. After dipping, the fiber is dried removing the excess of dichloromethane, leaving a coating of $[C_8MIM][PF_6]$. The dipping and evaporating cycle is repeated three times in order to achieve the desired coating thickness. The use of IL in this type of coating presents some advantages such as: (a) due to its liquid nature, the coating can be easily removed using appropriate solvents; (b) due to its liquid nature, the coating presents higher diffusion coefficients which involve faster sorption kinetics and (c) due to its high viscosity, the IL can cover easily the fiber. Using this approach, the determination of benzene, toluene, ethylbenzene and xylene in paints has been achieved. Taking into account that the IL is only physically coated into the fiber, the volume of IL is reduced and therefore the sorption capacity limited. Hsieh *et al.* have proposed the use of nafion membranes as inner coating in order to maximize the volume of IL (Hsieh *et al.*, 2006). This polymeric membrane presents some anionic groups that may interact with imidazolium cations increasing the volume of immobilized IL and providing a more stable coating. The proposed configuration used $[C_8MIM][Tfo]$ as extractant for the determination of ultra traces of PAHs in water samples. The physical coating can be also employed in coated fused silica capillary previously etched with an appropriate solution. Etching process produces a rough capillary surface which is able to immobilize a higher volume of IL compared to the original bare fused silica capillary (Huang *et al.*, 2009). The use of elastomers has been proposed to improve coating consistency. For this purpose a mixture of IL and elastomers in a volatile solvent is placed in a beaker and heated until a suitable viscosity is achieved. In this moment, a fused silica fiber is dipped in this solution. Later on, the fiber is drawn and dried until the solvent is completely evaporated. After a cycle of dipping and evaporation, a coated fiber can be obtained. The IL is trapped in the polymeric coated but it remains in its liquid form. A 50 µm in thickness polymeric coating containing $[C_4MIM][NTf_2]$ has been successfully applied to the determination of methamphetamine and amphetamine in forensic samples (He *et al.*, 2009b).

All the mentioned approaches correspond to a hybrid LPME-SPME strategy since typically SPME devices are employed while a liquid phase is used as extractant. The liquid nature of the extractant provides evident advantages (e.g faster sorption kinetics) but it limits the stability and reusability of the coatings. Polymeric ionic liquids (PILs) allow the development of classical SPME fibers providing better chemical and thermal stability compared to the above mentioned approaches (Zhao *et al.* 2008b). Moreover, PILs coatings present a longer life span (they can be used up to 150 times) and good reproducibility between fibers. This type of coatings are prepared with a dipping-evaporation process, similar to those described previously, using a solution of PILs in a volatile organic solvent. When the solvent is evaporated, a solid PIL-coating is obtained. SEM pictures of PILs coating are presented in Figure 2. PIL-based SPME fibers have been successfully employed for the extraction of a great variety of organic substances (López-Darias *et al.* 2010) or carbon dioxide (Zhao & Anderson, 2010).

ILs have been covalently bonded to a silica support in order to obtain a fiber coating (Wanigasekara *et al.* 2010). The IL is previously derivatized to obtain an appropriate reagent which is able to react with a silica subtract. As a result, the liquid state of the IL is lost but their inherent characteristics (such as low vapor pressure, chemical stability or affinity towards organic compounds) remain unaltered. This IL-silica coated fiber has been used for the extraction of polar compounds.

In the SPME context, ILs can be used as solvents for the synthesis of stationary phases for capillary microextraction (CME), a SPME mode based on the location of the sorptive phase in the inner volume of a fused- silica capillary. The IL may play a double role since it can act as a reaction solvent and/or porogen solvent. The use of IL as reaction solvent has some advantages such as: (a) the solvent can be recycled; (b) the solvent is thermo-stable in the temperature range in which the reaction takes place and (c) the reagents present a higher stability in ILs compared to other organic solvents. On the other hand, the use of ILs as porogens provides the final polymer with a higher porosity and therefore with a higher extraction efficiency. Moreover, the IL employed can be removed by a thermal treatment releasing a pure polymeric coating (Shearrow *et al.*, 2009a). The synthesis process is reproducible, both in run-to-run and capillary-to-capillary conditions. This approach has been successfully used for the fabrication of coatings focused on the extraction of polar substances, which is a challenge in the analytical chemistry context (Shearrow *et al.*, 2009b).

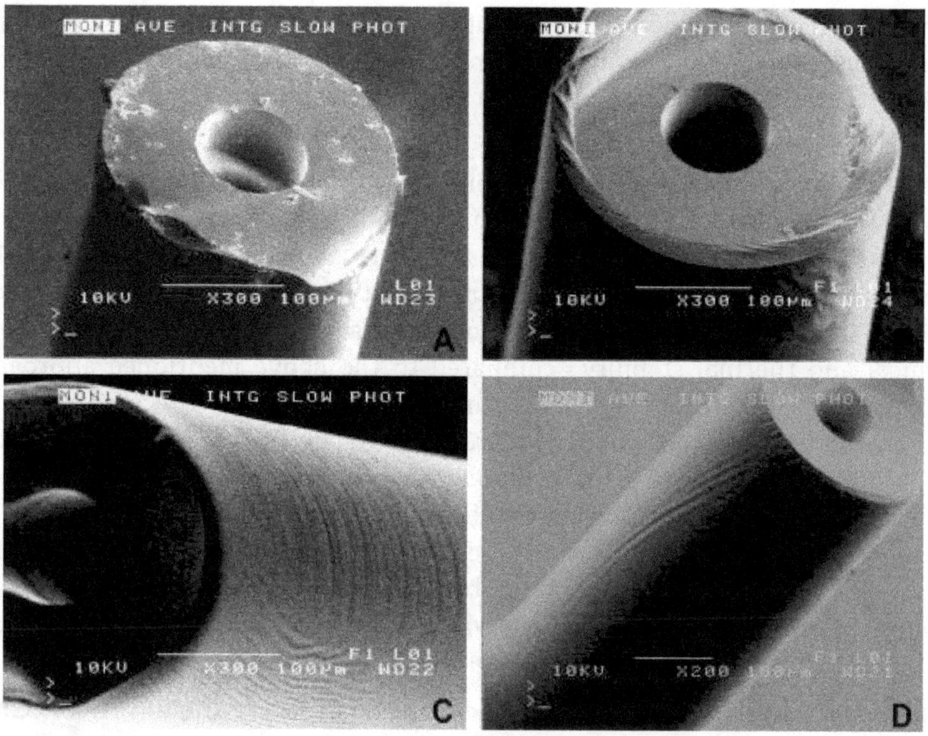

Fig. 2. Scanning electron micrographs of a 100-µm inner diameter bare fused silica support and various angles of the fused silica support coated with PIL (B–D). Figure from Zhao *et al.* (2008b) and reproduced with permission of Elsevier.

5. IONIC LIQUID-BASED SORBENTS FOR SOLID PHASE EXTRACTION

The use of a variety of sorbents in solid phase (micro)extraction techniques has several advantages compared to liquid phase (micro)extraction for sample clean-up and/or analyte(s) preconcentrations. In a previous section (see section 4.2) the main contributions of ILs to solid phase microextraction have been highlighted. However, ILs have also contributed to improve the sorbent capability of conventional sorbent materials in several ways, the most relevant of which are commented on.

The use of a soft functional material arising from the combination of ILs and carbon nanotubes, allowed by the specific interaction between imidazolium ion components of the IL and the p-electronic nanotube surface, has been described. The resultant gelatinous material, consisting of highly electroconductive nanowires and fluid electrolytes, has a high capacity to absorb/desorb analytes from a variety of liquid samples. This capacity is the result of the synergistic effect between ILs solvation power and the adsorption capacity of carbon nanotubes. This new sorbent has been successfully applied to the determination of traces of PAHs in river waters (Polo-Luque *et al.*, in press. a).

A conventional filter can be modified when a dissolution of carbon nanotubes in ILs is passed through it, leading to the stable retention of the nanotubes. This modified nanofilter has been applied to organic interference removal for the determination by non-aqueous capillary electrophoresis (NACE) of imidazolium and pyridinium-based ionic liquids in river water (Polo-Luque *et al.*, in press. b). In a similar way, typical C_{18} sorbents can be modified using a dissolution of carbon nanotubes in ILs; this sorbent has been used for the preconcentration and determination of sulphonamides in river water by capillary electrophoresis (Polo-Luque *et al.*, in press. c).

6. IONIC LIQUIDS IN SAMPLE DISSOLUTION

Dissolution is the classical beginning in the analysis of solid and solid-liquid samples (Kellner *et al.*, 2005). The general objectives of this substep of the preliminary operations of analytical processes are: (a) to make the next detection step possible; (b) to improve sample homogeneity and, thus, to enhance reproducibility (and quality) of the results; (c) to enhance sensitivity; and (d) to contribute to the simplification. In many real cases, it is not an easy task to distinguish between dissolution and extraction (see part 4 of this chapter). In this context, solvent changeover can also be considered.

The capacity of dissolution of ILs is undeniable (Weingärtner, 2008) (see part 3 of this article). They have been involved as solvents in both, physico-chemical studies and analytical processes. The most relevant applications in the field of chemical analysis are commented on below.

6.1 Ionic liquids as matrixes for mass spectrometric analysis

The application of ILs as replacements for the conventional liquid or solid matrixes in Matrix Assisted Laser Desorption/Ionization Mass Spectrometry (MALDI-MS) has recently been recognized (Baker *et al.*, 2005). MALDI-MS is a powerful technique for determining high-molecular-weight compounds (e.g. biomacromolecules, synthetic polymers) as well as low-molecular weight compounds. The requirements of an effective matrix for MALDI-MS are as follows (Liu *et al.*, 2009b):

- To dissolve (liquid matrix) or co-crystallize (solid matrix) with the sample;
- To contain a chromophore to absorb laser light and promote ionization of the analyte;
- To remain in the condensed phase under high-vacuum conditions;
- To minimize or avoid thermal and chemical sample degradation.

As can be seen in Figure 3, the applications of ILs as matrixes for MALDI-MS satisfy almost all of these requirements, but their ability to promote ionization of the analyte is not accomplished in all cases since it depends on the type of analyte and IL. In addition, the great variety of ILs makes not possible to extend their general features to their role as matrixes in MALDI-MS analysis. A recent review (Tholey & Heinzle, 2006) compiles the most relevant aspects and fields of application of the topic.

One of the first attempts to use ILs as matrixes in MALDI-MS was reported by Armstrong's group using peptides, proteins and poly(ethylene glycol) (PEG-2000) as model analytes (Armstrong *et al.*, 2001). All of the IL-based matrixes showed excellent solubilizing properties and vacuum stability compared to other frequently employed solid and liquid matrixes. There are other recent papers aimed to compare ILs with conventional solid and liquid matrixes for a variety of sample-analyte-pairs (Mank *et al.*, 2004) (Zabet-Moghaddam *et al.*, 2004). Besides the properties pointed out by Armstrong in 2001 (Armstrong *et al.*, 2001) (see Figure 3), it has been demonstrated that liquid consistency of IL-matrix sample preparations considerably enhances MALDI-MS analysis in terms of shot-to-shot reproducibility, which is a bottle-neck in this technique.

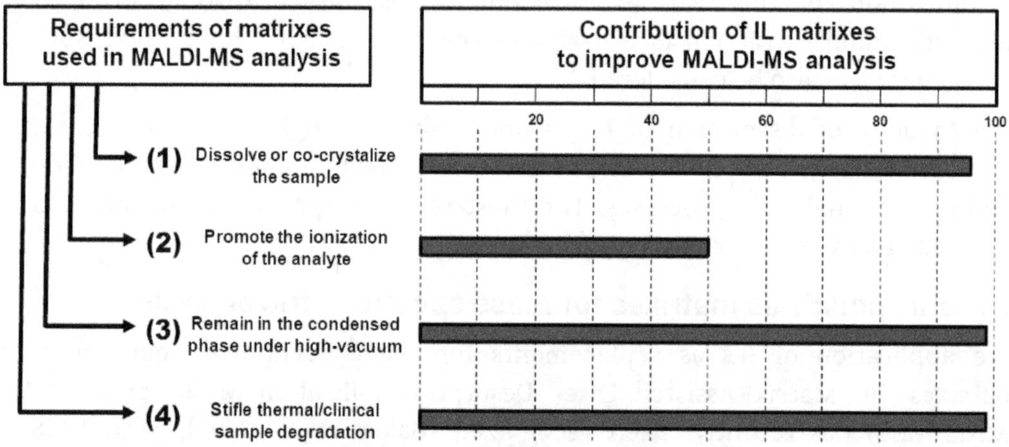

Fig. 3. Fulfillment by ILs of the basic requirements of matrixes in MALDI-MS analysis. For details, see text.

ILs as matrixes show also a synergy with the technique Matrix-Assisted Laser Desorption/Ionization-Time-of-Flight Mass Spectrometry (MALDI-TOF-MS) especially for screening purposes. Quantitative MALDI-TOF-MS using isotopic labeled internal standards and IL-based matrixes have been applied for the screening of 10 pyranose oxidase variants using aliquots of enzyme reaction mixtures without previous purification steps, which can be considered a relevant approach (Bungert et al., 2004). In some cases, this technique is combined with a previous separation step. Such is the case of the fast screening of a mixture of low molecular weight compounds (three alkaloids, two anesthetics and one antibiotic) based on Thin-Layer Chromatography (TLC) followed by direct on-shot MALDI-TOF-MS identification with nearly matrix-free mass spectra using a UV-absorbing IL matrix. This technique is fast and sensitive and it requires little sample preparation and manipulation (Santos *et al.*, 2004).

6.2 Ionic liquids as solvents in molecular spectroscopy

Since 1986, ionic liquids have been demonstrated to be able to solvate a wide range of chemical species including organic, inorganic, metal complexes and organometallic compounds (Appleby *et al.*, 1986). The main advantages of the use of ILs as solvents in spectroscopy are (Koel, 2005): (a) they offer stability to a variety of species, particularly metal complexes; (b) they circumvent problems arising from solvation and solvolysis; and (c) they permit reliable UV-visible and IR solution spectra. Enhanced resolution and significant spectra shifts compared to those of the standards published have been observed. The spectra of the substance of interest obtained in ILs resemble to those recorded in solvents of low dielectric constants such as dichloromethane. The majority of publications on this topic are related to physico-chemical studies, but analytical processes can have benefit from having the analyte dissolved in an appropriate solvent, which can be achieved using extraction (see part 4 of this chapter) or solvent changeover by dissolution of the solid residue after the dryness of the first solvent employing heating or under a N_2 stream.

6.3 Ionic liquids as solvents in headspace gas chromatography

Owing to the inherent characteristics of ILs, such as non-volatility and heat stability, they can be used advantageously as solvents in headspace GC. The study performed by André and co-workers (André *et al.*, 2005) is based on the use of three acidic, neutral and basic model analytes with boiling points above 200 °C which were dissolved in ILs; the analytes could be determined at the low-ppm level.

This approach has been successively applied to the determination of traces of low vapor pressure residual solvents in pharmaceutical drug products, a crucial aspect of their quality. Using the ionic liquid $[C_4MIM][BF_4]$ as the matrix medium, residual solvents such as N-methyl-2-pyrrolidone and dimethylformamide in commercially-available pharmaceutical preparations have been determined using static headspace GC (Liu & Jiang, 2007).

6.4 Ionic liquids as solvents for nanoparticles

ILs are suitable solvents for the dissolution (extraction) of a great variety of nanomaterials both organic (e.g. carbon nanoparticles) and inorganic (e.g. gold nanoparticles, quantum dots, etc.). Solubility of nanoparticles in ILs can have different foundations, the most relevant being salting out effects and cation exchange processes, among others.

ILs can be exploited as useful tools for the analysis of environmental and biological samples for the determination of the pollution caused by manufactured nanomaterials and for clarification of their toxicity in humans, animals and plants. Some specific developments in this context have been described in the last years (Wei *et al.*, 2004) (Huang *et al.*, 2006) (Nakashima & Kawai, 2005).

In addition, ILs can be used for the characterization of nanomaterials. The fact that they can be easily solubilized in ILs facilitates the analysis. This is the case of the characterization of a commercial mixture of single-walled carbon nanotubes that can be characterized by Raman spectrometry as a single component. Other alternative consists of the initial sample dissolution in an IL ([C_4MIM][BF_4]) using ultrasonic energy, and the subsequent encapsulation of the nanotubes in sodium dodecylsulphate micelles. An aliquot of the dissolution is introduced in a capillary electrophoretic equipment and an electropherogram of 8 peaks (in less than 4 minutes) is obtained, which demonstrates the non-homogeneity of the original sample of carbon nanotubes (López-Pastor *et al.*, 2008).

7. CONCISE SUMMARY

Due to the fact that the majority of the preliminary operations of the analytical process rely on the use of solvents (e.g. dissolution, extraction, solvent changeover, etc), any contribution to minimize or to avoid the negative connotations of the conventional ones, both organic (e.g. hexane) or inorganic (e.g. HNO_3) should be welcome. Such is the case of ILs which, since two decades ago, have open up new perspectives and promising approaches in this field, enhancing the analytical properties (top, basic and productive) through the improvement of the sample preparation steps (Valcárcel & Rios, 1993). The main advantages of ILs as solvents in comparison with the traditional ones can be justified by their unique properties that have been described in part 1 of this chapter, being the most relevant for such purpose: solvent capability and tunability, thermal stability and very low vapor pressure, as well as their viscosity, density, surface tension, immiscibility with other solvents, etc.

There are presently several drawbacks that limit the use of ILs in Analytical Sciences, the most relevant being the following: (a) hygroscopy, which can led to the hydrolysis of the anion components; (b) impurity, traces of other compounds (e.g.water) that change the expected properties of ILs to be used for a fixed fit-for-purpose; (c) instability, arising from the decomposition of ILs under heating, ultrasonication or microwave energy, frequently used in sample preparations; (d) alteration of the structure of the dissolved analytes (e.g. biomolecules); and (e) toxicity, despite the reputation of ILs as green solvents, recent studies evidence the relative toxicity of some of the ILs used in practice. The environmental fate and toxicity of ILs have been recently reviewed by Pham *et al* (Pham *et al.* 2010).

REFERENCES

Abdolmohammad-Zadeh, H. & Sadeghi, G. H. (2009). A novel microextraction technique based on 1-hexylpyridinium hexafluorophosphate ionic liquid for the preconcentration of zinc in water and milk samples. *Anal. Chim. Acta*, 649 (September, 2009) 211-217, ISSN: 0003-2670.

Abdolmohammad-Zadeh, H. & Sadeghi, G. H. (2010). Combination of ionic liquid-based dispersive liquid-liquid micro-extraction with stopped-flow spectrofluorometry for the pre-concentration and determination of aluminum in natural waters, fruit juice and food samples. *Talanta*, 81 (May, 2010) 778-785, ISSN: 0039-9140.

Abulhassani, J.; Manzoori, J. L. & Amjadi, M. (2010). Hollow fiber based-liquid phase microextraction using ionic liquid solvent for preconcentration of lead and nickel from environmental and biological samples prior to determination by electrothermal atomic absorption spectrometry. J. Hazard. Mater. 176 (April, 2010) 481-486, ISSN: 0304-3894.

Aguilera-Herrador, E.; Lucena, R.; Cárdenas, S. & Valcárcel, M. (2008a). Direct coupling of ionic liquid based single-drop microextraction and GC/MS. *Anal. Chem.* 80 (February, 2008) 793-800, ISSN: 0003-2700.

Aguilera-Herrador, E.; Lucena, R.; Cárdenas, S. & Valcárcel, M. (2008b). Ionic liquid-based single-drop microextraction/gas chromatographic/mass spectrometric determination of benzene, toluene, ethylbenzene and xylene isomers in waters. *J. Chromatogr. A*, 1201 (August, 2008), 106-111, ISSN: 0021-9673.

Aguilera-Herrador, E.; Lucena, R.; Cárdenas, S. & Valcárcel, M. (2008c). Determination of trihalomethanes in waters by ionic liquid-based single drop microextraction/gas chromatographic/mass spectrometry. *J. Chromatogr. A*, 1209 (October, 2008), 76-82, ISSN: 0021-9673.

Aguilera-Herrador, E.; Lucena, R.; Cárdenas, S. & Valcárcel, M. (2009). Ionic liquid-based single drop microextraction and room-temperature gas chromatography for on-site ion mobility spectrometric analysis. *J. Chromatogr. A*, 1216 (July, 2009), 5580-5587, ISSN: 0021-9673.

Anderson, J. L. & Armstrong. D. W. (2003). High-stability ionic liquids. A new class of stationary phases for gas chromatography. *Anal. Chem.*, 75 (September, 2003) 4851- 4858, ISSN: 0003-2700.

Andre, M.; Loidl, J.; Laus, G.; Schottenberger, H.; Bentivoglio, G.; Wurst, K. & Ongania, K. H. (2005). Ionic liquids as advantageous solvents for headspace gas chromatography of compounds with low vapor pressure. *Anal. Chem.*, 77 (January, 2005) 702-705, ISSN: 0003-2700.

Appleby, D.; Hyssey, C. L.; Seddon, K. R. & Turp, J. E. (1986). Room-temperature ionic liquids as solvents for electronic absorption spectroscopy of halide complexes. *Nature*, 323 (October, 1986) 614-615, ISSN: 0028-0836.

Armstrong, D. W.; He, L. & Liu Y. S. (1999). Examination of ionic liquids and their interaction with molecules, when used as stationary phases in gas chromatography. *Anal. Chem.* 71 (September, 1999) 3873-3876, ISSN: 0003-2700.

Armstrong, D. W.; Zhang, L. K; He, L. & Gross, M. L. (2001) Ionic liquids as matrixes for matrix-assisted laser desorption/ionization mass spectrometry. *Anal. Chem.* 73 (August, 2001) 3679-3686, ISSN: 0003-2700.

Bai, H.; Zhou, Q.; Xie, G. & Xiao, J. (2010). Temperature-controlled ionic liquid-liquid-phase microextraction for the pre-concentration of lead from environmental samples prior to flame atomic absorption spectrometry. *Talanta*, 80 (March, 2010) 1638-1642, ISSN: 0039-9140.

Baghdadi; M. & Sheminari, F. (2008). Cold-induced aggregation microextraction: a novel sample preparation technique based on ionic liquids. *Anal. Chim. Acta*, 613 (April, 2008) 56-63, ISSN: 0003-2670.

Baghdadi; M. & Sheminari, F. (2009). In situ solvent formation microextraction based on ionic liquids: A novel sample preparation technique for determination of inorganic species in saline solutions. *Anal. Chim. Acta*, 634 (February, 2009) 186-191, ISSN: 0003-2670.

Baker, G. A.; Baker, S. N.; Pandey, S. & Bright, F. V. (2005). An Analytical View of Ionic Liquids. *Analyst*, 130 (June, 2005) 800-808, ISSN: 0003-2654.

Basheer, C.; Alnedhary, A. A.; Madhava Rao, B. S.; Balasubramanian, R. & Lee, H. K. (2008). Ionic liquid supported three-phase liquid-liquid-liquid microextraction as a sample preparation technique for aliphatic and aromatic hydrocarbons prior to gas chromatography-mass spectrometry. *J. Chromatogr. A*, 1210 (November, 2008), 19-24, ISSN: 0021-9673.

Bennett, M. D. & Leo, D. J. (2004). Ionic liquids as stable solvents for ionic polymer transducers. *Sensor Actuator A. Phys.*, 115 (September, 2004) 79-90, ISSN 0924-4247.

Berladi, R. P. & Pawliszyn, J. B. (1989). The application of chemically modified fused silica fibers in the extraction of organics from water matrix samples and their rapid transfer to capillary columns. *Water Pollut. Res. J. Can.* 24 (1989) 179-191. ISSN: 0197-9140.

Berton, P.; Martinis, E. M.; Martinez, L. D. & Wuilloud, R. G. (2009). Room temperature ionic liquid-based microextraction for vanadium species separation and determination in water samples by electrothermal atomic absorption spectrometry. *Anal. Chim. Acta*, 640 (April, 2009) 40-46, ISSN: 0003-2670.

Berton, P. & Wuilloud, R. G. (2010a). Highly selective ionic liquid-based microextraction method for sensitive trace cobalt determination in environmental and biological samples. *Anal. Chim. Acta*, 662 (March, 2010) 115-162, ISSN: 0003-2670.

Berton, P.; Martinis, E. M. & Wuilloud R. G. (2010b). Development of an on-line temperature-assisted ionic liquid dispersive microextraction system for sensitive determination of vanadium in environmental and biological samples. *J. Hazard Mater.* 176 (April, 2010) 721-728, ISSN: 0304-3894.

Bungert, D.; Bastian, S.; Heckmann-Pohl, D. M.; Giffhorn, F.; Heinzle, E. & Tholey, A. (2004). Screening of sugar converting enzymes using quantitative MALDI-TOF mass spectrometry. *Biotech. Letters*, 26 (July, 2004) 1025-1030, ISSN: 0141-5492.

Buzzeo, M. C.; Hardacre, C. & Compton, R. C. (2004). Use of room temperature ionic liquids in gas sensor design. *Anal. Chem.*, 76 (July, 2004) 4583–4588, ISSN: 0003-2700.

Catherine L. Arthur, Janusz Pawliszyn. (1990). Solid phase microextraction with thermal desorption using fused silica optical fibers. *Anal. Chem.*, 62 (October, 1990) 2145– 2148, ISSN: 0003-2700.

Chiappe, C. & Pieraccini, D. (2005). Ionic liquids: solvent properties and organic reactivity. *J. Phys. Org. Chem.*, 18 (April, 2005) 275-297, ISSN: 0894-3230.

Chisvert, A.; Román, I. P.; Vidal, L. & Canals, A. (2009). Simple and commercial readily- available approach for the direct use of ionic liquid-based single-drop microextraction prior to gas chromatography determination of chlorobenzenes in real water samples as model analytical application. *J. Chromatogr. A*, 1216 (February, 2009), 1290-1295, ISSN: 0021-9673.

Couling, D. J.; Bernot, R. J.; Docherty, K. M.; Dixon, J. K. &. Maginn, E. J. (2006). Assessing the factors responsible for ionic liquid toxicity to aquatic organisms via quantitative structure–property relationship modeling. *Green Chem.* 8 (August, 2006) 82-90, ISSN: 1463-9262.

Cruz-Vera, M.; Lucena, R.; Cárdenas, S. & Valcárcel, M. (2008). Ionic liquid-based dynamic liquid-phase microextraction: application to the determination of anti- inflammatory drugs in urine samples. *J. Chromatogr. A*, 1202 (August, 2008) 1-7, ISSN: 0021-9673.

Cruz-Vera, M.; Lucena, R.; Cárdenas, S. & Valcárcel, M. (2009a). Determination of phenothiazine derivatives in human urine by using ionic liquid-based dynamic liquid-phase microextraction coupled with liquid chromatography. *J. Chromatogr. B*, 877 (January, 2009) 37-42, ISSN: 1570-0232.

Cruz-Vera, M.; Lucena, R.; Cárdenas, S. & Valcárcel, M. (2009b). One-step in-syringe ionic liquid-based dispersive liquid-liquid microextraction. *J. Chromatogr. A*, 1216 (September, 2010) 6459-6465, ISSN: 0021-9673.

Cruz-Vera, M.; Lucena, R.; Cárdenas, S. & Valcárcel, M. (2009c). Sorptive microextraction for liquidchromatographic determination of drugs in urine. *Trends Anal. Chem.*, 28 (month, 2009) 1164-1173, ISSN: 0165-9936.

Dadfarnia, S.; Shabani, A. M.; Bidabadi, M. S. & Jafari, A. A. (2010). A novel ionic liquid/micro-volume back extraction procedure combined with flame atomic absorption spectrometry for determination of trace nickel in samples of nutritional interest. *J Hazard Mater.*, 173 (January, 2010) 534-538, ISSN: 0304-3894.

Dai, S.; Ju, Y. H. & Barnes, C. E. (1999) Solvent extraction of strontium nitrate by a crown ether using room-temperature ionic liquids. *J. Chem. Soc. Dalton Trans.*, (month, 1999) 1201-1202, ISSN: 1477-9226.

Davis, J. H. (2004). Task specific ionic liquids. *Chem. Letters*, 33 (September, 2004) 1072-1077, ISSN: 0366-7022.

Fan, Y. C.; Hu, Z. L.; Chen, M. L.; Tu, C. S. & Zhu, Y. (2008). Ionic liquid based dispersive liquid–liquid microextraction of aromatic amines in water samples. *Chinese Chem. Lett.*, 19 (August, 2008), 985-987, ISSN 1001-8417.

Fan, Y.C.; Chen, M. L.; Tu, C. S. & Zhu, Y. (2009). A ionic liquid for dispersive liquid-liquid microextraction of phenols. *J. Anal. Chem.*; 64 (October, 2009) 1017-1022, ISSN: 1061-9348.

Fortunato, R.; Afonso, C. A. M.; Reis, M. A. M. & Crespo, J. G. (2004). Supported liquid membranes using ionic liquids: study of stability and transport mechanisms. *J. Membr. Sci.*, 242 (October, 2004) 197-2009, ISSN: 0376-7388.

Gharehbaghi, M.; Shemirani, F. & Farahani, M. D. (2009). Cold-induced aggregation microextraction based on ionic liquids and fiber optic-linear array detection spectrophotometry of cobalt in water samples. *J. Hazard. Mater.* 165 (June, 2009) 1049-1055, ISSN: 0304-3894.

He, Y. & Lee, H. K. (1997). Liquid-phase microextraction in a single drop of organic solvent by using a conventional microsyringe. *Anal. Chem.*, 69 (November, 1997) 4634-4640, ISSN: 0003-2700.

He, L.; Luo, X.; Xie, H.; Wang, C.; Jiang, X. & Lu, K. (2009a). Ionic liquid-based dispersive liquid-liquid microextraction followed high-performance liquid chromatography for the determination of organophosphorus pesticides in water sample. *Anal. Chim. Acta*, 655 (November, 2009) 52-59, ISSN: 0003-2670.

He, Y.; Pohl, J.; Engel, R.; Rothman, L. & Thomas, M. (2009b). Preparation of ionic liquid based solid-phase microextraction fiber and its application to forensic determination of methamphetamine and amphetamine in human urine. *J. Chromatogr. A.*, 1216 (June, 2009) 4824-4830, ISSN: 0021-9673.

Hirayama, N.; Deguchi, M.; Kawasumi, H. & Honjo, T. (2005). Use of 1-alkyl-3- methylimidazolium hexafluorophosphate room temperature ionic liquids as chelate extraction solvent with 4,4,4-trifluoro-1-(2-thienyl)-1,3-butanedione. *Talanta*, 65 (January, 2005) 255-260, ISSN: 0039-9140.

Hsieh, H.; Huang, P.; Sun, I.; Whang, T.; Hsu, C.; Huang, H. & Kuei, C. (2006). Nafion membrane-supported ionic liquid–solid phase microextraction for analyzing ultra trace PAHs in water samples. *Anal. Chim. Acta*, 557 (January, 2006) 321-328, ISSN: 0003-2670.

Huang, H. L.; Wang, H. P.; Wei, G. T.; Sun, I. W.; Huang, J. F. & Yang, Y. W. (2006). Extraction of nanosize copper pollutants with an ionic liquid. *Environ. Sci. Technol.*, 40 (August, 2006) 4761-4764, ISSN: 0013-936X.

Huang, K. P.; Wang, G. R.; Huang, B. Y. & Liu, C. Y. (2009). Preparation and application of ionic liquid-coated fused-silica capillary fibers for solid-phase microextraction. *Anal. Chim. Acta*, 645 (July, 2009) 42-47, ISSN: 0003-2670.

Jeannot, M. A. & Cantwell, F. (1996). Solvent microextraction into a single drop. *Anal. Chem.*, 68 (July, 1996) 2236-2240, ISSN: 0003-2700.

Jeannot, M. A. & Cantwell, F. (1997). Mass transfer characteristics of solvent extraction into a single drop at the tip of a syringe needle. *Anal. Chem.*, 69 (January, 1997) 235-239, ISSN: 0003-2700.

Kellner, R.; Mermet, J. M.; Otto, M.; Valcárcel, M. & Widmer, H. M. (2005). *Analytical Chemistry (2nd edition)*, Wiley-VCH, ISBN: 3-527-3059-4, Wheinheim.

Koel, M. (2005). Ionic Liquids in Chemical Analysis. *Crit. Rew. Anal. Chem.*, 35 (January, 2005) 177-192, ISSN: 1040-8347.

Li, Z.; Wei, Q.; Yuan, R.; Zhou, X.; Liu, H.; Shan, H. & Song, Q. (2007a). A new room temperature ionic liquid 1-butyl-3-trimethylsilylimidazolium hexafluorophosphate as a solvent for extraction and preconcentration of mercury with determination by cold vapor atomic absorption spectrometry. *Talanta*, 71 (January, 2007) 68-72, ISSN: 0039-9140.

Li, Z.; Lu, N.; Zhou, X. & Song, Q. (2007b). Extraction spectrophotometric determination of aluminum in dialysis concentrates with 3,5-ditertbutylsalicylfluorone and ionic liquid 1-butyl-3-trimethylsilylimidazolium hexafluorophosphate. *J. Pharm Biomed Anal.*, 43 (April, 2007) 1609-1614, ISSN: 0731-7085.

Li, S.; Cai, S.; Hu, W.; Chen, H. & Liu, H. (2009). Ionic liquid-based ultrasound-assisted dispersive liquid–liquid microextraction combined with electrothermal atomic absorption spectrometry for a sensitive determination of cadmium in water samples. *Spectrochim. Acta Part B*, 64, (July, 2009) 666-671, ISSN: 0584-8547.

Liang, C.; Yuan, C. Y.; Warmack, R. J.; Barnes, C. E. & Dai, S. (2002). Ionic liquids: a new class of sensing materials for detection of organic vapors based on the use of a quartz crystal microbalance. *Anal. Chem.*, 74, (May, 2002) 2172-2176, ISSN: 0003-2700.

Liu, H. & Dasgupta, P. K. (1996). Analytical Chemistry in a Drop. Solvent Extraction in a Microdrop. *Anal. Chem.*, 68, (June, 1996) 1817-1821, ISSN: 0003-2700.

Liu, J.; Jiang, G.; Chi, Y.; Cai, Y.; Zhou, Q. & Hu, J. (2003). Use of ionic liquids for liquid- phase microextraction of polycyclic aromatic hydrocarbons. *Anal. Chem.*, 75, (November, 2003) 5870-5876, ISSN: 0003-2700.

Liu, J. F.; Li, N.; Jiang, G. B.; Liu, J. M.; Jönsson, J. A. & Wen, M. J. (2005). Disposable ionic liquid coating for headspace solid-phase microextraction of benzene, toluene, ethylbenzene, and xylenes in paints followed by gas chromatography-flame ionization detection. *J. Chromatogr. A*, 1066(February, 2005) 27-32, ISSN: 0021-9673.

Liu, F. H. & Jiang, Y. (2007). Room temperature ionic liquid as matrix medium for the determination of residual solvents in pharmaceuticals by static headspace gas chromatography. *J. Chromatogr. A*, 1167 (October, 2007) 116-119, ISSN: 0021-9673.

Liu, Y.; Zhao, E.; Zhu, W.; Gao, H. & Zhou, Z. (2009a). Determination of four heterocyclic insecticides by ionic liquid dispersive liquid-liquid microextraction in water samples. *J. Chromatogr. A*, 1216, (February, 2009) 885-891, ISSN: 0021-9673.

Liu, R.; Liu, J. F.; Yin, Y. G.; Hu, X. L. & Jiang, G. B. (2009b). Ionic liquids in sample preparation. *Anal. Bioanal. Chem.*, 393 (February, 2009) 871-883, ISSN: 1618-2642.

López-Darias, J.; Pino, V.; Anderson, J. L.; Graham, C. M. & Afonso, A. M. (2010). Determination of water pollutants by direct-immersion solid-phase microextraction using polymeric ionic liquid coatings. *J. Chromatogr. A*, 1217, (February, 2010) 1236- 1243, ISSN: 0021-9673.

López-Pastor, M.; Domínguez-Vidal, A.; Ayora-Cañada, M. J.; Simonet, B. M.; Lendl, B. & Valcarcel, M. (2008). Separation of Single-Walled Carbon Nanotubes by Use of Ionic-Liquid-Aided Capillary Electrophoresis. *Anal. Chem.*, 80 (April, 2008) 2672-2679, ISSN: 0003-2700.

Lucena, R.; Cruz-Vera, M.; Cárdenas, S. & Valcárcel, M. (2009). Liquid phase microextraction in bioanalytical sample preparation. *Bioanalysis*, 1, (April, 2009) 135-149, ISSN 1757-6180.

Mahpishanian, S. & Shemirani, F. (2010). Preconcentration procedure using in situ solvent formation microextraction in the presence of ionic liquid for cadmium determination in saline samples by flame atomic absorption spectrometry. *Talanta*, 82 (July, 2010) 471-476, ISSN: 0039-9140.

Mank, M.; Stahl, B. & Boehm, G. (2004). 2,5-Dihydroxybenzoic acid butylamine and other ionic liquid matrixes for enhanced MALDI-MS analysis of biomolecules. *Anal. Chem.* 76 (May, 2004) 2938-2950, ISSN: 0003-2700.

Manzoori, J. L.; Amjadi, M. & Abulhassani, J. (2009a). Ionic liquid-based single drop microextraction combined with electrothermal atomic absorption spectrometry for the determination of manganese in water samples. *Talanta,* 77, (February, 2009) 1539-1544, ISSN: 0039-9140.

Manzoori, J. L.; Amjadi, M. & Abulhassani, J. (2009b). Ultra-trace determination of lead in water and food samples by using ionic liquid-based single drop microextraction- electrothermal atomic absorption spectrometry. *Anal. Chim. Acta,* 644, (June, 2009) 48-52, ISSN: 0003-2670.

Mao, T.; Hao, B.; He, J.; Li, W.; Li, S. & Yu, Z. (2009). Ultrasound assisted ionic liquid dispersive liquid phase extraction of lovastatin and simvastatin: a new pretreatment procedure. *J. Sep. Sci.* 32 (September, 2009) 3029-3033, ISSN: 1615-9306.

Martinis, E. M.; Bertón, P.; Altamirano, J. C.; Hakala, U. & Wuilloud, R. G. Tetradecyl(trihexyl) phosphonium chloride ionic liquid single-drop microextraction for electrothermal atomic absorption spectrometric determination of lead in water samples. *Talanta,* 80 (March, 2010) 2034-2040, ISSN: 0039-9140.

Mwongela, S. M.; Numan, A.; Gill, N. L.; Agbaria, R. A. & Warner, I. M. (2003). Separation of Achiral and Chiral Analytes Using Polymeric Surfactants with Ionic Liquids as Modifiers in Micellar Electrokinetic Chromatography. *Anal. Chem.* 75, (November, 2003) 6089-6096, ISSN: 0003-2700.

Nakashima, T. & Kawai, T. (2005). Quantum dots-ionic liquid hybrids: efficient extraction of cationic CdTe nanocrystals into an ionic liquid. *Chem. Commun.,* 28 (March, 2005) 1643-1645, ISSN: 1359-7345.

Palacio, M. & Bhushan, B. (2008). Ultrathin Wear-Resistant Ionic Liquid Films for Novel MEMS/NEMS Applications. *Adv. Mater.,* 20 (March, 2008) 1194-1198, ISSN: 0935-9648.

Pham, T. P.; Cho, C. W. & Yun, Y. S. (2010). Environmental fate and toxicity of ionic liquids: a review. *Water Res.* 44 (January, 2010) 352-372, ISSN: 0043-1354.

Pedersen-Bjergaard, S. & Rasmussen, K. E. (1999). Liquid-liquid-liquid microextraction for sample preparation of biological fluids prior to capillary electrophoresis. *Anal. Chem.* 71, (July, 1999) 2650-2656, ISSN: 0003-2700.

Pena, M. T.; Casais, M. C.; Mejuto, m. C. & Cela, R. (2009). Development of an ionic liquid based dispersive liquid-liquid microextraction method for the analysis of polycyclic aromatic hydrocarbons in water samples. *J. Chromatogr. A.,* 1216 (September, 2009) 6356-6364, ISSN: 0021-9673.

Pena-Pereira, F.; Lavilla, I.; Bendicho, C.; Vidal, L. & Canals, A. (2009). Speciation of mercury by ionic liquid-based single-drop microextraction combined with high performance liquid chromatography-photodiode array detection. *Talanta,* 78, (April, 2009) 537-541, ISSN: 0039-9140.

Peng, J.; Liu, J.; Jiang, G.; Tai, C. & Huang M. (2005). Ionic liquid for high temperature headspace liquid-phase microextraction of chlorinated anilines in environmental water samples. *J. Chromatogr. A,* 1072, (April, 2005) 3-6, ISSN: 0021-9673.

Peng, J.; Liu, J.; Hu, X. & Jiang, G. (2007). Direct determination of chlorophenols in environmental water samples by hollow fiber supported ionic liquid membrane extraction coupled with high-performance liquid chromatography. *J. Chromatogr. A,* 1139, (January, 2007) 165-170, ISSN: 0021-9673.

Polo-Luque, M. D.; Simonet, B. M. & Valcárcel, M. (in press. a). New soft materials of ionic liquids with carbon nanotubes as sorbents for SPE. *Anal. Chem.,* in press.

Polo-Luque, M. D.; Simonet, B. M. & Valcárcel, M. (in press. b). Combination of carbon nanotubes-modified filters for the NACE determination of trace levels of ionic- liquids in water. *J. Chromatogr. A.,* in press.

Polo-Luque, M. D.; Simonet, B. M. & Valcárcel, M. (in press. c). Solid-phase extraction- capillary electrophoretic determination of sulfonamide residues in surface water by using C18-carbon nanotubes as novel sorbent material. *Electrophoresis,* in press.

Poole, C. F. & Poole, S. K. (2010) Extraction of organic compounds with room temperature ionic liquids. *J. Chromatogr. A., 1217* (April, 2010) 2268-2286, ISSN: 0021-9673.

Qin, W. & Li, S. F. Y. (2002). An ionic liquid coating for determination of sildenafil and UK- 103,320 in human serum by capillary zone electrophoresis-ion trap mass spectrometry. *Electrophoresis*, 23 (December, 2002) 4110-4116, ISSN: 1522-2683.

Ravelo-Pérez, L. M.; Hernández-Borges, J.; Asensio-Ramos, M. & Rodríguez-Delgado, M. A. (2009a). Ionic liquid based dispersive liquid-liquid microextraction for the extraction of pesticides from bananas. *J. Chromatogr. A*, 1216 (October, 2009) 7336- 7345, ISSN: 0021-9673.

Ravelo-Pérez, L. M.; Hernández-Borges, J.; Herrera-Herrera, A. V. & Rodríguez-Delgado, M. A. (2009b). Pesticide extraction from table grapes and plums using ionic liquid based dispersive liquid-liquid microextraction. *Anal. Bioanal. Chem.*, 395 (December, 2009) 2387-2395, ISSN: 1618-2642.

Rezaee, M.; Assadi, Y.; Milani Hosseini, M. R.; Aghaee, E.; Ahmadi, F. & Berijani, S. (2006). Determination of organic compounds in water using dispersive liquid-liquid microextraction. *J. Chromatogr. A*, 1116, (May, 2006) 1-9, ISSN: 0021-9673.

Santos, L. S.; Haddad, R.; Höehr, N. F.; Pilli, R. A. & Eberlin, M. N. (2004). Fast screening of low molecular weight compounds by thin-layer chromatography and "on-spot" MALDI-TOF mass spectrometry. *Anal. Chem.*, 76 (April, 2004) 2144-2147, ISSN: 0003-2700.

Shearrow, A. M.; Harris, G. A.; Fang, L.; Sekhar, P. K.; Nguyen, L. T.; Turner, E. B.; Bhansali, S. & Malik, A. (2009a). Ionic liquid-mediated sol-gel coatings for capillary microextraction. . *J. Chromatogr. A*, 1216, (July, 2009) 5449-5458, ISSN: 0021-9673.

Shearrow, A. M.; Bhansali, S. & Malik, A. (2009b). Ionic liquid-mediated bis[(3- methyldimethoxysilyl) propyl] polypropylene oxide-based polar sol-gel coatings for capillary microextraction. *J. Chromatogr. A*, 1216, (September, 2009) 6349-6355, ISSN: 0021-9673.

Simoni, L. D.; Chapeaux, A.; Brennecke, J. F. & Stadtherr, M. A. (2010). Extraction of biofuels and biofeedstocks from aqueous solutions using ionic liquids. *Computers and chemical engineering*, 34 (September, 2010) 1406-1412, ISSN 0098-1354.

Tao, Y.; Liu, J.; Hu, X.; Li, H.; Wang, T. & Jiang, G. (2009). Hollow fiber supported ionic liquid membrane microextraction for determination of sulfonamides in environmental water samples by high-performance liquid chromatography. *J. Chromatogr. A*, 1216, (August, 2009) 6259-6266, ISSN: 0021-9673.

Tholey, A. & Heinzle, E. (2006) Ionic (liquid) matrices for matrix-assisted laser desorption/ionization mass spectrometry-applications and perspectives. *Anal. Bioanal. Chem.*, 386 (September, 2006) 24-37, ISSN: 1618-2642.

Valcárcel, M. & Ríos, A. (1993). The Hierarchy and Relationships of Analytical Properties. *Anal. Chem.*, 65 (September, 1993) 781A-787A, ISSN: 0003-2700.

Vidal, L.; Chisvert, A.; Canals, A. & Salvador, A. (2007a). Sensitive determination of free benzophenone-3 in human urine samples based on an ionic liquid as extractant phase in single-drop microextraction prior to liquid chromatography analysis. *J. Chromatogr. A*, 1174, (December, 2007) 95-103, ISSN: 0021-9673.

Vidal, L.; Psillakis, E.; Domini, C. E.; Grané, N.; Marken, F. & Canals, A. (2007b). An ionic liquid as a solvent for headspace single drop microextraction of chlorobenzenes from water samples. *Anal. Chim. Acta*, 584, (February, 2007) 189-195, ISSN: 0003-2670.

Vidal, L.; Domini, C. E.; Grané, N.; Psillakis, E. & Canals, A. (2007c). Microwave-assisted headspace single-drop microextration of chlorobenzenes from water samples. *Anal. Chim. Acta*, 592, (May, 2007) 9-15, ISSN: 0003-2670.

Vidal, L.; Chisvert, A.; Canals, A. & Salvador, A. (2010). Ionic liquid-based single-drop microextraction followed by liquid chromatography-ultraviolet spectrophotometry detection to determine typical UV filters in surface water samples. *Talanta*, 81, (April, 2010) 549-555, ISSN: 0039-9140.

Visser, A. E.; Swatloski, R. P.; Reichert, W. M.; Davis, J. H.; Rogers, R. D.; Mayton, R.; Sheff, S. & Wierzbicki, A. (2001). Task-specific ionic liquids for the extraction of metal ions from aqueous solutions. *Chem. Commun.*, (month, 2001), 135 - 136, ISSN: 1359-7345.

Wang, Y.; You, J.; Ren, R.; Xiao, Y.; Gao, S.; Zhang, H. & Yu, A. (2010). Determination of triazines in honey by dispersive liquid-liquid microextraction high-performance liquid chromatography. *J. Chromatogr. A.*, 1217 (June, 2010) 4241-4246, ISSN: 0021-9673.

Wanigasekara, E., Perera, S.; Crank, J. A.; Sidisky, L.; Shirey, R.; Berthod, A. & Armstrong, D. W. (2010). Bonded ionic liquid polymeric material for solid-phase microextraction GC analysis. *Anal. Bioanal. Chem.*, 396 (January, 2010) 511-524, ISSN: 1618-2642.

Wei, G. T.; Yang, Z. & Chen, C. J. (2003) Room temperature ionic liquid as a novel medium for liquid/ liquid extraction of metal ions. *Anal. Chim. Acta*, 488 (July, 2003) 183-192, ISSN: 0003-2670.

Wei, G. T.; Yang, Z.; Lee, C. Y.; Yang, H. Y. & Wang, C. R. (2004). Aqueous-organic phase transfer of gold nanoparticles and gold nanorods using an ionic liquid. *J. Am. Chem. Soc.*, 126 (April, 2004) 5036-5037, ISSN: 0002-7863.

Weingärtner, H. (2008). Understanding ionic liquids at the molecular level: facts, problems, and controversies. *Angew. Chem. Int. Ed.*, 47 (January, 2008) 654-670, ISSN: 1433-7851.

Xia, L.; Hu, B.; Jiang, Z.; Wu, Y.; Li, L. & Chen, R. (2005). 8-Hydroxyquinoline–chloroform single drop microextraction and electrothermal vaporization ICP-MS for the fractionation of aluminium in natural waters and drinks. *J. Anal. Atom. Spectrom., 20*, (May, 2005) 441-446, ISSN: 0267-9477.

Xia, L.; Li, X.; Wu, Y.; Hu, B. & Chen, R. (2008). Ionic liquids based single drop microextraction combined with electrothermal vaporization inductively coupled plasma mass spectrometry for determination of Co, Hg and Pb in biological and environmental samples. *Spectrochim. Acta Part B*, 63, (November, 2008) 1290-1296, ISSN: 0584-8547.

Yao, C. & Anderson, J. L. (2009). Dispersive liquid-liquid microextraction using an in situ metathesis reaction to form an ionic liquid extraction phase for the preconcentration of aromatic compounds from water. *Anal. Bioanal. Chem.* (November, 2009) 1491-1502, ISSN: 1618-2642.

Yao, C.; Twu, P. & Anderson, J. L. (2010). Headspace Single Drop Microextraction Using Micellar Ionic Liquid Extraction Solvents. *Chromatographia*, 72, (September, 2010) 393-402, ISSN 0009-5893.

Ye, C.; Zhou, Q. & Wang, X. (2006). Headspace liquid-phase microextraction using ionic liquid as extractant for the preconcentration of dichlorodiphenyltrichloroethane and its metabolites at trace levels in water samples. *Anal. Chim Acta.* 572 (July, 2006) 165-171, ISSN: 0003-2670.

Ye, C.; Zhou, Q.; Wang, X. & Xiao, J. (2007). Determination of phenols in environmental water samples by ionic liquid-based headspace liquid-phase microextraction coupled with high-performance liquid chromatography. *J. Sep. Sci.*, 30 (January, 2007) 42-47, ISSN: 1615-9306.

Yousefi, S. R. & Shemirani, F. (2010) Development of a robust ionic liquid-based dispersive liquid-liquid microextraction against high concentration of salt for preconcentration of trace metals in saline aqueous samples: application to the determination of Pb and Cd. *Anal. Chim. Acta*, 669 (June, 2010) 25-31, ISSN: 0003-2670.

Zabet-Moghaddam, M.; Heinzle, E. & Tholey, A. (2004) Qualitative and quantitative analysis of low molecular weight compounds by ultraviolet matrix-assisted laser desorption/ionization mass spectrometry using ionic liquid matrices. *Rapid Commun. Mass Spectrom.*, 18 (January, 2004) 141-148, ISSN: 0951-4198.

Zhang, J. & Lee, H. K. (2010). Headspace ionic liquid-based microdrop liquid-phase microextraction followed by microdrop thermal desorption-gas chromatographic analysis. *Talanta*, 81 (April, 2010) 537-542, ISSN: 0039-9140.

Zhao, H.; Xia, S. & Ma, P. (2005). Use of ionic liquids as green solvents for extractions. *J. Chem. Technol. Biotechnol.* 80 (October, 2005) 1089-1096, ISSN: 0268-2575.

Zhao, F.; Li, J. & Zeng, B. (2008a). Coupling of ionic liquid-based headspace single-drop microextraction with GC for sensitive detection of phenols. *J. Sep. Sci.*, 31 (September, 2008) 3045-3049, ISSN: 1615-9306.

Zhao, F.; Meng, Y. & Anderson, J. L. (2008b). Polymeric ionic liquids as selective coatings for the extraction of esters using solid-phase microextraction. *J. Chromatogr. A*, 1208 (October, 2008) 1-9, ISSN: 0021-9673.

Zhao, R.; Wang, X.; Sun J.; Wang, S.; Yuan, J. & Wang, X. (2010a). Trace determination of triclosan and triclocarban in environmental water samples with ionic liquid dispersive liquid-phase microextraction prior to HPLC–ESI-MS–MS. *Anal. Bioanal. Chem.*, 397 (June, 2010) 1627-1633, ISSN: 1618-2642.

Zhao, R. S.; Wang, X.; Sun, J.; Yuan, J. P.; Wang, S. S. & Wang, X. K. (2010b). Temperature- controlled ionic liquid dispersive liquid-phase microextraction for the sensitive determination of triclosan and triclocarban in environmental water samples prior to HPLC-ESI-MS/MS. *J. Sep. Sci.* 33 (June, 2010) 1842-1848, ISSN: 1615-9306.

Zhao, Q. & Anderson, J. L. (2010c). Selective extraction of CO2 from simulated flue gas using polymeric ionic liquid sorbent coatings in solid-phase microextraction gas chromatography. *J. Chromatogr. A*, 1217 (July, 2010) 4517- 4522, ISSN: 0021-9673.

Zhou, Q. & Ye, C. (2008). Ionic liquid for improved single-drop microextraction of aromatic amines in water samples. *Microchim. Acta*, 162 (July, 2008) 153-159, ISSN: 0026-3672.

Zhou, Q.; Bai, H.; Xie, G. & Xiao, J. (2008a). Temperature-controlled ionic liquid dispersive liquid phase micro-extraction. *J. Chromatogr. A*, 1177 (January, 2008) 43-49, ISSN: 0021-9673.

Zhou, Q.; Bai, H.; Xie, G. & Xiao, J. (2008b). Trace determination of organophosphorus pesticides in environmental samples by temperature-controlled ionic liquid dispersive liquid-phase microextraction. *J. Chromatogr. A*, 1188 (April, 2008) 148- 153, ISSN: 0021-9673.

Zhou, Q.; Bai, H.; Xie, G. & Xiao, J. (2009a). Temperature-controlled ionic liquid-dispersive liquid-phase microextraction for preconcentration of chlorotoluron, diethofencarb and chlorbenzuron in water samples. *J. Sep. Sci.* 32 (November, 2009) 3945-3950, ISSN: 1615-9306.

Zhou, Q.; Zhang, X. & Xiao, J. (2009b). Ultrasound-assisted ionic liquid dispersive liquid- phase micro-extraction: a novel approach for the sensitive determination of aromatic amines in water samples. *J Chromatogr A.* 1216 (May,2009) 4361-4365, ISSN: 0021-9673.

10
Chapter

COLD-INDUCED AGGREGATION MICROEXTRACTION (CIAME): A NOVEL IONIC LIQUID-BASED TECHNIQUE FOR COBALT PRECONCENTRATION IN WATER SAMPLES

1. INTRODUCTION: AN OVERVIEW

Despite advancements in contemporary analytical instruments, the direct quantification of trace analytes at low concentrations frequently poses challenges for analytical chemists, necessitating a sample-preparation phase. The sample preparation in the analytical process generally involves an extraction-separation step, which facilitates the isolation of target analytes from the matrix solution, thereby minimizing or potentially eliminating existing interferences, while also allowing for the preconcentration and detection of these analytes at very low levels. Liquid–liquid extraction (LLE), which relies on the partitioning of an analyte between two immiscible liquid phases, is a prevalent traditional sample preparation method. This procedure is laborious and time-intensive. Additionally, substantial quantities of high-purity solvents, which are costly and hazardous, are unavoidable in usage. The ongoing pursuit of innovative sample preparation techniques has resulted in the emergence of new methodologies, characterized by advantages such as increased speed and decreased solvent usage, including cloud point extraction (CPE), solid-phase microextraction (SPME), single drop microextraction (SDME), hollow fiber-liquid phase microextraction (HF-Liquid-phase microextraction (LPME) and dispersive liquid-liquid microextraction (DLLME), among others.

Cloud point extraction is regarded as an environmentally friendly extraction process, as it utilizes surfactants that often demonstrate less toxicity, volatility, and flammability in comparison to organic solvents employed in alternative extraction procedures. This method is extensively employed in the extraction of metals (Bezerra et al., 2005; Dallali et al., 2009), organic chemicals (Carabias-

Martinez et al., 2000), and proteins (Saitoh & Hinze, 1995). Despite the numerous advantages of utilizing CPE, elevated salt concentrations lead to an enhanced background due to the minimal aqueous sample included in the enrichment phase. Furthermore, when the concentration of water-miscible organic solvents, such as acetone, exceeds 3%, phase separation fails to occur, resulting in the degradation of the extraction system (Manzoori & Karim-Nezhad, 2003). The efficacy of this approach is constrained by temperature, pressure, pH, and the choice of certain surfactants (Paleologos et al., 2005).

Solid-phase microextraction (SPME) is a prevalent sample preparation process that is solvent-free, straightforward, and efficient, integrating extraction, preconcentration, and sample injection into a single step (Risticevic et al., 2009). Nevertheless, it is costly; its fiber is delicate and possesses a restricted lifespan. This procedure utilizes substantial quantities of eluents. Moreover, the carryover across extractions poses challenges when quantifying some analytes at minimal quantities.

Single-drop microextraction is a solvent microextraction (SME) technique executed by exposing a solitary drop of solvent to the headspace or directly into the sample matrix (Genfa & Dasgupta, 2000; He & Lee, 1997; Jeannot & Cantwell, 1996; Liu & Dasgupta, 1995). It was devised as a solvent-reduced sample preparation method that is cost-effective and beneficial. The minimum quantity of solvent employed results in negligible exposure to hazardous organic solvents (Ahmadi et al., 2006; Jeannot & Cantwell, 1997). Relative to solid phase microextraction, it offers advantages such as a broader selection of extraction solvents, more economical equipment, simplified operation, and the absence of analyte carryover. Nonetheless, comparatively minor quantities of extractant result in diminished sensitivity. Furthermore, it is labor-intensive, and at rapid stirring velocities, the organic droplet is fragmented, resulting in the formation of air bubbles.

Hollow-fiber liquid-phase microextraction is a sample preparation method wherein analytes are extracted into an organic layer contained within the pores of a suspended hollow fiber or subsequently re-extracted into an aqueous phase within the fiber, contingent upon the application of a two-phase or three-phase system (Pedersen-Bjergaard & Rasmussen, 1999; Rasmussen & Pedersen-Biergaard, 2004). The extraction efficiency achieved with this approach is superior to that of SDME due to the increased contact area between the analyte aqueous solution and the extraction phase. Nonetheless, the extraction process is laborious, and in many instances, equilibrium is not achieved within a little duration. Furthermore, carryover and contamination of the hollow fiber may occur.

Dispersive liquid-liquid microextraction (DLLME), a variant of solid-phase microextraction (SME) introduced in 2007, is characterized by its simplicity, rapidity, and cost-effectiveness. Nonetheless, the quantity of disperser solvent utilized is very substantial, which may result in a reduction in the extraction recovery of less hydrophobic species.

In recent years, room-temperature ionic liquids (RTILs) have garnered heightened interest and are increasingly utilized as extraction solvents, supplanting volatile solvents in sample preparation (Huddleston et al., 1998). This is attributed to their distinctive chemical and physical properties, including negligible vapor pressure, non-flammability, excellent extractability for diverse organic compounds and metal ions as neutral or charged complexes, along with adjustable viscosity and compatibility with water and organic solvents. Nonetheless, the majority of these studies were performed by liquid–liquid extraction, necessitating a substantial volume of ionic liquid, which is laborious and expensive (Smirnova et al., 2004; Xia et al., 2006). The microextraction utilizing ionic liquids was invented by Liu et al. (2003). Consequently, owing to the viscosity and cost of ionic liquids, we employed an ionic liquid (IL) at the microscale for the analytical extraction and quantification of mercury ions in water samples using spectrophotometric detection. (Gharehbaghi et al., 2009).

In 2008 Baghdadi and Shemirani developed a new method called cold-induced aggregation microextraction (CIAME), which was based on the use of ILs in homogeneous liquid–liquid microextraction (HLLME) (Baghdadi & Shemirani, 2008). In this method, a very small amount of hydrophobic IL as an extractant solvent is dissolved in the sample solution containing Triton X-114. Triton X-114 prevents IL from sticking onto the surface of a centrifuge tube wall, therefore we have named it "anti-sticking agent". One should be aware that there is no interface between the water and the extractant phases; as a result mass transfer from aqueous phase into separated phase has no important effect on the extraction step. Upon dissolution, the solution is chilled in an ice bath, resulting in a cloudy appearance due to the decreased solubility of the ionic liquid, leading to the formation of fine droplets. In the production of tiny droplets of the extractant phase, hydrophobic species are aggregated by the extractant molecules, culminating in the completion of the extraction process upon droplet formation. Upon centrifugation, the minute droplets of extractant phase accumulate at the base of the conical-bottom glass centrifuge tube.

This approach is straightforward, expeditious, secure, and resilient to elevated concentrations of salt and water-soluble organic solvents. In contrast to CPE, the background is excessively low for saline solutions due to the minimal solubility of water in ionic liquids. Furthermore, due to the elevated viscosity of ionic liquids, the extraction of the bulk aqueous phase is facilitated. This approach is more appropriate for extracting heat-sensitive species than CPE. CIAME offers a high recovery rate and exhibits little toxicity, as it utilizes minimal quantities of IL as a "green extraction solvent.".

2. IMPORTANCE

2.1 Cobalt; the occurrence, benefits and applications, exposure and toxicity and methods of determination

Cobalt is a natural element found throughout the environment, used in pigment manufacture and used to make superalloys (alloys that maintain their strength at high temperatures approaching their melting points). It can benefit or harm human health. Cobalt is an essential element in people's lives, as a constituent of vitamin B12. It has also been used as a treatment for anemia, because it stimulates red blood cell production. The significance of cobalt as a transition metal in its wide spectrum of applications is covering many frontier areas of study, particularly in medicine. We can be exposed to low levels of cobalt by breathing air, eating food or drinking water. Food and drinking water are the largest sources of exposure to cobalt for general population.

Toxicological effects of large amounts of cobalt include vasodilatation, flushing and cardiomyopathy in men and animals (Yuzefovsky et al., 1994). Even though cobalt is not considered to be as toxic as most heavy metals, it is an equally harmful element. Hence owing to the importance of cobalt, its determination from associated elements by extractive spectrophotometry has been of a considerable importance.

A wide variety of chelating agents has been reported for the spectrophotometric determination of cobalt (Appadoo & Bhagwat, 1994; Carvalho et al., 1996; Chaudhari & Sawant, 1993; Dasilva & Martins, 1992; Kalika et al., 1993; Kamburova et al., 1994; Khambekar & Sawant, 1997; Lokhande et al., 1996; Maheshwari & Balasubramanian, 1995; More & Sawant, 1992; Pillai & Shinde, 1995; R.A. Chaudhari & A.D. Sawant, 1993; Reddy & Sarma, 1994; Sharma & Dave, 1997; Shen et al., 1995; Taher & Puri, 1995; Toral et al., 1993; Umebayashi & Ishiguro, 1996). However, these methods suffer from limitations such as critical pH (Dasilva et al., 1992; Maheshwari et al., 1995; Pillai et al., 1995), long color development time (Umebayashi et al., 1996) and interference with some ions (Chaudhari et al., 1993; Kalika et al., 1993; Lokhande et al., 1996; Maheshwari et al., 1995; More et al., 1992; Pillai et al., 1995; Reddy et al., 1994; Toral et al., 1993; Umebayashi et al., 1996).

In addition, the preconcentration and determination of cobalt (II) has been studied in various extraction methods such as: liquid-liquid extraction (LLE), flow injection extraction (FIE), cloud point extraction (CPE), solid phase extraction (SPE) and dispersive liquid-liquid microextraction (DLLME).

New LLE methods based on liquid membranes (Kakoi et al., 1998; Ribeiro et al., 2004) and hollow fiber contactors (Soldenhoff et al., 2005) were used for extraction and preconcentration of cobalt. However, it is time-consuming, tedious and uses large amounts of high purity solvents, which are expensive and toxic. The continuous quest for novel sample preparation procedures has led to a development of new methods, among which main advantages belong their speed and negligible volume of used solvents. Initial efforts to address the problem of large solvent consumption have led to a development of the FIE method. Some of these methods have been applied for cobalt extraction (Andac et al., 2001; Cao et al., 1999; Cassella et al., 2001; Dadfarnia & Jafarzadeh, 1999; Dzherayan et al., 2002; Fujimoto et al., 1999; Li et al., 2006a; Li et al., 2006b; Nogueira et al., 1998; Shabani et al., 2003; Song et al., 2006; Tsakovski et al., 2002). FIE has some advantages compared with LLE like low cost, high extraction speed and reduced solvent and sample consumption. However, the amount of used solvent is still in the order of several hundred micro liters per analysis.

Also preconcentration of cobalt in some CPE methods has been reported (Bezerra et al., 2007; Chen & Teo, 2001; Donati et al., 2006; Ghaedi et al., 2008; Gil et al., 2008; Nascentes & Arruda, 2003; Safavi et al., 2004; Shemirani & Shokoufi, 2006; Shokoufi et al., 2007b). Despite many benefits from the use of CPE, in high content of salt, background is increased since the enrichment phase is composed of a little aqueous sample and in very high content of salt, the density of the sample becomes equal or even higher than that of micelles, therefore they can not be settled. Also in the presence of more than 3 % of water-miscible organic solvents such as acetone phase separation does not occur.

Several SPE preconcentration procedures for cobalt have been reported using various sorbents (Blitz-Raith et al., 2007; Ghaedi et al., 2007; Praveen et al., 2005; Yang et al., 2002).

DLLME is a type of solvent microextraction (SME), which has been developed in 2007. A combined method including fiber optic-linear array detection spectrophotometry (FO- LADS) and DLLME was developed for preconcentration and determination of cobalt (Shokoufi et al., 2007a). In this method 1,2-dichlorobenzene (a harmful solvent) was used as an extractant. Another DLLME method, which was applied to preconcentrate trace levels of cobalt as a prior step to its determination by spectrophotometric detection, was reported (Gharehbaghi et al., 2008). This method is simple, fast and inexpensive. However, the amount of used disperser solvent is relatively high, therefore it may happen that recoveries of relatively less hydrophobic species decrease.

Homogeneous liquid-liquid microextraction (HLLME) utilizes the phase separation phenomenon from a homogeneous solution and the target solutes are extracted into a separated phase. However, HLLME has some problems; for instance, sometimes it is not compatible with some instrumental analysis and it also requires an addition of a reagent such as acid, base, salt, etc. As a result, some interesting compounds are probably destroyed; moreover, the addition of the reagent causes a release of heat during the extraction.

2.2 Vantage points of cold-induced aggregation microextraction method

In this research the performance of CIAME was investigated with the determination of cobalt in water samples using FO-LADS. The effects of various experimental parameters on the extraction were investigated. This method is simple, rapid and viscosity independent for extraction and preconcentration of metal ions and organic compounds from water samples, and is robust against high content of salt and water-miscible organic solvents. Additionally, in comparison with the organic solvent extraction, it is much safer since only small amounts of surfactant and IL are used which are being considered as "green solvents" for various separation processes and in comparison with IL-based DLLME, the extraction recoveries of CIAME are relatively high (Gharehbaghi et al., 2009).

Another important capability of this method is that it can be operated in a continuous mode. If the solution is cooled during centrifuging, fine droplets of extractant phase are continuously formed and analytes are extracted and transferred to the bottom of the centrifuge tube. In order to have a good recovery in case of microextraction in which the volume of the extractant solvent is very small, a large distribution coefficient is necessary, unless the recovery is low and continuous microextraction is required. This project is currently being performed in our laboratory.

3. EXPERIMENTAL

3.1 Instrumentation

A UV–vis light source, optical fiber and spectrograph model 2048 were prepared from Avantes (Eerbeek, Netherlands) for FOLADS. A 50 μL quartz cylindrical micro-cell (Hellma, Mullheim, Germany) was used as a determination cell. A Universal 320R refrigerated centrifuge equipped with an angle rotor (6-place, 9000 rpm, Cat. No.1620A) and a Jeio Tech BW-05G water bath were obtained from Hettich (Kirchlengern, Germany).

3.2 Reagents

All used reagents were of analytical grade. Triply distilled water was used throughout the experiment. 1-(2-pyridylazo)-2-naphthol (PAN), acetone, ethanol, acetic acid, 1-hexyl-3- methylimidazolium hexafluorophosphate [Hmim][PF_6] and

1- hexyl-3-methylimidazolium bis (trifluoromethylsulfonyl) imide [Hmim][Tf_2N] and all used salts were obtained from Merck (Darmstadt, Germany). Triton X-100 and Triton X-114 were purchased from Fluka (Buchs, Switzerland). A stock solution of cobalt (II) (1000 mg L^{-1}) was prepared by dissolving appropriate amount of $CoCl_2{\cdot}6H_2O$ and working standard solutions were obtained by appropriate stepwise dilution of the stock standard solutions. All stock and working standard solutions were stored in plastic (polypropylene) bottles with leak proof screw cap, which were cleaned before being used by soaking in 10% nitric acid solution for at least 24 h and then rinsed thoroughly with triply distilled water. A 10^{-3} mol L^{-1} PAN solution was prepared by dissolving the appropriate amount in absolute ethanol. The viscosity of ILs is high and their handling is difficult, therefore working solutions ([Hmim][PF_6], 0.8 mg μL^{-1} and [Hmim][Tf_2N], 0.5 mg μL^{-1}) were prepared in acetone.

3.3 Cold-induced aggregation microextraction procedure

The sample or standard solution containing Co(II) in the range of 1.5–65 ng mL^{-1}, PAN ($2.5{\times}10^{-5}$ mol L^{-1}), Triton X-114 (0.05 %, w/v), sodium nitrate (0.25 %, w/v), oxalate ($4.0{\times}10^{-3}$ mol L^{-1}) was poured in a 10 mL conical-bottom glass centrifuge tube. After 1 or 2 min, while greenish complex of cobalt was formed, pH of the solution was adjusted to 1 by concentrated HCl. [Hmim][PF_6] (64mg) and [Hmim][Tf_2N] (5 mg) were added and the volume of the solution was adjusted to 10.0 mL by triply distilled water. After that, tube was kept in a thermostated bath at 35 °C for 4 min. After shaking, it was placed in an ice-water bath for 10 min and a cloudy solution was formed. Then, the mixture was centrifuged for 5 min at 5000 rpm. As a result, fine droplets of IL settled at the bottom of the centrifuge tube (about 8 μL). The bulk aqueous phase was removed simply by inverting the tubes. The extraction steps are illustrated in Fig. 1. Afterwards, IL-phase was dissolved in 50 μL of pure ethanol and transferred to quartz cylindrical micro-cell. The absorbance of the complex was measured at 578 nm.

4. RESULTS AND DISCUSSION

It is necessary to investigate the effect of all parameters that can probably influence the extraction performance. In this methodology these parameters are the kind and amount of IL and anti-sticking agent, ligand concentration, pH, salt concentration, temperature and centrifuge conditions, which were investigated and optimized in order to achieve a high recovery and enrichment factor. In all optimization steps concentration of cobalt was 30 ng mL^{-1}.

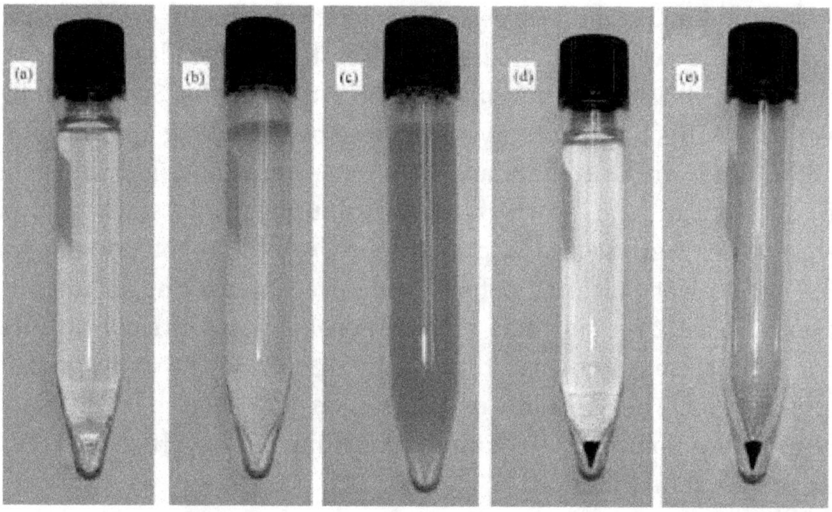

Fig. 1. Photography of different steps in CIAME: (a) after adding IL in the sample solution, (b) after shaking and dissolving the IL, (c) after cooling and phase separation, (d) after centrifuging and (e) after removing the bulk aqueous phase.

4.1 Selection of ionic liquid

For the selection of a suitable IL some considerations were recommended, which were mentioned in the previous work (Baghdadi et al., 2008). In this work, we focused on ILs containing imidazolium cation. Imidazolium-ILs containing PF_6^- as an anion are hydrophobe, relatively inexpensive and liquid in experimental conditions, therefore they are suitable for LLE. From [Bmim][PF_6], [Hmim][PF_6] and [Omim][PF_6] according to some physicochemical properties like density, viscosity and water solubility (Liu et al., 2005) and with regard to the point that the sample volume was 10 mL, [Hmim][PF_6] was chosen as an extractant. Thus, about 75 mg of this IL was solved in the aqueous sample solution.

4.2 Selection of anti-sticking agent

Like in the previous work, after centrifugation, it was observed that some of the IL-phase stick on the wall of the centrifuge tube. In order to overcome this problem, a non-ionic surfactant was added to the sample solution (Baghdadi et al., 2008). In the presence of non-ionic surfactant and during the phase separation, molecules of the surfactant surround the fine droplets of IL. Hence, interactions of IL with the wall of the centrifuge tube decrease and consequently, IL-phase does not stick on the wall of the centrifuge tube. The effects of two non-ionic surfactants (Triton X-114 and Triton X-100) were investigated and compared. In the presence of Triton X-100 absorbance decreased, while in the case of using Triton X-114 absorbance increased. Therefore, the Triton X-114 was chosen as the anti-sticking agent at the optimum concentration of 0.05 % (w/v).

4.3 Effect of pH and PAN concentration

In CIAME method pH plays a unique role in the metal-chelate formation and the subsequent extraction. The effect of pH on the extraction of cobalt from water samples was studied in the range of 0.5–6 after Co–PAN complex formation. Results reveal that the absorbance is slightly reduced by increasing pH. As well as at low pH, cations are less likely to precipitate; pH 1 seems a proper choice for extraction.

The effect of PAN concentration on the absorption is shown in Fig. 2. As it is well expected, in accordance with ML_2 stoichiometry of the complex the absorption is higher when increasing the PAN concentration. We investigated PAN concentration in the range of 1.5×10^{-7} to 4×10^{-5} mol L^{-1}. The maximum absorbance was obtained at a concentration of 2.5×10^{-5} mol L^{-1} of the ligand and after that, the absorbance stays approximately constant.

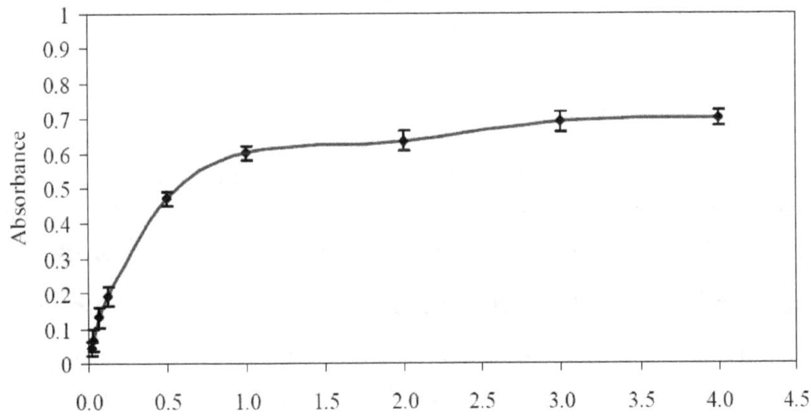

Fig. 2. Effect of PAN concentration on the absorbance of complex. Utilized conditions: Cobalt 30 ng mL^{-1}, pH = 1, $NaNO_3$ 0.2 % (w/v), Triton X-114 0.05 % (w/v), [Hmim][PF_6] 60 mg, [Hmim][Tf_2N] 5 mg, diluting agent 50 μL.

4.4 Effect of IL amount

In the presence of high content of salt, the solubility of [Hmim][PF6] increases and phase separation does not occur. But according to the common ion effect, solubility decreases in the presence of IL with a common ion (such as [Hmim][Tf_2N]). This is one of the interesting properties of ILs. The effect of [Hmim][PF_6] in the presence and absence of [Hmim][Tf_2N] is shown in Fig. 3. In the presence of [Hmim][Tf_2N], lower amount of [Hmim][PF_6] was required for the phase separation. The optimum amount of [Hmim][PF_6] for the complete extraction was 70 and 60 mg in the absence and presence of [Hmim][Tf_2N], respectively.

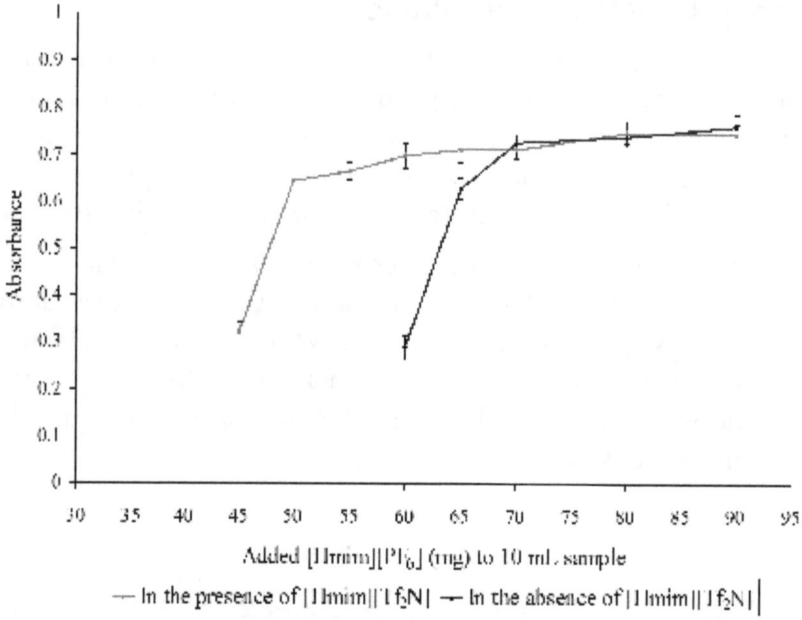

Fig. 3. Effect of the amount of [Hmim][PF_6] on the absorbance of Co-PAN complex. Utilized conditions: Cobalt 30 ng mL^{-1}, PAN 2.5×10^{-5} mol L^{-1}, pH = 1, $NaNO_3$ 0.2 % (w/v), Triton X- 114 0.05 % (w/v), [Hmim][Tf_2N] 5 mg, diluting agent 50 µL.

4.5 Effect of salt concentration and water-miscible organic solvents

$NaNO_3$ was chosen in order to study the salt effect in the range of 0–1 % (w/v). The absorbance slightly increased as a result of the salting out effect. A concentration of 0.2% (w/v) $NaNO_3$ was selected for subsequent experiments in order to increase recovery. Also to investigate the robustness of the method against salty solutions, various solutions with the concentration up to 10% were examined. As shown in Fig. 4, in the absence of [Hmim][Tf_2N], absorbance decreased by increasing the salt concentration as a result of increasing in solubility of [Hmim][PF_6], but in the presence of [Hmim][Tf_2N] phase separation occurred up to 10% $NaNO_3$.

Since acetone and ethanol were selected as the solvents for the IL and ligand, their effects were investigated, too. In the presence of acetone, the absorbance was nearly constant up to 10 % and afterwards, it decreased and in the presence of ethanol, the absorbance was almost constant up to 12 %. This is one of the advantages of CIAME compared to CPE. In case of CPE, the extraction system is destroyed in the presence of over 3% water-miscible organic solvents (Manzoori et al., 2003). Even in case of DLLME based on common organic solvents like chloroform, carbon tetrachloride, etc., the extraction efficiency was clearly reduced for solutions containing water-miscible organic solvents.

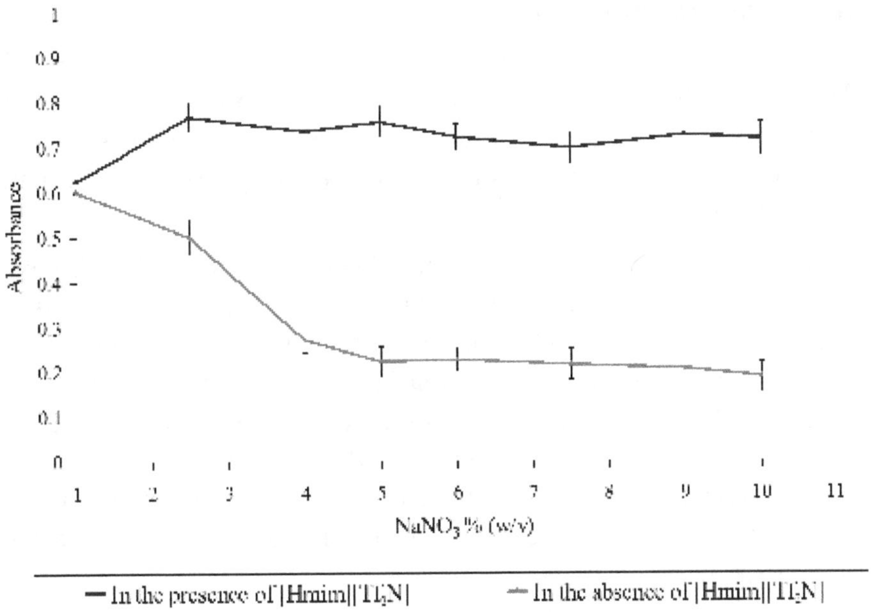

Fig. 4. Effect of NaNO$_3$ on the absorbance of Co-PAN complex in the range of 1-10 % (w/v). Utilized conditions: Cobalt 30 ng mL^{-1}, PAN 2.5×10^{-5} mol L^{-1}, pH = 1, Triton X-114 0.05 % (w/v), [Hmim][PF$_6$] 64 mg, [Hmim][Tf$_2$N] 5 mg, diluting agent 50 µL.

4.6 Effect of temperature

Before shaking the solutions containing IL, they were heated in the range of 20-60 °C. In this range the absorbance was constant; warming of the solution had no effect on the extraction process. This is another advantage of CIAME in comparison with CPE, especially if the species are heat-sensitive. Since ILs are dissolved more easily at the temperatures above 30 °C, a temperature of 35 °C was chosen for the heating step.

After dissolving of IL, samples were cooled in the temperature range of 0-25 °C. As the temperature decreased, the absorbance increased, as a result of decreasing IL solubility. Hence, a temperature of 0 °C was applied in all experiments by placing the centrifuge tubes in an ice bath for 10 min.

4.7 Effect of centrifuge conditions

The effect of centrifugation rate on the absorbance was studied in the range of 1000 - 7000 rpm. It was found, that over 4000 rpm IL-phase completely settled, therefore a rate of 5000 rpm was selected as the optimum point.

At the optimum rate, the absorbance was investigated as a function of the centrifugation time. Over 2 min, absorbance was constant, indicating complete transfer of IL-phase to the bottom of centrifuge tube. Therefore, the optimum centrifugation time was chosen as 3 min.

4.8 Effect of coexisting ions

The effects of coexisting ions in real water samples on the recovery of cobalt were also studied. In these experiments, 10mL solutions containing 30 ng mL^{-1} of cobalt and various amounts of interfering ions were treated. A given species was considered to interfere if it resulted in a ± 5% variation of the absorbance. The obtained results are given in Table 1.

Most of the examined cations and anions did not interfere with the extraction and determination of Co^{2+}. However, some of the tried species, such as Fe^{3+}, Cu^{2+} and Pd^{2+} interfered with the determination of Co^{2+} ion. These interferences were eliminated in the presence of proper masking agents, such as 1.5×10^{-4} mol L^{-1} acetylacetone for Pd^{2+}, 4.0×10^{-3} mol L^{-1} oxalate for Fe^{3+} (and Ca^{2+}) and 2.5×10^{-3} mol L^{-1} ascorbic acid and 4.0×10^{-3} mol L^{-1} iodide for Cu^{2+}. In the presence of the masking agents, no interference was observed for Ca^{2+} up to 1000, Fe^{3+} and Cu^{2+} up to 100 and Pd^{2+} up to 10 times relative to Co^{2+} concentration. Thus, a quantitative extraction of Co^{2+} was possible.

Table 1. Effect of interferents on the recovery of 30 µg L-1 Co(II) in water sample using CIAME

Interferent	Interferent/Co(II) (weight ratio)	Recovery (%)	Interferent	Interferent/Co(II) (weight ratio)	Recovery (%)
Ba^{2+}	1000	98	Cu^{2+}	100	-[c]
Mg^{2+}	1000	100		100	105[d]
Mn^{2+}	1000	99	Hg^{2+}	100	92
Cr^{3+}	1000	104		50	99
Cd^{2+}	1000	100	Pd^{2+}	10	188
Zn^{2+}	1000	103		10	95[e]
Ca^{2+}	1000	98[a]	I^-	1000	100
Ag^+	500	97	SO_4^{2-}	1000	101
Pb^{2+}	500	101	NO_3^-	1000	98
Al^{3+}	500	102	PO_4^{3-}	1000	98
Fe^{3+}	100	165	CH_3COO^-	1000	100
	100	98[b]	$NO2^-$	1000	98

[a]Ca^{2+} masked with 4.0×10^{-3} mol L^{-1} oxalate .

[b]Fe^{3+} masked with 4.0×10^{-3} mol L^{-1} oxalate.

[c]In the presence of this interferent the shape of spectrum is changed completely.

[d]Cu^{2+} masked with 2.8×10^{-3} mol L^{-1} ascorbic acid and 3.9×10^{-3} mol L^{-1} iodide. [e]Pd^{2+} masked with 1.9×10^{-4} mol L^{-1} acetylacetone.

4.9 Figures of merit

Table 2 summarizes analytical characteristics of the optimized method, including limit of detection, reproducibility and enhancement factor. The limit of detection (LOD = 0.14 ng mL^{-1}) was calculated as 3 S_b/m (S_b: standard deviation of the blank signals; m: slope of calibration curve after the preconcentration). A good correlation coefficient (r = 0.9997) was obtained and only small deviations between sequential determinations (RSD = 2.32 %) were found. The calibration curve was investigated up to 65 ng mL^{-1} which was linear. Enhancement factor (EF = 165) was obtained from the slope ratio of calibration curve after and before the preconcentration.

Table 2. Analytical characteristics of CIAME for determination of Co(II)

Parameter	Analytical Feature
Linear range (ng mL^{-1})	1.5-65
r^2	0.9995
Limit of detection (ng mL^{-1}) (3σ, n = 5)	0.14
Repeatability (RSD[a], %) (n = 5)	2.32
Enhancement factor[b] (EF)	165
Sample volume (mL)	10
Extraction time (min)	<20

[a]RSD was obtained for 30 µg L^{-1} concentration of Co(II) .

[b]Enhancement factor is the slope ratio of calibration graph after and before extraction.

The equation of the calibration curve after the preconcentration procedure is given in Eq. (1). Also the equation of the calibration curve before the preconcentration procedure is given at below in Eq. (2). (These standard samples of cobalt were chosen up to 200 ng mL^{-1} with execs PAN to measure their absorbance and obtain the equation of the calibration curve accurately).

$$A = 2.30 \times 10^{-2} C_{(Co)} + 0.0094 \ (r^2 = 0.9995, r = 0.9997) \quad (1)$$

$$A = 1.39 \times 10^{-4} C_{(Co)} + 0.0323 \ (r^2 = 0.9983, r = 0.9991) \quad (2)$$

4.10 Accuracy

For evaluating the accuracy of the method, a certified reference material (CRM) was analyzed. The Reference Standard “SCP-ES-L-1” (SCP SCIENCE Canada, ground water (Low level)) is a ground water that has been spiked with metals. The certified value and analytical results are presented in Table 3. The result indicates effectiveness and accuracy of the proposed method.

Table 3. Determination of cobalt in a standard reference material using DLLME

Certified reference material	Certified[a] (ng mL^{-1})	Found[b] (ng mL^{-1})	Found[c] (ng mL^{-1})	Recovery (%)
SCP-ES-L-1	51 ± 8	48.6 ± 0.2 [d]	49.5 ± 0.6 [d]	101.8

[a]Results after dilution 1 : 500.

[b]Determined by GFAAS.

[c]Determined by CIAME/FO-LADS.

[d]Mean value ± standard deviation based on five replicate measurements.

4.11 Analysis of natural water

The proposed methodology was applied for the determination of cobalt in different water samples. Damavand mineral water, Tajan river water and Caspian Sea water samples were collected from the north of Iran and analyzed by CIAME as a prior step to its determination by FO-LADS. No concentration of cobalt in the tap and mineral water samples was detected. Moreover, the robustness of the proposed method was checked by performing recovery tests on a saline serum and a synthetic sample (containing Mn^{2+}, Zn^{2+}, Mg^{2+}, Cd^{2+}, Cr^{3+} of 1000 ng mL^{-1} and Na^{+}, K^{+} of 3000 ng mL^{-1}). Each type of water was spiked with variable amounts of Co(II) to assess matrix effects. The results are shown in Table 4. The relative recoveries of cobalt from mentioned water samples at various spiking levels were between 96.0 and 103.0 %. These results demonstrated that matrices of these water samples, under presented conditions, had little effect on CIAME of cobalt.

Table 4. Determination of cobalt in real and synthetic samples

Sample	Spiked (ng mL^{-1})	Found (ng mL^{-1})[a]	Recovery (%)
Mineral water	–	N.D[b]	–
	10	10.1 ± 0.15	101.0
	20	20.3 ± 0.32	101.5
Tap water[c]	–	N.D[b]	–
	10	10.2 ± 0.22	102.0
	15	14.7 ± 0.35	98.0
River water	–	4.3 ± 0.39	–
	10	13.9 ± 0.23	96.0
	25	28.5 ± 0.33	96.8
Sea water	–	1.7 ± 0.27	–
	10	12.0 ± 0.32	103.0
	25	27.1 ± 0.55	101.6

Saline Serum[d]	25	25.6 ± 0.2	102.4
	45	43.9 ± 0.4	97.6
Synthetic sample	10	10.1 ± 0.18	101.0
	15.0	14.9 ± 0.23	99.3

[a]Mean ± S.D. (n = 5).

[b]Not detected.

[c]From drinking water system of Tehran, Iran.

[d]Produced by Daru Paksh Company for injection.

4.12 Comparison of CIAME with other methods

A comparison of the represented method with other reported preconcentration methods is given in Table 5. In comparison with other reported methods, CIAME has low LOD (0.14 ng mL^{-1}), high enrichment factor (165) and relative short extraction procedure (less than 20 min). All these results indicate that CIAME is a reproducible, simple and low cost technique that can be used for the preconcentration of metal ions like cobalt from water samples.

5. CONCLUSIONS AND FUTURE WORK

In this study, a new mode of HLLME based on IL, named cold-induced aggregation microextraction (CIAME), was developed for preconcentration of cobalt from water samples as a prior step to its determination by FO-LADS. Cobalt was employed as a test analyte and 1-(2-pyridylazo)-2-naphthol (PAN) as a complexing agent to assess the extraction procedure.

Table 5. Characteristic performance data obtained by using CIAME and other techniques for determination of cobalt

Method	LOD[a] (ng mL^{-1})	RSD.[b] (%)	Enhance-ment factor	Sample consumpt-ion (mL)	Calibration range (ng mL^{-1})	References
CPE/Spectrophotometry	7.5	2.2	10[c]	10	20–200	(Safavi et al., 2004)
CPE/FAAS	1.06	5.41	28.5	12.5	25-200	(Nascent es et al., 2003)
CPE/FAAS	0.24	2.1	57[d]	50	0-120	(Chen et al., 2001)
SPE/Spectrophotometry	10	2.23	100[e]	250	10-400	(Yang et al., 2002)
DLLME/FO- LADS	0.2	<4	165	10	1-70	(Shokoufi et al., 2007a)

DLLME/Spectrophotometry	0.5	2.5	125	50	2-60	(Gharehbaghi et al., 2008)
CPE/FAAS	5	1.71	20[f]	10	0-200	(Giokas et al., 2001)
SPE/ETAAS	0.004	4.5	87	11.5	0.01–0.25	(Sant'Ana et al., 2002)
CIAME/FO-LADS	0.14	2.32	165	10	1.5-65	This work

[a]Limit of detection.
[b]Relative standard deviation.
[c]Ratio of the aqueous phase to final volume of surfactant-rich phase.
[d]The enhancement factor, as the ratio of absorbance of preconcentrated sample to that obtained without preconcentration.
[e]Ratio of the aqueous phase to final volume of eluent phase.
[f]Preconcentration factor.

This method is simple, rapid, safe and robust against high content of salt and water-miscible organic solvents. In comparison with CPE, background is too little in case of saline solutions, because of very low solubility of water in ILs. Also owing to high viscosity of ILs removing the bulk aqueous phase is easier. This method is more suitable for extraction of heat-susceptible species in comparison with CPE. CIAME provides high recovery and has low toxicity since only very small amounts of an IL as a "green extraction solvent" is used. In addition, the proposed method offers good sensitivity in comparison with other combination methods, which were used, FAAS or conventional UV–vis spectrophotometer as the detection technique.

Our research group is presently engaged in continuous mode CIAME and aims to develop a microextraction process with a minimal volume of extractant solvent in future work. A substantial distribution coefficient is essential for effective recovery, unless continuous microextraction is employed, wherein fine droplets of the extractant phase are perpetually generated, facilitating the extraction and transport of analytes to the bottom of the centrifuge tube.

REFERENCES

Ahmadi, F.;Assadi, Y.;Hosseini, S., & Rezaee, M. (2006). Determination of organophosphorus pesticides in water samples by single drop microextraction and gas chromatography-flame photometric detector. *Journal of Chromatography A*, 1101(1-2), 307-312.

Andac, M.;Asan, A.;Isildak, I., & Cesur, H. (2001). Flow-injection spectrophotometric determination of cobalt(II) at low mu g 1(-1) levels with 4- benzylpiperidinedithiocarbamate. *Analytica Chimica Acta*, 434(1), 143-147.

Appadoo, C., & Bhagwat, V. W. (1994). Spectrophotometric determination of cobalt with m-(mercaptoacetamido) phenol. *Asian Journal of Chemistry*, 6, 703–705.

Baghdadi, M., & Shemirani, F. (2008). Cold-induced aggregation microextraction: A novel sample preparation technique based on ionic liquids. *Analytica Chimica Acta*, 613(1), 56-63.

Bezerra, M. A.;Maeda, S. M. D.;Oliveira, E. P.;de Carvalho, M. D. B., & Santelli, R. E. (2007). Internal standardization for the determination of cadmium, cobalt, chromium and manganese in saline produced water from petroleum industry by inductively coupled plasma optical emission spectrometry after cloud point extraction. *Spectrochimica Acta Part B-Atomic Spectroscopy*, 62(9), 985-991.

Bezerra, M. D.;Arruda, M. A. Z., & Ferreira, S. L. C. (2005). Cloud point extraction as a procedure of separation and pre-concentration for metal determination using spectroanalytical techniques: A review. *Applied Spectroscopy Reviews*, 40(4), 269-299.

Blitz-Raith, A. H.;Paimin, R.;Cattrall, R. W., & Kolev, S. D. (2007). Separation of cobalt(II) from nickel(II) by solid-phase extraction into Aliquat 336 chloride immobilized in poly(vinyl chloride). *Talanta*, 71(1), 419-423.

Cao, Q. E.;Zhao, Y. K.;Cheng, X. G.;Hu, Z. D., & Xu, Q. H. (1999). Sequential fluorescent determination of copper(II) and cobalt(II) in food samples by flow injection analysis. *Food Chemistry*, 65(3), 405-409.

Carabias-Martinez, R.;Rodriguez-Gonzalo, E.;Moreno-Cordero, B.;Perez-Pavon, J. L.;Garcia- Pinto, C., & Laespada, E. F. (2000). Surfactant cloud point extraction and preconcentration of organic compounds prior to chromatography and capillary electrophoresis. *Journal of Chromatography A*, 902(1), 251-265.

Carvalho, M. S.;Fraga, I. C. S.;Neto, K. C. M., & Silva, E. Q. (1996). Selective determination of cobalt using polyurethane foam and 2-(2-benzothiazolylazo)-2-p-cresol as a spectrophotometric reagent. *Talanta*, 43(10), 1675-1680.

Cassella, R. J.;Salim, V. A.;Jesuino, L. S.;Santelli, R. E.;Ferreira, S. L. C., & de Carvalho, M. S. (2001). Flow injection determination of cobalt after its sorption onto polyurethane foam loaded with 2-(2-thiazolylazo)-p-cresol (TAC). *Talanta*, 54(1), 61-67.

Chaudhari, R. A., & Sawant, A. D. (1993). Extraction and spectrophotometric determination of cobalt. *Asian Journal of Chemistry*, 5, 1-5.

Chen, J. R., & Teo, K. C. (2001). Determination of cobalt and nickel in water samples by flame atomic absorption spectrometry after cloud point extraction. *Analytica Chimica Acta*, 434(2), 325-330.

Dadfarnia, S., & Jafarzadeh, M. H. (1999). Online trace enrichment and determination of cobalt ion as an anionic complex by flow injection atomic absorption spectrometry. *Microchemical Journal*, 63(2), 226-234.

Dallali, N.;Zahedi, M. M.;Yamimi, Y., & Agrawal, Y. K. (2009). Simultaneous cloud point extraction and determination of zn, co, ni, and pb by flame atomic absorption spectrometry. *Reviews in Analytical Chemistry*, 28(2), 125-136.

Dasilva, J. F., & Martins, W. (1992). Extraction of Fe(III), Cu(II), Co(II), Ni(II) and Pb(II) with thenoyltrifluoroacetone using the ternary solvent system water ethanol methylisobutylketone. *Talanta*, 39(10), 1307-1312.

Donati, G. L.;Nascentes, C. C.;Nogueira, A. R. A.;Arruda, M. A. Z., & Nobrega, J. S. (2006). Acid extraction and cloud point preconcentration as sample preparation strategies for cobalt determination in biological materials by thermospray flame furnace atomic absorption spectrometry. *Microchemical Journal*, 82(2), 189-195.

Dzherayan, T. G.;Shkinev, V. M.;Shpigun, L. K.;Kamilova, P. M., & Geckeler, K. E. (2002). Water-soluble polymers for spectrophotometric and flow injection determination of cobalt with nitroso-R-salt. *Talanta*, 57(1), 7-13.

Fujimoto, T.;Teshima, N.;Kurihara, M.;Nakano, S., & Kawashima, T. (1999). Catalytic determination of cobalt at sub-nanogram levels using the oxidative coupling of 3-methyl-2-benzothiazolinone hydrazone with N-ethyl-N-(2-hydroxy-3-sulfopropyl)-3,5-dimethoxyaniline by manual and flow-injection methods. *Talanta*, 49(5), 1091-1098.

Genfa, Z., & Dasgupta, P. K. (2000). A continuous film-recirculable drop gas-liquid equilibration device. Measurement of trace gaseous ammonia. *Analytical Chemistry*, 72(14), 3165-3170.

Ghaedi, M.;Ahmadi, F., & Soylak, M. (2007). Preconcentration and separation of nickel, copper and cobalt using solid phase extraction and their determination in some real samples. *Journal of Hazardous Materials*, 147(1-2), 226-231.

Ghaedi, M.;Shokrollahi, A.;Ahmadi, F.;Rajabi, H. R., & Soylak, M. (2008). Cloud point extraction for the determination of copper, nickel and cobalt ions in environmental samples by flame atomic absorption spectrometry. *Journal of Hazardous Materials*, 150(3), 533-540.

Gharehbaghi, M.;Shemirani, F., & Baghdadi, M. (2008). Dispersive liquid-liquid microextraction and spectrophotometric determination of cobalt in water samples. *InternationalJournal ofEnvironmental Analytical Chemistry*, 88(7), 513-523.

Gharehbaghi, M.;Shemirani, F., & Baghdadi, M. (2009). Dispersive liquid-liquid microextraction based on ionic liquid and spectrophotometric determination of mercury in water samples. *International Journal of Environmental Analytical Chemistry*, 89(1), 21-33.

Gil, R. A.;Gasquez, J. A.;Olsina, R.;Martinez, L. D., & Cerutti, S. (2008). Cloud point extraction for cobalt preconcentration with on-line phase separation in a knotted reactor followed by ETAAS determination in drinking waters. *Talanta*, 76(3), 669- 673.

Giokas, D. L.;Paleologos, E. K.;Tzouwara-Karayanni, S. M., & Karayannis, M. I. (2001). Single-sample cloud point determination of iron, cobalt and nickel by flow injection analysis flame atomic absorption spectrometry - application to real samples and certified reference materials. *Journal of Analytical Atomic Spectrometry*, 16(5), 521-526.

He, Y., & Lee, H. K. (1997). Liquid phase microextraction in a single drop of organic solvent by using a conventional microsyringe. *Analytical Chemistry*, 69(22), 4634-4640.

Huddleston, J. G.;Willauer, H. D.;Swatloski, R. P.;Visser, A. E., & Rogers, R. D. (1998). Room temperature ionic liquids as novel media for 'clean' liquid-liquid extraction. *Chemical Communications*(16), 1765-1766.

Jeannot, M. A., & Cantwell, F. F. (1996). Solvent microextraction into a single drop. *Analytical Chemistry*, 68(13), 2236-2240.

Jeannot, M. A., & Cantwell, F. F. (1997). Mass transfer characteristics of solvent extraction into a single drop at the tip of a syringe needle. *Analytical Chemistry*, 69(2), 235-239.

Kakoi, T.;Ura, T.;Kasaini, H.;Goto, M., & Nakashio, F. (1998). Separation of cobalt and nickel by liquid surfactant membranes containing a synthesized cationic surfactant. *Separation Science and Technology*, 33(8), 1163-1180.

Kalika, K. C.;Kumar, A.;Rajeev, P., & Rajeev, S. (1993). Simultaneous spectrophotometric estimation of cobalt(II) and nickel(II) with phenylazobenzaldoxime. *Asian Journal of Chemistry*, 5, 559–562.

Kamburova, M.;Alexandrov, A., & Trifonov, K. (1994). Extraction-spectrophotometric determination of cobalt in soils and plants using tetrazolium violet. *Chemia Analityczna*, 39(5), 639-643.

Khambekar, A. M., & Sawant, A. D. (1997). Extractive and spectrophotometric determination of Co(II) using p-nitroisonitrosoacetophenone. *Indian Journal of Chemistry Section a-Inorganic Bio-Inorganic Physical Theoretical & Analytical Chemistry*, 36(5), 459-460.

Li, B. X.;Wang, D. M.;Lv, H., & Zhang, Z. J. (2006a). Chemometrics-assisted simultaneous determination of cobalt(II) and chromium(III) with flow-injection chemiluminescence method. *Spectrochimica Acta Part a-Molecular and Biomolecular Spectroscopy*, 65(1), 67-72.

Li, B. X.;Wang, D. M.;Lv, H. G., & Zhang, Z. J. (2006b). Flow-injection chemiluminescence simultaneous determination of cobalt(II) and copper(II) using partial least squares calibration. *Talanta*, 69(1), 160-165.

Liu, J.;Jiang, G. B.;Chi, Y. G.;Cai, Y. Q.;Zhou, Q. X., & Hu, J. T. (2003). Use of ionic liquids for liquid-phase microextraction of polycyclic aromatic hydrocarbons. *Analytical Chemistry*, 75(21), 5870-5876.

Liu, J. F.;Jonsson, J. A., & Jiang, G. B. (2005). Application of ionic liquids in analytical chemistry. *Trac-Trends in Analytical Chemistry*, 24(1), 20-27.

Liu, S. R., & Dasgupta, P. K. (1995). Liquid droplet - a renewable gas sampling interface. *Analytical Chemistry*, 67(13), 2042-2049.

Lokhande, R. S.;Jaywant, A. S., & Barhate, V. D. (1996). Extractive spectrophotometric determination of cobalt(II) with p-methylisonitrosoacetophenonehydrazone. *Asian Journal of Chemistry*, 8(4), 610-612.

Maheshwari, V., & Balasubramanian, N. (1995). Spectrophotometric determination of cobalt with nitrosochromotropic acid. *Fresenius Journal of Analytical Chemistry*, 351, 333- 335.

Manzoori, J. L., & Karim-Nezhad, G. (2003). Selective cloud point extraction and preconcentration of trace amounts of silver as a dithizone complex prior to fame atomic absorption spectrometric determination. *Analytica Chimica Acta*, 484(2), 155- 161.

More, P. S., & Sawant, A. D. (1992). Extractive and spectrophotometric determination of Co(II) at trace levels using isonitroso-4-methyl-2-pentanone. *Indian Journal of Chemistry Section a-Inorganic Bio-Inorganic Physical Theoretical & Analytical Chemistry*, 31(12), 984-985.

Nascentes, C. C., & Arruda, M. A. Z. (2003). Cloud point formation based on mixed micelles in the presence of electrolytes for cobalt extraction and preconcentration. *Talanta*, 61(6), 759-768.

Nogueira, A. R. A.;Souza, G. B.;Batista, L. A. R.;Arruda, M. A. Z., & Miyazawa, M. (1998). Catalytic determination of cobalt by flow injection spectrophotometry: application in animal faeces analysis using hydrochloric acid as extracting solution. *Animal Feed Science and Technology*, 74(1), 79-84.

Paleologos, E. K.;Giokas, D. L., & Karayannis, M. I. (2005). Micelle-mediated separation and cloud-point extraction. *Trac-Trends in Analytical Chemistry*, 24(5), 426-436.

Pedersen-Bjergaard, S., & Rasmussen, K. E. (1999). Liquid-liquid-liquid microextraction for sample preparation of biological fluids prior to capillary electrophoresis. *Analytical Chemistry*, 71(14), 2650-2656.

Pillai, V. P., & Shinde, V. M. (1995). Synergistic extraction of copper(II), cobalt(II) and nickel(II) with trifluoroacetylacetone and pyridine. *Indian Journal of Chemistry Section a-Inorganic Bio-Inorganic Physical Theoretical & Analytical Chemistry*, 34(5), 407-409.

Praveen, R. S.;Daniel, S., & Rao, T. P. (2005). Solid phase extraction preconcentration of cobalt and nickel with 5,7-dichloroquinone-8-ol embedded styrene-ethylene glycol dimethacrylate polymer particles and determination by fame atomic absorption spectrometry (FAAS). *Talanta*, 66(2), 513-520.

R.A. Chaudhari & A.D. Sawant (1993). Extraction and spectrophotometric determination of cobalt. *Asian Journal of Chemistry*, 5, 1-5.

Rasmussen, K. E., & Pedersen-Biergaard, S. (2004). Developments in hollow fibre-based, liquid-phase microextraction. *Trac-Trends in Analytical Chemistry*, 23(1), 1-10.

Reddy, B. R., & Sarma, P. (1994). Extractive spectrophotometric determination of cobalt using cyanex-272. *Talanta*, 41(8), 1335-1339.

Ribeiro, C. P.;Costa, A. O. S.;Lopes, I. P. B.;Campos, F. F.;Ferreira, A. A., & Salum, A. (2004). Cobalt extraction and cobalt-nickel separation from a simulated industrial leaching liquor by liquid surfactant membranes using Cyanex 302 as carrier. *Journal of Membrane Science*, 241(1), 45-54.

Risticevic, S.;Niri, V. H.;Vuckovic, D., & Pawliszyn, J. (2009). Recent developments in solid- phase microextraction. *Analytical and Bioanalytical Chemistry*, 393(3), 781-795.

Safavi, A.;Abdollahi, H.;Nezhad, M. R. H., & Kamali, R. (2004). Cloud point extraction, preconcentration and simultaneous spectrophotometric determination of nickel and cobalt in water samples. *Spectrochimica Acta Part a-Molecular and Biomolecular Spectroscopy*, 60(12), 2897-2901.

Saitoh, T., & Hinze, W. L. (1995). Use of surfactant-mediated phase-separation (cloud point extraction) with affinity ligands for the extraction of hydrophilic proteins. *Talanta*, 42(1), 119-127.

Sant'Ana, O. D.;Oliveira, L. G.;Jesuino, L. S.;Carvalho, M. S.;Domingues, M. D. F.;Cassella, R. J., & Santelli, R. E. (2002). A novel sampling strategy for cobalt determination by ETAAS using polyurethane foam. *Journal of Analytical Atomic Spectrometry*, 17(3), 258-262.

Shabani, A. M. H.;Dadfarnia, S., & Dehghan, K. (2003). On-line preconcentration and determination of cobalt by chelating microcolumns and flow injection atomic spectrometry. *Talanta*, 59(4), 719-725.

Sharma, R. R., & Dave, D. P. (1997). Solvent extraction and spectrophotometric determination of cobalt(II) with isonitrosopropiophenone. *Indian Journal of Chemistry Section a-Inorganic Bio-Inorganic Physical Theoretical & Analytical Chemistry*, 36(7), 629-630.

Shemirani, F., & Shokoufi, N. (2006). Laser induced thermal lens spectrometry for cobalt determination after cloud point extraction. *Analytica Chimica Acta*, 577(2), 238-243.

Shen, H. X.;Tang, Y. P.;Xiao, X. L.;Zhang, S. F., & Liu, R. X. (1995). Synthesis and studies on the analytical functions of a highly selective spectrophotometric reagent 2-(8- quinolylazo)-5-n,n'-dimethylaminobenzoic acid. *Analyst*, 120(5), 1599-1602.

Shokoufi, N.;Shemirani, F., & Assadi, Y. (2007a). Fiber optic-linear array detection spectrophotometry in combination with dispersive liquid-liquid microextraction for simultaneous preconcentration and determination of palladium and cobalt. *Analytica Chimica Acta*, 597(2), 349-356.

Shokoufi, N.;Shemirani, F., & Memarzadeh, E. (2007b). Fiber optic-linear array detection spectrophotometry in combination with cloud point extraction for simultaneous preconcentration and determination of cobalt and nickel. *Analytica Chimica Acta*, 601(2), 204-211.

Smirnova, S. V.;Torocheshnikova, II;Formanovsky, A. A., & Pletnev, I. V. (2004). Solvent extraction of amino acids into a room temperature ionic liquid with dicyclohexano- 18-crown-6. *Analytical and Bioanalytical Chemistry*, 378(5), 1369-1375.

Soldenhoff, K.;Shamieh, M., & Manis, A. (2005). Liquid-liquid extraction of cobalt with hollow fiber contactor. *Journal of Membrane Science*, 252(1-2), 183-194.

Song, Z. H.;Yue, Q. L., & Wang, C. N. (2006). Flow injection chemiluminescence determination of femtogram-level cobalt in egg yolk, fish tissue and human serum. *Food Chemistry*, 94(3), 457-463.

Taher, M. A., & Puri, B. K. (1995). Column preconcentration of cobalt with the ion-pair of 2- nitroso-1-naphthol-4-sulfonic acid tetradecyldimethylbenzylammonium chloride supported on naphthalene using 2nd-derivative spectrophotometry. *Analyst*, 120(5), 1589-1592.

Toral, M. I.;Richter, P.;Silva, L., & Salinas, A. (1993). Determination of mixtures of cobalt and iron by 1st derivative spectrophotometry. *Microchemical Journal*, 48(2), 221-228.

Tsakovski, S.;Benkhedda, K.;Ivanova, E., & Adams, F. C. (2002). Comparative study of 8-hydroxyquinoline derivatives as chelating reagents for flow-injection preconcentration of cobalt in a knotted reactor. *Analytica Chimica Acta*, 453(1), 143-154.

Umebayashi, Y., & Ishiguro, S. I. (1996). Spectrophotometric study of thiocyanato complexation of cobalt(II) and nickel(II) ions in micellar solutions of a nonionic surfactant Triton X-100. *Journal of Solution Chemistry*, 25(8), 731-746.

Xia, Z.;Li, Z.;Rui, Y., & Liu, H. (2006). A novel room temperature ionic liquid extraction spectrophotometric determination of trace germanium in natural water with methybenzeneazosalicylfluorone. *Analytical Letters*, 39(4/6), 863-877.

Yang, G. Y.;Huang, Z. J.;Hu, Q. F., & Yin, J. Y. (2002). Study on the solid phase extraction of Co(II)-QADEAB chelate with C-18 disk and its application to the determination of trace cobalt. *Talanta*, 58(3), 511-515.

Yuzefovsky, A. I.;Lonardo, R. F.;Wang, M. H., & Michel, R. G. (1994). Determination of ultra-trace amounts of cobalt in ocean water by laser-excited atomic fluorescence spectrometry in a graphite electrothermal atomizer with semi online flow-injection preconcentration. *Journal of Analytical Atomic Spectrometry*, 9(11), 1195-1202.

11 Chapter

UTILIZATION OF IONIC LIQUIDS IN THE SEPARATION OF AZEOTROPIC MIXTURES

1. INTRODUCTION: AN OVERVIEW

The growing apprehension regarding environmental issues, coupled with the implementation of new rules, has recently focused the scientific community's attention on innovative methods utilizing greener technologies. In numerous industrial sectors, the accumulation of solvent mixes arises from the challenges associated with recycling. It is essential to separate these combinations into their pure components for potential reuse. Nevertheless, the majority of solvent mixes contain azeotropes, rendering their separation via simple distillation unfeasible. Extractive distillation is the predominant separation technique employed to eliminate a component from an azeotropic system. The introduction of a new solvent (entrainer) in this process enhances interaction with one component of the previous mixture, hence modifying their relative volatilities. This evident advantage is limited by the substantial energy expenses required to attain a fluid phase system. In this context, liquid-liquid separation, predicated on the immiscibility of two liquid phases at ambient temperature, presents an advantageous alternative for diminishing energy usage and environmental effect.

Ionic liquids (ILs) have emerged as a prominent "green" medium for engineers, attributable to its exceptional physicochemical features and recyclability. The primary reason for interest in these modern solvents is their minimal vapor pressure at ambient temperature (Earle et al., 2006), which mitigates the risk of worker exposure and solvent evaporation into the atmosphere. Furthermore, ionic liquids can be customized for particular applications by the precise selection of cations and anions (Huddleston et al., 2001). This characteristic is highly appealing to the industry, as the fine-tuning of solvent properties enables the optimization of chemical engineering requirements for efficiency and cost-effectiveness of processes.During the last years, ILs have exhibited the ability to separate azeotropes including ethanol + water (Jork et al., 2004; Seiler et al., 2004), tetrahydrofuran (THF) + water (Jork et al., 2004; Seiler et al., 2004; Hu et al., 2006), benzene + heptane (Letcher & Deenadayalu, 2003; Gonzalez et al., 2009), ethyl acetate + ethanol

(Zhang et al., 2008), and ethyl tert-butyl ether (ETBE) + ethanol (Arce et al., 2007). Although the increasing number of publications addressing azeotropic separations with ILs, these studies only analyze the liquid-liquid equilibria (LLE), vapor-liquid equilibria or simulation of the extractive distillation process. The present work considers the use of ILs formed by 1-alkyl-3-methylimidazolium cation ($[C_nMIM^+]$) and several anions as solvents in a bench-scale extraction process. One of the selected anions was hexafluorophosphate $[PF_6^-]$ because it has been commonly investigated and considered historically one of the most important anion families; despite the fact that this anion can undergo hydrolysis producing hydrofluoric acid when in contact with water and at high temperatures (Swatloski et al., 2003; Najdanovic-Visak et al., 2002). Consequently, its industrial application has been restricted to those processes under water free conditions and moderate temperatures. Other anions chosen are methyl sulfate $[MeSO_4]$ and ethyl sulfate $[EtSO_4]$, since they display the most promising potential for application in industrial processes (Holbrey et al., 2002) because they can be easily synthesized in an halide-free manner at reasonable cost, they are chemically and thermally stable, and they have low melting points and relatively low viscosities. Taking into account all these features, these ILs were considered good candidates to be tested as extracting solvents or entrainers in the separation of azeotropic mixtures.

To assess this possibility, the liquid–liquid equilibria of various azeotropic mixes with ionic liquids at 298.15 K and atmospheric pressure were conducted. The experimental data were effectively correlated using the NRTL equation (Renon & Prausnitz, 1968), hence enabling their application in computerized systems. The efficacy of the chosen ionic liquids as azeotrope disruptors in liquid extraction procedures was assessed using selectivity and solute distribution ratio analysis. This capability was compared with various extracting solvents mentioned in the literature.

Based on the examination of extraction capacities, the systems yielding the most favorable findings were chosen to conduct lab-scale extraction operations that include a solvent recycling phase. The operational parameters of the lab-scale extraction process were optimized with HYSYS software. The optimized conditions were evaluated in a laboratory-scale packed column, and the extraction efficiencies of the processes within the column were quantified.

2. EXPERIMENTAL

2.1 Chemicals

1-Ethyl-3-methylimidazolium ethyl sulfate ([C2MIM] [EtSO4]) was purchased from Solvent Innovation with a purity ≥ 99 wt%. The others ionic liquids were synthesized according to procedure described in previous research (Pereiro et al., 2006a; Pereiro et al., 2007a; Pereiro et al., 2007b). NMR and positive FAB mass spectra were performed and the results are in agreement with literature. The ionic liquids were always used directly following the reduction of its water content to a mass fraction < 0.03%, determined by Karl Fischer titration. Hexane (from Aldrich, ≥ 99.0 wt%), heptane (from Aldrich, ≥ 99.0 wt%), 2- butanone (from Merck, 99.5 wt%), 2-propanol (from Merck, 99.7 wt%), ethyl acetate (from Fluka, 99.8 wt%), cyclohexane (from Fluka, 99.8 wt%) and ethanol (from Merck, ≥ 99.8 wt%) were dried over 4 Å molecular sieves (supplied by Aldrich) for several weeks before use, and the purities were verified by means of gas chromatography.

2.2 Equipment and techniques

2.2.1 Experimental liquid- liquid equilibrium procedure

Ternary liquid-liquid equilibrium data were determined in a glass cell (Fig. 1a-b) containing a magnetic stirrer and thermostatted by a water jacket connected to a bath controlled to ± 0.01 K. The temperature in the cell was measured with an ASL F200 digital thermometer with an uncertainty of ± 0.01 K (Fig. 1c).

Two techniques were used to study the ternary liquid-liquid equilibrium. The first based on the experimental determination of the binodal curve, estimating the immiscible area. The second is grounded on the determination of the tie-lines, which calculates the composition of each layer.

The binodal curve was determined by adding known quantities of the three components corresponding to the immiscible area into the equilibrium cell. Then, we slowly added known quantities of solute maintaining the stirring until the "cloud point" disappears (Fig. 1a).

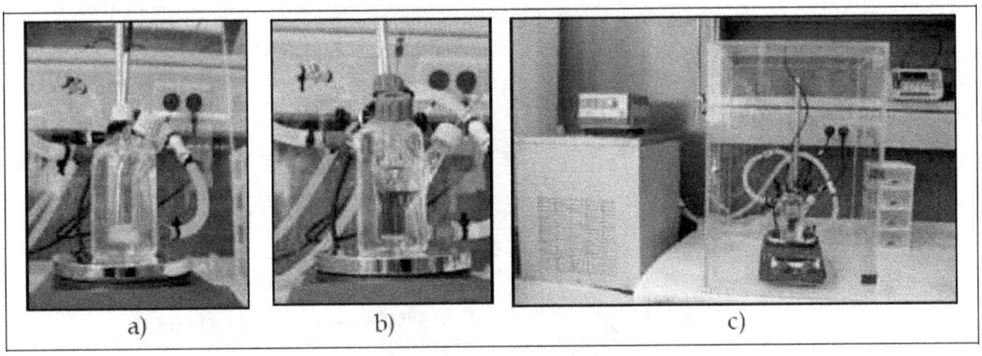

Fig. 1. (a)-(b) Liquid-liquid equilibria cells. (c) Experimental equipment used to determine of the liquid-liquid equilibria.

For the determination of the tie-lines, 30 ml of ternary mixture of known composition was added to the cell, the temperature was brought to 298.15 ± 0.01 K, and the mixture was stirred vigorously for 1 h and left to settle for 4 h (Fig.

1b). Then, samples of both layers were taken with a syringe and the densities and refractive indices were determined. Lastly, the phase compositions were inferred by means of calibration curves which had been previously constructed at 298.15 K. These curves were obtained by fitting the composition on the binodal curve by means of refractive indices and densities at 298.15 K. The uncertainty of the phase composition is ± 0.004 in mass fraction. All weight measurements were performed in a Mettler AX-205 Delta Range balance with an uncertainty of ± 10^{-4} mass fraction. Also, densities were measured with an Anton Paar DSA-48 digital vibrating tube densimeter with an uncertainty of ± 2×10^{-4} g cm^{-3}. Finally, refractive indices were calculated via a Dr. Kernchen ABBEMAT WR automatic refractometer with an uncertainty of ± 4×10^{-5}.

2.2.2 Packed-column experiments

The practical performance of ionic liquids as azeotrope breakers in extraction processes was researched by using ionic liquids for continuous countercurrent separation of the azeotropic mixture in a 54 × 1585 mm glass extraction column packed with 8 × 8 mm Raschig rings to a height of 1475 mm (Fig. 2).

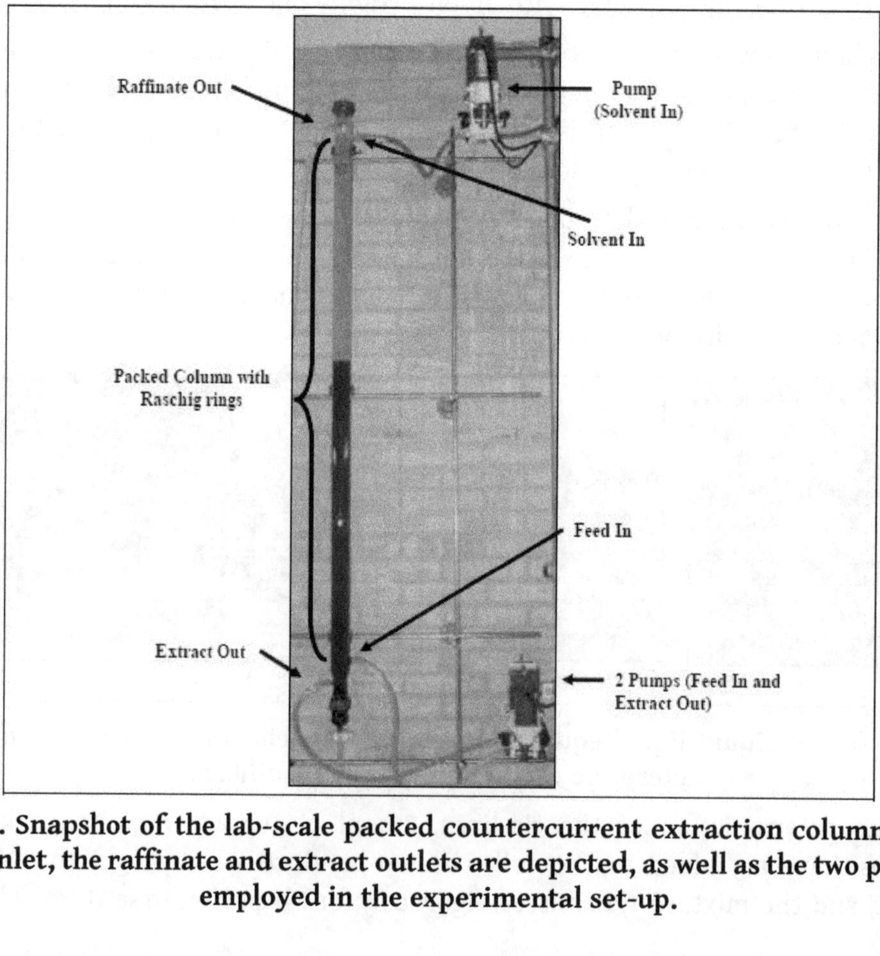

Fig. 2. Snapshot of the lab-scale packed countercurrent extraction column. The feed inlet, the raffinate and extract outlets are depicted, as well as the two pumps employed in the experimental set-up.

The experiment was carried out at room temperature under steady-state conditions. First, the azeotropic mixture and solvent streams were pumped in, the extract stream was pumped out (with FMI QV laboratory pumps) and the raffinate stream came out of the column under gravity. Then, samples from the top and bottom of the column were taken periodically and the composition was determined by analyzing their density and refractive index. At last, the ionic liquid was recovered from the extract stream and was recycled into the packed column after regeneration.

2.2.3 Regeneration of the Ionic Liquids

The ionic liquids used during these experiments were recovered and purified from the extract stream by removing the rest of its components in a Büchi R 3000 rotary evaporator with a vacuum controller. This operation is straightforward due to the fact that vapor pressure of ionic liquids is lower than the one of the rest of the components. The purity of the recovered ILs was verified by comparing its density at 298.15 K and NMR with the density and NMR of the freshly synthesized ones.

3. RESULTS AND DISCUSSION

3.1 Liquid-liquid separation of the azeotropic mixtures

The implementation of a separation process requires the accurate knowledge and careful control of the thermodynamic properties of the mixture, especially the phase boundaries. In this study, ternary liquid-liquid equilibria of different azeotropic mixtures with ionic liquids as extraction agents (Table 1) were measured at 298.15 K and atmospheric pressure. The selected azeotropic mixtures are commonly used in different processes of coating industry, petrochemical and food industry. The separation of each azeotropic mixture is discussed in the following sections.

Table 1. Azeotropic mixture + ionic liquid systems analyzed by liquid-liquid equilibria at 298.15 K and atmospheric pressure.

Azeotropic mixture	Ionic Liquids	Azeotropic mixture	Ionic Liquids
Ethanol + 2-Butanone	$[C_1MIM][MeSO_4]$	Hexane+Ethanol	$[C_1MIM][MeSO_4]$
	$[C_4MIM][PF_6]$		$[C_4MIM][MeSO_4]$
2-Propanol+2-Butanone	$[C_1MIM][MeSO_4]$		$[C_2MIM][EtSO_4]$
	$[C_4MIM][PF_6]$		$[C_6MIM][PF_6]$
2-Propanol+Ethyl acetate	$[C_1MIM][MeSO_4]$		$[C_8MIM][PF_6]$
	$[C_4MIM][PF_6]$	Heptane+Ethanol	$[C_1MIM][MeSO_4]$
	$[C_6MIM][PF_6]$		$[C_4MIM][MeSO_4]$
Hexane+Ethyl acetate	$[C_6MIM][PF_6]$		$[C_2MIM][EtSO_4]$
	$[C_8MIM][PF_6]$		$[C_6MIM][PF_6]$
Cyclohexane+2-Butanone	$[C_6MIM][PF_6]$		$[C_8MIM][PF_6]$
	$[C_8MIM][PF_6]$		

Values of solute distribution ratio, β, and selectivity, S, are widely used parameters in assessing the solvent feasibility the in liquid–liquid extraction. The solute distribution ratio supplies the amount of ionic liquid required for the process, related to the capacity of the IL and the selectivity evaluate the efficiency of the ionic liquid used as solvent. These parameters are defined as follows:

$$\beta = \frac{x_2^{II}}{x_2^{I}} \tag{1}$$

$$S = \left(\frac{x_1^{I}}{x_1^{II}}\right) \cdot \left(\frac{x_2^{II}}{x_2^{I}}\right) \tag{2}$$

where x is the mole fraction, subscripts 1 and 2 refer to inert and solute, respectively, and superscripts I and II indicate the organic (raffinate) and ionic liquid (extract) phases, respectively.

In order to perform simulation studies and process design, the experimental data were fitted with NRTL equation. The parameters were adjusted to minimize the difference between the experimental and calculated mole fraction defined as:

$$\text{O.F.} = \sum_{i=1}^{n} \left[\left(x_{1i}^{I} - x_{1i}^{I}(\text{calc})\right)^2 + \left(x_{2i}^{I} - x_{2i}^{I}(\text{calc})\right)^2 + \left(x_{1i}^{II} - x_{1i}^{II}(\text{calc})\right)^2 + \left(x_{2i}^{II} - x_{2i}^{II}(\text{calc})\right)^2 \right] \tag{3}$$

where x_{1i}^{I}, x_{2i}^{I}, x_{1i}^{II}, x_{2i}^{II} are the experimental mole fraction; $x_{1i}^{I}(\text{calc})$, $x_{2i}^{I}(\text{calc})$, $x_{1i}^{II}(\text{calc})$ and $x_{2i}^{II}(\text{calc})$ are the calculated mole fraction; and superscripts I and II indicate the organic (raffinate) and ionic liquid (extract) phases, respectively.

The deviations were calculated by applying the following expression:

$$\sigma = \left(\frac{\sum_{i} \left(x_{ilm}^{\text{exp}} - x_{ilm}^{\text{calc}}\right)^2}{6k} \right)^{1/2} \tag{4}$$

where x is the mole fraction and the subscripts i, l and m provide the component, the phase and the tie - line, respectively. The k value refers to the number of experimental tie-lines.

3.1.1 Separation of alcohols + 2-Butanone

Traditionally, 2-butanone has been used as a solvent in paints and resin adhesives. A mixture of different alcohols with this ketone that form azeotropes is a very common product. Given the wide diversity of alcohol + ketone mixtures, study of the binary mixtures ethanol or 2-propanol with 2-butanone has been considered (Pereiro & Rodriguez, 2007c; Pereiro & Rodriguez, 2007d).

At present, the separation of these azeotropic mixtures is made by azeotropic distillation (Berg, 1995; Berg, 1999) using as entrainers: amyl acetate, methyl formate, 2,2-dimethyl butane or 2,3-dimethyl butane for the azeotrope ethanol + 2-butanone; and 3- methylpentane, amyl ether or acetonitrile for the azeotrope 2-propanol + 2-butanone.

The evaluation of [C_4MIM][PF_6] and [C_1MIM][$MeSO_4$] as potential solvents in liquid -liquid extraction for the recovery of alcohols from the azeotropes was carried out through the analysis of liquid-liquid equilibrium data. Binodal curves and the tie lines were obtained for the mixtures of ethanol + 2-butanone + [C_4MIM][PF_6], 2-propanol + 2-butanone + [C_4MIM][PF_6], 2-butanone + ethanol + [C_1MIM][$MeSO_4$] and 2-butanone + 2-propanol + [C_1MIM][$MeSO_4$] at 298.15 K and atmospheric pressure.

The values of the selectivity for the studied ternary systems as a function of the solute composition in the organic phase are plotted in Figure 3. A comparison with conventional organic extractive solvents (Katayama et al., 1998; Katayama & Amano, 2005) was made and also depicted in the Figure 3.

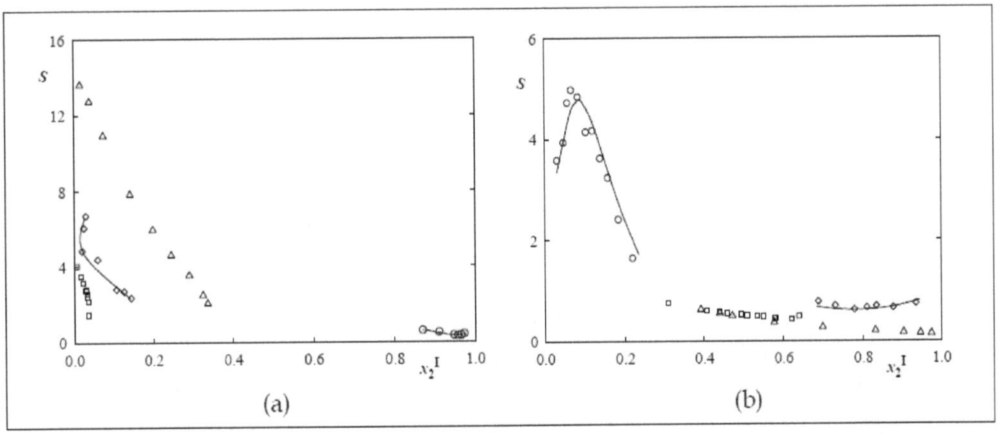

Fig. 3. Selectivity (S) of the systems presenting azeotrope: (a) ethanol (1) + 2-butanone (2); (b) 2-propanol (1) + 2-butanone (2) where: O, [C_4MIM][PF_6] this work; O, [C_1MIM][$MeSO_4$] this work; Δ, glycerol (Katayama et al., 1998); D, water (Katayama & Amano, 2005); solid line, NRTL correlation as a function of the solute mole fraction (ethanol or 2-butanone) in the organic phases at 298.15 K.

All the selectivity values for all areas of the binodal curves are higher than unity, from which it can be inferred that the extraction of the solute from the azeotropic system is indeed possible. A comparison between the selectivity values for the ternary systems shows that the [C_1MIM][$MeSO_4$] obtains higher values than [C_4MIM][PF_6] for the ethanol + 2- butanone and the opposite behavior is observed for 2-propanol + 2-butanone. In Figure 3a, the selectivity values at low concentration of ethanol for the ethanol + 2-butanone separation are similar in the $[C_1MIM][MeSO_4]$ and glycerol and higher than they are in water and $[C_4MIM][PF_6]$. On the other hand, for 2-propanol + 2-butanone, the selectivity values of the $[C_4MIM][PF_6]$ are higher than for the other solvents and ionic liquid.

3.1.2 Separation of 2-Propanol + Ethyl Acetate

The azeotropic mixture ethyl acetate + 2-propanol is present in the solvent extraction of edible oils (Bera et al., 2006). The liquid-liquid equilibrium data were obtained for the ternary mixtures of ethyl acetate + 2-propanol + $[C_1MIM][MeSO_4]$, 2-propanol + ethyl acetate + $[C_4MIM][PF_6]$ and 2-propanol + ethyl acetate + $[C_6MIM][PF_6]$ at 298.15 K and atmospheric pressure (Pereiro & Rodriguez, 2007e). The binodal curve for ternary mixture 2- propanol + ethyl acetate + $[C_8MIM][PF_6]$ was also determined (Fig. 4).

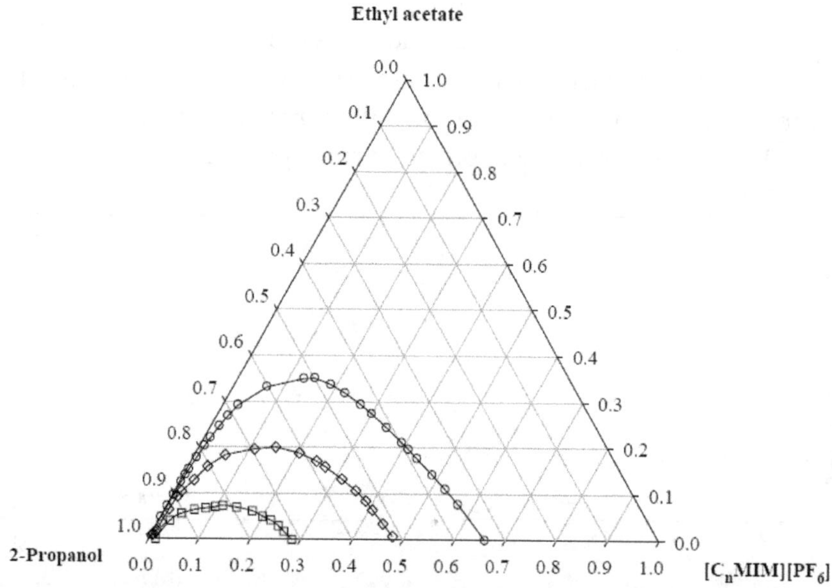

Fig. 4. Experimental binodal curves of the ternary systems 2-propanol (1) + ethyl acetate (2) + $[C_nMIM][PF_6]$ (3) at 298.15 K, where: O, $[C_4MIM][PF_6]$; O, $[C_6MIM][PF_6]$; and D, $[C_8MIM][PF_6]$.

In Figure 4, it is observed how the immiscibility region decreases if the length of the alkyl chain in the imidazolium ring increases when the $[C_nMIM][PF_6]$ was used as solvent in the liquid-liquid extraction.

The analysis of the LLE data (Pereiro & Rodriguez, 2007e) indicates that the alkyl chain length of the imidazolium ring plays a negative role in the capability of the $[C_nMIM][PF_6]$ to purify the alcohol. This negative role may be caused mainly by the hydrophobic steric effect of the alkyl group which reduces the polar character of the secondary –OH group of 2- propanol. From the LLE data it is also verified that the use of $[C_1MIM][MeSO_4]$ as solvent leads to higher values of solute distribution ratio and selectivity than the ionic liquids involving PF_6 as the anion due to the fact that the $MeSO_4$ ionic liquid contains just two $-CH_3$ groups and the steric alkyl effect is reduced.

3.1.3 Separation of Hexane + Ethyl Acetate

The ILs $[C_6MIM][PF_6]$ and $[C_8MIM][PF_6]$ have been chosen for the separation of the azeotropic mixture hexane + ethyl acetate. This azeotrope is present in the process for purifying grafted polyolefins (Gupta & Carey, 2006), and its separation is made by azeotropic batch distillation (Rodriguez-Donis et al., 2005) with heterogeneous entrainers such as methanol, acetonitrile, water, and nitromethane. The evaluation of these two ILs as extraction solvents for the recovery of ethyl acetate from its mixture with hexane was carried out through the analysis of LLE (Pereiro & Rodriguez, 2008a).

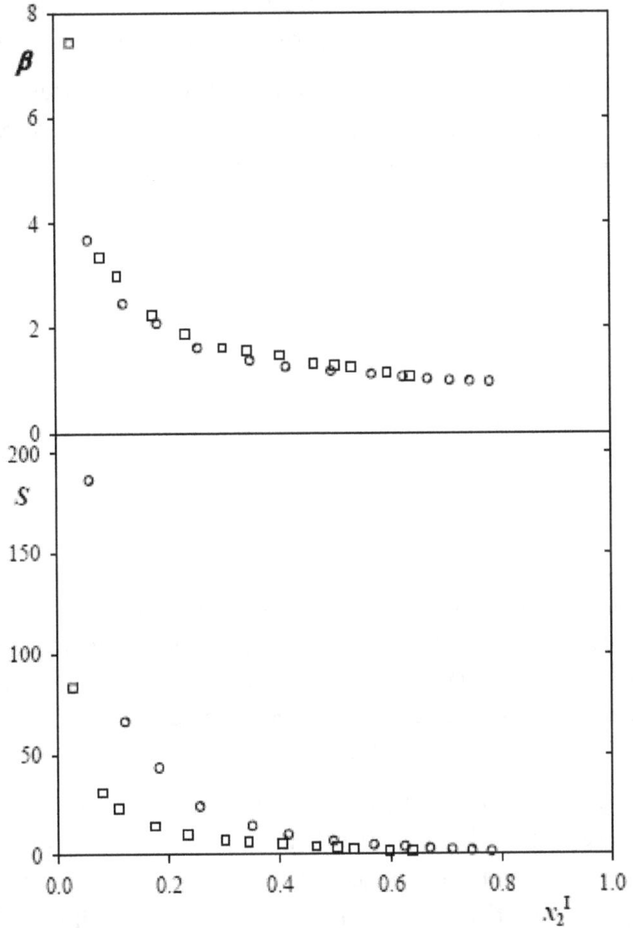

Fig. 5. Solute distribution ratio (β) and selectivity (S) of the systems presenting azeotrope hexane (1) + ethyl acetate (2) with: O, $[C_6MIM][PF_6]$ and D, $[C_8MIM][PF_6]$ versus ethyl acetate mole fraction in the organic phases at 298.15 K.

The solute distribution ratio (Eq. 1) provides the solvent capacity of the ionic liquid, related to the amount of ionic liquid required for the process. The selectivity (Eq. 2) is an important parameter to assess the efficiency of the ionic liquid used as solvent in the selective extraction of the solute from the azeotropic system. The values of the solute distribution ratio and selectivity for the studied ternary systems as a function of the solute composition in the organic phase are plotted in Figure 5.

A comparison between the selectivity values for the ternary systems shows that the $[C_6MIM][PF_6]$ obtains better values than $[C_8MIM][PF_6]$ for the removal of hexane from its azeotropic mixture with ethyl acetate.

3.1.4 Separation of Cyclohexane + 2-Butanone

The azeotropic mixture cyclohexane + 2-butanone occurs in the process for purifying grafted polyolefins (Gupta & Carey, 2006) and its separation is made by membrane in conjunction with a dephlegmation (Wijmans et al., 2005).

Liquid-liquid equilibrium data were obtained for the mixtures of cyclohexane + 2- butanone + $[C_6MIM][PF_6]$ and cyclohexane + 2-butanone + $[C_8MIM][PF_6]$ at 298.15 K and atmospheric pressure (Pereiro & Rodriguez, 2008b).

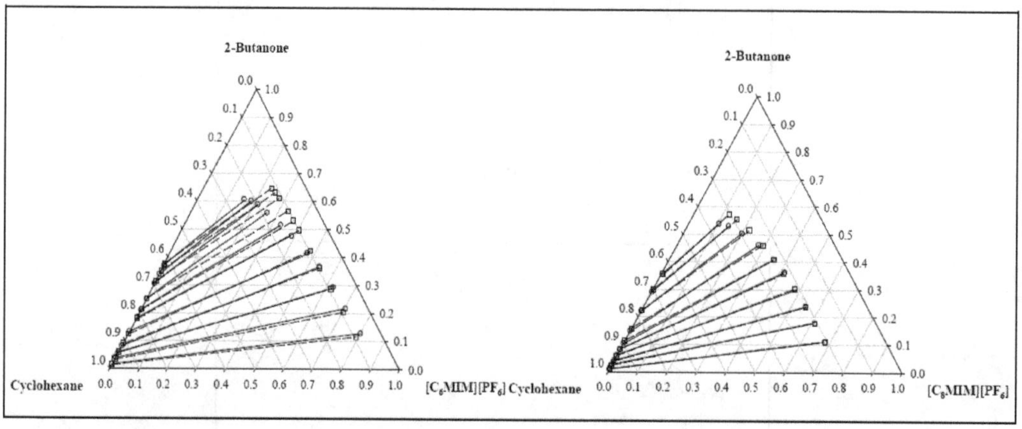

Fig. 6. Experimental tie-lines of the ternary systems at 298.15 K: (a) cyclohexane (1) + 2- butanone (2) + $[C_6MIM][PF_6]$ (3); (b) cyclohexane (1) + 2-butanone (2) + $[C_8MIM][PF_6]$ (3) where: O and solid line, experimental data; D and dashed line, NRTL correlation.

The corresponding triangular diagrams with the experimental tie-lines for the studied systems are shown in Figure 6. An examination of this figure indicates a clear idea of the shape and the size of the immiscibility region of the systems. The positive slope shows that the solute goes preferentially to the solvent-rich phase. Another significant aspect is the fact that the ionic liquid does not enter in the organic-rich phase, *i.e.* the presence of $[C_6MIM][PF_6]$ and $[C_8MIM][PF_6]$ was not detected in the upper phase.

The use of $[C_6MIM][PF_6]$ as solvent leads to higher values of selectivity (Pereiro & Rodriguez, 2008b) than the $[C_8MIM][PF_6]$, indicating that it would be a better choice as solvent for this azeotropic separation. The analysis of the data indicate that the increase of alkyl chain length of the imidazolium decreases the capability of the $[C_nMIM][PF_6]$ to purify the cyclohexane from the azeotropic mixture.

3.1.5 Separation of Alkanes + Ethanol

As a result of the reduction of lead in gasoline, a growing number of processes in which alkanols and alkanes co-exist to produce oxygenated additives for gasolines are under development or have already reached the industrial production stage (Pucci, 1989). The azeotropic mixtures of either hexane or heptane with ethanol are chosen due to the difficulty that lies in separating them. The liquid–liquid separation leads to an environmentally friendly extraction process of these azeotropic mixtures as an alternative to azeotropic distillation (Laroche et al., 1991; Marwil, 1984), pervaporation (Okada & Matsuura, 1988) and reverse osmosis (Laatikainen & Lindstrom, 1986) which are procedures used for the separation of these azeotropes.

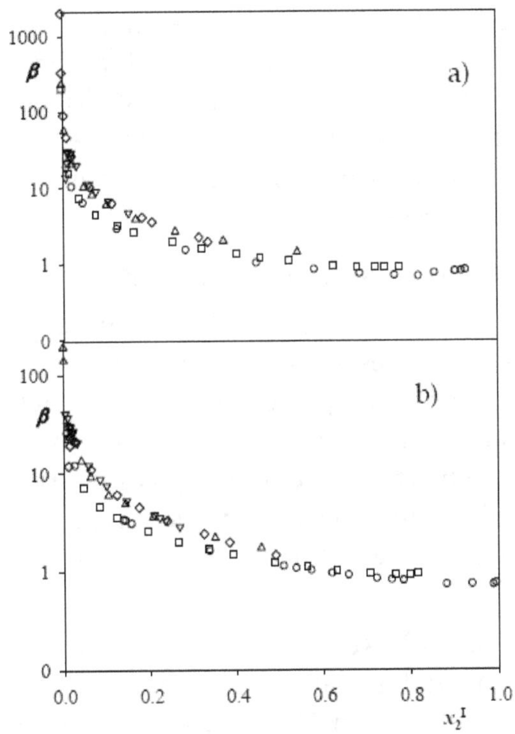

Fig. 7. Solute distribution ratio (β) of the systems presenting azeotrope: (a) hexane (1) + ethanol (2); (b) heptane (1) + ethanol (2) where: o, $[C_6MIM][PF_6]$; D, $[C_8MIM][PF_6]$; D., $[C_1MIM][MeSO_4]$; v, $[C_4MIM][MeSO_4]$; and o, $[C_2MIM][EtSO_4]$ as a function of the solute mole fraction (ethanol) in the organic phase at 298.15 K.

The $[C_1MIM][MeSO_4]$, $[C_4MIM][MeSO_4]$, $[C_2MIM][EtSO_4]$, $[C_6MIM][PF_6]$ and $[C_8MIM][PF_6]$ ILs have been selected to act as azeotrope breakers for the alkanes + ethanol separation. Ternary liquid-liquid equilibria of theses mixtures were determined at 298.15 K and atmospheric pressure (Pereiro et al., 2006b; Pereiro & Rodriguez, 2008c; Pereiro & Rodriguez, 2008d;Pereiro & Rodriguez, 2009a; Pereiro & Rodriguez, 2009b; Pereiro & Rodriguez, 2009c; Pereiro et al., 2010).

The corresponding values for the solute distribution ratio of the studied ternary systems are plotted in Figure 7 as a function of the ethanol mass fraction in the organic-rich phase. High values of these parameters are desired and all ionic liquids are suitable for extraction processes. Although, by analysing the selectivity values (Pereiro & Rodriguez, 2009c) we can conclude that shorter alkyl chain on the imidazolium cation increases selectivity, being favourable for alkane/ethanol separation.

3.2 Lab-Scale extraction process

From the analysis of all extraction capacities, the systems composed by alkanes (hexane or heptane) + ethanol + ionic liquids with the methyl sulfate anion ($[C_1MIM][MeSO_4]$ and $[C_4MIM][MeSO_4]$) were selected to carried out the lab-scale extraction process incorporating a solvent recycling stage (Pereiro & Rodriguez, 2008d; Pereiro & Rodriguez, 2009a; Pereiro & Rodriguez, 2009b; Pereiro & Rodriguez, 2009c). The use of alkylsulfate ILs as azeotrope breakers amplifies the selectivities and solute distribution ratio, encouraging us to test these ILs in our lab-scale extraction process.

3.2.1 Selection of column operation conditions

Operating conditions for simulations and packed column experiments were selected in order to lessen cost while respecting the requirement of an elevated purity of the raffinate. Both cost and purity rise when the solvent/feed flow ratio in the column and/or the purity of the solvent stream increase. Moreover, the cost of solvent recovery grows when the purity is as high as desired. For that reason, this study has been carried out for solvent purities (ILs) of 70–100%. The restrictions applied on solvent stream were raffinate purities >85 wt% and solvent/feed ratio lower than 2.

Figure 8 shows plots of raffinate purity alongside solvent/feed ratio calculated from the experimental tie line data for the ternary system heptane + ethanol + $[C_1MIM][MeSO_4]$ as illustrated in Figure 9, where is showed an example of a tie-line. Each point in Figure 8 corresponds to the crossing between the tie-line and the line that combine the feed (azeotrope) with solvent stream (IL stream with different purities). More details on this procedure can be found elsewhere (Pereiro & Rodriguez, 2008d).

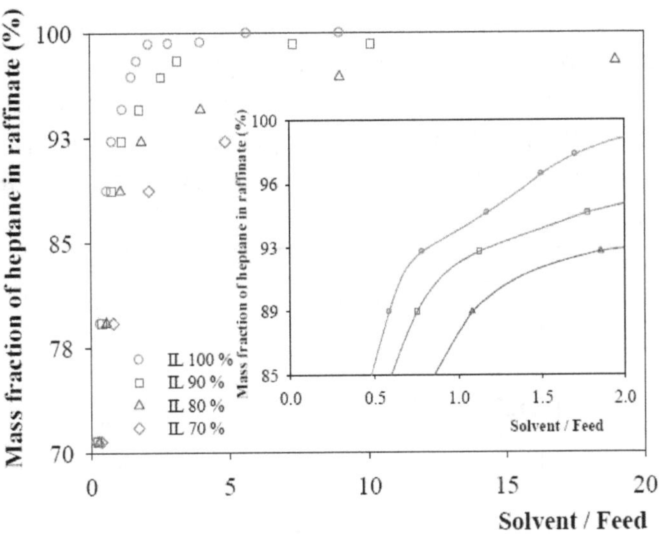

Fig. 8. Effect of solvent purity ([C_1MIM][$MeSO_4$]) on the raffinate purity for the ternary system heptane + ethanol + [C_1MIM][$MeSO_4$] at 298.15 K.

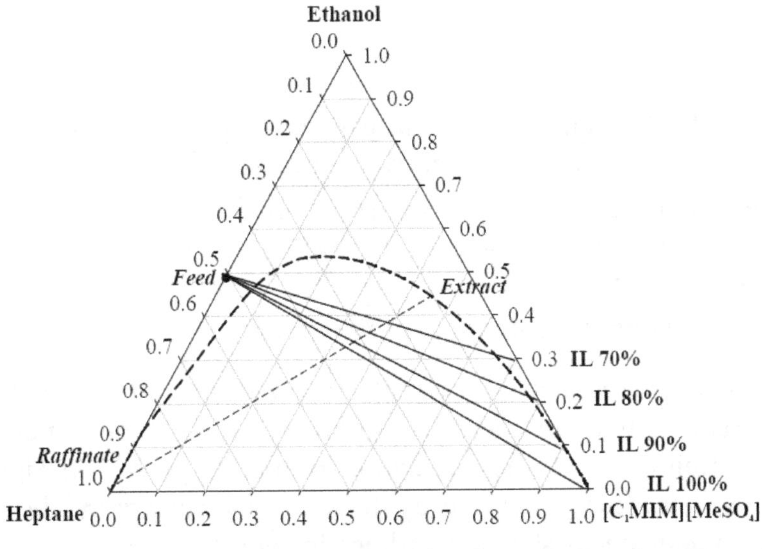

Fig. 9. Ternary diagram depicting the phase compositions for the *Raffinate* and *Extract* streams when an azeotropic mixture of heptane and ethanol (*Feed*) is mixed with [C_1MIM][$MeSO_4$] containing: 0% (100% IL), 10% (90% IL), 20% (80% IL), or 30% (70% IL) of ethanol in ratios corresponding to the crossing with the tie-line (dashed line).

3.2.2 Simulation results

The design of the extraction process was accomplish via HYSYS v.3.2 (Aspen Technology Inc., Cambridge, MA, USA) with the NRTL equation fitted to experimental tie-line data. The simulation model is illustrated schematically in

Figure 10, where a liquid–liquid extractor with one equilibrium stage models the packed column and a short-cut distillation process models solvent recovery. As simulation constraints, the solvent and feed compositions were kept constant (in the region of the chosen the theoretical operation conditions, as mentioned in the above sub-section), and the flow rates were optimized in order to maximize raffinate purity.

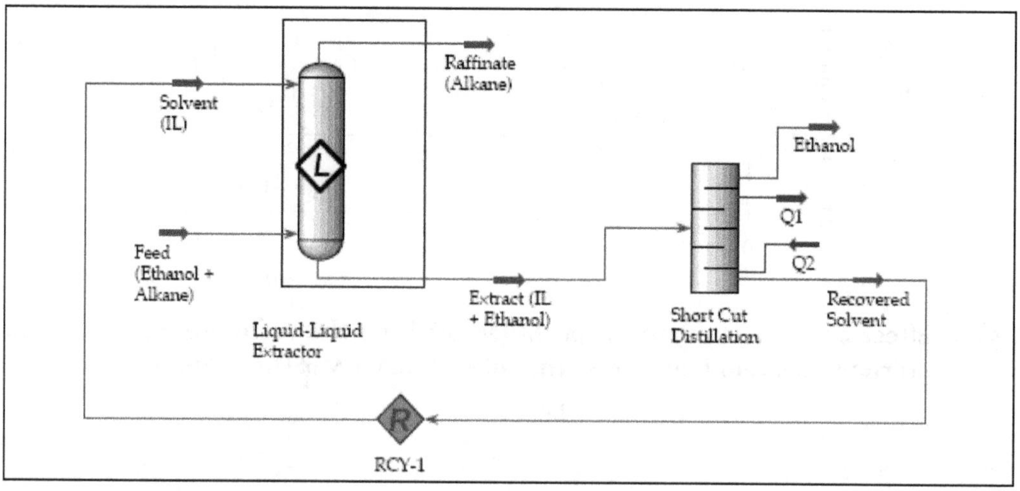

Fig. 10. Flowsheet simulation for the extraction process of ethanol + alkanes azeotropic mixtures using ionic liquid as solvent.

The ability of ionic liquids to act as azeotrope breakers in liquid–liquid extraction processes for the separation of the mixture ethanol + alkanes has been clearly proven by the performed simulations.

3.2.3 Results of packed-column experiments

The evolution of the alkane content observed in raffinate stream of the countercurrent column extraction experiment, carried out under operating conditions attained from the simulation results, is plotted in Figure 11. The amount of time needed to attain steady state, raffinate purity and the extract mass composition under steady state are outlined in Table 2.

The extraction of alkane with a purity ≥ 98 wt. % was feasible by using the packed extraction column and the above mentioned ILs ([C_1MIM][$MeSO_4$] and [C_4MIM][$MeSO_4$]) as solvent. A comparison between packed column experimental data and theoretical data from experimental tie-lines is also depicted in Table 2. The performance of the experimental column exceeds the theoretical (from LLE) and simulation-based expectations, certainly due to the superior mixing between the feed and the solvent.

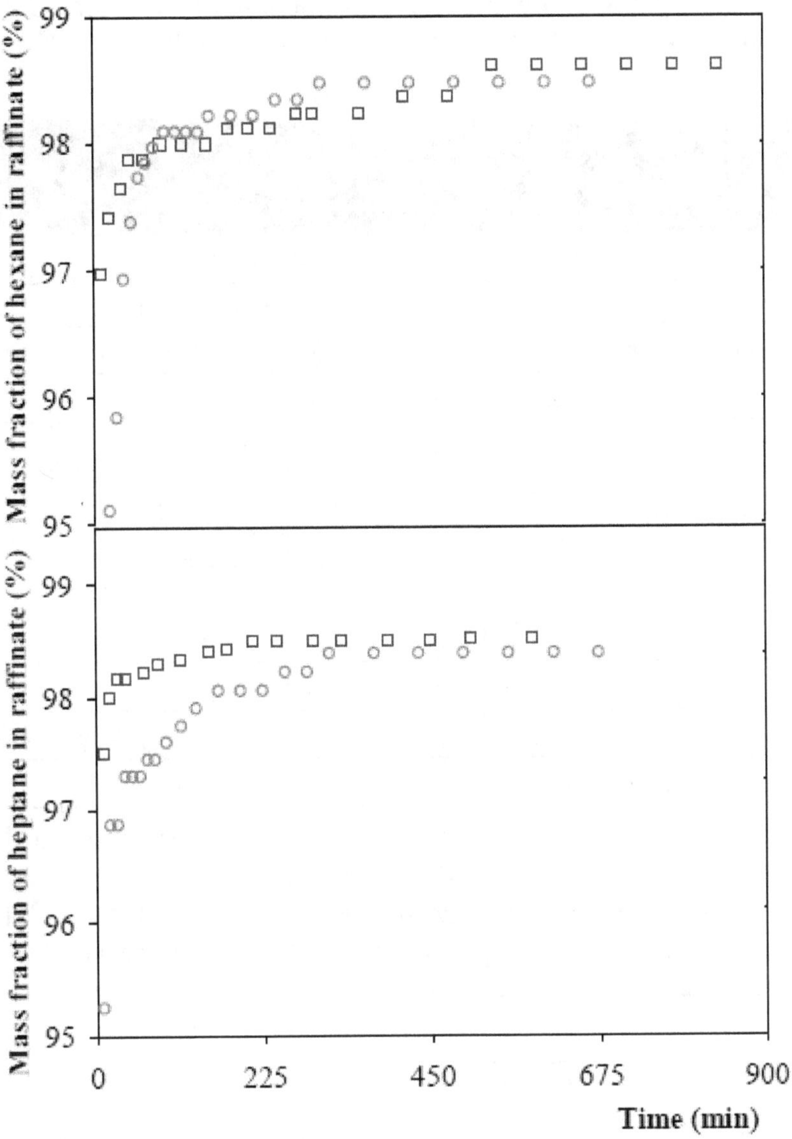

Fig. 11. Time dependence of the mass fractions of heptane in raffinate following start-up of the extraction column of Fig. 2 for: □, [C_1MIM][$MeSO_4$]; and ○, [C_4MIM][$MeSO_4$].

Figure 12 compares the compositions of the initial feed and the two outlet streams of the countercurrent packed column for the two ionic liquids (([C_1MIM][$MeSO_4$] and [C_4MIM][$MeSO_4$]). The results indicate clearly that the [C_1MIM][$MeSO_4$] has a lower amount of alkane in the extract stream.

Table 2. Comparison of experimental data under steady state for the extraction processes with theoretical data (from tie-lines) for the studied ternary systems.

	Raffinate				Extract		
	$\% w^{I}_{1}$	$\% w^{I}_{2}$	$\% w^{I}_{3}$		$\% w^{II}_{1}$	$\% w^{II}_{2}$	$\% w^{II}_{3}$
Hexane + Ethanol + $[C_1MIM][MeSO_4]$							
t = 305 min	98.5	1.5	0.0	t = 485 min	6.2	36.8	57.0
one equilibrium stage	89.8	9.9	0.3	one equilibrium stage	3.7	32.6	63.7
Hexane + Ethanol + $[C_4MIM][MeSO_4]$							
t = 535 min	98.6	1.4	0.0	t = 655 min	9.4	36.4	54.2
one equilibrium stage	96.2	3.5	0.3	one equilibrium stage	10.0	38.1	51.9
Heptane + Ethanol + $[C_1MIM][MeSO_4]$							
t = 315 min	98.4	1.6	0.0	t = 495 min	3.6	38.4	58.0
one equilibrium stage	88.7	10.6	0.7	one equilibrium stage	4.7	40.1	55.2
Heptane + Ethanol + $[C_4MIM][MeSO_4]$							
t = 4 min	98.1	1.9	0.0	t = 8 min	11.9	46.0	42.1
one equilibrium stage	92.8	7.2	0.0	one equilibrium stage	10.5	44.0	45.5

In order to contrast the two ionic liquids for extraction processes in the packed column, the extraction efficiency, *E*, was calculated. This parameter indicates the ability of extraction solvent ($[C_1MIM][MeSO_4]$ and $[C_4MIM][MeSO_4]$) to remove solute (ethanol) from the azeotropic mixtures (ethanol + alkanes) in the extraction column. This parameter is defined as follows:

$$E = \frac{w_1^{F} - w_1^{R}}{w_1^{F} - w_1^{Eq}} \tag{5}$$

where *w* is the mass fraction, subscript 1 refer to inert component (alkane) and superscripts F, R and Eq indicate the feed stream, raffinate stream and one equilibrium stage (theoretical data), respectively.

Table 3. Extraction efficiencies for the separation of the azeotrope alkanes + ethanol in the packed column.

Azeotropic mixture	$[C_1MIM][MeSO_4]$	$[C_4MIM][MeSO_4]$
Hexane+Ethanol	1.81	1.14
Heptane+Ethanol	1.26	1.13

In the separations of the azeotropes alkanes (hexane and heptane) + ethanol, the extraction efficiencies are listed in Table 3. The values obtained for [C_1MIM][$MeSO_4$] are better in comparison with those for [C_4MIM][$MeSO_4$]. Moreover, [C_1MIM][$MeSO_4$] has both lower viscosity and synthesized cost.

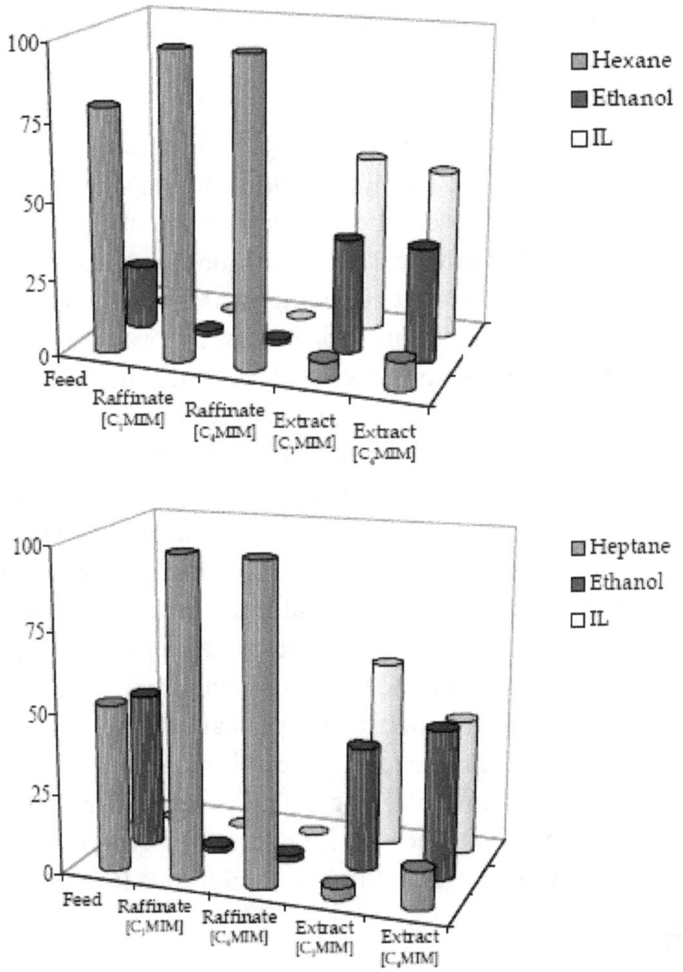

Fig. 12. Mass composition (%) of the feed, raffinate and extract streams of the packed column (Figure 2) under steady state conditions for [C_1MIM][$MeSO_4$] and [C_4MIM][$MeSO_4$].

4. CONCISE SUMMARY

The efficacy of various ionic liquids as solvents for azeotropic separation by liquid-liquid extraction was illustrated. The initial strategy for building any extraction process must consider numerous factors: i) Determination of liquid-liquid extraction (LLE) data for ternary systems involving azeotropic mixtures and ionic liquids (ILs), ii) Correlation of experimental data using theoretical equations,

iii) Assessment of extraction capacity through critical parameters such as solute distribution ratios and selectivities, iv) Computational simulation to optimize operational conditions, v) Experimental lab-scale extraction process.

Alkyl sulfate-based ionic liquids are the most promising alternatives as solvents for the separation of the azeotropes examined in this study. The elongation of the alkyl side chain in the imidazolium ring adversely affects the extraction efficiency of the ionic liquids.

The separation operations for the azeotropes of ethanol and alkanes were conducted using methyl sulfate-based ionic liquids ([C1MIM][MeSO4] and [C4MIM][MeSO4]). The LLE data facilitated the determination of theoretically suitable operating conditions for the countercurrent continuous extraction process at room temperature, incorporating a solvent recycling phase. The operational conditions were later refined through simulation approaches. Laboratory-scale experiments with a packed column under steady-state conditions attained a raffinate purity of 98 wt %. The laboratory-scale trials further validated the feasibility of immediate online recovery of specific ionic liquids. he extraction efficiency was determined to compare the various extraction procedures in the packed column. The values obtained for [C1MIM][MeSO4] surpass those for [C4MIM][MeSO4]. Furthermore, [C1MIM][MeSO4] is regarded as the optimal candidate because to its reduced viscosity and affordability. Considering the raffinate purity of the alkane produced during the extraction process, scaling up for industrial application appears feasible. Upcoming concerns to be resolved include the scalability of the process in a pilot plant to assess the feasibility of industrial application. The application of ionic liquids as azeotrope disruptors is expected to mitigate the impact of traditional solvents (VOCs) on global climate change. This strategy presents a significant potential to attain the elevated objectives for decreasing greenhouse gas emissions set forth at the recent United Nations Climate Conference (COP15).

REFERENCES

Arce, A.; Rodriguez, H.; Soto, A. (2007). Use of a green and cheap ionic liquid to purify gasoline octane boosters. *Green Chem.*, 9, 3, March 2007 247–253, 1463-9262

Bera, D.; Lahiri, D.; De Leonardis, A.; De, K. B.; Nag, A. (2006). A Novel Azeotropic Mixture for Solvent Extraction of Edible Oils. *Agricultural Engineering International: the CIGR Ejournal.*, Manuscript FP 06 005, 8, April 2006 1–6, 1682-1130

Berg, Lloyd. (1995). Separation of 2-butanone from isopropanol by azeotropic distillation. *U.S. Patent* 5423954, June 1995

Berg, Lloyd. (1999). Separation of methyl ethyl ketone from ethanol by azeotropic distillation. *U.S. Patent* 5868907, February 1999

Earle, M. J.; Esperança, J. M. S. S.; Gilea, M. A.; Lopes, J. N. C.; Rebelo, L. P. N.; Magee, J. W.; Seddon, K. R.; Widegren, J. A. (2006). The distillation and volatility of ionic liquids. *Nature*, 439, 7078, February 2006 831–834, 0028-0836

Gonzalez, E. J.; Calvar, N.; Gonzalez, B.; Dominguez, A. (2009). (Liquid plus liquid) equilibria for ternary mixtures of (alkane plus benzene plus [EMpy] [ESO_4]) at several temperatures and atmospheric pressure. *J. Chem. Thermodyn.*, 41, 11, November 2009 1215-1221, 0021-9614

Gupta, V. P.; Carey, E. P. (2006). Extraction of Impurities from Grafted Polyolefins. *U.S. Patent* 7060787, June 2006

Holbrey, J. D.; Reichert, W. M.; Swatloski, R. P.; Broker, G. A.; Pitner, W. R.; Seddon, K. R.; Rogers, R. D. (2002). Efficient, halide free synthesis of new, low cost ionic liquids: 1,3-dialkylimidazolium salts containing methyl- and ethyl-sulfate anions. *Green Chem.* 4, 5, October 2002, 407–413, 1463-9262

Hu, X.; Yu, J.; Liu, H. (2006). Separation of THF and water by room temperature ionic liquids. *Water Sci. Tech.*, 53, 11, 2006 245–249, 0273-1223

Huddleston, J. G.; Visser, A. E.; Reichert, W. M.; Willauer, H. D.; Broker, G. A.; Rogers, R. D. (2001). Characterization and comparison of hydrophilic and hydrophobic room temperature ionic liquids incorporating the imidazolium cation. *Green Chem.*, 3, 4, August 2001 156–164, 1463-9262

Jork, C.; Seiler, M.; Beste, Y.-A.; Arlt, W. (2004). Influence of ionic liquids on the phase behavior of aqueous azeotropic systems. *J. Chem. Eng. Data*, 49, 4, July-August 2004 852–857, 0021-9568

Katayama, H.; Hayakawa, T.; Kobayashi, T. (1998). Liquid-liquid equilibria of three ternary systems: 2-propanone-glycerol-methanol, 2-butanone-glycerol-ethanol, and 2-butanone-glycerol-2-propanol in the range of 283.15 to 303.15 K. *Fluid Phase Equilib.*, 144, 1-2, February 1998 157-167, 0378-3812

Katayama, H.; Amano, K. (2005). Liquid-liquid equilibria of three ternary systems: Water plus 2-butanone plus aliphatic alcohols (Ethanol, 2-propanol and 1-propanol). *J. Chem. Eng. Japan*, 38, 7, July 2005 459-464, 0021-9592

Laatikainen, M.; Lindstrom, M. (1986). Separation of methanol-ethanol and ethanol-normal- heptane mixtures by reverse-osmosis and pervaporation. *Acta Polytech Scandinavica Chem. Tech. Metall. Series*, 175, 1986 1-61, 1239-0518

Laroche, L.; Bekiaris, N.; Andersen, H. W.; Morari, M. (1991). Homogeneous azeotropic distillation - comparing entrainers. *Can. J. Chem. Eng.*, 69, 6, December 1991 1302-1319, 0008-4034

Letcher, T. M.; Deenadayalu, N. (2003). Ternary liquid-liquid equilibria for mixtures of 1- methyl-3-octyl-imidazolium chloride plus benzene plus an alkane at T-298.2 K and 1 atm. *J. Chem. Thermodyn.*, 35, 1, January 2003 67-76, 0021-9614

Marwil, S. J. (1984). Separation of hydrocarbon and alcohol azeotropic mixtures by distillation with anhydrous ammonia. *U.S. Patent* 4437941, March 1984

Najdanovic-Visak, V.; Esperanca, J. M. S. S.; Rebelo, L. P. N.; Nunes da Ponte, M.; Guedes, H. J. R.; Seddon, K. R.; Szydlowski, J. (2002). Phase Behaviour of Room Temperature Ionic Liquid Solutions: an Unusually Large Co-Solvent Effect in (Water + Ethanol). *Phys. Chem. Chem. Phys.*, 4, 10, 2002 1701–1703, 1463-9076

Okada, T.; Matsuura, T. (1988). A study on the pervaporation of ethyl alcohol/heptane mixtures by porous cellulose membranes. *Proc. Int. Conf. Pervaporation Processes Chem. Ind.* 3, pp. 224–230

Pereiro, A. B.; Santamarta, F.; Tojo, E.; Rodriguez, A.; Tojo, J. (2006a). Temperature dependence of physical properties of ionic liquid 1,3-dimethylimidazolium methyl sulfate. *J. Chem. Eng. Data*, 51, 3, May 2006 952-954, 0021-9568

Pereiro, A. B.; Tojo, E.; Rodriguez, A.; Canosa, J.; Tojo, J. (2006b). $HMImPF_6$ ionic liquid that separates the azeotropic mixture etanol + heptane. *Green Chem.*, 8, 3, March 2006 307-310, 1463-9262

Pereiro, A. B.; Legido, J. L.; Rodriguez, A. (2007a). Physical properties of ionic liquids based on 1-alkyl-3-methylimidazolium cation and hexafluorophosphate as anion and temperature dependence. *J. Chem. Thermodyn.*, 39, 8, August 2007 1168-1175, 0021-9614

Pereiro, A. B.; Verdia, P.; Tojo, E.; Rodriguez, A. (2007b). Physical properties of 1-butyl-3-methylimidazolium methyl sulfate as a function of temperature. *J. Chem. Eng. Data*, 52, 2, March-April 2007 377-380, 0021-9568

Pereiro, A. B.; Rodriguez, A. (2007c). Ternary Liquid-Liquid Equilibria Ethanol + 2-Butanone + 1-Butyl-3-Methylimidazolium Hexafluorophosphate, 2-Propanol + 2-Butanone + 1-Butyl-3-Methylimidazolium Hexafluorophosphate, and 2-Butanone + 2-Propanol + 1,3-Dimethylimidazolium Methyl Sulfate at 298.15 K. *J. Chem. Eng. Data,* 52, 6, November 2007 2138-2142, 0021-9568

Pereiro, A. B.; Canosa, J.; Rodriguez, A. (2007d). Liquid-liquid equilibria of 1,3- dimethylimidazolium methyl sulfate with ketones, dialkyl carbonates and acetates. *Fluid Phase Equilib.,* 254, 1-2, June 2007 150-157, 0378-3812

Pereiro, A. B.; Rodriguez, A. (2007e). Ternary (liquid + liquid) equilibria of the azeotrope (ethyl acetate + 2 – propanol) with different ionic liquids at T = 298.15 K. *J. Chem. Thermodyn.,* 39, 12, December 2007 1608-1613, 0021-9614

Pereiro, A. B.; Rodriguez, A. (2008a). Phase Equilibria of the Azeotropic Mixture Hexane + Ethyl Acetate with Ionic Liquids at 298.15 K. *J. Chem. Eng. Data,* 53, 6, June 2008 1360-1366, 0021-9568

Pereiro, A. B.; Rodriguez, A. (2008b). Measurement and correlation of liquid – liquid equilibria of the azeotrope (cyclohexane + 2 – butanone) with different ionic liquids at T = 298.15 K. *J. Chem. Thermodyn.,* 40, 8, August 2008 1282-1289, 0021-9614

Pereiro, A. B.; Rodriguez, A. (2008c). A study on the liquid-liquid equilibria of 1-alkyl-3-methylimidazolium hexafluorophosphate with ethanol and alkanes. *Fluid Phase Equilib.,* 270, 1-2, August 2008 23-29, 0378-3812

Pereiro, A. B.; Rodriguez, A. (2008d). Azeotrope-breaking using [BMIM] [$MeSO_4$] ionic liquid in an extraction column. *Sep. Purif. Technol.,* 62, 3, September 2008 733-738, 1383-5866

Pereiro, A. B.; Rodriguez, A. (2009a). Separation of Ethanol-Heptane Azeotropic Mixtures by Solvent Extraction with an Ionic Liquid. *Ind. Eng. Chem. Res.,* 48, 3, February 2009 1579-1585, 0888-5885

Pereiro, A. B.; Rodriguez, A. (2009b). Purification of hexane with effective extraction using ionic liquid as solvent. *Green Chem.,* 11, 3, March 2009 346-350, 1463-9262

Pereiro, A. B.; Rodriguez, A. (2009c). Effective extraction in packed column of ethanol from the azeotropic mixture ethanol + hexane with an ionic liquid as solvent. *Chem. Eng. J.,* 153, 1-3, November 2009 80-85, 1385-8947

Pereiro, A. B.; Deive, F. J.; Esperança, J. M. S. S.; Rodriguez, A. (2010). Alkylsulfate-based ionic liquids to separate azeotropic mixtures. *Fluid Phase Equilib.,* 291, 1, April 2010 13-17, 0378-3812

Renon, H.; Prausnitz, J. M. (1968). Local compositions in thermodynamic excess functions for liquid mixtures. *AIChE J.,* 14, 1, 1968 135-144, 0001-1541

Pucci, A. (1989). Phase-equilibria of alkanol alkane mixtures in new oil and gas process- development. *Pure Appl. Chem.,* 61, 8, August 1989 1363-1372, 0033-4545

Rodriguez-Donis, I.; Acosta-Esquijarosa, J.; Gerbaud, V.; Pardillo-Fondevila, E.; Joulia, X. (2005). Separation of n-hexane-ethyl acetate mixtures by azeotropic batch distillation with heterogeneous entrainers. *Chem. Eng. Process.,* 44, 1, January 2005 131-137, 0255-2701

Seiler, M.; Jork, C.; Kavarnou, A.; Arlt, W.; Hirsch, R. (2004). Separation of azeotropic mixtures using hyperbranched polymers or ionic liquids. *AIChE J.,* 50, 10, October 2004 2439–2454, 0001-1541

Swatloski, R. P.; Rogers, J. D.; Rogers, R. D. (2003). Ionic Liquids are not always Green: Hydrolysis of 1-Butyl-3-methylimidazolium Hexafluorophosphate. *Green Chem.,* 5, 4, August 2003 361–363, 1463-9262

Wijmans, J. G.; Baker, R. W.; Mairal, A. P. (2005). Separation of organic mixtures using gas separation or pervation and dephlegmation. *U.S. Patent* 6899743, May 2005

Zhang, D. L.; Deng, Y. F.; Li, C. B.; Chen, J. (2008). Separation of ethyl acetate-ethanol azeotropic mixture using hydrophilic ionic liquids. *Ind. Eng. Chem. Res.,* 47, 6, March 2008 1995–2001, 0888-5885

12 Chapter

APPLICATION OF IONIC LIQUIDS IN LIQUID CHROMATOGRAPHY

1. Introduction: An Overview

Ionic liquids (ILs) are salts composed of large, asymmetrical organic cations containing nitrogen or phosphorus, such as alkyl-imidazolium, pyridinium, pyrrolidinium, ammonium, or phosphonium, paired with various organic anions like triflate, dicyanamide, acetate, trifluoroacetate, trifluoromethylsulfate, or inorganic anions such as bromide, chloride, nitrate, perchlorate, chloroaluminate, tetrafluoroborate, or hexafluorophosphate. The number of potential combinations of anion-cation rises to 10^18, rendering this class of compounds one of the largest in the chemical repertoire. The characteristics of ionic liquids are contingent upon their molecular structure. To achieve an ionic liquid with optimal qualities, including a specified melting temperature, viscosity, density, and miscibility with water and organic solvents, appropriate ions must be selected. Ionic liquids are frequently referred to be "designer" solvents for this reason.

This class of chemical compounds possesses distinct common properties. They possess the capability to dissolve both inorganic and organic substances, including enzymes, while simultaneously preserving their activity. They exhibit stability both chemically and thermally. The difference between the boiling temperature and the melting point is 300 °C. Ionic liquids exhibit low vapor pressure, hence preventing the generation of volatile environmental pollutants. The defining characteristic of these compounds is their low melting point, which is below 100 °C. If the melting point is below room temperature (25 °C), the ionic liquid is referred to as a room-temperature ionic liquid (RTIL).

Ionic liquids have been utilized as novel non-molecular solvents with distinctive features throughout various domains of chemistry, including organic synthesis, electrochemistry, extraction, spectroscopy, and mass spectrometry. The increasing interest in ionic liquids (ILs) has also been noted in chromatographic techniques: gas chromatography (GC), capillary electrophoresis (CE), and liquid chromatography (LC). From the perspective of separation techniques, an essential characteristic of ionic liquids (ILs) is their capacity to facilitate robust donor-acceptor interactions and engage in Coulombic interactions (chromatography), exhibit high electrical conductivity (capillary electrophoresis), possess significant viscosity and surface tension, maintain thermal stability (gas chromatography), and demonstrate superior solvent properties (extraction). It is important to highlight critical assessments that underscore the special advantages of ionic liquids in chromatographic and electromigration procedures. (Buszewski & Studzińska, 2008; Marszałł & Kaliszan, 2007; Berthod et al., 2008; Pandey, 2006; Koel, 2005; Stalcup & Cabovska, 2004; Poole, 2007; Sun & Armstrong, 2010; Poole, 2004).

A major problem associated with using ILs, as solvents, in liquid chromatography is their high viscosity, many times bigger than one of methanol or acetonitrile. A completely different situation concerns using ionic liquids as small additives (2-60 mM) to water- organic mobile phases. In such minute concentration, ionic liquids lose their unique properties and they simply become salts. Apart from that, small concentration of the additive doesn't change the eluent viscosity drastically and can be applied in conditions both of high pressure chromatography (HPLC) as well as thin-layer chromatography (TLC). The addition of ionic liquids is particularly beneficial in analysis of basic compounds on the reversed-phase columns obtained on the basis of silica, where silanol interactions were usually a disadvantage. Activity of reducing of silanol interactions through ionic liquids exceeds commonly used tertiary amines as well as bivalent cations, frequently applied for this reason.

Modification of the stationary phase with the help of ILs relies on covalent bonding of silanol groups of silica gel with the cation of an ionic liquid. Obtained phases change its properties into ion-exchanging able to separate inorganic as well as organic ionic analytes.

This review focuses on application of ILs in liquid chromatography (HPLC and TLC), and electromigration techniques. Control of ILs by these techniques is also discussed according to increases of their popularity in industry.

2. PHYSICOCHEMICAL PROPERTIES OF IONIC LIQUIDS

Physicochemical properties of ILs depend on their components. The asymmetric organic cations and large polyatomic anions are responsible for lowering of the lattice energy and finally lowering of the melting point of resulting compound. For instance the melting point of archetypical sodium chloride equals 800 °C. Displacement of sodium by imidazolium cation lowers the melting point ten times. Increasing popularity of ILs in chemical industry is connected with their low vapor pressure, which is lower than the one of water, mercury and nonacosane - least volatile among alkanes. In the room temperature, the pressure achieves non-measurable values thus they appear to be friendly for environment. Besides that, they exhibit a wide range of solubility, viscosity and density depending on the kind of anion and cation in their structure. To find proper composition for exact analytical application, physicochemical properties of almost all existing ILs are collected in data banks available on the following websites: Dortmund Data Bank, Germany (Data Bank of Ionic Liquids, DECHEMA, DETHERM on the WEB, Thermophysical Properties of Pure Substances & Mixtures, http://www.ddbst.de/new/Default.htm or another one: NIST Boulder Colorado, USA (Ionic Liquids Database (IL Thermo), Thermodynamics Research Center, National Institute of Standards and Technology, Boulder, CO, USA, http://ilthermo.boulder.nist.gov/ILThermo/mainmenu.uix.

One of the advantages of ILs is their high thermal robustness. They possess thermal stability in the range approximately from -40°C to 200°C, which is much wider than other solvents like water, ammonium, acetone, methanol, benzene or nitrobenzene. ILs could be hydrophobic or hydrophilic depending on the properties of their ions. Hydrophilic ILs similarly to regular salts dissociate into ions which are hydrated in aqueous solutions. Ions exerting strong interactions with water molecules cause increase of the structuring of water and they are known as ''structure-makers" or 'kosmotropes'. Hydrophobic ions partially or completely immiscible with water which decrease the structuring of water are known as 'chaotropes. meaning disorder-makers'. These terms originated from the Hofmeister series, which orders the ion abilities of proteins precipitation. Comparing several thermodynamic parameters like the viscosity *B*-coefficients obtained from the Jones-Dole empirical equation, structural entropies (ΔS^0), structural volumes and ion mobility expressed as the limiting Walden product $\Lambda_m\eta$ quantification of the ILs ions kosmotropicity/chaotropicity can be evaluated (Zhao, 2006). According to Zhao investigation, it appears that smaller organic cations including $[MMIM]^+$, $[EMIM]^+$, $[MePy]^+$, $[EtPy]^+$ and $[Me_4N^+]$ are chaotropes, while those of medium size (such as $[BuPy]^+$ and $[Et_4N^+]$ are borderline ions, in turn larger ions such as $[BMIM]^+$, $[Pr_4N^+]$, $[Bu_4N^+]$, and $[Pe_4N^+]$ possess kosmotropic properties. The anions can be kosmotropes or chaotropes depending on the position in Hofmeister series of salts.

The properties of a modern ILs could be characterized by the following features (Johnson, 2007):viscosity: < 100 cP, dielectric constant: < 30, specific conductivity: < 10 $mScm^{-1}$, molar conductivity: < 10 $Scm^2\ mol^{-1}$, electrochemical window > 2V.

Comparing properties of ILs and other liquids according to their molarities and molar conductance, it is clear that modern ILs consist of not only ions but also un-dissociated molecules in a form of ion-pairs. From a point of view of liquid chromatography such properties as viscosity, refractivity, solubility, E_T^N - Reichardt polarity index are the most important. Table 1 contains properties of ILs common in chromatographic techniques.

Table 1. Physicochemical properties of selected ionic liquids (Berthod et al., 2008).

Symbol	Name	Melting point (°C)	Density (25 °C)	Refrac. index	Visc. (25 °C) (cP)	Sol. in water	E_T^N (×100)
EMIM BF_4	1-Ethyl-3-methyl imidazolium tetrafluoroborate	6	1.248		66	s	
BMIM BF_4	1-Butyl-3-methyl imidazolium tetrafluoroborate	−82	1.208	1.429	233	s	67
BMMIM BF_4	1-Butyl-2,3-dimethyl imidazolium tetrafluoroborate	37	1.2		780	s	
HMIM BF_4	1-Hexyl-3-methyl imidazolium tetrafluoroborate	−82	1.208		310	p	
OMIM BF_4	1-Octyl-3-methyl imidazolium tetrafluoroborate	−79	1.11		440	p	54
DMIM BF_4	1-Decyl-3-methyl imidazolium tetrafluoroborate	−25	1.072		930		63
BMIM PF_6	1-Butyl-3-methyl imidazolium hexafluorophosphate	10	1.373	1.411	400	18 g L^{-1}	67
HMIM PF_6	1-Hexyl-3-methyl imidazolium hexafluorophosphate	−61	1.304		800	n	
OMIM PF_6	1-Octyl-3-methyl imidazolium hexafluorophosphate	−40	1.2		810	n	
EMIM Cl	1-Ethyl-3-methyl imidazolium chloride	89	1.12*	Solid	Solid	s	
HMIM Cl	1-Hexyl-3-methyl imidazolium chloride	−75	1.05		7500	s	
OMIM Cl	1-Octyl-3-methyl imidazolium chloride	0	1		16,000	s	
$(C_7)_4$ N Cl	Tetraheptyl ammonium chloride	−9	0.882		598	s	
BMIM Cl	1-Butyl-3-methyl imidazolium chloride	65	1.10*	Solid	Solid	s	

BMIM Br	1-Butyl-3-methyl imidazolium bromide					s	
BMIM I	1-Butyl-3-methyl imidazolium iodide	−72			1110		
EMIM Act	1-Ethyl-3-methyl imidazolium acetate	−20	1.03		91	s	
BMIM Act	1-Butyl-3-methyl imidazolium acetate	−20	1.06		525	s	
EA NO_3	Ethylammonium nitrate	12.5	1.122	1.4537	32	s	95
PA NO_3	Propylammonium nitrate	4	1.157	1.4561	67	s	92
TBA NO_3	Tributylammonium nitrate	21.5	0.918	1.4627	640	s	80
EMIM MS	1-Ethyl-3-methyl imidazolium methylsulfate	5	1.24		80	s	
BMIM MS	1-Butyl-3-methyl imidazolium methyl sulfate	−20	1.2		180	s	

3. CHROMATOGRAPHY

ILs were applied firstly in gas chromatography as new liquid stationary phases owing to their low volatility, high viscosity and non-flammability (Barber et al., 1959). Barber and co-workers proposed GC stationary phases modified with molten stearates of Mn, Co, Ni, Cu, and Zn for separation of alcohols and amines. Later on, GC columns based on ethylammonium nitrate and ethylpyridinium bromide and tetraalkylphosphonium salts were prepared and tested (Pachloec et al., 1982; Pacholec & Poole, 1983). ILs-based stationary phases exhibited selectivity towards polar and hydrogen bond forming analytes. However, they possessed some limitations concerning narrow liquid range, poor thermal stabilities, poor efficiency and lack of selectivity according to n-hydrocarbons. Since imidazoium ILs: 1-Butyl-3-methylimidazolium hexafluorophosphate [BMIM][PF_6] and chloride [BMIM][Cl] salts started to be coated onto silica capillaries, GC systems have achieved satisfactory efficiency useful for separation either nonpolor or polar compounds (Armstrong et al., 1999; Berthod et al., 2001; Anderson & Armstrong, 2003).

In liquid chromatography, ILs have not been widely applied so far. However, the number of publications dealing with this subject has recently visibly increased. Generally, ILs are too viscous to be used as solvents and exhibit poor transmittance of low-wavelength UV light, worsening detection limits. However, when ILs are diluted, properties characteristic for the neat ones are lost and ILs become common organic salts. Among whole group, only pyridinium and imidazolium hexafluorophosphates or tetrafluoroborates are the most commonly applied in solution as mobile phase modifiers. In this case ILs can play different roles: coating residual silanols, modifying the stationary phase and acting as ion-pairing reagents.

3.1 Thin-layer chromatography

The first paper concerning application of ILs in thin-layer chromatography (TLC) was published by Kaliszan's group (Kaliszan et al., 2004). The imidazolium classes of tetrafluoroborate ILs covering [EMIM] [BF_4], [HxMIM] [BF_4], [Hx-HpOMIM] [BF_4] were tested as mobile phase modifiers in analysis of several basic compounds on an octadecyl bound-silica stationary phase. It appeared that proposed modifiers were more effective in chromatographic separation of chosen basic analytes in comparison to other standard additives such as triethylamine (TEA), dimethyloctylamine (DMOA) or ammonia. Whereas TEA, DMOA and ammonia showed negligible effect on investigated compounds retention, ILs appeared to be the most effective suppressor of attractive interactions between acidic free silanols present on the stationary phase surface and analyzed basic drugs causing drastically retention decreases, improvement of selectivity and spots shape. Effects of investigated additives on chromatographic results are compared in Fig.1.

Dependences of ionic liquid concentration in eluent system versus obtained experimental retardation factors (R_F) exhibit typical Langmuir shapes. Independently on the kind of stationary phase (silica or the octadecyl silica-covered plates), mobile phase composition (pure acetonitrile, water-acetonitrile) and IL used, saturation level was achieved at ca. 0.5% v/v of ionic liquid content in eluent system. Comparison of the effectiveness of investigated silanol screening agents on the basis of reciprocal of retardation factor vs. reciprocal of the additive concentration relationship (Fig.2), proved the most effective adsorption of exemplary ionic liquid [EMIM] [BF_4].

Addition of imidazolium tetrafluoroborates to mobile phase improved also linearity of the classical Snyder-Soczewiński relationship of R_M (log1-R_F/R_F) versus organic modifier concentration in the mobile phase. This suggests that chromatographic systems modified with ILs could be applied for much more precise determination of chromatographic lipophilicity parameters (R_M^0, φ_0) by extrapolation and interpolation graphical techniques. Using of IL as the mobile phase modifier is useful also in normal-phase TLC systems. Kaliszan and his group (Bączek et al., 2005) described behavior of a homologous series of peptides analysed on a silica support by the use of the eluent system enriched with [EMIM] [BF_4]. It appeared that experimental dependences of the retention coefficient, R_M on the volume percentage of organic modifier in the mobile phase containing ionic liquid were nonlinear and could be described with a third-degree polynomial function. Whereas quadratic equation was well fitted to dependences obtained for chromatographic systems without addition of IL.

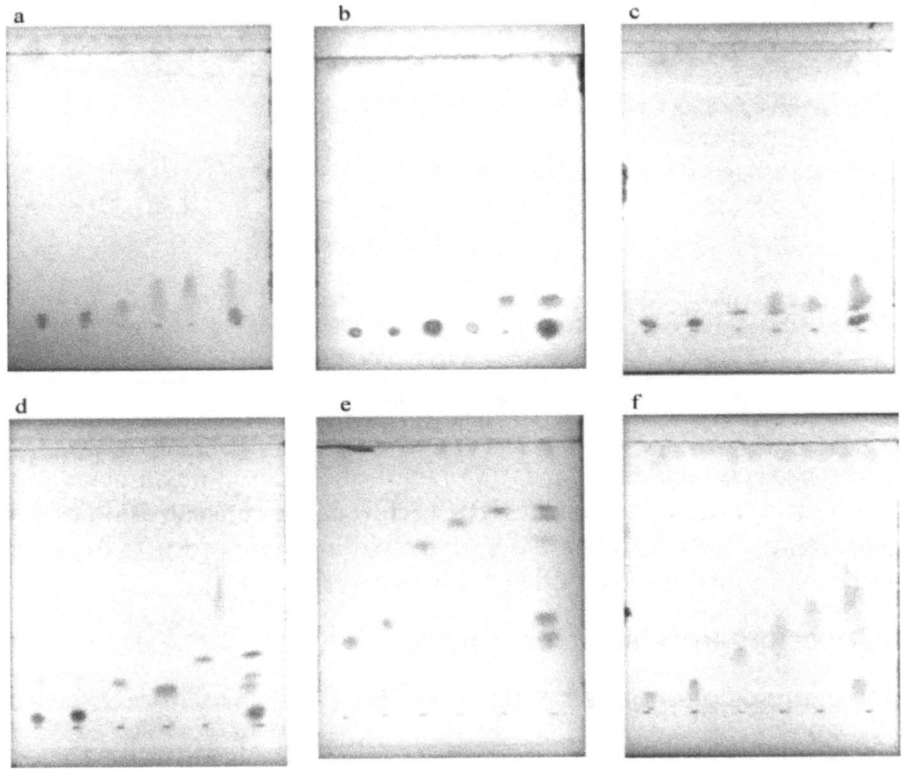

Fig. 1. Chromatograms of thioridazine, trifluoropromazine, phenazoline, naphazoline, tiamenidine and the mixture of the drugs, as spotted from left to right, on RP-18 F254 plates, developed with water-acetonitrile 40:60 v/v eluent either pure or containing 1.5% v/v of various additives: a-no additive; b-NH4OH; c-DMOA; d-TEA; e-[EMIM][BF4]; f-buffer of pH 2.87. (Kaliszan et al., 2004).

RTILs were examined also as matrixes for online separation and identification of peptides, proteins, and poly(ethylene glycol) (PEG-2000) with matrix-assisted laser desorption/ionization-mass spectrometry (MALDI-MS) (Armstrong at el., 2001; Mehl & Hercules, 2000). The method described hybrid TLC-MALDI plates. Authors concluded that tested ILs, owing to good solubilizing properties and vacuum stability, in comparison to other commonly used liquids and solid matrixes, should be further investigated as potential MALDI matrixes.

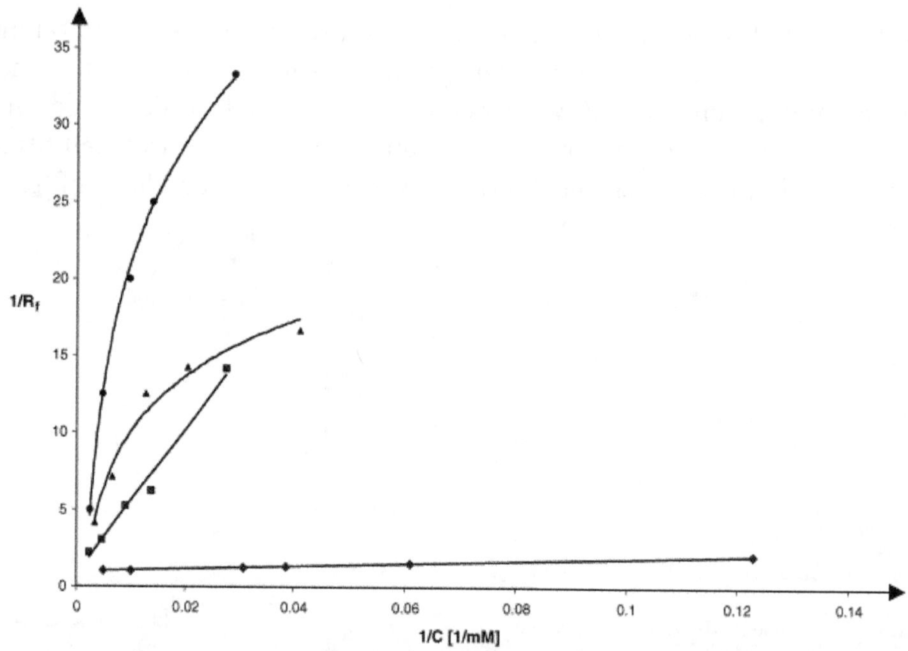

Fig. 2. Plots of reciprocal of retardation factor of tiamenidine on octadecylsilica plates with acetonitrile as eluent vs. the reciprocal of concentration of additive in the mobile phase. Additives are denoted as follows: IL 1 (✦), TEA (▪), DMOA (▴) and NH_4OH ( ).(Kaliszan et al., 2004).

3.2 High-performance liquid chromatography

3.2 1 Ionic liquids as mobile phase solvents

In the 80s of the last century, Poole et al. (Poole et al., 1986) elaborated on the synthesis of six alkylammonium salts of the general formula $R_{n-a}NH^{+}_{a}X^{-}$, where R = ethyl, propyl, butyl or sec.-butyl, X^{-} = nitrate or thiocyanate, $n < 4$ and a = 0, 1,2 or 3. Authors measured their physical properties including melting point, refractive index, UV spectra, miscibility with common organic solvents, aqueous solution pH, density and viscosity and concluded that obtained liquid organic salts forming immiscible solvent pairs with non-polar organic solvents appear to be suitable for liquid—liquid chromatography and liquid—liquid extraction. Some of them, such as the alkylammonium nitrate or the thiocyanate salts mixed with a second solvent of low viscosity such as water, methanol, acetonitrile, tetrahydrofuran and dichloromethane could be used as mobile phases in liquid chromatography. Studies of their solvent properties in microcolumn RPLC demonstrated that the back pressure on the column was too high, in spite of the fact of a small flow of 0.05 ml/min, a small internal diameter of the column (1mm) and the least viscous of ILs applied (Shetty et al., 1987). Additionally, applied aqueous solutions of thiocyanates corroded metal parts of the apparatus and demonstrated poor baseline stability.

In turn tetra-n-alkylammonium sulfonates were too basic and caused silica-based columns degradation process. Therefore, the project was abandoned.

A couple of years after that, Waichigo applied a few monoalkyloammonium nitrates and acetates adding them, as modifiers, to water in concentrations of 20-60 % using classical C18 columns or the copolymer-styrene-divinylobenzene (Waichigo et al., 2005; Waichigo & Danielson, 2006 a; Waichigo & Danielson, 2006 b; Waichigo et al., 2007). It appears that ethylammonium nitrate possesses similar solvent strength to tetrahydrofuran. Elongation of alkyl chain in ammonium salt structure is connected with increasing of elution strength of IL in chromatographic system. In turn, the solvent strength of ethylammonium acetate is similar to methanol. Waichigo observed the lower number of theoretical plates (20%) and the twice higher back pressure on the column compared with the conventional arrangement of reversed phases and water-organic eluent. One should emphasize it that majority of authors applying alkylammonium salts achieved worse effects comparing with classical RP- HPLC as for: the transparency in UV, viscosities triggering back pressure, worsened efficiency. Taking into account the fact that alkylammonium salts possess the lowest viscosity in comparison to other ILs, it is difficult to expect that they would displace currently applied organic modifiers like methanol or acetonitrile. However, the non- molecular nature of ILs could give meaningful improvement of selectivities (Fig.3)

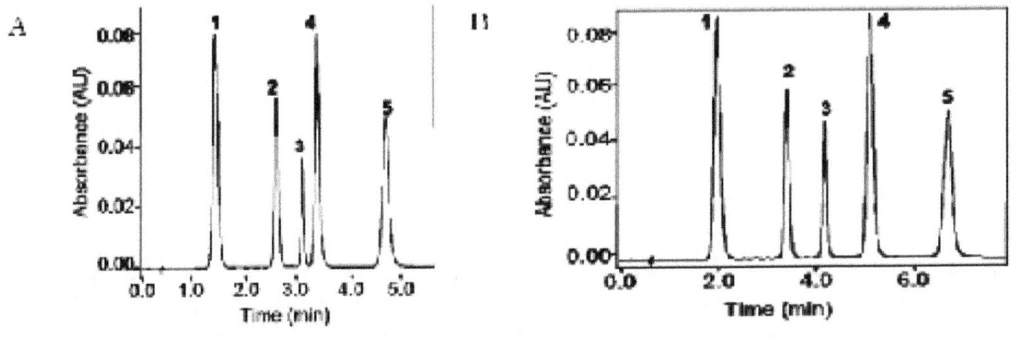

Fig. 3. Chromatogram of 1-ascorbic acid, 2-nicotinic acid, 3-thiamine, 4-pyridoxine, and 5- niacinamide using the Aquasep C8 column with a)30%methanol-70%water. a)30% ionic liquid ethylammonium acetate-70%acetate. Flowrate:1.0mLmin-1. UV detection at 254 nm. (Waichigo et al., 2005).

3.2 2 Ionic liquids as mobile phase modifiers.

ILs were found to be more useful when added to the mobile phase at millimolar concentration, however all their unique properties were lost and they were just salts adjusting the solvent strength. By dilution of IL with water or organic-aqueous solvent or by increasing the column temperature, the viscosity of ionic liquids can be drastically reduced. Several advantages associated with the use of ionic liquids as mobile phase additives should be emphasized. First of all, these compounds don't damage the silica based columns, do not change radically pH of the mobile phase, and could replace the less advantageous alkylamines with aim of silanol screening.

Many papers demonstrate the usefulness of ILs at these conditions in terms of improvement of efficiency and peak symmetry (He et al., 2003; Zhang et al., 2003; Zhang et al., 2004; Marszałł et al., 2005,; Marszałł et al., 2006 a; Ruiz-Angel et al., 2006). Fig.4 presents comparison of prochlorperazine peaks depending on the kind of modifier added to the mobile phase (Flieger, 2009).

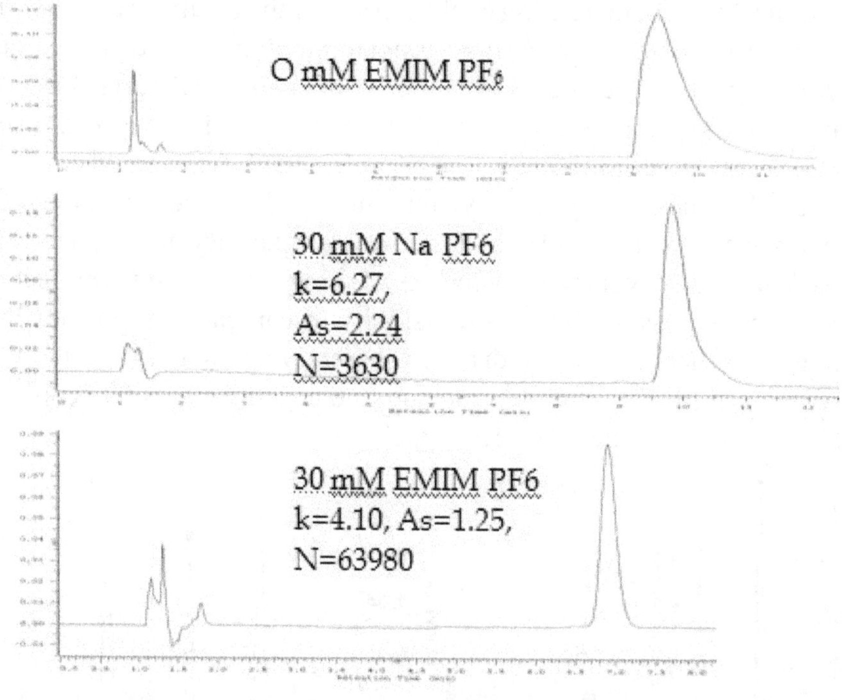

Fig. 4. The effect of different mobile phase additives on chromatographic parameters of prochlorperazine. Eluent: 40%ACN/10mM phosphate buffer (pH2.8), column: Zorbax Extend-C18 (Flieger, 2009).

To understand the influence of the particular ions of added reagents, the eluent systems containing the following components: ACN/phosphate buffer pH=2.8, ACN/phosphate buffer (pH=2.8)/$NaPF_6$ and ACN/phosphate buffer (pH=2.8)/ [EMIM] [PF_6] were compared. As it can be seen, unacceptable parameters in system containing only organic- aqueous mobile phase acidified by phosphate buffer being a consequence of the ionic nature of the analyte, cause the creation of tailing peaks with not satisfactory symmetry. Addition of chaotropic additive $NaPF_6$ to the mobile phase improves the peak symmetry and efficiency of the system, nevertheless A_s factor is still out of an acceptable range. The best parameters of peak symmetry and efficiency were obtained for ionic liquid [EMIM] [PF_6] as the mobile phase additive. Change of Na^+ cation, which is not adsorbed into hydrophobic surface into the enlarged imidazole cation possessing the structure enabling to penetrate the space between the hydrocarbon chains, appears to be the most advantageous.

However, special attention should be devoted to appropriate concentration of IL in the mobile phase to avoid dissymmetry and multiplification of peak as a result of existing analyte in different forms: ion-pair and free protonated base. Fig.5 shows how peaks profile of nortryptyline can be improved according to increasing concentration of [BMIM][PF_6] (Flieger & Czajkowsa-Żelazko, 2010).

Improving peak symmetry, efficiency expressed as theoretical plates number and visible in decreasing peaks tailing and reducing band broadening, are connected with changing of the retention mechanism in the presence of ILs in a chromatographic system. Both ions contribute to the retention of analytes. They can affect obtained results either antagonistically or synergistically.

Many authors confirm suppression of silanophilic interaction by ILs which coat the surface and interact with free silanol groups of silica based supports (Marszałł et al.,2006 a; Marszałł et al.,2005; Ruiz-Angel et al.,2006; Flieger, 2009). As new masking agents enabling reduction of deleterious silanol effect, ILs appeared to be more advantageous in comparison to traditional additives like amino quenchers (triethylamine (TEA), dimethyloctylamine (DMOA), cyclohexylamine, quaternary ammonium ions) applied so far for this reason. It was demonstrated that either anions can be adsorbed on hydrophobic stationary phase according to their chaotropicity (Pan et al., 2004; Gritti & Guiochon, 2004) or ionic liquids' cations following their lyotropy related to the lengths of the alkyl chains (Berthod et al.,2005; Zhao, 2006). Ions can also move together with the mobile phase and interact with ionic analytes. Protonated basic analytes could interact with anions creating neutral ion-pairs with longer retention. In this case ion-associated complexes are formed only with that ones exhibiting chaotropic properties ($H_2PO_4 < HCOO < CH_3SO_3 < Cl < NO_3 < CF_3COO < BF_4 < ClO_4 < PF_6$) (LoBrutto et al., 2001; Flieger, 2006). So far, adsorption of ILs on adsorbent surface and ion-pairing mechanism has been proposed to explain retention mechanism in chromatography applying ILs as modifiers.

Firstly Jiang (Jiang et al., 2004) described addition of 2–50 mM IL to aqueous mobile phases as silanol suppressing agents in the separation of basic and strong hydrophilic analytes ephedrines, catecholamines, and amino acids. Later on Marszałł et al. evaluated the silanol- suppressing potency of ILs (Marszałł et al., 2006 a) applying dual retention model proposed by Nahum and Horvath:

$$\frac{[A]}{k_0 - k} = \frac{1}{k_2 K_A} + \frac{[A]}{k_2} \qquad (1)$$

where k_0 indicates retention factor at the absence of any additives and k is the retention factor obtained at the concentration [A] of the silanol suppressor.

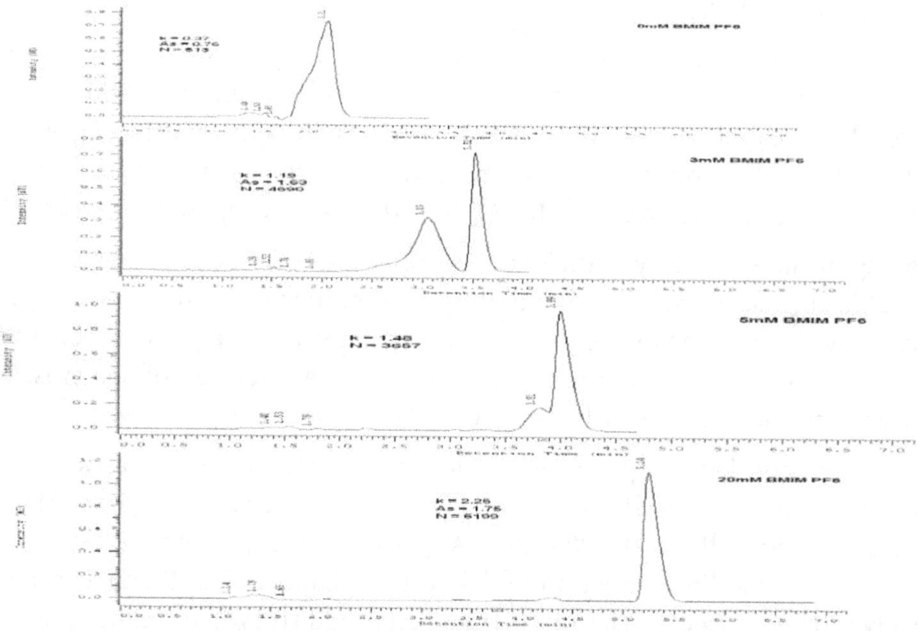

Fig. 5. The effect of different concentration of BMIM PF$_6$ on retention, peak symmetry and efficiency of nortryptyline. Column: on Zorbax Eclipse- XDB-C18, Eluent system: 40% ACN/20 mM phosphate buffer pH = 2.9 (Flieger & Czajkowska-Żelazko, 2010).

The effect of different imidazolium based ILs was established on the basis of calculated binding constant of the silanol complex (K_A). It appears that the higher values were obtained for chloride and bromide ionic liquids than for tetrafluoroborates. Additionally, it should be emphasized that observed screening effect was achieved at lower concentration of these additives. Determined stability constant of the silanol complexes together with recommended additive concentration are presented in Table 2.

Advantages connected with suppression of silanophilic interaction appear simultaneously with the changes in analytes' retention factor. The increasing K_A values of alkylimidazolium halides, reflecting the silanol-masking potency, are connected with decreases of the basic analytes retention and selectivity of their separation.

In the light of conducted experiments, it is obvious that the cation of ionic liquid is responsible for the silanol screening effect, whereas the retention factor primarily depends on the properties of its anion. Only more chaotropic character of ILs anion, meaning not only hydrophobic nature but also entirely ion-pairing properties, leads to drastic retention increase. When anion of IL does not create ion-associated complex with anlayte in ionic form, decrease in retention is observed. Summarizing, the retention mechanism in chromatographic system modified with ILs additives is complex covering the creation of ion pairs as well as ion exchange and hydrophobic partitions.

Generally, the retention of the basic compounds increases with the increase of the ionic liquids with chaotropic anion concentration in the mobile phase in reversed-phase systems (Flieger, 2009). However, these dependences differ in the shapes especially at their initial sectors. Fig. 6 illustrates the effect of concentration of 1-ethyl-3-methyl-imidazolium hexafluorophospate [EMIM][PF_6], 1-butyl-3-methyl-imidazolium hexafluorophosphate [BMIM][PF_6] added to acetonitrile-aqueous eluent system on retention of basic compounds belonging to phenothiazine derivatives on an octadecyl bounded-silica stationary phase.

Table 2. Stability constant of the silanol complex, K_A and the recommended additive concentration, C_A, in 50:50 v/v acetonitrile/water pH 3 when using 1-alkyl-3- methylimidazolium ionic liquids. (Marszałł et al., 2006a).

Investigated ionic liquids	Benzyltriethylammonium hydrochloride		Thioridazine hydrochloride		Chlorpromazine hydrochloride	
	$K_A \times 10^{-3}$ [M^{-1}]	C_A^* [mM]	$K_A \times 10^{-3}$ [M^{-1}]	C_A^* [mM]	$K_A \times 10^{-3}$ [M^{-1}]	C_A^* [mM]
[EMIM][BF_4]	2.15	4.65	7.11	1.41	7.40	1.35
[PMIM][BF_4]	2.37	4.22	7.67	1.30	8.68	1.15
[BMIM][BF_4]	2.38	4.20	7.92	1.26	9.45	1.06
[HMIM][BF_4]	2.71	3.69	9.39	1.07	17.95	0.56
[OMIM][BF_4]	3.12	3.20	13.52	0.74	18.43	0.54
TEA	1.13	8.86	0.52	19.31	0.61	16.49
[MMIM][MSO_4]	2.45	4.07	8.53	1.17	14.31	0.69
[EMIM][ESO_4]	2.61	3.82	6.99	1.43	7.09	1.41
[BMIM][OSO_4]	1.90	5.26	1.05	9.52	0.94	10.63
[EMIM][Br]	3.57	3.34	10.31	0.96	8.24	1.21
[EMIM][Cl]	2.99	2.80	9.7	1.03	7.90	1.26
[EMIM][Ts]	0.26	38.46	5.44	1.83	6.36	1.57
[BMIM][Br]	2.59	3.86	7.08	1.41	7.05	1.41
[BMIM][Cl]	2.89	3.46	6.54	1.52	6.94	1.43
[OMIM][Cl]	5.58	1.79	13.81	0.72	16.28	0.61

This class of isotherms possesses characteristic Langmuir shape related to horizontal orientation of the adsorbate molecules on the adsorbent surface.

At higher concentration of ionic liquids, the shape of the isotherms differs, depending on the type of the additive. Quite different are also the concentrations provided to column saturation. The addition of [BMIM] [PF_6] ensures obtaining of the plateau at 2mM, and the shape of this dependence is a typical one of the monolayer adsorption (the basic type of Langmuir adsorption). In case of [EMIM] [PF_6], the shape of the above mentioned dependence suggests the incidence of a multi-layer adsorption. Quite different is the shape of the isotherm for the additive containing anion of a neutral character [BMIM] [Cl]. In this case, retention time of bases decreases with increasing ionic liquid concentration. This effect could be a consequence of repulsion forces between the adsorbed imidazolium cation on the adsorbent surface and protonated analytes.

Xiaohua obtained similar results (Xiaohua et al., 2004) analyzing the series of amines: benzylamine, benzidine, N-ethylaniline, N, N'-dimethylaniline and *k* o-,m-,p-phthalic acids in the presence of [BMIm][BF4] in the mobile phase. Decreases of basic and acidic compounds retention proved that the part of ionic liquids coat on the reversed-phase surface and simultaneously anion of ionic liquid added does not interact with analytes. Kaliszan and his group also observed the retention decrease of neuroleptic basic drugs on octadecylsilica support and mobile phase comprising acetonitrile: phosphoric buffer (pH 3) 70/30 (v/v) modified with 1-ethyl-3-methylimidazolium tetrafluoroborate and 1,3- dimethylimidazolium methyl sulfate (Marszałł et al., 2005).

Owing to addition of ILs at minute concentration to the mobile phase we can improve also separation selectivity. Xiaohua and co-workers (Xiaohua et al., 2004) showed improvement of amines mixture separation covering benzylamine, benzydine, N,N-dimethyloaniline, N- ethyloanilinecan on C18 column and water with addition of 1-ethyl-3-methylimidazolium tetrafluoroboran as an eluent system.

In my latest studies, some new aspects devoted to application of ionic liquids as mobile phase modifier were revealed (Flieger & Czajkowska-Żelazko, 2010). Analyzing chosen basic drugs in reversed-phase systems on two columns: Zorbax XDB-C18 and Zorbax SB- Phenyl in acetonitrile aqueous mobile phase acidified and modified with [BMIM] [Cl] or [BMIM][PF_6], visible differences in chromatographic behavior were detected. Fig.7 presents the retention dependences of pindolol on the concentration of investigated ionic liquids in 10%ACN buffered eluent system.

As it could be seen, pindolol shows an increase of retention with increasing concentration of ionic liquid possessing chaotropic anion [PF_6], as a result of ion-pair creation between protonated basic molecule [BH^+] and the counter-anion [PF_6]. This process achieves saturation level after which ionic liquid concentration does not influence the retention. In turn, the presence of neutral anion [Cl^-] causes deceasing of retention factor. Moreover, at this graph unexpected retention increase on Zorbax SB-Phenyl at low concentration (1 mM) of [BMIM] [Cl] in 10% ACN mobile phase could be observed. This deviation is the consequence of differences in structure of interphases created in compared systems.

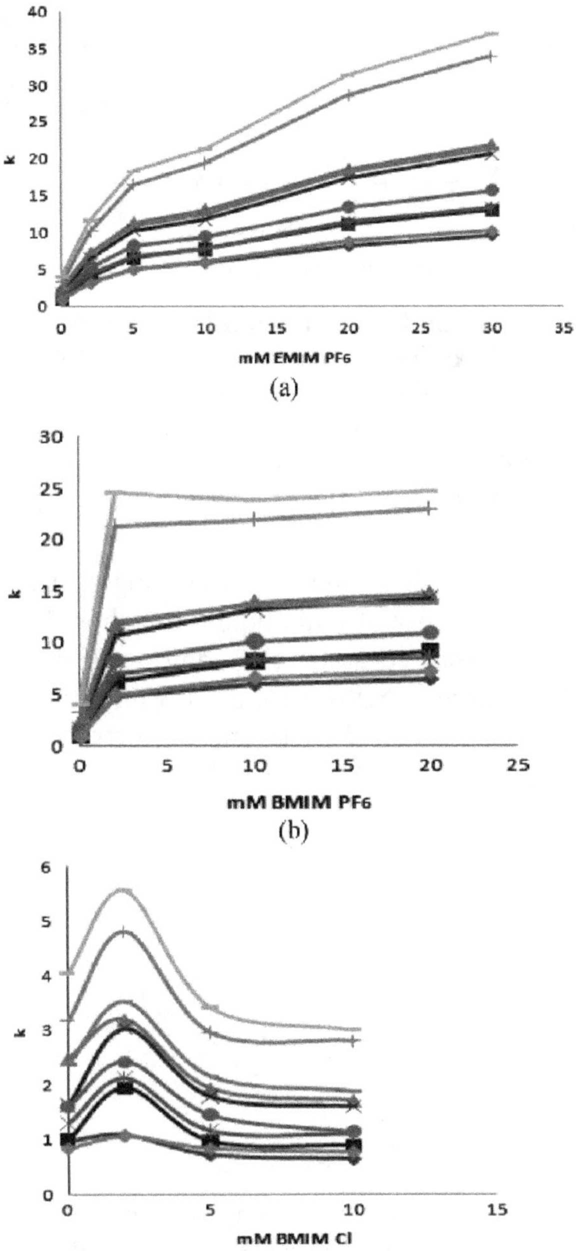

Fig. 6. Experimental retention factor for the investigated neuroleptic drugs vs. concentration of ILs in mobile phase containing 30% ACN/10 mM phosphate buffer (pH=2.8). A-EMIM PF_6, B-BMIM PF_6, C-BMIM Cl. (Flieger, 2009).

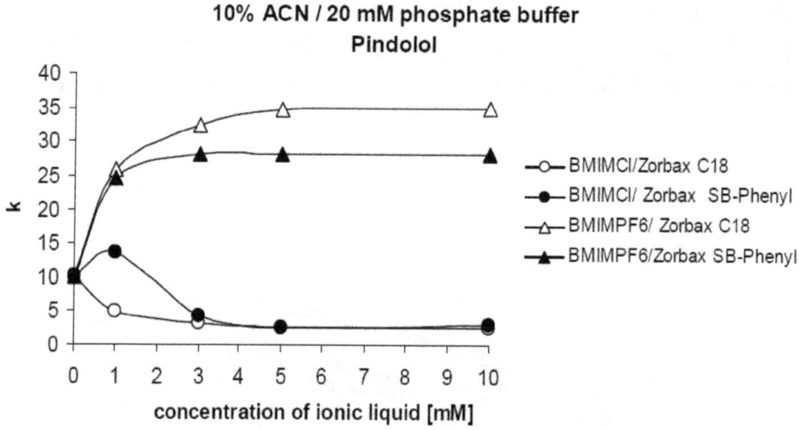

Fig. 7. Experimental retention factors for pindolol versus ionic liquids (1-butyl-3-methyl-imidazolium hexafluorophosphate-[BMIM] [PF_6], 1-butyl-3-methyl-imidazolium chloride- [BMIM] [Cl] concentration in 10% ACN / 20 mM phosphate buffer pH = 2.9 obtained for different hydrophobic columns: Zorbax Eclipse-XDB-C18 and Zorbax SB-Phenyl. Flow rate: 0.9 ml/min., temperature: 20°C, UV detection at 254 nm; sample: 3 µl injection. (Flieger & Czjkowska-Żelazko, in press).

Multilayer of adsorbed acetonitrile molecules forming on the adsorbent surface is a good medium for solubility of ionic liquids as a liophilic species. Assuming that free silanols are in ionized state, the addition of even small amount of ionic liquid (1 mM for Zorbax XDB- C18 or 1-3 mM for Zorbax SB-Phenyl) could be enough to reduce free silanols and to increase of stationary phase hydrophobicity owing to electrostatic interactions occurring accordingly to the following scheme:

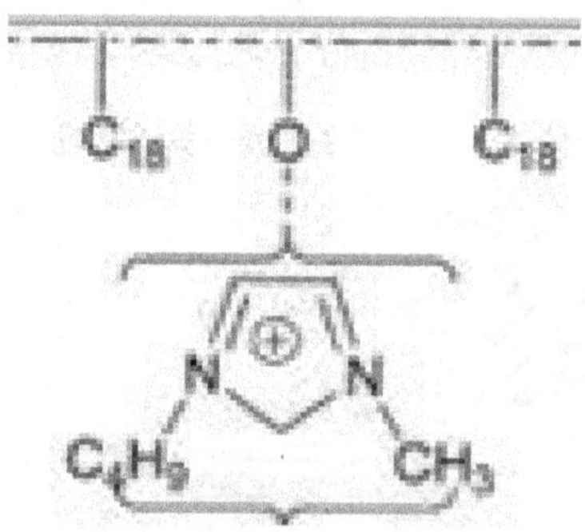

This effect could be useful in practice for improvement of separation selectivity by retention enhancement. Fig.8 presents example of improvement of 1,4 benzodiazepine separation by addition of 3 mM [BMIM][Cl] to acetonitrile-water mobile phase on Zorbax SB-Phenyl column.

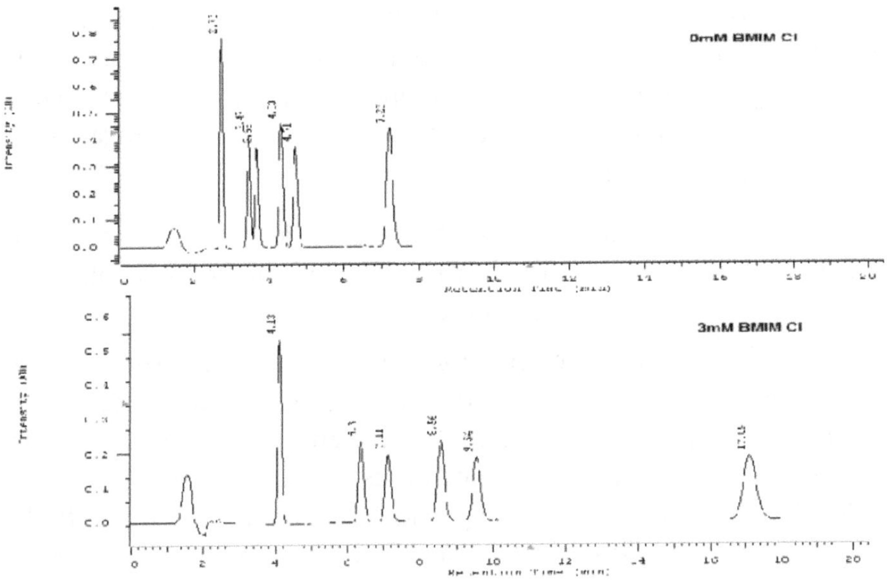

Fig. 8. Separation of 1,4-benzodiazepines mixture on Zorbax SB-Phenyl column and eluent system: 50% ACN / 20 mM phosphate buffer pH = 2.9 with addition of 3 mM [BMIM] [Cl]. For HPLC conditions see Experimental part. The peaks order: dechlordiazepam, oxazepam, lorazepam, nitrazepam, demethyldiazepam, diazepam.

Further increase of [BMIM] [Cl] concentration in the mobile phase causes decrease in retention of protonated amines probably owing to excess adsorption of liophilic ionic liquid cation with polar part oriented toward polar eluent environment in agreement with the following scheme:

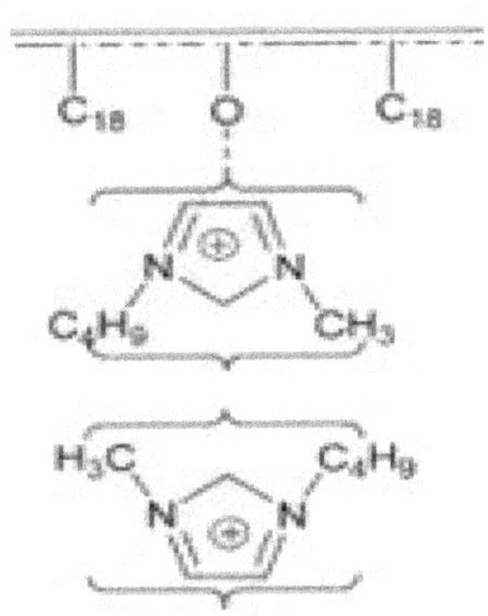

According to this scheme, positive charge on the surface is responsible for retention decrease of protonated amines at higher concentration of ionic liquid chloride in the mobile phase.Though, some papers (Hu et al., 2009; Stepnowski et al., 2006) demonstrated adsorption of ILs on hydrophobic surface, nobody has shown before real reflectance of these speculations in retention factor and connection with ILs concentration in the mobile phase.

Addition of ILs to the mobile phase in RP-HPLC systems appears to be especially advantageous for determination of the lipophilicity parameters of ionogenic basic analytes owing to supression of silanophilic interactions and improvement of linearity of log k vs. volume fraction of organic modifier in eluent system:

$$\log k = -S\varphi + \log k_w \quad (2)$$

where log k indicates the isocratic capacity factor, φ the volume fraction of organic modifier in the mobile phase, log k_w the retention factor extrapolated to pure water, S the slope of the regression curve.

The correlation between the reference lipiophilicity parameter from the n-octanol-water partition system (log P_{oct}) and chromatographically determined lipophilicity parameters (S, log k_w, φ_0) is statistically significant mainly for congeneric group of neutral solutes. Application of ionic liquids additives improved evidently this correlation for compounds exhibiting acid–base behavior. Kaliszan (Kaliszan et al 2004) proposed [EMIM] [BF_4] for this purpose. Fig.9 illustrates improvement of correlations between the chromatographic lipophilicity parameters, log k_w, and log P_{oct} obtained for a series of basic drugs by gradient HPLC at the presence of [EMIM] [BF_4] in the mobile phase.

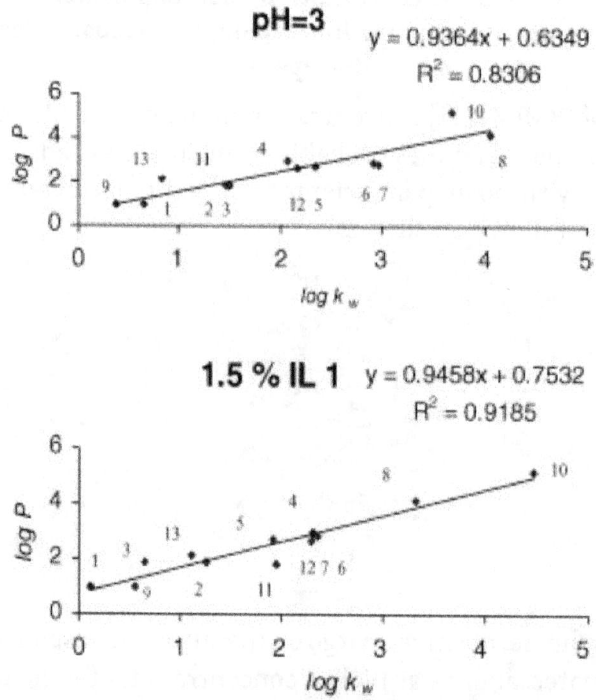

Fig. 9. Relationships between log P and the lipophilicity parameter log k_w, determined by gradient HPLC with a buffered at pH 2.87 water-methanol eluent not comprising (upper plot) and comprising (bottom plot) 1.5% (v/v) of IL (Kaliszan et al., 2004).

For acid compounds, the mechanism for separation and the effects of ionic liquids appearto be different. Hu et al (Hu et al., 20) applied 1-butyl-3-methylimidazolium chloride [C_4MIM][Cl], 1-octyl-3-methylimidazolium chloride [C_8MIM][Cl], and 1-decyl-3- methylimidazolium chloride [C_{10}MIM][Cl] as mobile phase additives in reversed-phase separation phenoxy acid herbicides and phenols at neutral pH on conventional C18 column. It was demonstrated that [BMIM][Cl], was the most efficient. The retention time of the target acids was shortened with the increase of the alkyl chain length and the concentrations of ionic liquids. The delocalization of the positive charge on the imidazolium cation, the repulsion between chlorine ions of ionic liquids and the acid compounds, as well as the stereo-hindrance effect are probably responsible for the observed effects.

3.2.3 Ionic liquids as liquid chromatography stationary phases

The newest approach in liquid chromatography is stationary phase modification by the use of ILs. Initial experiments were performed by the dynamic coating of the silica support using alkylammonium nitrate ILs in chlorinated hydrocarbons or ethers in hexane containing up to 75% (v/v) of the polar organic solvent as the mobile phase (Shetty et al. 1990). In this way, ILs formed the so-called supported liquid phases generating liquid-liquid separation mechanism. Obtained biphasic systems were effective for identifying systems for liquid–liquid extraction of complex mixtures and determination of partition coefficients. This method, owing to favorable features including speed, accuracy and minimal consumption of reagents, appears to be suitable in spite of some disadvantages concerning UV absorbance limitations caused by ILs. This idea found application in countercurrent chromatography (CCC) requiring two immiscible liquid phases as separation medium. Berthod et al. (Berthod et al. 2004) constructed biphasic liquid system consisting of water and [BMIM] [PF_6] with addition of acetonitrile reducing viscosity for determination of the distribution constant of various aromatic compounds containing basic, acidic and neutral substituents. A preparative high-speed counter-current chromatography method for isolation and purification of flavonoid compounds from *Oroxylum indicum* was successfully established by using ionic liquids as the modifier of the two-phase solvent system. Two flavonoid compounds including baicalein-7-O-diglucoside and baicalein-7-O-glucoside were purified from the crude extract of *O. indicum* by using ethyl acetate–water–[BMIM][PF_6] (5:5:0.2, v/v) as two-phase solvent system (Liu et al., 2010).

Liu and his group was among the first who examined new LC stationary phases based on covalently bonded ILs. They immobilized a vinyl-hexylimidazolium tetrafluorobarate IL at the surface of porous silica particles activated with a silane-coupling agent (mercaptopropyl trimethoxyxilane). The obtained column was applied for separation of an ephedrine mixture (Jiang et al. 2004). The initial results weren't satisfactory. However, the next study of the same team concerning covalent bonding between silanol groups and the cation of the ILs turned out to be a considerable success (Qiu et al., 2006 a; Qiu et al., 2006 b). Owing to the reaction of silanised with 3-chloropropyltrimethoxysilane gel with imidazolium salt, stationary phase possessing anion-exchanging properties able to separate inorganic as well as organic anions was obtained. A schematic diagram of SilprMim synthesis and separation results is shown in Fig.10 and Fig.11.

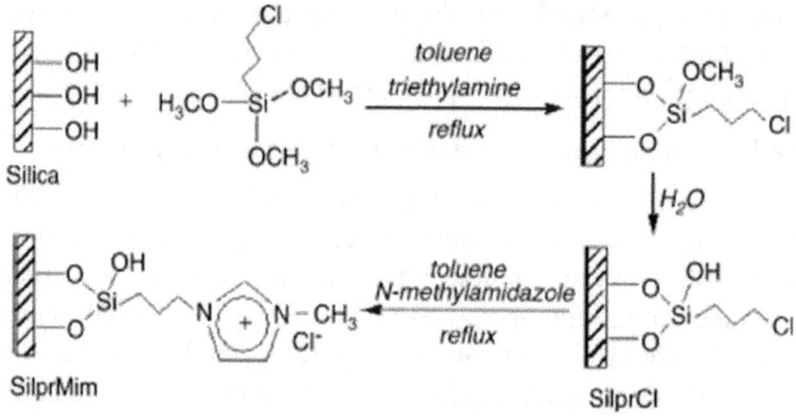

Fig. 10. Preparation of N-methylimidazolium functionalized silica. (Qiu et al. 2006 b)

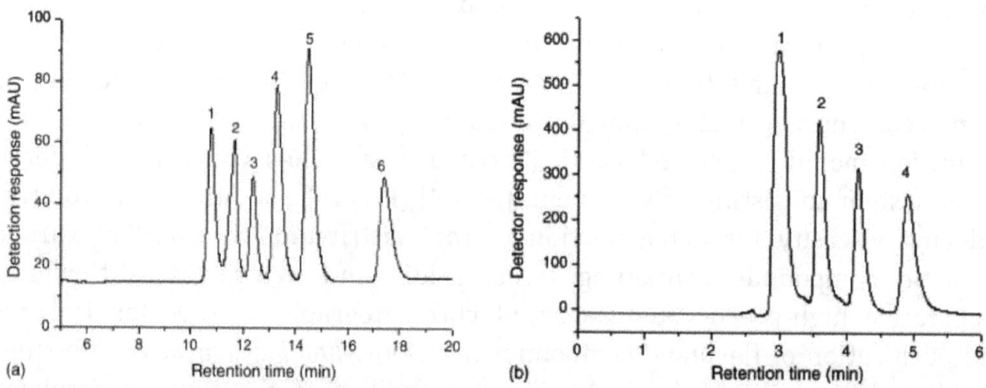

Fig. 11. Chromatogram of separation of test mixture composed of: (a) iodate(1) chloride(2); bromide(3); nitrate(4); iodide(5); thiocyanate(6). Mobile phase 20mmoll^{-1}KH_2PO_4 (pH 4.6); (b) N-ethylaniline(1); N,N-dimethylaniline(2); o-nitroaniline(3);andbenzidine(4); - Mobile phase was 50mmoll^{-1} KH_2PO_4 (pH 4.6) (Qiu et al. 2006 b).

Next, Stalcup and co-workers (Sun et al. 2005) received new column thanks to the reaction of silica gel with 1-butyl-3 heptylimidazolium bromide. IL's cation was covalently immobilized through an n-alkyl ether chain. Properties of obtained column were studied under reversed-phase conditions for the separation of a group of neutral aromatic solutes applying the linear solvation energy relationship approach. Retention properties were remarkably similar to conventional phenyl-based stationary phases despite the presence of significant charge on the surface. The slope of the log $k_{butylimidazolium}$ versus log k_{phenyl} is close to unity (0.958) indicating additionally that residual alkyl linker does not play a large role in the retention process (Fig.12).

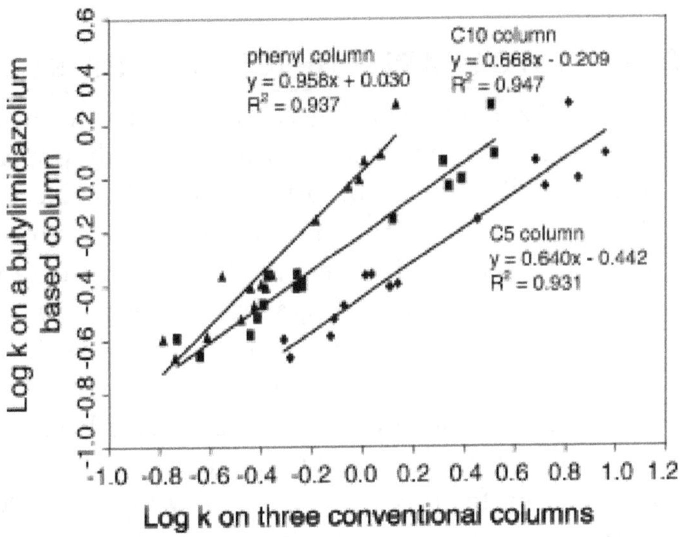

Fig. 12. Plot of log *k* obtained using 60% MeOH on the butylimidazolium column versus log *k* obtained on conventional aliphatic reversed phase columns under the same mobile phase conditions (Sun et al. 2005).

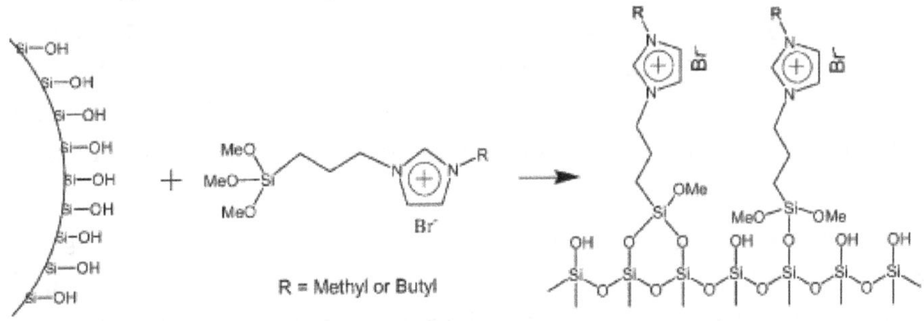

Fig. 13. Schematic procedure of ILs bonded stationary phase synthesis (Wang et al. 2006).

Recently, consecutive study describing other stationary phase has been published (Wang et al. 2006). Trimethoxysilane "ionosilane" derivatives of RTILs based on alkylimidazolium bromides were bonded to 3 μm diameter silica particles. The ionic liquid moiety was predominantly attached to the silica surface through two siloxane bonds of the ionosilane derivative. The obtained structure was confirmed by thermogravimetric analysis (TGA) and ^{13}C and ^{29}Si NMR spectroscopies. Properties of the column were studied by using aromatic carboxylic acids as model compounds. It appears that the separation mechanism involves multiple interactions including ion exchange, hydrophobic interaction, and other electrostatic interactions. Authors of the study emphasize it that a clear advantage of the new columns is a possibility of reducing the content of organic solvent in the mobile phase and a diversified selectivity depending on the spatial orientation of bounded ligands (Fig.13).

4. LIQUID CHROMATOGRAPHY OF IONIC LIQUIDS

In the recent years, ILs have gained great popularity in chemical industry. Their determination is very important because of high probability of their migration into the environment. Unfortunately, this subject has not been extensively explored so far. There are only several papers dealing with analytical methods suitable for ILs determination. Among them chromatographic techniques, especially reversed-phase liquid chromatography, appear to be the most suitable for analysis of ILs as well as determination of their properties or identifying systems suitable for their liquid-liquid extraction. Different stationary phases such as cholesterol ligands bonded chemically to silica, alkylamide, butyl, octyl, octadecyl, phenyl, aryl, mixed (containing octyl, octadecyl, phenyl, aminopropyl, cyanopropyl) were tested for this purpose (Buszewski et al., 2006). Due to the non-molecular nature of ILs, the retention mechanism is not as simple as the one of molecular species. The retention mechanism combines hydrophobic and ionic interactions (ion-pairing, ion-exchange). In agreement with hydrophobic nature of retention mechanism, retention of ILs increases with increasing hydrophobicity of the stationary phase. However, for packings containing protonized functional groups, the lowering of the retention factor was observed owing to electrostatic repulsion forces between entities and charged surface. The main interactions, predominant in the case of aryl stationary phase, are dispersive ones and of π-π type.

The best results according to separation selectivity towards ILs were achieved on the butyl and silica monolith with bonded octadecyl groups probably owing to the lowest heterogeneity of these ligands (Buszewski et al., 2006). Routinely, liquid chromatography with the use of an octyl stationary phase could be applied for isolation, identification and quantification of 1-alkyl and 1-aryl-3-methylimidazilium based ILs (Stepnowski et al., 2003). It should be stressed that when a salt is added to the mobile phase, only ILs differing with the cation can be separated (Ruiz-Angel & Berthod, 2006). To improve IL peak shape and enhance retention time, $NaPF_6$ or $NaClO_4$ could be added to the mobile phase. These anionic additives, possessing chaotropic properties, form ion-pairs with the IL cations. In turn, mobile phases containing no salts are able to separate ILs differing with the anion. Their elution order depends on the lyotropy of the anions. Berthod investigated eleven 1-alkyl-3- methyl imidazolium ionic liquid salts in reversed phase mode with a Kromasil C_{18} column. It was shown that it is possible to separate different ILs depending on the mobile phase composition. Additionally, working with salt-free mobile phases, fronting peaks for all ILs were observed, associated with the IL/C_{18} stationary phase interactions responsible for concave isotherms. With mobile phase containing 0.01 M of salt, tailing IL peaks corresponding to convex adsorption isotherms were obtained, as could be seen in Fig.14.

Taking into account the fact that short chain ILs, as the early eluting compounds in reversed-phase systems, are not satisfactorily resolved, Stepnowski and Mrozik (Stepnowski & Mrozik, 2005) applied the SCX stationary phase with strong cation exchange properties. They investigated the chromatographic behavior of 8 Ils: 7 homologues of 1-alkyl-3- methylimidazolium and 4-methyl-*N*-butylpyridinium. Depending on the mobile phase composition, different interactions such as cation exchange, nonspecific hydrophobic interactions, and adsorption affected the retention process. In comparison to the chromatographic behavior of Ils in the classical reversed phase system much higher selectivity of highly hydrophilic compounds was noticed.

In order to take advantage of the potential π–π interactions, an ether-linked phenyl stationary phase with polar endcapping was also proposed in the separation of imidazolium and pyridinium ILs (Stepnowski et al., 2006). With phenyl-bonded phases, very good separations of the cations were achieved however; gradient elution was required to achieve separation at appropriate time (Fig.15). Once again special attention was paid to the short- chain hydrophilic entities, known to be poorly separated on conventional reversed-phase columns. Authors declare the applicability of the proposed method for monitoring degraded cations of 1-butyl-3-methylimidazolium salts obtained in the Fenton process.

The aim of this study was also investigation of the nature of interactions occurring in the chromatographic system. Possible interaction between imidazolium IL and phenyl-bonded stationary phase is illustrated schematically in Fig.16. In case of acetonitrile, which is able to form electrono-donor-acceptor complexes, suppression of π–π interactions between the solute and the aromatic moiety of the stationary phase could be assumed. Thus at higher acetonitrile concentration in the mobile phase only hydrophobic and hydrogen-bonding interactions would be responsible for the separation mode. Recently, Studzińska and Buszewski proved that ILs are bonded much stronger to variety of stationary phase surfaces when pH of mobile phase is less acidic (Studzińska & Buszewski 2010).

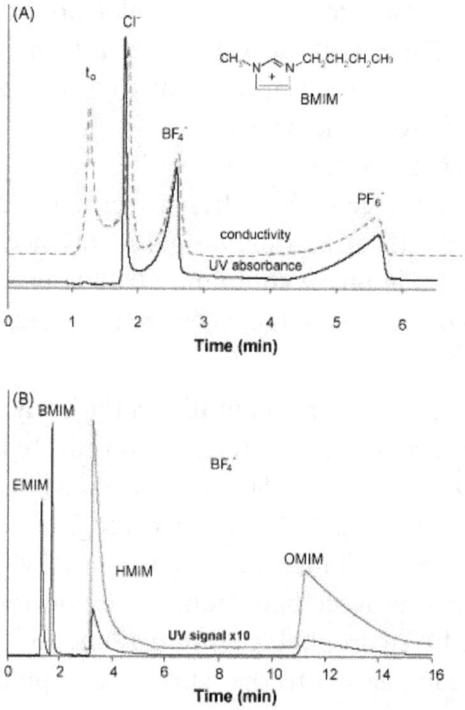

Fig. 14. RPLC separation of ionic liquids. (A) Separation of three BMIM ILs; continuous line: UV 254 nm detection; dotted line: conductimetric detection; salt-free mobile phase: acetonitrile/water 10/90% (v/v), 1 mL/min. (B) Separation of EMIM Cl, BMIM BF_4, HMIM BF_4 and OMIM PF_6; mobile phase: acetonitrile/water 0.01 M NaCl 30/70% (v/v), 1 mL/min; UV detection 254 nm; column Kromasil C_{18} 150 mm × 4.6 mm i.d. (Ruiz-Angel & Berthod, 2006).

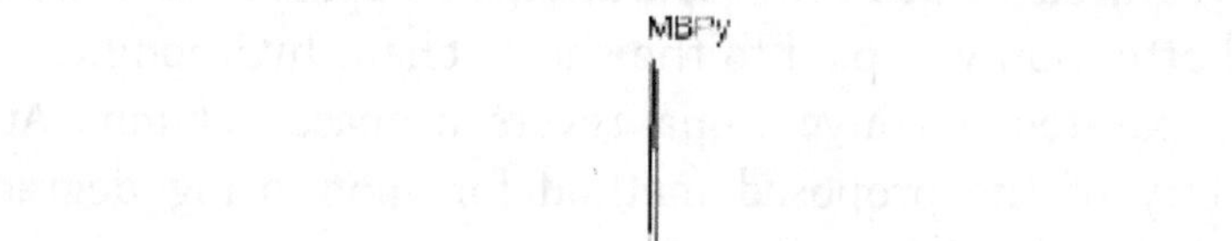

Fig. 15. Gradient separation of selected imidazolium and pyridinium ionic liquids under the conditions investigated. Mobile phase: methanol (2–20%)-water (5mM K_2HPO_4).pH 3.0.Column: Synergi Polar RP (Stepnowski et al., 2006).

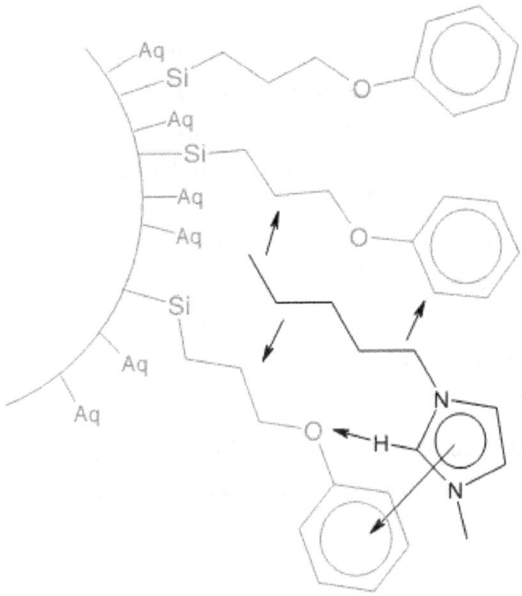

Fig. 16. Possible interactions of imidazolium ionic liquid cation and phenyl-bonded stationary phase. Arrows indicate hydrogen-bonding, hydrophobic and π–π interactions (Stepnowski et al., 2006).

Chromatographic behavior of ionic liquid cations in view of quantitative structure-retention relationship (QSRR) was studied by Molíková and co-workers (Molíková et al., 2010). The most significant QSRR equation between experimentally determined log k_w values and the structural descriptors of test analytes from molecular modeling (the solvent accessible surface: *SAS*, the differences in the energies of the highest occupied and the lowest unoccupied molecular orbitals: *diffHL*) was obtained in case of columns with the highest carbon loads and larger specific surface areas, i.e. Symmetry C18 and Symmetry C8. In turn, the column ACE 5 C18 allowed to obtain the best separations of the ILs according to cations kind. As the authors declare, identification of QSRRs can provide the knowledge regarding hydrophobicity and retention mechanism of ILs. It could be further helpful in prediction of their biological effects and the way of these chemicals into the environment.

5. ELECTROMIGRATION TECHNIQUES

Ils, due to good electrical conductivity and appropriate viscosity, could be used as electrolytes, additives in the run buffer or reagents for dynamic coating or covalent attachment to the capillary walls in capillary electrophoresis (CE). The main disadvantage of this chromatographic technique is an electroosmotic flow (EOF) provoked by silanol groups on the inner surface of silica capillaries.

To reverse this process ILs can be covalently bonded to a fused-silica capillary surface. Qin and Li (Qin &Li, 2002; Qin et al. 2003) covalently bonded propyl methyl imidazolium chloride [PMIM][Cl] to a silica capillary. Due to the positive charge of the bonded moieties and reduction of interaction between analytes and the capillary wall, separation of sildefanil and its metabolites in human serum and DNA analysis with MS detection were possible. The capillary wall modified with 1-ethyl-3- methylimidazolium was used for separation of positively charged inorganic ions: NH_4^+ and K^+in human urine, the separation of K^+, Na^+, Li^+, Ca^{2+}, Mg^{2+} and Ba^{2+} ions in red wine. The main advantage of those capillaries appears to be increases in the working pH range of the running buffer with good efficiency and the compatibility with mass spectrometry (Qin & Li, 2002; Qin & Li, 2003; Qin et al., 2003).

Fused-silica capillaries could be chemically modified with an analogue of the imidazole- based IL (1-butyl-3-(3-triethoxysilylpropyl)-4,5-dihydroimidazolium tetrafluoroborate) as well as zwitterionic salt (*N* -3-(3-triethoxysilylpropyl)-4,5-dihydroimidazole) (Borissowa et al., 2007).

It appeared that the electroosmotic flow in the capillary coated with an ionic liquid analogue was found to be anodic and dependent on the pH of the separation buffer. Whereas in the case of a zwitterionic capillary, the electroosmotic flow was cathodic and its velocity remained almost constant in the pH range of 4–7. Schematic illustration of ionic liquid bonding to the capillary wall is presented in Fig.17.

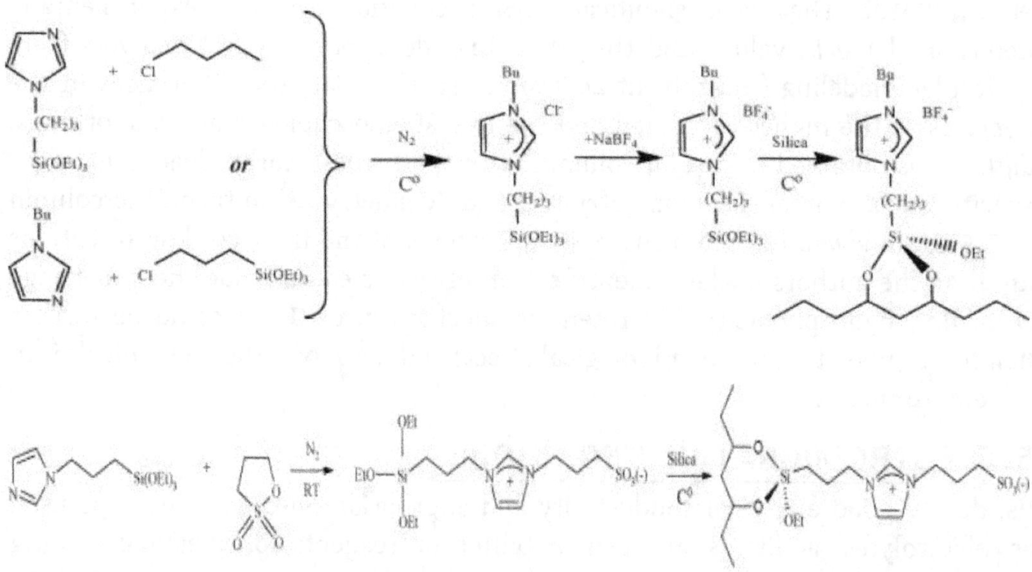

Fig. 17. Scheme representing the IL and zwitterionic salt coating procedure (Borissowa et al., 2007).

ILs could be applied also as background electrolytes in aqueous as well as non-aqueous elecytophoresis (NACE). In NACE, the EOF does not change its direction by addition of IL indicating that the dynamic coating process is meaningless. That is why interactions between the solute and ILs ions and significant changes in the electrophoretic mobility are mainly responsible for enhancement of separation selectivity. Effect of ionic liquids on EOF in NACE was monitored by contactless conductivity detection (Seiman et al., 2008). Dialkylimidazolium ILs were used in NACE for separation of water insoluble dyes (Vaher et al., 2000), phenols and aromatic acids (Vaher et al., 2002a), polyphenols (Vaher et al., 2002b).

In turn, ILs used as the electrolytes in aqueous capillary electrophoresis, enable the dynamic coating of the capillaries wall. IL cations, as well as anions, influence the migration behavior of analytes but the resolving power of the system is much more affected by activity of the IL cation. Analytes can interact either with the ILs cations adsorbed on the capillary wall or with the free IL cations in the electrolyte solution. Not meaningless is the suppression of the EOF and changing eletrophoretic properties of capillary wall by IL's cationic molecule. It makes the separation mechanism more complex. Fig.18 illustrates mechanism of polyphenol separation using 1-alkyl-3-methylimidazolium based IL in the background electrolyte (Yanes et al., 2001).

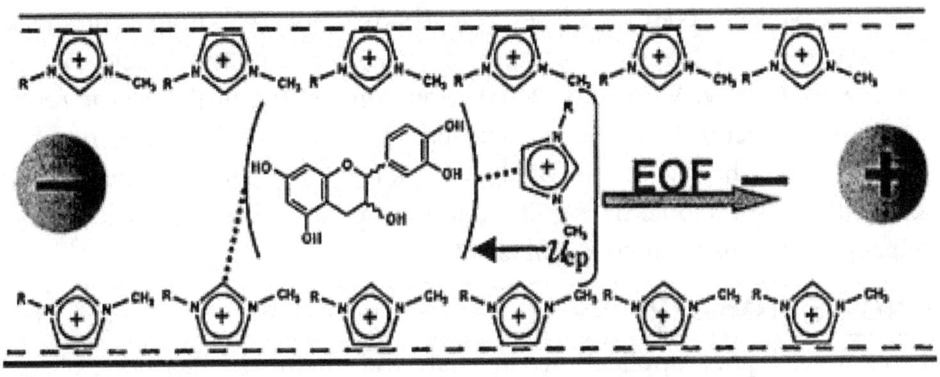

Fig. 18. Mechanism of polyphenols' separation using 1-alkyl-3-methylimidazolium-based ionic liquids. (Yanes et al., 2001).

ILs as dynamic coating materials were used for separation of different groups of compounds. For instance, dimethyldinonylammonium bromide was used for determination of carboxylates as copper complexes (Laamanen et al.,2005), 1-alkyl-3-methylimidazolium- based ionic liquids were used as the running electrolytes for separation of acidic and basic proteins (Jiang et al.,2003; Wu et al., 2008) and flavone derivatives (Qi et al.,2006), nicotinic, isonicotinic and picolinic acids (Marszałł et al., 2006 b), monocationic and dicationic phosphonium-based ILs were tested for separation of inorganic and small organic anions (Krizek et al.,2009).

Dicationic ion paring reagent (*N,N'*-dibutyl 1,1′- pentylenedipyrrolidinium) was applied for determination of several anions in the positive mode using CE-ESI-MS method achieving greater sensitivity, in comparison to the direct MS detection in the negative mode (Lin et al., 2009).

Separation of both neutral and charged analytes is possible owing to micellar capillary electrophoresis (MCE). In this chromatographic technique, long chain alkylimidazolium ILs could be used as surfactants and short chain ones as modifiers. 1-butyl-3- methylimidazolium tetrafluoroborate [BMIM][BF_4] was added as modifier to the micellar system containing SDS resulting in the complete separation of active components of lignans in medicinal herbs (Tian et al., 2005). As the authors claimed, ILs affected the size and shape of the micelles, altering the separation. Long chain alkylimidazolium ILs added to the aqueous buffer were used as surfactants for separation of neutral methylresorcinol isomers and benzene derivatives (Borissova et al., 2008). 1-Butyl-3-methylimidazolium dodecanesulfonate (BAS) was applied in microchip MCE for the separation of proteins (Xu et al., 2008).

Chiral ILs could be used for enantiomeric separation in electromigation techniques. S-[3- (chloro-2-hydroxypropyl)trimethylammonium] NTf_2 was used as a chiral selector to separate various enantiomeric pharmaceutical compounds in CE (Tran& Mejac, 2008), undecenoxycarbonyl-L -pryrrolidinol bromide, and undecenoxycarbonyl-L -leucinol bromide to separate enantiomeric acidic analytes were used in MCE (Rizvi & Shamasi 2006). Sometimes ILs chiral selector requires another chiral selector to achieve enantioseparation. For instance, two chiral ILs: (ethyl- and phenylcholine of bis(trifluoromethylsulfonyl)imide) acting synergistically with cyclodextrins were able to improve the enantioselectivity for 2- arylpropionic acids (Francois et al.,2007).

6. CONCISE SUMMARY

Growth of ionic liquids applications in analytical chemistry has been observed during the previous years. It seems justifiable, given a broad spectrum of their physico-chemical properties. As environmentally friendly solvents, they may have numerous industrial applications in the future. Therefore, it is imperative to elaborate on novel sensitive methods for their determination in environmental and industrial samples. Considering chromatographic techniques, the most potential advancements associated with the advantageous solvent properties of ionic liquids, their low volatility, and electrical conductivity are evident in liquid chromatography and electromigration procedures. The subsequent generation of ionic liquids possessing chiral characteristics will be created and will discover more applications. Special precautions must be observed about the toxicity, stability, and purity of these chemicals. The long-term stability of ionic liquids with perfluorinated anions must be considered in relation to the breakdown products that may harm silica-based adsorbents. The characteristics of ionic liquids in chromatographic systems remain ambiguous. Certain writers contend that ionic liquids should not serve as substitutes for conventional salts, as the differentiation between liquid and solid is diminished in solution, resulting in ionic liquids exhibiting properties akin to those of organic salts. Other writers effectively illustrate the distinctive behavior of ionic liquids in chromatographic systems, emphasizing the unique modification of the stationary phase and beneficial interactions with the analyzed substances. The interpretation of separation mechanisms using ionic liquids remains a topic of interest within the scientific community.

REFERENCES

Armstrong, D.W., He, L. & Liu, Y.S.(1999). Examination of ionic liquids and their interactions with molecules, when used as stationary phase in gas chromatography, *Anal.Chem.* Vol. 71 (No. 17): 3873-3876.

Armstrong, D.W., Zhang, Li-K., He, L. & Gross, M.L. (2001). Ionic Liquids as Matrixes for Matrix-Assisted Laser Desorption/Ionization Mass Spectrometry, *Anal. Chem.* Vol.73 (No.15): 3679–3686.

Anderson, J.L. & Armstrong, D.W. (2003). High-stability ionic liquids. A new class of stationary phases for gas chromatography, *Anal.Chem.* Vol. 75 (No. 19): 4851-4858.

Bączek, T., Marszałł, M.P., Kaliszan, R., Walijewski, Ł., Makowiecka, W., Sparzak, B., Grzonka, Z., Wiśniewski, K.& Juszczyk, P. (2005). Behavior of peptides and computer-assisted optimization of peptides separations In a normal-phase thin- layer chromatography system with and without the addition of ionic liquid In the eluent, *Biomed.Chromatogr.* Vol. 19 *(No. 1): 1-8.*

Barber, D.W., Phillips, C.S.G., Tusa G. F. & Verdin, A. (1959). The chromatography of gases and vapours. Part VI. Use of the stearates of bivalent manganese, cobalt, nickel, copper, and zinc as column liquids in gas chromatography, *J. Chem. Soc.*18-24.

Berthod, A., He, L. & Armstrong, D,W. (2001). Ionic liquids as stationary phases solvents for methylated cyclodextrines in gas chromatography, *Chromatographia* Vol. 53 (No.1- 2): 63-68.

Berthod, A. & Carda-Broch, S. (2004). Use of the ionic liquid 1-butyl-3-methylimidazolium hexafluorophosphate in countercurrent chromatography, *Anal. Bioanal. Chem.* Vol. 380 (No.1): 168-177.

Berthod, A., Ruiz-Angel, M.J.& Huguet, S.(2005). Nonmolecular Solvents in Separation Methods: Dual Nature of Room Temperature Ionic Liquids, *Anal. Chem.* Vol. 77 (No. 13): 4071-4080.

Berthod, A., Ruiz-Angel, M.J. & Carda-Broch, S. (2008). Ionic liquids in separation techniques, *Journal of Chromatography A* Vol. 1184 (No. 1-2): 6-18.

Borissova, M.; Vaher, M.; Koel, M.& Kaljurand, M. (2007). Capillary zone electrophoresis on chemically bonded imidazolium based salts, *J. Chromatogr. A* Vol. 1160 (No. 1- 2):320–325.

Borissova, M., Palk, K.& Koel, M. (2008). Micellar electrophoresis using ionic liquids, *J. Chromatogr. A* Vol. 1183 (No. 1-2):192-195.

Buszewski,B., Kowalska, S.& Stepnowski, P. ((2006). Influence of stationary phase properties on the separation of ionic liquid cations by RP-HPLC, *J.Sep.Sci.* Vol. 29 (No. 8): 1116-1125.

Buszewski, B. & Studzińska, S. (2008). A review of ionic liquids in chromatographic and electromigration techniques, *Chromatographia* Vol. 68 (No. 1-2): 1-10.

Flieger, J. (2006). The effect of chaotropic mobile phase additives on the separation of selected alkaloids in reversed-phase high-performance liquid chromatography, *J. Chromatogr. A* Vol. 1113 (No. 1-2): 37-44.

Flieger, J. (2009). Effect of ionic liquids as mobile phase additives on chromatographic parameters of neuroleptic drugs in reversed-phase high-performance liquid chromatography, *Anal..Lett.* Vol. 42 (No. 11): 1632–1649.

Flieger, J.& Czajkowska-Żelazko, A. (2010). Ionic liquids as mobile phase additives in reversed phase high performance liquid, submitted to the *Journal of Separation science.*

Francois, Y., Varenne, A., Juillerat, E., Villemin, D. & Gareil P. (2007). Evaluation of chiral ionic liquids as additives to cyclodextrins for enantiomeric separations by capillary electrophoresi,s *J. Chromatogr. A* Vol. 1155 (No. 2):134-141.

Gritti, F.& Guiochon, G. (2004). Retention of Ionizable Compounds in Reversed-Phase Liquid Chromatography. Effect of the Ionic Strength of the Mobile Phase and the Nature of the Salts Used on the Overloading Behavior, *Anal.Chem.* Vol. 76 (No. 16): 4779-4789.

He, L., Zhang, W., Zhao, L., Liu, X. & Jiang, S. (2003). Effect of 1-alkyl-3-methylimidazolium- based ionic liquids as the eluent on the separation of ephedrines by liquid chromatography, *J. Chromatogr. A* Vol. 1007 (No. 1-2):39-45.

Hu, X., PengJ., Huang, Y., Yin, D.& LiuJ. (2009). Ionic liquids as mobile phase additives for high-performance liquid chromatography separation of phenoxy acid herbicides and phenols, *J.Sep. Sci.* Vol.32 (No.23-24):4126-4132.

Jiang, T., Gu, Y., Liang, B., Li, J., Shi, Y.& Ou, Q. (2003). Dynamically coating the capillary with 1-alkyl-3-methylimidazolium-based ionic liquids for separation of basic proteins by capillary electrophoresis, *Anal. Chim. Acta* Vol. 479 (No. 2):249-254.

Johnson, K.E. (2007). What's an ionic liquid?, *The Electrochemical Society Interface Spring 2007:38-41.* www.electrochem.org/dl/interface/spr/spr07/spr07_p38.pdf

Kaliszan, R., Marszałł, M.P., Markuszewski, M.J., Bączek, T. & Pernak, J. (2004). Suppresion of deleterious effects of free silanols in liquid chromatography by imidazolium tetrafuoroborate ionic liquids. *J. Chromatogr. A* Vol. 1030 (No. 1-2): 263-271.

Koel, M., (2005). Ionic liquids in chemical analysis, *Critical Review in Analytical Chemistry* Vol.35 (No.3): 177-192.

Krizek, T., Breitbach, Z.S., Armstrong, D.W., Tesarova, E.& Coufal, P. (2009). Separation of inorganic and small organic anions by CE using phosphonium-based mono- and dicationic reagent, *Electrophoresis* Vol. 30 (No.22):3955-3963.

Laamanen, P., Busi, S., Lahtinen, M.& Matilainen, R. (2005). A new ionic liquid dimethyldinonylammonium bromide as a flow modifier for the simultaneous determination of eight carboxylates by capillary electrophoresis, *J. Chromatogr. A* Vol. 1095 (No. 1-2):164-171.

Lin, X., Gerardi, A.R., Breitbach, Z.S., Armstrong, D.W.& Colyer, C.L. (2009). CE-ESI-MS analysis of singly charged inorganic and organic anions using a dicationic reagent as a complexing agent, *Electrophoresis* Vol. 30 (No.22):3918-3925.

Liu, S.J., Zhou, F., Xiao, X.H., Zhao, L., Liu, X. & Jiang, S.X. (2004). Surface confined ionic liquid - A new stationary phase for the separation of ephedrines in high- performance liquid chromatography, *Chin. Chem. Lett.* Vol. 15 (No. 9): 1060-1062.

Liu, R., Xu, L., Li,A. & Sun,A. (2010). Preparative isolation of flavonoid compounds from *Oroxylum indicum* by high-speed counter-current chromatography by using ionic liquids as the modifier of two-phase solvent system, *J.Sep.Sci.*Vol.33 (No.8):1058-1063.

LoBrutto, R., Jones, A. & Kazakevich, Y.V. (2001). Effect of counter-anion concentration on retention in high-performance liquid chromatography of protonated basic analytes, *J. Chromatogr. A* Vol. 913 (No. 1-2): 189-196.

Marszałł, M.P., Baczek, T.& Kaliszan, R. (2005). Reduction of silanophilic interactions in liquid chromatography with the use of ionic liquids, *Anal. Chim. Acta* Vol. 547 (No. 2): 172-178.

a. Marszałł, M.P., Baczek, T.& Kaliszan, R. (2006). Evaluation of the silanol-suppressing potency of ionic liquids, *J. Sep. Sci.* Vol. 29 (No. 8):1138-1145.

b. Marszałł, M.P., Markuszewski, M.J.& Kaliszan, R. (2006). Separation of nicotinic acid and its structural isomers using 1-ethyl-3-methylimidazolium ionic liquid as a buffer additive by capillary electrophoresis, *J. Pharm. Biomed. Anal.* Vol. 41 (No. 1):329–332.

Marszałł, M.P.& Kaliszan, R., (2007). Application of ionic liquids In liquid chromatography, *Critical Reviews in Analytical Chemistry* Vol. 37 (No.14): 127-140.

Mehl, J.T.& Hercules, D.M., (2000). Direct TLC-MALDI coupling using a hybrid plate. *Anal.Chem.*Vol. 72 (No. 1): 68-73.

Molíková,M., Markuszewsk,i M.J., Kaliszan,R.& Jandera,P.(2010). Chromatographic behaviour of ionic liquid cations in view of quantitative structure-retention relationship, *J.Chromatogr. A* Vol.1217 (No.8):1305-1312.

Pacholec, F., Butler, H.T.& Poole, C.F. (1982). Molten organic salt phase for gas-liquid chromatography, *Anal. Chem.*, Vol. 54 (No. 12): 938-1941.

Pacholec, F.& Pool,e C.F. (1983). Stationary phase properties of the organic molton salt ethylpyridinium bromie In gas chromatohraphy, *Chromatographia* Vol. 17 (No. 7): 370-374.

Pan, L., LoBrutto, R., Kazakevish, Y.V.& Thomson, R. (2004). Influence of inorganic mobile phase additives on the retention, efficiency and peak symmetry of protonated basic compounds in reversed-phase liquid chromatography, *J. Chromatogr. A* Vol. 1049 (No.1-2):63-73.

Pandey S (2006). Analytical applications of room-temperature ionic liquids: A review of recent efforts, *Anal. Chim. Acta* Vol. 556 (No.1):38–45.

Poole, C.F., Kersten, B.R., Ho,S.S.J., Coddens, M.E.,& Furton, K.G. (1986). Organic salts, liquid at room temperature, as mobile phases in liquid chromatography, *J.Chromatogr.A* Vol. 352 (No. 1-2): 407-425.

Poole, C.F. (2004). Chromatographic and spectroscopic methods for the determination of solvent properties of room temperature ionic liquids, *Journal of Chromatography A* Vol. 1037 (No.1-2): 49-82.

Poole, C.F., (2006). Applications of Ionic Liquids in Extraction, Chromatography, and Electrophoresis, in Grushka, E., Grinberg, N., (ed.), *Advances in Chromatography,* Vol. 45, CRC Press, Boca Raton, Fl, pp. 89-125.

Qi, S., Li, Y., Deng, Y., Cheng, Y., Chen, X. & Hu, Z. (2006). Simultaneous determination of bioactive flavone derivatives in Chinese herb extraction by capillary electrophoresis used different electrolyte systems—Borate and ionic liquids, *J. Chromatogr. A* Vol. 1109 (No. 2):300-306.

Qin, W. & Li, S.F.Y. (2002).An ionic liquid coating for determination of sildenafil and UK- 103,320 in human serum by capillary zone electrophoresis-ion trap mass spectrometry, *Electrophoresis* Vol. 23 (No. 24):4110-4116.

Qin, W.& Li, .SF.Y. (2003). Electrophoresis of DNA in ionic liquid coated capillary, *Analyst* Vol.128 (No. 1): 37–41.

Qin, W., Wei, H. & Li, S.F.Y. (2003). 1,3-Dialkylimidazolium-based room-temperature ionic liquids as background electrolyte and coating material in aqueous capillary electrophoresis, *J. Chromatogr. A* Vol. 985 (No. 1-2): 447–454.

a. Qiu H., Jiang S., Liu X., (2006). *N*-Methylimidazolium anion-exchange stationary phase for high-performance liquid chromatography *J. Chromatogr. A*, Vol. 1103, (No. 2), p. 265-270.

b. Qiu H., Jiang S., Liu X., Zhao L.,(2006). Novel imidazolium stationary phase for high- performance liquid chromatography *J. Chromatogr. A* Vol. 1116, (No.1-2):46-50.

Rizvi, S.A.A.& Shamsi, S.A. (2006). Synthesis, Characterization, and Application of Chiral Ionic Liquids and Their Polymers in Micellar Electrokinetic Chromatography, *Anal. Chem.* Vol. 78 (No. 19):7061-7069.

Ruiz-Angel, M.J.& Berthod, A.(2006). Reversed phase liquid chromatography of alkyl- imidazolium ionic liquids, *J. Chromatogr. A* Vol. 1113 (No. 1-2):101- 108.

Ruiz-Angel, M.J., Carda-Broch, S.& Berthod, A. (2006). Ionic liquids versus triethylamine as mobile phase additives in the analysis of β-blockers, *J. Chromatogr. A* Vol. 1119 (No. 1-2):202-208.

Seiman, A., Vaher, M.& Kaljurand, M. (2008). Monitoring of the electroosmotic flow of ionic liquid solutions in non-aqueous media using thermal marks, *J. Chromatogr. A* Vol. 1189 (No. 1-2):266-273.

Shetty, P.H., Poole, S.K. & Poole. C.F. (1990). Applications of ethylammonium and propylammonium nitrate solvents in liquid-liquid extraction and chromatography, *Anal. Chim. Acta* Vol. 236 (No.1):51-61.

Shetty, P.H., Youngberg, P.J., Kersten, B.R.& Poole, C.F., (1987). Solvent properties of liquid organic salts used as mobile phases in microcolumn reversed-phase liquid chromatography, *J. Chromatogr.* Vol. 411 (No.1-2): 61-79.

Stalcup, A.M. & Cabovska, B. (2004). Ionic liquids in chromatography and capillary electrophoresis, *Journal of Liquid Chromatography and Related Technologies* Vol. 27 (No. 7-9): 1443-1459.

Stepnowski, P., Müller, A., Behrend, P., Ranke, J., Hoffmann, J.& Jastorff, B. (2003). Reversed-phase liquid chromatographic method for the determination of selected room-temperature ionic liquid cations, *J.Chromatogr.A* Vol. 993 (No.1-2): 173-178.

Stepnowski, P. & Mrozik, W. (2005). Analysis of selected ionic liquid cations by ion exchange chromatography and reversed-phase high performance liquid chromatography, *J.Sep.Sci.*, Vol. 28 (No. 2): 149-154.

Stepnowski, P., Nichthauser, J.,Mrozik, W. & Buszewski, B.(2006). Usefulness of π...π aromatic interactions in the selective separation and analysis of imidazolium and pyridinium ionic liquid cations, *Analytical & Bioanalytical Chemistry* Vol. 385 (No.8): 1483-1491.

Studzińska, S.& Buszewski, B. (2010). Study of retention mechanism of imidazolium-based ionic liquids in HPLC, *J.Sep.Sci.* Vol.33 (No.9): 1264-1273.

Sun,Y., Cabovska B., Evans C.E., Ridgway T.H., Stalcup A.M., (2005). Retention characteristics of a new butylimidazolium-based stationary phase, *Anal. Bioanal. Chem.* Vol.382 (No.3), p. 728-734.

Sun, P., Armstrong, D.W. (2010). Ionic liquids in analytical chemistry, *Anal.Chim.Acta* Vol.661 (No. 1): 1-16.

Tian, K., Qi, S., Cheng, Y., Chen, X.. & Hua, Z. (2005). Separation and determination of lignans from seeds of Schisandra species by micellar electrokinetic capillary chromatography using ionic liquid as modifier, *J. Chromatogr. A* Vol.1078 (No. 1- 2):181-187.

Tran, C.D.& Mejac, I.. (2008). Chiral ionic liquids for enantioseparation of pharmaceutical products by capillary electrophoresis, *J. Chromatogr. A* Vol. 1204 (No.2): 204-209.

Vaher, M., Koel ,M.& Kaljurand, M. (2000). Non-aqueous capillary electrophoresis in acetonitrile using lonic-liquid buffer electrolytes, *Chromatographia* Vol.53 (No.5- 6):S302–S306.

a. Vaher, M., Koel, M.& Kaljurand, M. (2002). Application of 1-alkyl-3-methylimidazolium- based ionic liquids in non-aqueous capillary electrophoresis, *J. Chromatogr. A* Vol. 979 (No.1-2):27-32.

b. Vaher, M., Koel, M.& Kaljurand, M. (2002). Ionic liquids as electrolytes for nonaqueous capillary electrophoresis, *Electrophoresis* Vol.23 (No.3) :426-430.

Waichigo M.M., Riechel T.L. & Danielson N.D., (2005). Ethylammonium Acetate as a Mobile Phase Modifier for Reversed Phase Liquid Chromatography, *Chromatographia* Vol. 61 (No. 1-2):17-23.

a. Waichigo, M.M.& Danielson, N.D. (2006). Ethylammonium Formate as an Organic Solvent Replacement for Ion-Pair Reversed-Phase Liquid Chromatography, *J. Chromatogr. Sci.*, Vol. 44 (No.10): 607-614.

b. Waichigo, M.M.& Danielson, N.D., (2006). Comparison of ethylammonium formate to methanol as a mobile-phase modifier for reversed-phase liquid chromatography, *J. Sep. Sci.* Vol. 29 (No5):599-606.

Waichigo, M.M., Hunter, B.M., Riechel, T.L.& Danielson, N.D. (2007). Alkylammonium Formate Ionic Liquids as Organic Mobile Phase Replacements for Reversed-Phase Liquid Chromatography, *J. Liq. Chromatogr. Rel. Technol.* Vol. 30 (No. 2):165-184.

Wang Q., Baker G.A., Baker S.N., Colón L.A. (2006). Surface confined ionic liquid as a stationary phase for HPLC, *Analyst*, Vol.131 (No.9)1000-1005.

Wu, X., Wei, W., Su, Q., Xu, L.& Chen, G. (2008).Simultaneous separation of basic and acidic proteins using 1-butyl-3-methylimidazolium-based ion liquid as dynamic coating and background electrolyte in capillary electrophoresis *Electrophoresis* Vol. 29 (No. 11):2356-2362.

Xiaohua, X., Liang, Z., Xia, L., & Shengxiang, J. (2004). Ionic liquids as additives in high performance liquid chromatography: Analysis of amines and the interaction mechanism of ionic liquids, *Anal. Chim. Acta* Vol. 519 (No. 2): 207–211.

Xu, Y., Li, J. & Wang, E. (2008). Eukaryotic expression and antimicrobial spectrum determination of the peptide tachyplesin II, *J. Chromatogr. A* Vol. 1207 (No. 2):175- 183.

Yanes E.G., Gratz S.R., Baldwin M.J., Robison S.E.& Stalcup A.M. (2001). Capillary Electrophoretic Application of 1-Alkyl-3-methylimidazolium-Based Ionic Liquids, *Anal. Chem.* Vol.73 (No. 16): 3838-3844.

Zhang, W., He, L., Gu, Y., Liu, X. & Jiang, S.(2003). Archaeal lipid biomarkers and isotopic evidence of anaerobic methane oxidation associated with gas hydrates in the Gulf of Mexico, *Anal. Lett.* Vol. 36 (No. 6): 827-836.

Zhang, W., He, L., Liu, X. & Jiang, S. (2004). Generalized nonlinear models and variance function estimation, *Chin. J. Chem.* Vol. 22 (No. 3): 549-570.

Zhao, H., (2006). Are ionic liquids kosmotropic or chaotropic? An evaluation of available thermodynamic parameters for quantifying the ion kosmotropicity of ionic liquids, *J.Chem.Technol.Biotechnol.* Vol. 81 (No.6): 877-891.

13 Chapter

SELECTION OF IONIC LIQUID SOLVENTS FOR CHEMICAL SEPARATIONS UTILIZING THE ABRAHAM MODEL

1. INTRODUCTION: AN OVERVIEW

Room temperature ionic liquids (RTILs) have garnered significant attention over the past decade due to their distinctive physical and chemical characteristics. The annual quantity of published applications utilizing RTILs as solvent media has risen. Next-generation RTILs are a favored solvent option for manufacturing applications related to nanomaterials and novel pharmaceutical drug molecules, serving as high-temperature lubricants for metal-to-metal interactions, reservoirs for the controlled release of drug molecules in pharmaceutical formulations, chromatographic stationary phases for gas chromatographic separations, gas absorption agents, and extraction solvent systems for the removal of aromatic nitrogen and sulfur compounds from coal and petroleum feedstocks. RTILs are typically synthesized by amalgamating a weakly coordinating cation with an anion to produce a highly polar ionic liquid. Ionic liquids are frequently immiscible with supercritical carbon dioxide, saturated linear hydrocarbons, and other acyclic organic solvents. Liquid immiscibility renders room temperature ionic liquids particularly advantageous for synthetic techniques utilizing biphasic catalysis. Synthetic methods now exist for the preparation of over 300 distinct ionic liquids (ILs), encompassing dicationic and tricationic bis/tris-imidazolium-based ILs, polymeric ionic liquids (PILs), and chiral ionic liquids. Techniques have been established for the incorporation of polar functional groups at the terminus of an alkyl CH2-chain. The comprehensive physical and solubilizing characteristics of ionic liquids (ILs) stem from the combined attributes of their cation and anion. The anion typically governs the degree of miscibility of the RTIL with water. The cation of an ionic liquid often has a voluminous organic framework (such as alkylimidazolium, alkylpyridinium, alkylpyrrolidinium, tetraalkylphosphonium, and tetraalkylammonium – refer to Figure 1 for molecular structures) characterized by low molecular symmetry. The kind and dimensions/symmetry of the cation influence the melting point temperature of the ionic liquid. The melting point is significant since it indicates the lower limit of liquidity, and when considered alongside thermal stability, it delineates the temperature range within which the ionic liquid can function as a liquid solvent. Experimental studies have demonstrated that ionic liquids containing dications exhibit a broader temperature range of thermal stability compared to their monocation equivalents.

The solubilizing characteristics of an ionic liquid can be altered by modifying the cation-anion pairing. Recent advancements in our comprehension of ionic liquids' characteristics have enabled us to propose potential ionic liquid solvents for specific chemical separations. A particular use entails the incorporation of an ionic liquid (as an entrainer) into an azeotropic system composed of components that cannot be separated by conventional distillation techniques. The introduced ionic liquid entrainer engages with the constituents of the azeotropic system, thereby modifying the relative volatilities of the components. Verma and Banerjee (2010) investigated various ionic liquid combinations comprising 10 cations (imidazolium, pyridinium, and quinolium) and 24 anions as entrainers for water + ethanol, water + 2-propanol, and water + tetrahydrofuran systems, utilizing the 1,3-dimethylimidazolium cation, [M2Im]+, in conjunction with the acetate, [Ac]-, chloride, [Cl]-, and bromide, [Br]-, anions, which yielded the highest relative volatility. Chemicals in azeotropic mixtures have also been separated using liquid-liquid extraction utilizing an ionic liquid solvent, as schematically seen in Figure 2. Pereiro et al. (2010) investigated 1-ethyl-3-methylimidazolium ethyl sulfate ionic liquid as a solvent for extracting ethanol from ethanol + heptane and ethanol + hexane azeotropic mixtures. Ionic liquid entrainers have been effectively employed in the separation of alkanes from alkenes (propane and propene (Mokrushina et al., 2010), hexane and 1-hexene (Lei et al., 2006), alkanes from aromatic compounds (cyclohexane and benzene (Zhu et al., 2004), and acetylene from ethylene (Palgunadi et al., 2010). Recent review papers (Soukup-Hein et al., 2009; Lei et al., 2003; Pandey, 2006; Poole and Poole, 2010) have examined the advancements in chemical separations utilizing ionic liquid solvents.

Gas-liquid chromatography (GLC) and high-performance liquid chromatography (HPLC) provide efficient techniques for the separation of organic chemicals in mixtures before quantifying the composition of the mixture. The ionic liquid can function as the stationary phase, or in the context of high-performance liquid chromatography, it may be utilized as an organic modifier incorporated into the mobile phase to influence the solute's partitioning properties. The solute distributes between the stationary and mobile phases while traversing the chromatographic column.. The elution time is governed by the partition coefficient, which is defined as

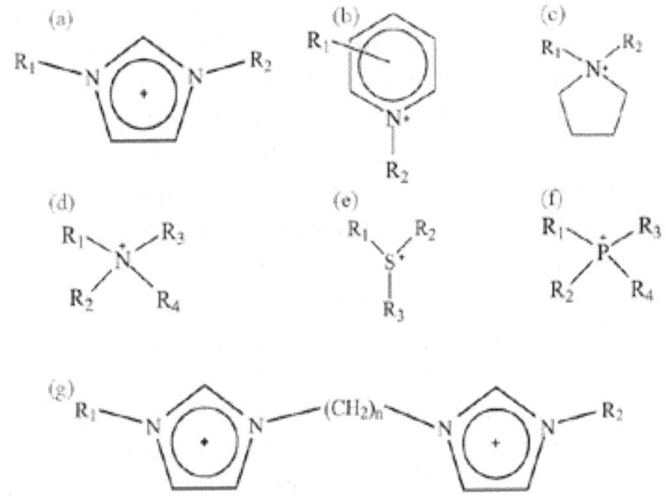

Fig. 1. Molecular Structures for 1,3-disubstituted imidazolium (a), substituted pyridinium (b), substituted pyrrolidinium (c), tetraalkylammonium (d), trialkylsulfonium (e), tetraalkylphosphonium (f), and *bis*(1,3-disubstituted imidazolium) cations (g). The substitutes are alkyl or functionalized alkyl chains and are denoted as R_1, R_2, R_3 and R_4.

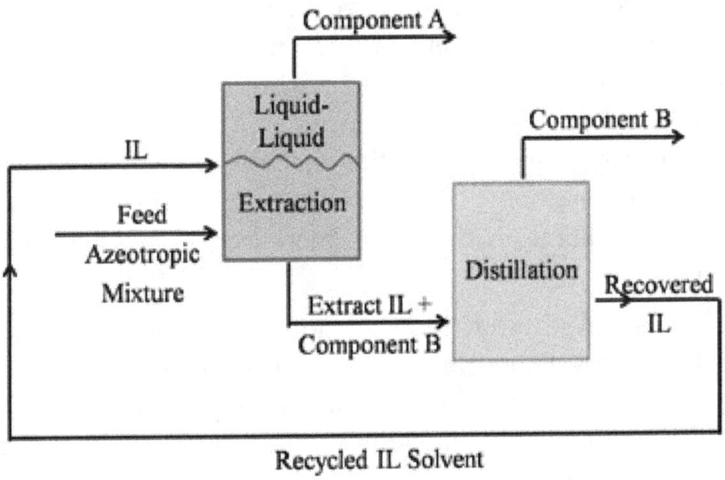

Fig. 2. Process for the separation of Component A and Component B from an azeotropic mixture using an IL solvent

$$P = \frac{C_{solute,mobile\ phase}}{C_{solute,stationary\ phase}} \tag{1}$$

the ratio of the molar concentration of the solute in the respective mobile and stationary phases. The partition coefficient is a relative measure of the affinity that the solute exhibits for each phase. Each solute is expected to interact differently with the two phases.

Numerous publications have reported using IL solvents in conjunction with chromatographic analyses (Soukup-hein *et al.*, 2009; Pandey, 2006; Poole and Poole, 2010). For example, Seely *et al.* (2008) studied chemical separations of complex mixtures on a two- dimensional gas chromatographic column. The system consisted of a trihexyl(tetradecyl)phoshonium *bis*(trifluoromethane)-sulfonamide IL primary column coupled to a (5 % diphenyl + 95 % dimethyl) polysiloxane secondary column. The authors found that the two-dimensional system separated the major components of diesel fuel into three narrow bands: saturated hydrocarbons, monoaromatics and diaromatics. Yao *et al.* (2009) employed various ILs containing the *tris*(pentafluoroethyl)trifluorophosphate anion. $[FAP]^-$, as extraction solvents for direct immersion single drop micro-extraction studies prior to hplc analysis. Ionic liquids tend to form larger and more stable microdroplets than do the more traditional organic solvents, and their elution time in hplc is very short and does not affect the chemical separation. The largest enrichment factors for compounds with high molar masses and fused rings were obtained with trihexyl(tetradecyl)phoshonium *bis*(pentafluoroethyl) trifluorophosphate. 1-Methyl-3-hexyl-imidazolium [FAP] gave the better results for the smaller molecules studied.

Headspace micro-extraction methods (Zhao *et al.* 2008 and 2009; Aguilera-Herrador *et al.* 2008; Liu *et al.*, 2003) have been developed for the partitioning of volatile and semivolatile compounds into an exposed IL drop. In the micro-extraction setup depicted in Figure 3 a syringe needle is inserted through the sample vial septum. The syringe needle is slowly depressed to expose the IL microdrop to the vapor above the aqueous sample. The volatile solutes then partition into the ionic liquid solvent from the vapor phase. After a predetermined equilibration time, the microdrop is retracted back into the syringe needle.

The syringe is removed and inserted into the injection port of a gas chromatograph. The syringe plunger is depressed in order to expose, but not inject, the IL microdrop. The volatile solutes thermally desorb from the IL drop at the higher temperature. Solute concentrations are quantified by gas chromatographic analysis.

The fore-mentioned experimental studies document the application of IL solvents to separation problems encountered by chemicals and engineers. Our knowledge of the solubilizing properties of ionic liquids has now matured to the point where researchers can make educated guesses concerning which IL solvent is most likely to give the desired separation. This chapter will discuss predictions based upon the Abraham general solvation model (Abraham, 1993a,b; Abraham *et al.*, 2004).

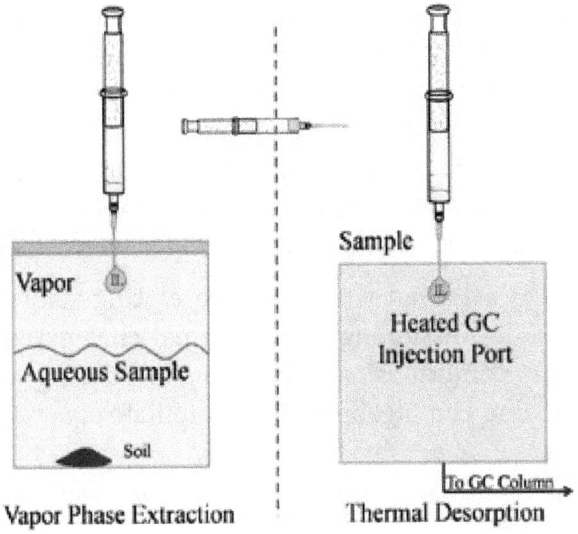

Fig. 3. Simple headspace ionic liquid-based microextraction setup for extracting volatile organic compounds from a contaminated soil sample. Organic contaminates on the soil dissolve into the aqueous phase. The organic vapors above the aqueous solution partition into the ionic liquid droplet.

2. THERMODYNAMIC PROPERTIES AND CHEMICAL SEPARATIONS

The thermodynamic basis for chemical separations using IL solvents is governed by solute partitioning between two immiscible or partly miscible phases. In the case of gas-liuqid chromatography the measured adjusted retention time, t_r', is related to the solute's infinite dilution activity coefficient, γ_{solu}^{∞}, (Mutelet *et al.*, 2010)

$$\ln \gamma_{1,2}^{\infty} = \ln\left(\frac{n_{solvent}.R.T}{V_N.P_1^0}\right) - P_1^0.\frac{B_{11}-V_1^0}{RT} + \frac{2.B_{13}-V_1^{\infty}}{RT}.J.P_0 \qquad (2)$$

$$V_N = \frac{3}{2}x\frac{[(\frac{P_{inlet}}{P_{outlet}})^2-1]}{[(\frac{P_{inlet}}{P_{outlet}})^3-1]}U_o\, t_r{}'x\frac{T_{col}}{T_{room}}(1-\frac{P_{water}}{P_{outlet}}) \qquad (3)$$

and gas-to-liquid partition coefficient, K,

$$\log K = \log\left(\frac{RT}{\gamma_{solute}^{\infty} P_{solute}^{o} V_{solvent}}\right) \qquad (4)$$

through standard thermodynamic relationships. In Eqns. 2 – 4, T_{col} is the column temperature, U_o is the flow rate of the carrier gas (mobile phase) measured at ambient room temperature (T_{room}) with a soap-bubble flowmeter, and P_{water} is the vapor pressure of water at T_{room}. P_{inlet} and P_{outlet} denote inlet and outlet pressures, respectively. Rigorous calculations do account for vapor phase nonideality, B_{11} is the second virial coefficient of the solute (component 1) in the gaseous state at temperature T, B_{13} is the mutual coefficient between the solute and mobile phase carrier gas, and P_1 is the solute's vapor pressure at temperature T, R denotes the universal gas constant, $V_{solvent}$ refers to the molar volume of the solvent at temperature T, and $n_{solvent}$ is the number of moles of solvent inside the column.

Chemical separation is achieved whenever two eluting solutes have sufficiently different retention times. Assuming an isothermal chromatographic separation, and that the column conditions remain constant during the course of the chemical separation, one can algebraically manipulate equations 2-4 to obtain the following expression

$$\frac{t_{r,solute\,B}{}'}{t_{r,solute\,A}{}'} = (\frac{\gamma_A^{\infty}}{\gamma_B^{\infty}})x(\frac{P_A^{o}}{P_B^{o}}) \quad (5)$$

relating the ratio of adjusted retention times and the infinite dilution activity coefficients. The ratio of vapor pressures represents the separation (separation factor is $\alpha = \gamma_A^{\infty} P_A^{o} / \gamma_B^{\infty} P_B$) that would be expected from the vapor pressure differences for the two solutes. The infinite dilution activity coefficient measures the enhanced separation that one could get from solute interactions with the liquid phase solvent. Solution models that accurately predict infinite dilution activity coefficients facilitate the design of manufacturing processes and analytical methods employing azeotropic distillations, gas stripping and gas chromatographic separations.

Chemical separations can also be achieved through solute partitioning between two condensed phases. Practical examples include liquid-liquid extraction and hplc. These partitioning processes are described by a partition coefficient as well

$$P = \frac{C_{solute,\,phase\,\alpha}}{C_{solute,\,phase\,\beta}} \quad (6)$$

For notation purposes gas-to-liquid partition coefficients are denoted as K in the chapter, while condensed phase-to-condensed phase partition coefficients are referred to as P. The notational distinction is needed because the Abraham solvation parameter model uses different linear free energy relationships (LFER) to predict log K and log P values.

From thermodynamic considerations it can be shown that the solute's partition coefficient is

$$P = \frac{\gamma_{solute,phase\,\alpha}^{\infty} V_{phase\,\alpha}}{\gamma_{solute,phase\,\beta}^{\infty} V_{phase\,\beta}} \quad (7)$$

a ratio of the product of the solute's infinite dilution activity coefficient in each phase ($\gamma_{solute, phase\ \alpha}^{\infty}$ or $\gamma_{solute, phase\ \beta}^{\infty}$) times the molar volumes of each respective phase corrected for solvent α - solvent β mutual saturation. The solute activity coefficients in Eqn. 7 pertain to the equilibrated phases in physical contact with one another, and thus are not necessarily the values that would be obtained for the solute dissolved in each pure solvent.

Practical partition coefficients should not be confused with indirect, hypothetical partitioning processes. The latter are also thermodynamic transfer processes. The Gibbs energy of transfer is

$$\Delta G_{trans} = - RT \ln \text{Partition coefficient} \quad (8)$$

It is possible to calculate ΔG_{trans} values for solute transfer even when the two condensed phases are not in direct physical contact with each other. A hypothetical transfer process can be set up

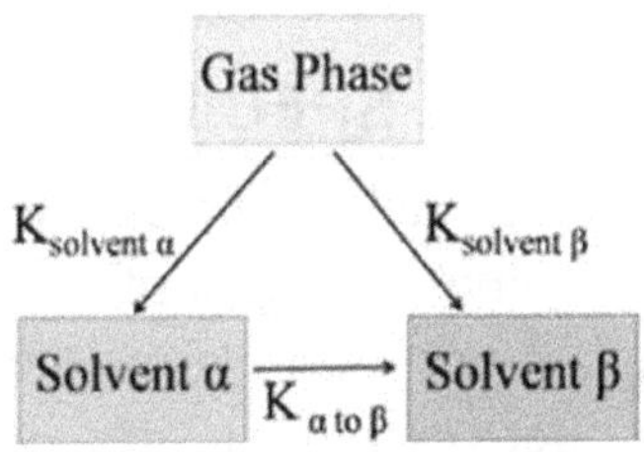

where the "hypothetical" solvent α-to-solvent β partition coefficient can be calculated from

$$\log P_{\alpha\text{-}\beta} = \log K_{solvent\ \beta} - \log K_{solvent\ \alpha} \quad (9)$$

the gas-to-solvent α and gas-to-solvent β partition coefficients. Abraham model correlations have been developed for predicting "hypothetical" water-to-anhydrous IL solvents. Even though hypothetical these predicted log P values for water-to-anhydrous ILs are useful. The predicted values can be converted back to γ_{solute}^{∞} values for solutes dissolved in the anhydrous IL using Eqns. 4 and 10.

$$\log P = \log K_{IL} - \log K_W \quad (10)$$

In Eqn. 10 K_W is the solute's gas-to-water partition coefficient.

Most of the Abraham model correlations that have been developed for predicting partition coefficients of solutes into IL solvents pertain to 298 K (Acree and Abraham, 2006; Abraham and Acree, 2006; Proctor *et al.*, 2008; Sprunger *et al.*, 2007b; Sprunger *et al.*, 2008; Sprunger *et al.*, 2009a,b,c; Sprunger *et al.*, 2010; Abraham *et al.*, 2009). Not all chemical separations take place at 298 K or even under isothermal conditions. There is a need to convert estimated partition coefficients for 298 K to other temperatures. From a thermodynamic standpoint, the gas-to-condensed phase partition coefficient, K, and water-to-organic solvent partition coefficient,

P, can be estimated at other temperatures from measured partition coefficient at 298.15 K and the solute's enthalpy of solvation, ΔH_{solv}, or the enthalpy of transfer, ΔH_{trans}, between the two condensed phases from Eqn. 11 or Eqn. 12.

$$\log K(at\,T) - \log K(at\,298.15K) = \frac{-\Delta H_{Solv}}{2.303R}(1/T - 1/298.15) \quad (11)$$

and

The enthalpy of transfer needed in Eqn. 12 is defined as

$$\Delta H_{trans} = \Delta H_{Solv,Org} - \Delta H_{Solv,W} \quad (13)$$

the difference in the enthalpy of solvation of the solute in the specified organic solvent minus its enthalpy of solvation in water. The above equations assume zero heat capacity changes. Abraham model correlations have been developed for estimating ΔH_{solv} for organic vapors and gases in IL solvents using ion-specific equation coefficients.

3. THE ABRAHAM MODEL: IL-SPECIFIC CORRELATIONS

The Abraham general solvation model is one of the more useful approaches for the analysis and prediction of Gibbs energies of solute transfer in chemical and biochemical systems. Published applications include the partitioning of solutes into organic and IL solvents both from the gas phase and from water (Acree and Abraham, 2006; Abraham and Acree, 2006; Proctor *et al.*, 2008; Sprunger *et al.*, 2007b; Sprunger *et al.*, 2008; Sprunger *et al.*, 2009a,b,c; Sprunger *et al.*, 2010; Abraham *et al.*, 2009), partitioning of volatile organic compounds and drug molecules between human/rat blood and select body organs/tissues (Abraham *et al.*, 2006; Abraham *et al.*, 2007; Abraham *et al.*, 2008), partitioning of solutes into humic acid (Mintz *et al.*, 2008a), sorption of gases and organic solutes onto polydimethylsiloxane solid- phase microextraction surfaces, (Sprunger *et al.*, 2007c) and the distribution of solutes between water and sodium dodecyl sulfate (SDS) micelles (Sprunger *et al.*, 2007a).

The method relies on two linear free energy relationships (lfers), one for transfer processes occurring within condensed phases (Abraham, 1993a,b; Abraham *et al.*, 2004):

$$SP = c + e \cdot \mathbf{E} + s \cdot \mathbf{S} + a \cdot \mathbf{A} + b \cdot \mathbf{B} + v \cdot \mathbf{V} \quad (14)$$

and one for processes involving gas-to-condensed phase transfer

$$SP = c + e \cdot \mathbf{E} + s \cdot \mathbf{S} + a \cdot \mathbf{A} + b \cdot \mathbf{B} + l \cdot \mathbf{L} \quad (15)$$

The dependent variable, SP, is some property of a series of solutes in a fixed phase, which in the present study will be the logarithm of solute partition coefficient between two immiscible (or partly miscible) phases or the enthalpy of solvation.

The independent variables, or descriptors, are solute properties as follows: **E** and **S** refer to the excess molar refraction and dipolarity/-polarizability descriptors of the solute, respectively, **A** and **B** are measures of the solute hydrogen-bond acidity and basicity, **V** is the McGowan volume of the solute and **L** is the logarithm of the solute gas phase dimensionless Ostwald partition coefficient into hexadecane at 298 K. The first four descriptors can be regarded as measures of the tendency of the given solute to undergo various solute-solvent interactions. The latter two descriptors, **V** and **L**, are both measures of solute size, and so will be measures of the solvent cavity term that will accommodate the dissolved solute. General dispersion interactions are also related to solute size, hence, both **V** and **L** will also describe the general solute-solvent interactions. Solute descriptors are available for more than 4,000 organic, organometallic and inorganic solutes. No single article lists all of the numerical values; however, a large compilation is available in one published review article (Abraham *et al.*, 1993a), and in the supporting material that has accompanied several of our published papers (Abraham *et al.*, 2006; Abraham *et al.*, 2009; Mintz *et al.*, 2007). Solute descriptors can be obtained by regression analysis using various types of experimental data, including water-to-solvent partitions, gas-to-solvent partitions, solubility data and chromatographic retention data as discussed elsewhere (Abraham *et al.*, 2010; Zissimos *et al.*, 2002a,b). There are also commercial software packages (Pharma Algorithms, 2006) and several published estimation schemes (Mutelet and Rogalski, 2001; Arey *et al.*, 2005; Platts *et al.*, 1999; Abraham and McGowan, 1987) for calculating the numerical values of solute descriptors from molecular structural information if one is unable to find the necessary partition, chromatographic and/or solubility data. For any fully characterized system/process (those with calculated values for the equation coefficients) further values of SP can be estimated for solutes with known values for the solute descriptors.

The usefulness of Eqns. 14 and 15 in the characterization of solvent phases is that he coefficients *e, s, a, b, l* and *v* are not just curve-fitting constants. The coefficients reflect particular solute-solvent interactions that correspond to chemical properties of the solvent phase. The excess molar refraction, **E**, is defined from the solute refractive index, and hence the *e* coefficient gives a measure of general solute-solvent dispersion interactions. The **V** and **L** descriptors were set up as measures of the endoergic effect of disrupting solvent-solvent bonds. However, solute volume is always well correlated with polarizability and so the *v* and *l* coefficients will include not only an endoergic cavity effect but also exoergic solute- solvent effects that arise through solute polarizability. The **S** descriptor is a measure of dipolarity and polarizability and hence the *s* coefficient will reflect the ability of a solvent to undergo dipole- and dipole-induced dipole interactions with the solute. The **A** descriptor is a measure of solute hydrogen bond acidity, and hence the *a* coefficient will reflect the complementary solvent hydrogen bond basicity. Similarly the *b* coefficient will be a measure of solvent hydrogen bond acidity. All this is straightforward for gas-to-solvent partitions because there are no interactions to consider in the gas phase. For partition between solvents, the coefficients in Eqn. 14 then refer to differences between the properties of the two phases.

Listed in Tables 1 and 2 are the Abraham model equation coefficients that have been reported (Grubbs *et al.*, 2010; Revelli *et al.*, 2009; Mutelet *et al.*, 2010; Revelli *et al.*, 2010) or calculated for the chapter for describing solute transfer from the gas phase (log K) and from water (log P) into 1-methyl-3-ethylimidazolium *bis*(trifluoromethyl-sulfonyl)imide, ($[MEIm]^+[(Tf)_2N]^-$), 1-methyl-3-butylimidazolium *bis*(trifluoromethylsulfonyl)-imide, ($[MBIm]^+[(Tf)_2N]^-$), 1-methyl-3-hexyl-imidazolium *bis*(trifluoromethylsulfonyl)imide, ($[MHIm]^+[(Tf)_2N]^-$), trimethylbutylammonium *bis*(trifluoromethyl-sulfonyl)imide, ($[M_3BAm]^+[(Tf)_2N]^-$), hexyltrimethylammonium *bis*(trifluoromethylsulfonyl)imide, ($[HexM_3Am]^+[(Tf)_2N]^-$), 1,3-dimethoxyimidazolium *bis*((trifluoromethyl)sulfonyl) imide ($[(Meo)_2Im]^+[(Tf)_2N]^-$), 1-ethanol-3-methylimidazolium *bis*(trifluoromethyl) sulfonylimide, ($[EtOHMIm]^+[(Tf)_2N]^-$), trihexyltetradecylphosphonium *bis*(trifluoromethyl-sulfonyl)imide, ($[H_3TdP]^+[(Tf)_2N]^-$), 1-methylethylether-3-methyl-imidazolium *bis*((trifluoromethyl)sulfonyl)imide, ($[MeoeMIm]^+[(Tf)_2N]^-$), 1-methyl-3-butylimidazolium tetrafluoroborate, ($[MBIm]^+[BF_4]^-$), 1-methyl-3-octylimidazolium tetrafluoroborate, ($[MOIm]^+[BF_4]^-$), 1-methyl-3-butyl-imidazolium hexafluorophosphate, ($[MBIm]^+[PF_6]^-$), 1-methyl-3-ethylimidazolium ethylsulfate, ($[MEIm]^+[EtSO_4]^-$), 1-methyl-3-butylimidazolium octylsulfate, ($[MBIm]^+[OtSO_4]^-$), 1-methyl-3-butylimidazolium trifluoromethanesulfonate, ($[MBIm]^+[Trif]^-$), 1-methyl-3-ethyl-imidazolium dicyanamide, ($[MEIm]^+[C(CN)_2]^-$), and 1-(3- cyanopropyl)-3-methylimidazolium dicyanamide, ($[CNPrMIm]^+[C(CN)_2]^-$). The IL-specific equation coefficients were obtained by regressing infinite dilution activity coefficients and Henry's law constants into the respective IL in accordance with Eqns. 14 and 15. Solubility data for gaseous solutes are often reported in the literature as Henry's law constants, which are related to log K by

$$\log K = \log \left(\frac{RT}{K_{Henry}\ V_{solvent}}\right) \tag{16}$$

The statistics of each derived correlation are quite good as reflected by the low standard deviations (SD) and near unity values for the squared correlation coefficients (R^2). The standard error in each equation coefficient is denoted in parenthesis immediately below the coefficient. The remaining statistical information is as follows: N denotes the number of experimental data points for the given IL data set.

Table 1. Abraham model correlation equation coefficients for the transfer of solutes from gas to the anhydrous (dry) ionic liquid solvents (log K correlation).

Solvent	c	e	s	a	b	l	N[a]	SD	R^2
([MBIm]+[BF4]-)	-0.600	0.356	2.534	3.312	0.284	0.604	66	0.099	0.997
	(0.026)	(0.075)	(0.068)	(0.100)	(0.074)	(0.011)			
([EtOHMIm]+[(Tf)2N]-)	-0.793	0.139	2.404	2.587	1.353	0.581	81	0.100	0.993
	(0.047)	(0.061)	(0.065)	(0.074)	(0.077)	(0.011)			
([MBIm]+[PF6]-)	-0.460	-0.191	2.747	2.228	0.363	0.663	91	0.154	0.994
	(0.033)	(0.081)	(0.086)	(0.097)	(0.101)	(0.015)			
([MBIm]+[(Tf)2N]-)	-0.394	0.089	1.969	2.283	0.873	0.696	104	0.111	0.994
	(0.029)	(0.068)	(0.085)	(0.104)	(0.113)	(0.009)			
([MHIm]+[(Tf)2N]-)	-0.348	-0.240	2.060	2.184	0.561	0.754	77	0.117	0.993
	(0.032)	(0.102)	(0.102)	(0.136)	(0.113)	(0.010)			
([MOIm]+[BF4]-)	-0.409	-0.049	1.562	2.911	0.803	0.778	61	0.140	0.987
	(0.050)	(0.115)	(0.135)	(0.125)	(0.155)	(0.013)			
([MEIm]+[(Tf)2N]-)	-0.486	0.068	2.296	2.278	0.988	0.651	65	0.094	0.996
	(0.041)	(0.058)	(0.052)	(0.052)	(0.126)	(0.067)			
([M3BAm]+[(Tf)2N]-)	-0.457	0.000	2.188	2.375	0.663	0.668	58	0.120	0.990
	(0.048)		(0.150)	(0.198)	(0.197)	(0.013)			
([MBIm]+[Trif]-)	-0.666	0.179	2.264	3.856	0.567	0.698	51	0.105	0.989
	(0.057)	(0.124)	(0.130)	(0.140)	(0.118)	(0.018)			
([MeomMIm]+[(Tf)2N]-)	-0.508	0.000	2.635	2.379	0.418	0.602	52	0.108	0.981
	(0.076)		(0.087)	(0.129)	(0.093)	(0.018)			
([Meo)2Im]+[(Tf)2N]-)	-0.762	-0.013	2.557	2.427	1.157	0.584	48	0.084	0.989
	(0.063)	(0.046)	(0.072)	(0.106)	(0.077)	(0.015)			
([CNPrMIm]+[C(CN)2]-)	-1.489	-0.418	3.089	4.807	0.626	0.644	45	0.121	0.987
	(0.098)	(0.117)	(0.115)	(0.163)	(0.135)	(0.025)			
([MEIm]+[C(CN)2]-)	-0.888	0.361	2.833	4.690	0.364	0.596	74	0.109	0.989
	(0.055)	(0.082)	(0.075)	(0.104)	(0.074)	(0.014)			
([HexM3Im]+[(Tf)2]-)	-0.462	0.000	2.073	2.022	0.637	0.684	50	0.123	0.968
	(0.090)		(0.081)	(0.150)	(0.103)	(0.021)			
([H3TdP]+[(Tf)2]-)	-0.406	-0.576	1.602	2.358	-0.009	0.959	59	0.112	0.982
	(0.058)	(0.123)	(0.125)	(0.156)	(0.147)	(0.018)			
([MBIm]+[OtSO4]-)	-0.228	-0.287	1.940	4.862	-0.302	0.880	56	0.116	0.984
	(0.081)	(0.093)	(0.103)	(0.168)	(0.140)	(0.023)			
([MEIm]+[EtSO4]-)	-0.677	0.000	2.557	5.327	0.000	0.588	53	0.125	0.986
	(0.062)		(0.072)	(0.072)		(0.015)			

[a] Statistical information: N is the number of data points, SD is the standard deviation, and R^2 is the squared correlation coefficient.

Table 2. Abraham model correlation equation coefficients for the transfer of solutes from water to the anhydrous (dry) ionic liquid solvents (log P correlation).

Solvent	c	e	s	a	b	v	N[a]	SD	R^2
($[MBIm]^+[BF_4]^-$)	-0.082	0.454	0.541	-0.427	-4.583	2.961	66	0.132	0.992
	(0.049)	(0.092)	(0.095)	(0.134)	(0.099)	(0.057)			
($[EtOHMIm]^+[(Tf)_2N]^-$)	-0.402	0.304	0.470	-1.082	-3.510	2.977	79	0.133	0.990
	(0.078)	(0.078)	(0.091)	(0.099)	(0.103)	(0.063)			
($[MBIm]^+[PF_6]^-$)	-0.056	0.193	0.737	-1.351	-4.526	3.109	86	0.154	0.988
	(0.046)	(0.080)	(0.087)	(0.100)	(0.102)	(0.059)			
($[MBIm]^+[(Tf)_2N]^-$)	-0.018	0.416	0.153	-1.312	-4.187	3.347	101	0.131	0.994
	(0.044)	(0.084)	(0.105)	(0.131)	(0.139)	(0.039)			
($[MHIm]^+[(Tf)_2N]^-$)	-0.065	0.010	0.260	-1.476	-4.313	3.587	75	0.115	0.996
	(0.042)	(0.105)	(0.103)	(0.135)	(0.111)	(0.039)			
($[MOIm]^+[BF_4]^-$)	-0.115	0.210	0.000	-0.511	-4.338	3.617	59	0.159	0.994
	(0.076)	(0.130)		(0.143)	(0.177)	(0.063)			
($[MEIm]^+[(Tf)_2N]^-$)	0.029	0.351	0.202	-1.684	-3.585	3.059	64	0.119	0.993
	(0.048)	(0.103)	(0.121)	(0.163)	(0.137)	(0.055)			
($[M_3BAm]^+[(Tf)_2N]^-$)	0.047	-0.051	0.356	-1.262	-4.400	3.209	57	0.120	0.996
	(0.060)	(0.133)	(0.154)	(0.202)	(0.198)	(0.049)			
($[MBIm]^+[Trif]^-$)	-0.217	0.195	0.497	0.062	-4.310	3.282	51	0.130	0.992
	(0.091)	(0.143)	(0.166)	(0.176)	(0.145)	(0.084)			
($[MeomMIm]^+[(Tf)_2N]^-$)	-0.140	-0.067	0.818	-1.185	-4.523	3.105	48	0.120	0.993
	(0.120)	(0.097)	(0.122)	(0.172)	(0.116)	(0.117)			
($[(Meo)_2Im]^+[(Tf)_2N]^-$)	-0.412	-0.104	0.761	-1.124	-3.776	3.055	46	0.130	0.993
	(0.133)	(0.123)	(0.124)	(0.171)	(0.118)	(0.106)			
($[CNPrMIm]^+[C(CN)_2]^-$)	-0.928	0.373	1.224	1.042	-4.307	3.046	44	0.150	0.988
	(0.156)	(0.146)	(0.154)	(0.207)	(0.166)	(0.125)			
($[MEIm]^+[C(CN)_2]^-$)	-0.404	0.344	0.945	0.987	-4.526	2.957	70	0.126	0.992
	(0.105)	(0.095)	(0.100)	(0.133)	(0.088)	(0.087)			
($[HexM_3Im]^+[(Tf)_2]^-$)	-0.275	0.000	0.407	-1.478	4.320	3.510	48	0.140	0.994
	(0.136)		(0.115)	(0.176)	(0.114)	(0.111)			
($[H_3TdP]^+[(Tf)_2]^-$)	-0.155	-0.163	-0.029	-1.271	-5.042	4.246	59	0.136	0.996
	(0.092)	(0.146)	(0.157)	(0.192)	(0.179)	(0.085)			
($[MBIm]^+[OtSO_4]^-$)	-0.050	0.198	0.179	1.146	-5.154	4.008	55	0.179	0.986
	(0.155)	(0.132)	(0.175)	(0.296)	(0.228)	(0.135)			
($[MEIm]^+[EtSO_4]^-$)	-0.079	-0.021	0.554	1.491	-4.944	2.895	53	0.147	0.992
	(0.090)	(0.158)	(0.183)	(0.228)	(0.211)	(0.068)			

[a] Statistical information: N is the number of data points, SD is the standard deviation, and R^2 is the squared correlation coefficient.

The log P correlations do pertain to solute transfer from water-to-anhydrous IL solvent. As noted above these represent a hypothetical partitioning processes, and calculated values based on the equation coefficients listed in Table 2 may be different than the experimental log P values determined by direct partitioning of the solute between water and the IL solvent. At the present time there has been few practical water-to-IL systems studied. Abraham *et al.* (2003) reported log P correlations for 1-butyl-3-methylimidazoium hexafluorophosphate, ($[BMIm]^+[PF_6]^-$)

$$\text{Log P} = -0.17 + 0.45\,\mathbf{E} + 0.23\,\mathbf{S} - 1.76\,\mathbf{A} - 1.83\,\mathbf{B} + 2.15\,\mathbf{V} \quad (17)$$

and for 1-hexyl-3-methylimidazolium hexafluorophosphate, ($[HMIm]^+[PF_6]^-$)

$$\text{Log P} = -0.13 + 0.05\,\mathbf{E} + 0.40\,\mathbf{S} - 1.48\,\mathbf{A} - 2.11\,\mathbf{B} + 2.30\,\mathbf{V} \quad (18)$$

Room temperature ionic liquids have been used as a solvent media for organic synthesis. Most (if not all) of the classic synthetic methods have been performed in IL solvents. Much higher product yields and shorter reaction times have been reported for ILs than for the more conventional organic solvents. Product removal from the IL reaction media can often be accomplished by carbon dioxide supercritical fluid extraction. Many imidazolium-based ionic liquids are nearly insoluble in supercritical carbon dioxide, even at very high pressures. Roth (2009) recently reviewed the partitioning of organic compounds between IL solvents and supercritical fluids, with particular emphasis on supercritical carbon dioxide. There have been very few attempts to correlate/predict the partition coefficients of organic solutes in IL solvents – supercritical fluid carbon dioxide systems. Planeta and coworkers (Planeta and Roth 2005; Planeta *et al.*, 2007; Planeta *et al.*, 2009) correlated the relative partition coefficients of solutes in the ($[BMIm]^+[BF_4]^-$)/CO_2, ($[BMIm]^+[MeSO4]^-$)/CO_2 and ($[ThtdP]^+[Cl]^-$)/CO_2 with the Abraham model

$$\log \frac{P_{solute,A}}{P_{naphthalene}} = eE_A + sS_A + aA_A + bB_A + vV_A \quad (19)$$

The authors used naphthalene as the reference solute, and the descriptors in Eqn. 19 pertain to solute A. The derived correlations did provide a satisfactory description of the experimental partition coefficient data, suggesting that the Abraham model could be used in the preliminary design stage of real-world extraction processes employing IL/CO2 systems. The solubilizing properties of supercritical carbon dioxide are temperature and pressure- dependent. Correlations derived for one set of experimental conditions might not apply at other operating temperatures and processes.

4. ABRAHAM MODEL: ION-SPECIFIC EQUATION COEFFICIENTS

The physical and chemical properties of ILs can be modified by changing the cation-anion combination. There are currently synthetic procedures for making more than 300 different ILs and the number is likely to grow in number in future years as more applications are discovered. It would be impractical to develop an Abraham model correlation for each known IL. To address this concern, Sprunger and coworkers (Sprunger *et al.*, 2007b; Sprunger *et al.*, 2008; Spurnger *et al.*, 2009c; Grubbs *et al.*, 2010) modified the basic Abraham solvation parameter model for the gas-to-IL partition coefficient

$$\log K = c_{cation} + c_{anion} + (e_{cation} + e_{anion})\,\mathbf{E} + (s_{cation} + s_{anion})\,\mathbf{S} + (a_{cation} + a_{anion})\,\mathbf{A} + (b_{cation} + b_{anion})\,\mathbf{B} + (l_{cation} + l_{anion})\,\mathbf{L} \quad (20)$$

and water-to-IL partition coefficient

$$\log P = c_{cation} + c_{anion} + (e_{cation} + e_{anion})\,\mathbf{E} + (s_{cation} + s_{anion})\,\mathbf{S} + (a_{cation} + a_{anion})\,\mathbf{A} + (b_{cation} + b_{anion})\,\mathbf{B} + (v_{cation} + v_{anion})\,\mathbf{V} \quad (21)$$

by rewriting each of the six solvent equation coefficients as the summation of their respective cation and anion contribution. In separating the equation coefficients it was assumed that the soltue's interaction with a given cation is not influenced by the surrounding IL anion, and that the solute's interaction with a given anion is unaffected by the surrounding IL cation. Once calculated, the ion-specific equation coefficients can be combined to build the Abraham model correlation for the desired cation-anion pair.

During the three years since the modified version was first suggested we have periodically updated the numerical values of the coefficients as new experimental data became available. For this chapter we have reanalyzed our large log K and log P databases for solutes dissolved in ILs to yield the following correlations

$$\log K = \sum_{cation} (c_{cation} + e_{cation} E + s_{cation} S + a_{cation} A + b_{cation} B + l_{cation} L) + \sum_{anion} (c_{anion} + e_{anion} E + s_{anion} S + a_{anion} A + b_{anion} B + l_{anion} L) \quad (22)$$

$(N = 2084, R^2 = 0.998, R^2_{adj} = 0.998, SD = 0.112, F = 5810)$

$$\log P = \sum_{cation} (c_{cation} + e_{cation} E + s_{cation} S + a_{cation} A + b_{cation} B + v_{cation} V) + \sum_{anion} (c_{anion} + e_{anion} E + s_{anion} S + a_{anion} A + b_{anion} B + v_{anion} V) \quad (23)$$

$(N = 2054, R^2 = 0.997, R^2_{adj} = 0.996, SD = 0.139, F = 2720)$

The calculated cation-specific and anion-specific equation coefficients are listed in Tables 3 and 4, along with their respective standard errors. For the most part, the larger standard errors are noted in the equation coefficients for those ions for which experimental data is limited. The number of data points for the individual ions range from a minimum of 22 log K values for the $[B(CN)_4]^-$ anion to more than 400 log K values for the $[BMIm]^+$ and $[EMIm]^+$ cations and $[(Tf)_2N]^-$ anion, which is more than sufficient for the regression analysis. The 23 cation-specific equation coefficients and 13 anion-specific equation coefficients that have been reported in Tables 3 and 4 can be combined to give predictive log K and log P correlations for a total of 299 (23 times 13) IL solvents. Predicted log K and log P values can be converted to infinite dilution activity coefficients, γ_{solute}^{∞}, through Eqns.4 and 10.

Sprunger *et al.* found that very little (if any) loss in descriptive/predictive ability resulted from splitting the equation coefficients into the individual ion contributions.

One note regarding the computation methodology that we used in calculating of the ion- specific equation coefficients. The cation-specific and anion-specific coefficients are paired in that each cation-specific coefficient goes together with its anion-specific counterpart to make up a summed value that the five solute descriptors are multiplied by. If one were to perform a regression analysis on Eqns. 20 - 23 the statistical software would generate numerical equation coefficients based on some reference point. The reference point would likely depend on the particular database used and the software's built-in convergence routine. Calculation of additional ion values at some later time would be difficult as there would be no guarantee that the next regression analyses would find the same reference point. In accordance with the computation methodology suggested by Sprunger *et al.* (2007b) the anion-specific equation coefficients of $[(Tf)_2N]^-$ were set equal to zero. Setting fixed reference points is not uncommon in thermodynamics. For example in calculating the chemical potentials of individual ions, the chemical potential of the hydrogen ion in water is defined to be zero and the values of all other ions are computed relative to this defined thermodynamic reference state.

Table 3. Cation-Specific and Anion-Specific Equation Coefficients for the Abraham Model Gas-to-RTIL Correlation (Eqn.22)

Ion	cion	eion	sion	aion	bion	lion
CATIONS[a]						
$[MEIm]^+$	-0.502	0.050	2.421	2.266	0.875	0.650
(N=420)[b]	(0.026)	(0.073)	(0.088)	(0.112)	(0.100)	(0.009)
$[BMIm]^+$	-0.407	0.064	2.071	2.279	0.761	0.699
(N=402)	(0.023)	(0.053)	(0.062)	(0.077)	(0.078)	(0.007)
$[MHIm]^+$	-0.339	-0.135	2.034	2.365	0.505	0.757
(N=238)	(0.027)	(0.078)	(0.093)	(0.108)	(0.100)	(0.008)
$[MOIm]^+$	-0.229	-0.246	1.292	1.808	1.146	0.836
(N=114)	(0.032)	(0.090)	(0.114)	(0.108)	(0.127)	(0.010)

$[M_3BAm]^+$	-0.457	-0.005	2.188	2.375	0.663	0.668
(N=58)	(0.045)	(0.123)	(0.141)	(0.187)	(0.185)	(0.012)
$[M_2EIm]^+$	-0.611	0.188	2.380	2.101	0.899	0.667
(N=39)	(0.072)	(0.134)	(0.142)	(0.175)	(0.138)	(0.022)
$[4\text{-}BMPy]^+$	-0.479	0.141	2.293	2.364	0.672	0.713
(N=76)	(0.055)	(0.115)	(0.132)	(0.138)	(0.121)	(0.017)
$[3\text{-}BMPy]^+$	-0.263	0.000	2.368	2.265	0.271	0.688
(N=36)	(0.115)		(0.260)	(0.253)	(0.246)	(0.034)
$[NEP]^+$	-0.668	0.246	2.399	2.403	0.936	0.672
(N=31)	(0.076)	(0.192)	(0.180)	(0.209)	(0.150)	(0.028)
$[PM_2Im]^+$	-0.822	0.780	2.357	3.432	0.926	0.526
(N=34)	(0.120)	(0.169)	(0.196)	(0.227)	(0.207)	(0.035)
$[HexM_3Am]^+$	-0.459	-0.039	2.096	2.021	0.624	0.684
(N=50)	(0.083)	(0.102)	(0.095)	(0.138)	(0.100)	(0.020)
$[HexdMIm]^+$	0.019	-0.452	0.821	1.810	0.523	0.997
(N=31)	(0.142)	(0.157)	(0.165)	(0.245)	(0.187)	(0.044)
$[HxomMIm]^+$	-0.463	-0.394	2.478	2.428	0.337	0.786
(N=34)	(0.109)	(0.212)	(0.272)	(0.205)	(0.217)	(0.033)
$[(Hxom)_2Im]^+$	-0.314	-0.479	2.076	2.376	0.287	0.835
(N=34)	(0.109)	(0.212)	(0.272)	(0.205)	(0.217)	(0.033)
$[(Meo)_2Im]^+$	-0.762	-0.013	2.557	2.427	1.154	0.584
(N=48)	(0.083)	(0.104)	(0.096)	(0.140)	(0.101)	(0.020)
$[EtOHMIm]^+$	-0.843	0.098	2.438	2.684	1.402	0.578
(N=151)	(0.039)	(0.059)	(0.062)	(0.073)	(0.074)	(0.010)
$[H_3TdP]^+$	-0.406	-0.576	1.602	2.338	-0.009	0.959
(N=59)	(0.058)	(0.124)	(0.126)	(0.157)	(0.149)	(0.018)
$[MeoeMIm]^+$	-0.507	-0.015	2.644	2.378	0.413	0.602
(N=52)	(0.078)	(0.020)	(0.094)	(0.135)	(0.100)	(0.019)
$[BMPyrr]^+$	-0.570	-0.075	2.687	2.338	0.570	0.711
(N=31)	(0.076)	(0.300)	(0.391)	(0.273)	(0.282)	(0.025)
$[CNPrMIm]^+$	-1.103	0.108	2.677	2.383	1.136	0.699
(N=45)	(0.110)	(0.155)	(0.159)	(0.216)	(0.177)	(0.029)
$[PrOHPy]^+$	-0.895	-0.065	3.006	3.013	1.483	0.673
(N=38)	(0.117)	(0.202)	(0.250)	(0.240)	(0.247)	(0.034)
$[E_3S]^+$	-0.606	-0.196	2.992	2.444	0.355	0.690
(N=31)	(0.112)	(0.278)	(0.368)	(0.241)	(0.254)	(0.033)
$[MO_3Am]^+$	-0.387	-0.130	1.460	2.327	0.501	0.927
(N=32)	(0.064)	(0.222)	(0.353)	(0.336)	(0.463)	(0.020)
ANIONS[c]						

$[(Tf)_2N]^-$	0.000	0.000	0.000	0.000	0.000	0.000
(N=860)						
$[BF_4]^-$	-0.203	0.217	0.370	1.093	-0.408	-0.068
(N=307)	(0.027)	(0.066)	(0.073)	(0.091)	(0.086)	(0.009)
$[PF_6]^-$	-0.033	-0.184	0.647	0.007	-0.362	-0.085
(N=169)	(0.029)	(0.068)	(0.075)	(0.090)	(0.092)	(0.010)
$[EtSO_4]^-$	-0.176	-0.035	0.123	3.048	-0.862	-0.063
(N=53)	(0.062)	(0.143)	(0.162)	(0.205)	(0.189)	(0.016)
$[Trif]^-$	-0.301	0.030	0.329	1.712	-0.321	0.011
(N=163)	(0.044)	(0.114)	(0.133)	(0.147)	(0.128)	(0.015)
$[F_3Ac]^-$	-0.305	-0.191	0.499	3.265	-0.336	0.019
(N=32)	(0.055)	(0.233)	(0.343)	(0.686)	(0.725)	(0.019)
$[OtSO_4]^-$	0.277	-0.232	-0.150	2.402	-1.015	0.120
(N=58)	(0.072)	(0.099)	(0.117)	(0.175)	(0.156)	(0.018)
$[SCN]^-$	-0.727	0.284	0.783	2.607	-0.288	-0.006
(N=91)	(0.069)	(0.153)	(0.202)	(0.153)	(0.167)	(0.020)
$[C(CN)_2]^-$	-0.387	0.310	0.412	2.424	-0.510	-0.055
(N=119)	(0.063)	(0.113)	(0.118)	(0.156)	(0.126)	(0.017)
$[E_2PO_4]^-$	0.089	0.145	-0.184	5.186	-0.966	0.064
(N=38)	(0.105)	(0.175)	(0.182)	(0.237)	(0.184)	(0.035)
$[FAP]^-$	0.179	-0.015	0.063	-1.314	0.238	-0.053
(N=144)	(0.050)	(0.099)	(0.117)	(0.129)	(0.136)	(0.015)
$[B(CN)_4]^-$	0.243	-0.347	0.445	0.034	-0.225	-0.005
(N=23)	(0.157)	(0.469)	(0.607)	(0.839)	(0.775)	(0.047)
$[NO_3]^-$	-0.211	0.624	0.577	2.578	-0.788	-0.191
(N=27)	(0.224)	(0.200)	(0.203)	(0.328)	(0.192)	(0.067)

[a] Cation abbreviations: $[4\text{-BMPy}]^+$ is 4-methyl-N-butylpyridinium cation; $[3\text{-BMPy}]^+$ is 3-methyl-N- butylpyridinium cation; $[MEIm]^+$ is 1-methyl-3-ethylimidazolium cation ; $[M_2EIm]^+$ is 1,2-dimethyl-3- ethylimidazolium cation; $[MBIm]^+$ is 1-methyl-3-butylimidazolium cation; $[MHIm]^+$ is1-methyl-3- hexylimidazoliumcation; $[M_3BAm]^+$ is trimethylbutylammonium cation; $[MOIm]^+$ is 1-methyl-3-octylimidazoliumcation; $[NEPy]^+$ isN-ethylpyridiniumcation; $[PM_2Im]^+$ is1-propyl-2,3- dimethylimidazolium cation; $[(Meo)_2Im]^+$ is 1,3-dimethoxy-imidazolium cation; $[EtOHMIm]^+$ is 1- ethanol-3-methylimidazolium cation; $[H_3TdP]^+$ is trihexyltetradecylphosphonium cation; $[MeoeMIm]^+$ is 1-methylethylether-3-methyl-imidazolium cation; $[HxomMIm]^+$ is 1-hexyloxymethyl-3- methylimidazolium; $[(Hxom)_2Im]^+$ is 1,3-dihexyloxymethylimidazolium cation; $[PrOHPy]^+$ is 1-(3- hydroxypropyl)pyridinium; $[BMPyrr]^+$ is 1-butyl-1-methylpyrrolidinum cation; $[CMPrMIm]^+$ is 1-(3- cyanopropyl)-3-methylimidazolium cation; $[HexM_3Am]^+$ is hexyltrimethylammonium cation; $[HexdMIm]^+$ is 1-hexadecyl-3-methylimidazolium cation; $[E_3S]^+$ is triethylsulfonium cation; and $[MO_3Am]^+$ is methyl(trioctyl)ammonium cation.

[b] Number of experimental data points associated with the specified ion.

[c] Anion abbreviations: $[(Tf)_2N]^-$ is *bis*(trifluoromethylsulfonyl)imide anion; $[BF_4]^-$ is tetrafluoroborate anion; $[PF_6]^-$ is hexafluorophosphate anion; $[EtSO_4]^-$ is ethylsulfate anion; $[Trif]^-$ is the trifluoromethanesulfonate anion; $[F_3Ac]^-$ is the trifluoroacetate anion; $[OtSO_4]^-$ is octylsulfate anion; $[SCN]^-$ is thiocyanate anion; $[C(CN)_2]^-$ is dicyanamide anion; $[E_2PO_4]^-$ is diethylphosphate anion; $[B(CN)_4]^-$ is tetracyanoborate anion; $[NO_3]^-$ is nitrate anion; and $[FAP]^-$ is *tris*(pentafluoroethyl)trifluorophosphate anion.

Table 4. Cation-Specific and Anion-Specific Equation Coefficients for the Abraham Model Water-to-RTIL Correlation (Eqn.23)

Ion[a]	c_{ion}	e_{ion}	s_{ion}	a_{ion}	b_{ion}	v_{ion}
CATIONS						
$[MEIm]^+$	-0.037	0.152	0.554	-1.422	-4.011	3.150
(N=415)[b]	(-0.043)	(0.091)	(0.113)	(0.142)	(0.125)	(0.045)
$[BMIm]^+$	-0.028	0.448	0.180	-1.399	-4.160	3.332
(N=392)	(0.036)	(0.065)	(0.079)	(0.099)	(0.098)	(0.034)
$[MHIm]^+$	-0.040	0.252	0.165	-1.263	-4.375	3.586
(N=236)	(0.045)	(0.098)	(0.116)	(0.136)	(0.124)	(0.041)
$[MOIm]^+$	-0.011	0.009	-0.150	-1.641	-3.98	3.872
(N=112)	(0.055)	(0.111)	(0.143)	(0.137)	(0.159)	(0.052)
$[M_3BAm]^+$	0.047	-0.051	0.356	-1.262	-4.400	3.209
(N=57)	(0.071)	(0.156)	(0.180)	(0.236)	(0.231)	(0.058)
$[M_2EIm]^+$	-0.095	0.292	0.443	-1.681	-4.024	3.174
(N=39)	(0.115)	(0.156)	(0.187)	(0.224)	(0.172)	(0.103)
$[4\text{-}BMPy]^+$	-0.128	0.207	0.591	-1.349	-4.283	3.433
(N=76)	(0.087)	(0.133)	(0.169)	(0.176)	(0.151)	(0.079)
$[3\text{-}BMPy]^+$	0.134	0.170	0.678	-1.204	-4.934	3.324
(N=36)	(0.151)	(0.309)	(0.317)	(0.325)	(0.258)	(0.173)
$[NEP]^+$	-0.322	0.323	0.552	-1.234	-3.951	3.370
(N=31)	(0.130)	(0.222)	(0.231)	(0.273)	(0.189)	(0.140)
$[PM_2Im]^+$	-0.499	0.685	0.568	-0.238	-4.006	2.907
(N=34)	(0.174)	(0.192)	(0.256)	(0.284)	(0.258)	(0.151)
$[HexM_3Am]^+$	-0.278	0.013	0.401	-1.476	-4.315	3.512
(N=48)	(0.139)	(0.127)	(0.129)	(0.176)	(0.123)	(0.111)
$[HexdMIm]^+$	0.167	-0.218	-0.448	-1.884	-4.597	4.461
(N=31)	(0.212)	(0.182)	(0.227)	(0.311)	(0.234)	(0.190)
$[HxomMIm]^+$	-0.039	-0.645	1.184	-1.374	-4.779	3.609
(N=34)	(0.163)	(0.255)	(0.335)	(0.256)	(0.275)	(0.142)
$[(Hxom)_2Im]^+$	0.107	-0.628	0.747	-1.441	-4.808	3.750
(N=34)	(0.163)	(0.255)	(0.335)	(0.256)	(0.275)	(0.142)
$[(Meo)_2Im]^+$	-0.412	-0.104	0.761	-1.124	-3.776	3.055
(N=34)	(0.142)	(0.131)	(0.132)	(0.183)	(0.126)	(0.113)
$[EtOHMIm]^+$	-0.404	0.229	0.517	-1.026	-3.493	2.931
(N=148)	(0.061)	(0.070)	(0.080)	(0.092)	(0.091)	(0.052)
$[H_3TdP]^+$	-0.155	-0.164	-0.055	-1.317	-4.985	4.250
(N=59)	(0.093)	(0.148)	(0.154)	(0.189)	(0.168)	(0.086)
$[MeoeMIm]^+$	-0.161	0.014	0.658	-1.282	-4.262	3.125
(N=50)	(0.131)	(0.026)	(0.127)	(0.173)	(0.124)	(0.104)

$[BMPyrr]^+$	-0.182	-0.760	2.188	-1.449	-5.039	3.423
(N=31)	(0.127)	(0.368)	(0.487)	(0.344)	(0.355)	(0.120)
$[CNPrMIm]^+$	-0.596	0.172	0.724	-1.497	-3.509	3.284
(N=44)	(0.179)	(0.194)	(0.210)	(0.277)	(0.220)	(0.148)
$[PrOHPy]^+$	-0.302	-0.299	1.476	-0.503	-3.760	3.125
(N=38)	(0.175)	(0.249)	(0.311)	(0.299)	(0.310)	(0.151)
$[E_3S]^+$	-0.062	-1.347	2.716	-1.550	-5.274	3.242
(N=31)	(0.170)	(0.348)	(0.458)	(0.302)	(0.325)	(0.148)
$[MO_3Am]^+$	-0.092	0.000	-0.146	-1.129	-4.609	4.114
(N=32)	(0.103)		(0.135)	(0.116)	(0.274)	(0.095)
ANIONS						
$[(Tf)_2N]^-$	0.000	0.000	0.000	0.000	0.000	0.000
(N=843)						
$[BF_4]^-$	-0.084	0.129	0.240	1.108	-0.401	-0.296
(N=305)	(0.045)	(0.080)	(0.094)	(0.116)	(0.108)	(0.043)
$[PF_6]^-$	0.000	-0.299	0.544	0.088	-0.321	-0.264
(N=164)		(0.083)	(0.096)	(0.115)	(0.114)	(0.048)
$[EtSO_4]^-$	-0.042	-0.173	0.000	2.913	-0.933	-0.256
(N=53)	(0.095)	(0.174)		(0.258)	(0.236)	(0.078)
$[Trif]^-$	-0.193	-0.339	0.506	1.520	-0.293	-0.047
(N=162)	(0.072)	(0.136)	(0.166)	(0.187)	(0.161)	(0.068)
$[F_3Ac]^-$	-0.288	-0.771	0.891	2.255	0.332	0.049
(N=32)	(0.094)	(0.270)	(0.426)	(0.854)	(0.904)	(0.096)
$[OtSO_4]^-$	0.013	-0.258	0.000	2.537	-1.001	0.678
(N=56)	(0.126)	(0.121)		(0.249)	(0.201)	(0.109)
$[SCN]^-$	-0.566	-0.757	1.878	2.390	-0.843	-0.120
(N=91)	(0.105)	(0.191)	(0.251)	(0.193)	(0.212)	(0.190)
$[C(CN)_2]^-$	-0.287	0.179	0.351	2.350	-0.525	-0.255
(N=116)	(0.105)	(0.139)	(0.156)	(0.200)	(0.158)	(0.094)
$[E_2PO_4]^-$	0.059	0.136	-0.120	5.217	-1.029	0.196
(N=38)	(0.164)	(0.194)	(0.242)	(0.302)	(0.231)	(0.156)
$[FAP]^-$	0.132	-0.171	0.121	-1.314	0.244	-0.107
(N=144)	(0.079)	(0.122)	(0.148)	(0.162)	(0.170)	(0.069)
$[B(CN)_4]^-$	0.346	-1.011	1.195	-0.105	-0.676	-0.070
(N=23)	(0.237)	(0.618)	(0.766)	(1.053)	(0.952)	(0.205)
$[NO_3]^-$	-0.037	0.322	0.376	2.539	-0.801	-0.676
(N=27)	(0.303)	(0.215)	(0.283)	(0.403)	(0.243)	(0.268)

[a]Cation and anion abbreviations are given in Table 3.

[b]Number of experimental data points associated with the specified ion.

Both LFERs are statistically very good, and describe experimental log K and log P databases that cover a 12.5 log unit and 8.7 log unit range to within standard deviations of 0.112 log units (Eqn. 22) and 0.139 log units (Eqn. 23) as shown in Figures 4 and 5.

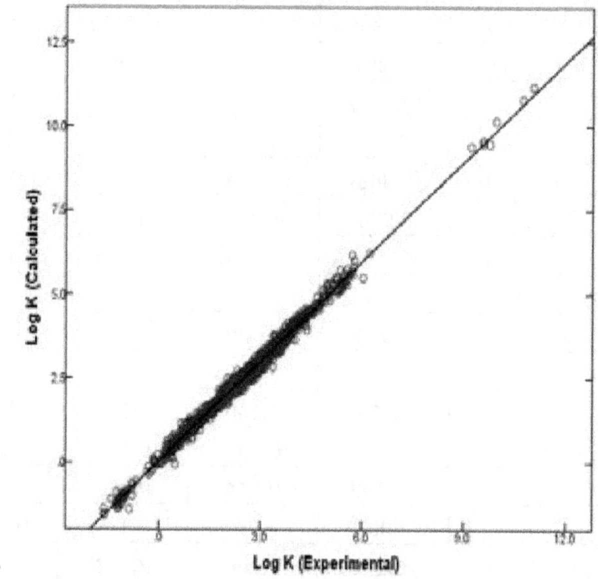

Fig. 4. Comparison between the 2084 experimental log K data points and predicted values based on Eqn. 22.

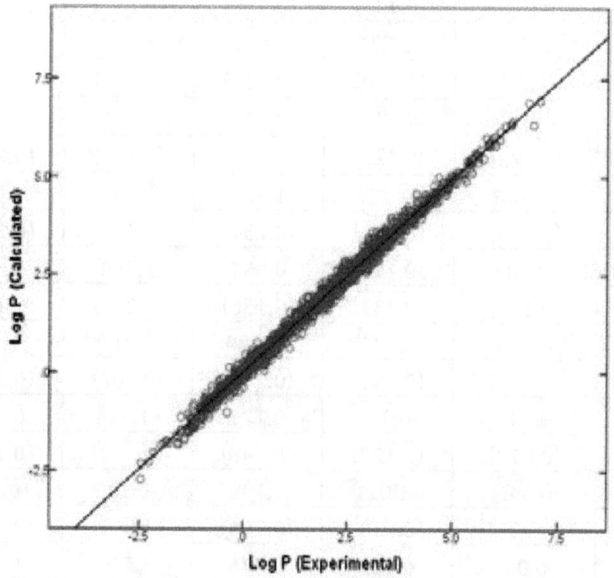

Fig. 5. Comparison between the 2054 experimental log P data points and predicted values based on Eqn. 23.

Based on the limited number of replicate independent activity coefficient measurements that have been performed for solutes dissolved in ILs we believe that 0.05 to 0.15 log units would be a reasonable guesstimate of the uncertainty associated with the experimental log K values. Slightly larger standard deviations are expected for the log P correlation which also includes the experimental uncertainties in the log K_w data used to convert the log K values to log P. Careful examination of the individual residuals between the calculated and observed values revealed that Equation 22 described 70.1 % (1461 of 2084 values) of the gas-to-IL partition coefficient data to within 0.1 log units, 92.0 % (1918 of 2084 values) to within 0.2 log units, and 98.1 % (2045 of 2084 values) to within 0.3 log units of observed values. Only 1.9 % of the predicted log K values fell more than 0.3 log units from the experimental value, with the largest residual being -0.55 log units. Similar results were noted for Eqn. 23; 56.8 % of the back-calculated water-to-IL partition coefficients differed from the observed value by less than 0.1 log units, 85.8 % differed by less than 0.2 log units, and 98.7 % differed by less than 0.3 log units. Less than 1.3 % of the predicted log P values were more than 0.3 log units from the observed value. The largest residual for the log P calculations is -0.61 log units. These values should reflect the predictive ability that Eqns. 22 and 23 would exhibit in terms of predicting partition coefficients for new compounds dissolved in ILs containing the 23 cations and 13 anions given in Tables 3 and 4, provided that the solute descriptors of the compounds fall within the area of predictive chemical space defined by the calculated equation coefficients: **E** = 0.000 to **E** = 1.500; **S** = 0.000 to **S** = 1.720; **A** = 0.000 to **A** = 1.030; **B** = 0.000 to **B** = 1.280; **V** = 0.109 to **V** = 1.799; and **L** = -1.200 to **L** = 7.833. A few of the ion-specific data sets spanned a slightly smaller range of solute descriptors. Predicted activity coefficients can be converted to infinite dilution activity coefficients, γ_{solut}^{∞}, through Eqns.4 and 10.

The major advantage of splitting the equation coefficients into individual cation-specific and anion-specific contributions is that one can make predictions for more RTILs. Normally one needs partition coefficient data for 40-50 solutes dissolved in a given RTIL to develop a RTIL-specific Abraham model correlation. By combining all of the experimental data for a RTIL containing lets say either a 1-methyl-3-octylimidazolium cation, $[MOIm]^+$, and trifluoromethane-sulfonate anion, $[Trif]^-$, it may be possible to calculate $[MOIm]^+$-specific and $[Trif]^-$-specific equation coefficients where there was too few data points for the $[MOIm]^+[Trif]^-$ ionic liquid to develop a meaningful correlation. Revelli *et al.* (2010a) further extended the Abraham model's predictive capability for IL solvents by further splitting the cation-specific equation coefficient into functional group values

$$\log K = \sum_{group} c_i n_i + \sum_{group} e_i n_i E + \sum_{group} s_i n_i S + \sum_{group} a_i n_i A + \sum_{group} b_i n_i B + \sum_{group} l_i n_i L + (c_{anion} + e_{anion} E + s_{anion} S + a_{anion} A + b_{anion} B + l_{anion} L) \quad (24)$$

and

$$\log P = \sum_{group} c_i n_i + \sum_{group} e_i n_i E + \sum_{group} s_i n_i S + \sum_{group} a_i n_i A + \sum_{group} b_i n_i B + \sum_{group} v_i n_i V + (c_{anion} + e_{anion} E + s_{anion} S + a_{abion} A + b_{anion} B + v_{anion} V) \quad (25)$$

where n_i is the number of times a particular functional group appears in the cation. The authors were able to mathematically describe the 1450 available gas-to-IL partition coefficients (log K values) and 1410 water-to-IL partition coefficients (log P values) with 21 groups: 12 functional groups characterizing the cations (CH_3, CH_2, N_{cyclic}, C_{cyclic}, *etc.*) and 9 individual anions ($[PF_6]^-$, $[BF_4]^-$, $[SCN]^-$, *etc.*) to within 0.15 and 0.17 log units, respectively. The number of functional group values is likely to increase in future years as more experimental data becomes available for functionalized IL solvents. In selecting a predictive method, we suggest that the first choice by the IL-specific Abraham equation if available, followed by the ion-specific Abraham model, and then the Abraham group contribution model of Revelli *et al.* For many IL solvents the group contribution model will be the only option available at the present time.

5. ABRAHAM MODEL: ENTHALPY OF SOLVATION CORRELATIONS

Abraham model correlations constructed from the ion-specific equation coefficients in Tables 3 and 4 pertain to 298.15 K. A method is needed for extrapolating the predicted log K and log P values to other temperatures for applications involving azeotropic distillation and temperature-programmed gas-liquid chromatographic separation. The majority of experimental data used in the log K and log P regressions came from published gas-liquid chromatographic retention measurements. As part of the experimental studies the authors determined the infinite dilution coefficients of volatile solutes in IL solvents at several temperatures. For the majority of measurements the median temperature corresponded to 323 ± 5 K. Solutes studied include inert gases, diatomic gas molecules, linear and cyclic alkanes and alkenes (up to dodecane), alkylbenzenes, linear and branched alcohols, linear and cyclic monoethers (plus 1,4-dioxane), chlorinated methanes and a few of the smaller aldehydes and ketones. The solute descriptor space defined by these compounds would be: **E** = 0.000 to **E** = 0.850; **S** = 0.000 to **S** = 0.900; **A** = 0.000 to **A** = 0.430; **B** = 0.000 to **B** = 0.650; **V** = 0.109 to **V** = 1.800; and **L** = -1.200 to **L** = 5.700. The predictive area of chemical space spanned by this set of solutes is smaller than the range covered by the log K (Eqn. 22) and log P (Eqn. 23) correlations.

The solute's molar enthalpy of solution in the IL is calculated from the variation of γ_{solut}^{∞} with temperature, *i.e.*, $\Delta H^{ex,\infty} = R\, \partial \ln \gamma_{solute}^{\infty}/\partial(1/T)$. Enthalpies of solution determined in this fashion assume that $\Delta H^{ex,\infty}$ is independent of temperature over the range of the experimental measurements. Enthalpies of solvation are calculated by

$$\Delta H_{Solv} = \Delta H^{ex,\infty} - \Delta H_{Vap} \quad (26)$$

$$\Delta H_{Solv} = \Delta H^{ex,\infty} - \Delta H_{Sub} \quad (27)$$

the solute's enthalpy of vaporization (ΔH_{Vap}, liquid solute) or enthalpy of sublimation (ΔH_{Sub}, solid solute) from the molar enthalpy of solution. Large compilations of ΔH_{Vap} and ΔH_{Sub} data are available in the published literature (Chickos and Acree, 2002; Chickos and Acree, 2003) for use in converting $\Delta H^{ex,\infty}$ to ΔH_{Solv}, and *vice versa.*

Mintz *et al.* (Mintz *et al.*, 2007; Mintz *et al.*, 2008b; Mintz *et al.*, 2009) and Sprunger *et al.* (2009a) demonstrated that the Abraham model does provide an accurate mathematical description of enthalpies of solvation of organic vapors and gases in water, $\Delta H_{Solv,W}$,

$$\Delta H_{Solv,W}\ (\text{kJ/mole}) = -13.310(0.457) + 9.910(0.814)\mathbf{E} + 2.836(0.807)\mathbf{S} - 32.010(1.102)\mathbf{A} - 41.816(0.781)\mathbf{B} - 6.354(0.200)\mathbf{L}$$

(with N = 368, SD = 3.68, $R^2 = 0.964$, $R^2_{adj} = 0.964$, F = 1950.5) (28)

and

$$\Delta H_{Solv,W}\ (\text{kJ/mole}) = -6.952(0.651) + 1.415(0.770)\,\mathbf{E} - 2.859(0.855)\,\mathbf{S} - 34.086(1.225)\,\mathbf{A} - 42.686(0.850)\,\mathbf{B} - 22.720(0.800)\,\mathbf{V}$$

(with N = 369, SD = 4.04, $R^2 = 0.959$, $R^2_{adj} = 0.958$, F = 1688.2) (29)

as well as into organic solvents and ionic liquids. From a thermodynamic standpoint Eqn. 28 is the enthalpic temperature derivative of the Abraham model's gas-to-condensed phase transfer equation. Equation 29 might be more useful in some predictive applications in instances where the L-descriptor is not known. Equation 29 uses the McGowan volume, V- descriptor, which is easily calculable from the individual atomic sizes and numbers of bonds in the molecule (Abraham and McGowan, 1987).

To provide as complete set of ion-specific equation coefficients as possible we have reanalyzed our enthalpy of solvation data for solutes dissolved in IL solvents at 323 K in accordance with the Eqns. 14 and 15 of the Abraham model. The derived correlations

$$\Delta H_{solv} = \sum_{cation} (c_{cation} + e_{cation}E + s_{cation}S + a_{cation}A + b_{cation}B + l_{cation}L) + \sum_{anion} (c_{anion} + e_{anion}E + s_{anion}S + a_{anion}A + b_{anion}B + l_{anion}L) \quad (30)$$

(N = 977, $R^2 = 0.998$, $R^2_{adj} = 0.998$, SD = 1.602, F = 3058) and

$$\Delta H_{solv} = \sum_{cation} (c_{cation} + e_{cation} E + s_{cation} S + a_{cation} A + b_{cation} B + v_{cation} V) + \sum_{anion} (c_{anion} + e_{anion} E + s_{anion} S + a_{anion} A + b_{anion} B + v_{anion} V) \quad (31)$$

(N = 977, R^2 = 0.998, R^2_{adj} = 0.997, SD = 1.655, F = 2860)

describe the observed enthalpy of solvation to within standard deviations of SD = 1.602 kJ/mole and SD = 1.655 kJ/mole using the ion-specific equation coefficients tabulated in Tables 5 and 6. Standard errors in the equation coefficients are given in parenthesis immediately below the respective coefficient. For the most part, the larger standard errors were noted in the equation coefficients for those ions for which experimental data were limited. Large standard errors were similarly noted in the ion-specific equation coefficients for our initial log K and log P Abraham model correlations. The standard errors did decrease in magnitude when additional log *K* and log P values were added to the database. We expect the same to happen with the ΔH_{solv} correlations.

Comparsions of experimental ΔH_{Solv} data versus back-calculated values based on the two Abraham model correlations are graphically depicted in Figures 6 and 7. Each calculation uses only the values for the cation and anion in the given RTIL. For example, to predict the enthalpies of solvation of organic vapors and gases in 1-(3-hydroxypropyl)pyridinium *tris*(pentafluoroethyl)trifluorophosphate one would use only the six equation coefficients for the $[PrOHPy]^+$ and the six equation coefficients for the $[FAP]^-$ anion. The 13 cation-specific equation coefficients and 10 anion-specific equation coefficients listed in Tables 5 and 6 can be used to predict enthalpies of solvation of organic vapors and gases in 130 (13 x 10) different IL solvents. The number of ionic liquids that one can make ΔH_{Solv} predictions for is far smaller than the 299 ILs for which log K and log P predictions can be made.

Predicted values of enthalpies of solvation based on Eqns. 30 and 31 can be used to correct predicted log K and log P values at 298.15 K to other temperatures. For the log P corrections one will need an experimental or predicted value for the enthalpy of solvation of the solute in water, $\Delta H_{Solv,W}$, as well (see Eqns. 12 and 13). The Abraham model correlation derived by Mintz *et al.* can be used to estimate $\Delta H_{Solv,W}$ if an experimental value is not available. An error/uncertainty of ± 2 kJ/mole in the enthalpy of solvation results in an error of slightly less than 0.04 log units in extrapolating a log K value measured at 298.15 K to a temperature of 313.15. This level of predictive error will be sufficient for most practical chemical separation applications. For gas-liquid chromatographic separations Abraham model correlations have been reported (Breitbach and Armstrong, 2008; Anderson *et al.*, 2002; Baltazar *et al.*, 2008; Zhao *et al.*, 2009) for predicting solute retention factors on a few select mono-cationic and di-cationic IL stationary phases at higher temperatures.

Table 5. Cation-Specific and Anion-Specific Equation Coefficients for the Abraham Model Enthalpy of Solvation Correlation (Eqn.30)

Ion[a]	c_{ion}	e_{ion}	s_{ion}	a_{ion}	b_{ion}	l_{ion}
CATIONS						
$[MEIm]^+$	-6.007	3.914	-15.247	-20.601	-7.607	-7.153
(N=257)[b]	(0.821)	(1.563)	(1.749)	(2.147)	(1.704)	(0.249)
$[BMIm]^+$	-6.223	-1.307	-8.699	-20.815	-9.869	-7.072
(N=179)	(0.642)	(1.407)	(1.926)	(2.169)	(2.328)	(0.173)
$[MHIm]^+$	-5.543	0.111	-6.842	-19.484	-10.780	-7.795
(N=101)	(0.685)	(1.577)	(2.013)	(2.362)	(2.357)	(0.187)
$[MOIm]^+$	-7.758	1.088	-4.778	-21.886	-6.993	-7.448
(N=86)	(0.907)	(2.153)	(2.631)	(2.686)	(2.588)	(0.274)
$[M_3BAm]^+$	-7.255	1.352	-9.405	-22.715	-7.191	-6.761
(N=51)	(0.969)	(2.041)	(2.724)	(3.067)	(3.151)	(0.237)
$[M_2EIm]^+$	-0.592	4.122	-15.849	-23.339	-7.193	-8.667
(N=37)	(1.570)	(2.097)	(2.221)	(2.629)	(2.015)	(0.442)
$[4\text{-}BMPy]^+$	-2.796	5.802	-19.743	-27.272	-2.956	-8.378
(N=69)	(1.127)	(2.094)	(2.554)	(2.343)	(2.210)	(0.334)
$[3\text{-}BMPy]^+$	-6.188	5.800	-18.460	-20.302	-2.050	-7.565
(N=36)	(1.992)	(3.406)	(4.291)	(4.378)	(4.375)	(0.589)
$[E_3S]^+$	-2.913	15.064	-32.880	-15.755	-5.551	-7.845
(N=28)	(1.642)	(4.601)	(7.428)	(10.973)	(12.526)	(0.495)
$[BMPyr]^+$	-5.800	10.249	-25.046	-16.107	-3.629	-7.191
(N=30)	(1.710)	(4.440)	(5.683)	(4.725)	(4.739)	(0.539)
$[HxomMIm]^+$	-3.748	13.318	-24.942	-27.250	1.260	-8.645
(N=34)	(1.589)	(3.132)	(4.019)	(3.056)	(3.335)	(0.479)
$[(Hxom)_2Im]^+$	-4.547	13.240	-21.422	-28.533	0.734	-8.833
(N=34)	(1.589)	(3.132)	(4.019)	(3.056)	(3.235)	(0.479)
$[PrOHPy]^+$	-25.941	-15.090	28.322	-15.026	-33.898	-2.429
(N=35)	(2.861)	(7.406)	(9.520)	(12.678)	(11.759)	(0.844)
ANIONS						
$[(Tf)_2N]^-$	0.000	0.000	0.000	0.000	0.000	0.000
(N=394)						
$[BF_4]^-$	-0.520	-3.920	3.391	1.969	-1.162	0.494

(N=136)	(0.751)	(1.805)	(2.224)	(2.303)	(2.144)	(0.227)
$[PF_6]^-$	-5.420	12.391	-24.092	13.322	14.997	0.806
(N=54)	(0.813)	(2.647)	(3.369)	(3.463)	(3.616)	(0.254)
$[EtSO_4]^-$	-3.021	-2.344	6.454	-16.911	1.412	1.106
(N=42)	(1.551)	(2.634)	(3.036)	(3.747)	(3.276)	(0.379)
$[Trif]^-$	2.171	2.471	-5.152	-19.614	6.577	-0.696
(N=132)	(1.253)	(2.224)	(2.796)	(3.195)	(3.132)	(0.378)
$[F_3Ac]^-$	5.481	9.027	-6.060	-12.187	-16.266	-1.763
(N=27)	(1.849)	(4.864)	(7.649)	(11.189)	(12.648)	(0.561)
$[NO_3]^-$	-3.591	0.499	-0.226	-14.195	5.620	0.520
(N=28)	(3.254)	(2.953)	(2.980)	(4.528)	(2.842)	(0.970)
$[SCN]^-$	7.978	14.895	-23.862	-26.264	12.733	-2.220
(N=84)	(1.103)	(2.444)	(3.255)	(2.689)	(2.961)	(0.316)
$[B(CN)_4]^-$	20.462	5.485	-24.845	-39.242	21.911	-4.524
(N=22)	(2.434)	(7.100)	(9.553)	(13.453)	(12.973)	(0.729)
$[FAP]^-$	25.824	17.977	-48.802	-4.930	10.119	-4.938
(N=58)	-2.405	(6.922)	(8.926)	(12.306)	(11.333)	(0.713)

[a]Cation and anion abbreviations are given in the footnote to Table 3.

[b]Number of experimental data points associated with the specified ion.

Table 6. Cation-Specific and Anion-Specific Equation Coefficients for the Abraham Model Enthalpy of Solvation Correlation (Eqn.31)

Ion[a]	c_{ion}	e_{ion}	s_{ion}	a_{ion}	b_{ion}	v_{ion}
CATIONS						
$[MEIm]^+$	-0.522	-8.710	-15.755	-24.770	-7.602	-25.280
(N=257)[b]	(1.027)	(1.469)	(1.838)	(2.270)	(1.765)	(0.914)
$[BMIm]^+$	0.671	-10.742	-13.978	-23.150	-10.347	-26.260
(N=179)	(0.794)	(1.409)	(2.004)	(2.238)	(2.395)	(0.646)
$[MHIm]^+$	0.103	-12.846	-7.474	-16.018	-15.782	-27.380
(N=101)	(0.851)	(1.584)	(2.083)	(2.421)	(2.412)	(0.697)
$[MOIm]^+$	-3.094	-14.937	-1.923	-21.742	-10.249	-25.306
(N=86)	(1.147)	(2.122)	(2.745)	(2.799)	(2.672)	(1.021)
$[M_3BAm]^+$	-1.158	-7.091	-15.020	-25.483	-6.221	-24.671

(N=51)	(1.197)	(2.059)	(2.867)	(3.202)	(3.259)	(0.891)
$[M_2EIm]^+$	4.006	-5.855	-21.604	-28.854	-3.145	-29.304
(N=37)	(1.881)	(1.962)	(2.463)	(2.829)	(2.100)	(1.576)
$[4\text{-}BMPy]^+$	1.375	-11.008	-16.169	-28.921	-5.933	-27.874
(N=69)	(1.372)	(2.024)	(2.654)	(2.454)	(2.280)	(1.195)
$[3\text{-}BMPy]^+$	1.324	-6.735	-20.274	-23.121	-4.582	-28.299
(N=36)	(2.435)	(3.415)	(4.442)	(4.505)	(4.529)	(2.107)
$[E_3S]^+$	4.009	-10.891	-17.029	-23.710	-12.059	-28.593
(N=28)	(2.124)	(4.789)	(7.631)	(11.458)	(12.920)	(1.860)
$[BMPyr]^+$	2.509	-6.665	-20.110	-20.750	-9.554	-27.793
(N=30)	(2.141)	(4.169)	(5.749)	(4.885)	(4.842)	(1.961)
$[HxomMIm]^+$	2.168	-5.927	-19.612	-27.547	-4.531	-30.045
(N=34)	(1.977)	(3.112)	(4.111)	(3.166)	(3.397)	(1.732)
$[(Hxom)_2Im]^+$	1.063	-6.379	-15.883	-28.570	-5.017	-30.303
(N=34)	(1.977)	(3.112)	(4.111)	(3.166)	(3.397)	(1.732)
$[PrOHPy]^+$	-24.316	-6.193	7.321	-15.580	-23.480	-8.586
(N=35)	(3.552)	(8.005)	(9.947)	(13.201)	(11.998)	(3.057)
ANIONS						
$[(Tf)_2N]^-$	0.000	0.000	0.000	0.000	0.000	0.000
(N=394)						
$[BF_4]^-$	0.525	3.258	-5.021	-0.272	3.289	0.404
(N=136)	(0.968)	(1.774)	(2.346)	(2.427)	(2.225)	(0.862)
$[PF_6]^-$	-3.055	14.791	-27.034	9.240	18.214	0.108
(N=54)	(1.074)	(2.662)	(3.515)	(3.608)	(3.733)	(1.006)
$[EtSO_4]^-$	-1.724	5.034	-1.145	-17.689	4.040	2.461
(N=42)	(1.964)	(2.584)	(3.309)	(4.005)	(3.406)	(1.449)
$[Trif]^-$	0.513	-0.513	-2.324	-16.775	5.462	-0.738
(N=132)	(1.525)	(2.158)	(2.908)	(3.281)	(3.220)	(1.340)
$[F_3Ac]^-$	7.964	-7.797	13.304	-17.312	-24.326	-7.359
(N=27)	(2.404)	(5.003)	(7.884)	(11.705)	(13.049)	(2.124)
$[NO_3]^-$	-6.235	4.613	-3.847	-10.664	8.528	3.247
(N=28)	(3.616)	(2.622)	(3.381)	(4.656)	(2.911)	(3.178)
$[SCN]^-$	9.842	1.469	-11.336	-31.443	8.565	-8.113
(N=84)	(1.422)	(2.512)	(3.355)	(2.831)	(3.130)	(1.205)
$[B(CN)_4]^-$	23.574	-15.339	-4.703	-39.309	10.023	-15.787
(N=22)	(3.047)	(7.629)	(9.887)	(14.040)	(13.192)	(2.647)
$[FAP]^-$	29.378	-4.429	-27.315	-4.750	-3.066	-17.347
(N=58)	(3.001)	(7.540)	(9.341)	(12.818)	(11.546)	(2.585)

[a] Cation and anion abbreviations are given in the footnote to Table 3.

[b] Number of experimental data points associated with the specified ion.

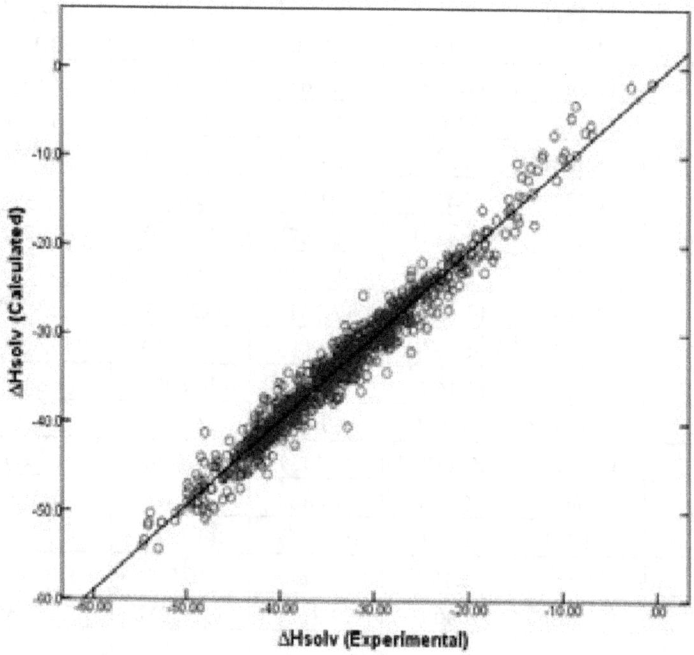

Fig. 6. Comparison between the 977 experimental ΔH_{Solv} data points and predicted values based on Eqn. 30.

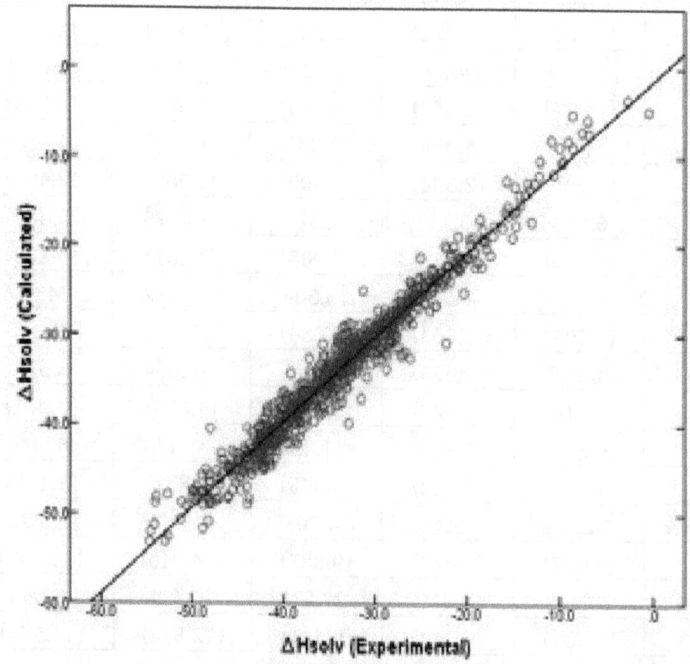

Fig. 7. Comparison between the 977 experimental ΔH_{Solv} data points and predicted values based on Eqn. 31

6. CONCISE SUMMARY

The Abraham universal solvation model offers a very precise mathematical representation of the thermodynamic characteristics that dictate solute transfer into anhydrous ionic liquid solvents from both aqueous and gaseous phases. Expressions derived from the Abraham model facilitate the estimation of log K and log P values at 298 K for various solute-IL systems, utilizing published ion-specific equation coefficients and/or group contribution values. The comparison of experimental and predicted values for over 30 IL solvents indicates that the predicted log P and log K values typically align within 0.15 log units of the observed values. The model includes mechanisms for adjusting the anticipated log K and log P values for temperatures close to 298 K.

REFERENCES

Abraham, M. H. & McGowan, J. C. (1987) The use of characteristic volumes to measure cavity terms in reversed phase liquid chromatography. *Chromatographia* 23 (4) 243- 246.

Abraham, M. H. (1993a) Scales of solute hydrogen-bonding: their construction and application to physicochemical and biochemical processes. *Chemical Society Reviews* 22 (2),73-83.

Abraham, M. H. (1993b) Application of solvation equations to chemical and biochemical processes.*Pure and Applied Chemistry* 65 (12), 2503-12.

Abraham, M. H.; Acree, W. E., Jr. (2006) Comparative analysis of solvation and selectivity in room temperature ionic liquids using the Abraham linear free energy relationship. *Green Chemistry* 8 (10), 906-915.

Abraham, M. H.; Zissimos, A. M.; Huddleston, J. G.; Willauer, H. D.; Rogers, R. D. & Acree, W. E., Jr.(2003) Some Novel Liquid Partitioning Systems: Water-Ionic Liquids and Aqueous Biphasic Systems. *Industrial & Engineering Chemistry Research* 42, 413-418.

Abraham, M. H.; Ibrahim, A. & Zissimos, A. M. (2004) Determination of sets of solute descriptors from chromatographic measurements. *Journal of Chromatography, A* 1037 (1-2), 29-47.

Abraham, M. H.; Ibrahim, A.; Zhao, Y.; Acree, W. E., Jr. (2006) A data base for partition of volatile organic compounds and drugs from blood/plasma/serum to brain, and an LFER analysis of the data. *Journal of Pharmaceutical Sciences* 95 (10), 2091-2100.

Abraham, M. H.; Ibrahim, A. & Acree, W. E. Jr. (2007) Air to liver partition coefficients for volatile organic compounds and blood to liver partition coefficients for volatile organic compounds and drugs . *European Journal of Medicinal Chemistry* 42 (6), 743- 751.

Abraham, M. H.; Ibrahim, A. & Acree, W. E. Jr. (2008) Air to lung partition coefficients for volatile organic compounds and blood to lung partition coefficients for volatile organic compounds and drugs. *European Journal of Medicinal Chemistry* 43 (3), 478- 485.

Abraham, M. H.; Acree, W. E., Jr. & Cometto-Muniz, J. E. (2009) Partition of compounds from water and from air into amides. *New Journal of Chemistry* 33 (10), 2034-2043.

Abraham, M. H.; Smith, R. E.; Luchtefeld, R.; Boorem, A. J.; Luo, R.; Acree, W. E., Jr. (2010) Prediction of solubility of drugs and other compounds in organic solvents. Journal of Pharmaceutical Sciences 99 (3), 1500-1515.

Acree, W. E., Jr. & Abraham, M. H. (2006) The analysis of solvation in ionic liquids and organic solvents using the Abraham linear free energy relationship. *Journal of Chemical Technology and Biotechnology* 81 (8), 1441-1446.

Aguilera-Herrador, E.; Lucena, R.; Cardenas, S.; Valcarcel, M. (2008) Ionic liquid-based single-drop microextraction/gas chromatographic/mass spectrometric determination of benzene, toluene, ethylbenzene and xylene isomers in waters. *Journal of Chromatography, A* 1201 (1), 106-111.

Anderson, J. L.; Ding, J.; Welton, T. & Armstrong, D. W. (2002) Characterizing ionic liquids on the basis of multiple solvation interactions. *Journal of the American Chemical Society* 124 (47), 14247-14254.

Arey, J. S.; Green, W. H., Jr. & Gschwend, P. M. (2005) The electrostatic origin of Abraham's solute polarity parameter. *J. Phys. Chem. B 109*, 7564-7573.

Baltazar, Q. Q.; Leininger, Suzette K.; Anderson, J. L. (2008) Binary ionic liquid mixtures as gas chromatography stationary phases for improving the separation selectivity of alcohols and aromatic compounds. *Journal of Chromatography, A*, 1182 (1), 119-127.

Breitbach, Z. S.; Armstrong, D. W. (2008) Characterization of phosphonium ionic liquids through a linear solvation energy relationship and their use as GLC stationary phases. *Analytical and Bioanalytical Chemistry* 390 (6), 1605-1617.

Chickos, James S.; Acree, William E., Jr. (2003) Enthalpies of vaporization of organic and organometallic compounds, 1880-2002. *Journal of Physical and Chemical Reference Data* 32 (2), 519-878.

Chickos, J. S. & Acree, W. E., Jr. (2002) Enthalpies of sublimation of organic and organometallic compounds. 1910-2001. *Journal of Physical and Chemical Reference Data* 31 (2), 537-698.

Grubbs, L. M.; Saifullah,M.; De La Rosa, N. E.; Acree, W. E., Jr.; Abraham, M. H.; Zhao, Q. & Anderson, J. L. Cation-specific and anion-specific Abraham model correlations for solute transfer into ionic liquids. *Global Journal of Physical Chemistry*, 1 (1), 1-19.

Lei, Z.; Li, C. & Chen, B. (2003) Extractive distillation: a review. *Separation and Purification Reviews* 32 (2), 121-213.

Lei, Z.; Arlt, W.; Wasserscheid, P. (2006) Separation of 1-hexene and n-hexane with ionic liquids. *Fluid Phase Equilibria*, 241 (1-2), 290-299.

Liu, J.-F.; Jiang, G.-B.; Chi, Y.-G.; Cai, Y.-Q.; Zhou, Q.-X.; Hu, J.-T. (2003) Use of ionic liquids for liquid-phase micro-extraction of polycyclic aromatic hydrocarbons. *Analytical Chemistry* 75 (21), 5870-5876.

Mintz, C. & Acree, W. E Jr. (2007) Partition coefficient correlations for transfer of solutes from gas phase and water to room temperature ionic liquids. *Physics and Chemistry of Liquids* 45 (3), 241-249.

Mintz, C.; Clark, M.; Acree, W. E., Jr. & Abraham, M. H. (2007) Enthalpy of solvation correlations for gaseous solutes dissolved in water and in 1-octanol based on the Abraham model. *Journal of Chemical Information and Modeling* 47 (1), 115-121.

Mintz, C.; Ladlie, T.; Burton, K.; Clark, M.; Acree, W. E., Jr. & Abraham, M. H. (2008a) Characterization of the partitioning of gaseous solutes into humic acid with the Abraham model and temperature-independent equation coefficients. *QSAR & Combinatorial Science* 27 (4), 483-491.

Mintz, C.; Ladlie, T.; Burton, K.; Clark, M.; Acree, W. E., Jr. & Abraham, M. H. (2008b) Enthalpy of solvation correlations for gaseous solutes dissolved in alcohol solvents based on the Abraham model. *QSAR & Combinatorial Science*, 27 (5), 627-635.

Mintz, C.; Burton, K.; Ladlie, T.; Clark, M.; Acree, W.E.; Abraham, M. H. (2009) Enthalpy of solvation correlations for organic solutes and gases dissolved in N,N- dimethylformamide and *tert*-butanol. *Journal of Molecular Liquids* 144 (1-2), 23-31.

Mokrushin, V.; Assenbaum, D.; Paape, N.; Gerhard, D.; Mokrushina, L.; Wasserscheid, P.; Arlt, W.; Kistenmacher, H.; Neuendorf, S. & Goeke, V. (2010) Ionic liquids for propene-propane separation. *Chemical Engineering & Technology*, 33 (1), 63-73.

Mutelet, F. & Rogalski, M. (2001) Experimental determination and prediction of the gas- liquid n-hexadecane partition coefficients. *J. Chromatogr., A 923*, 153-163.

Mutelet, F.; Revelli, A.-L.; Jaubert, J.-N.; Sprunger, L. M.; Acree, W. E., Jr. & Baker, G. A. (2010) Partition Coefficients of Organic Compounds in New Imidazolium and Tetralkyl-ammonium Based Ionic Liquids Using Inverse Gas Chromatography. *Journal of Chemical & Engineering Data* 55 (1), 234-242.

Palgunadi, J.; Kim, H. S.; Lee, J. M. & Jung, S. (2010) Ionic liquids for acetylene and ethylene separation: Material selection and solubility investigation. *Chemical Engineering and Processing* 49 (2), 192-198.

Pandey, S. (2006) Analytical applications of room-temperature ionic liquids: A review of recent efforts. *Analytica Chimica Acta* 556 (1), 38-45.

Pereiro, A. B.; Deive, F. J.; Esperança, J. M. M. S. & Rodríguez, A. (2010) Alkylsulfate-based ionic liquids to separate azeotropic mixtures. *Fluid Phase Equilibr.*, 291 (1), 13-17.

Pharma Algorithms, (2006) ADME Boxes, Version 3.0; Pharma Algorithms Inc.: Toronto, Ontario, Canada, 2006.

Planeta, J. & Roth, M. (2005) Solute Partitioning between the Ionic Liquid 1-n-Butyl-3- methylimidazolium Tetrafluoroborate and Supercritical CO2 from Capillary- Column Chromatography. *Journal of Physical Chemistry B* 109 (31), 15165-15171.

Planeta, J.; Karasek, P. & Roth, M (2007) Limiting partition coefficients of solutes in biphasic trihexyltetradecyl-phosphonium chloride ionic liquid-supercritical CO2 system: measurement and LSER-based correlation. *Journal of Physical Chemistry B* 111 (26), 7620-7625.

Planeta, J.; Karasek, P. & Roth, M. (2009) Distribution of organic solutes in biphasic 1-n- butyl-3-methylimidazolium methyl sulfate-supercritical CO2 system. *Journal of Physical Chemistry B*, 113 (28), 9520-9526.

Platts, J. A.; Butina, D.; Abraham, M. H.; Hersey, A. (1999) Estimation of molecular linear free energy relation descriptors using a group contribution approach. *J. Chem. Inf. Comput. Sci.* 1999, 39, 835-845.

Poole, C. F. & Poole, Salwa K. (2010) Extraction of organic compounds with room temperature ionic liquids. *Journal of Chromatography, A* 1217(16), 2268-2286.

Proctor, A. Sprunger, L. M.; Acree, W. E., Jr.; Abraham, M. H. (2008) LFER correlations for the solubilising characterisation of room temperature ionic liquids containing trifluoromethanesulfonate and trifluoroacetate anions. *Physics and Chemistry of Liquids* 46 (6), 631-642.

Revelli, A.-L.; Sprunger, L. M.; Gibbs, J.; Acree, W. E.; Baker, G. A. & Mutelet, F. (2009) Activity coefficients at infinite dilution of organic compounds in trihexyl(tetradecyl)phosphonium bis(trifluoromethylsulfonyl)imide using inverse gas chromatography. *Journal of Chemical and Engineering Data* 54 (3), 977-985.

Revelli, A.-L.; Mutelet, F. & Jaubert, J.-N. (2010a) Prediction of partition coefficients of organic compounds in ionic liquids: use of a linear solvation energy relationship with parameters calculated through a group contribution method. *Industrial & Engineering Chemistry Research* 49 (8), 3883-3892.

Revelli, A.-L.; Mutelet, F.; Jaubert, J.-N.; Garcia-Martinez, M.; Sprunger, L. M.; Acree, W. E., Jr. & Baker, Gary A. (2010b) Study of ether-, alcohol-, or cyano-functionalized ionic liquids using inverse gas chromatography. *Journal of Chemical & Engineering Data*, 55 (7), 2434-2443.

Roth, M. (2009) Partitioning behaviour of organic compounds between ionic liquids and supercritical fluids. *Journal of Chromatography*, A 1216 (10), 1861-1880.

Seeley, J. V.; Seeley, S. K.; Libby, E. K.; Breitbach, Z. S. & Armstrong, D. W. (2008) Comprehensive two-dimensional gas chromatography using a high-temperature phosphonium ionic liquid column. *Analytical and Chemistry* 390(1), 323-332.

Soukup-Hein, R. J.; Warnke, M. M. & Armstrong, D. W. (2009) Ionic liquids in analytical chemistry. *Annual Review of Analytical Chemistry* 2, 145-168.

Sprunger L.; Acree W. E., Jr. & Abraham M. H. (2007a) Linear free energy relationship correlation of the distribution of solutes between water and sodium dodecyl sulfate (SDS) micelles and between gas and SDS micelles. *Journal of Chemical Information and Modeling* 47 (5), 1808-17.

Sprunger, L.; Clark, M.; Acree, W. E., Jr. & Abraham, M. H. (2007b) Characterization of room-temperature ionic liquids by the Abraham model with cation-specific and anion-specific equation coefficients. *Journal of Chemical Information and Modeling* 47 (3), 1123-1129.

Sprunger, L.; Proctor, A.; Acree, W. E. Jr. & Abraham, M. H. (2007c) Characterization of the sorption of gaseous and organic solutes onto polydimethyl siloxane solid-phase microextraction surfaces using the Abraham model. *Journal of Chromatography, A* 1175 (2), 162-173.

Sprunger, L. M.; Proctor, A.; Acree, W. E., Jr. & Abraham, M. H. (2008) LFER correlations for room temperature ionic liquids: Separation of equation coefficients into individual cation-specific and anion-specific contributions. *Fluid Phase Equilibria* 265 (1-2), 104-111.

Sprunger, L. M.; Achi, S. S.; Acree, W. E. & Abraham, M. H. (2009a) Linear Free Energy Relationship Correlations for Enthalpies of Solvation of Organic Solutes into Room- Temperature Ionic Liquids Based on the Abraham Model with Ion-Specific Equation Coefficients.. *Industrial and Engineering Chemistry Research* 48 (18), 8704- 8709.

Sprunger, L. M.; Gibbs, J.; Baltazar, Q. Q.; Acree, W. E., Jr.; Abraham, M. H.; Anderson, J. L. (2009b) Characterisation of room temperature ionic liquid chromatographic stationary phases by combining experimental retention factor and partition coefficient data into a single model. *Physics and Chemistry of Liquids* 47 (1), 74-83.

Sprunger, L. M.; Gibbs, J.; Proctor, A.; Acree, W. E., Jr.; Abraham, M. H.; Meng, Y.; Yao, C. & Anderson, J. L. (2009c) Linear free energy relationship correlations for room temperature ionic liquids: revised cation-specific and anion-specific equation coefficients for predictive applications covering a much larger area of chemical space. *Industrial and Engineering Chemistry Research* 48 (8), 4145-4154.

Verma, V. K. & Banerjee, T. (2010) Ionic liquids as entrainers for water+ethanol, water+2- propanol, and water+THF systems: A quantum chemical approach. *Journal of Chemical Thermodynamics* 42 (7), 909-919.

Yao, C.; Pitner, W. R. & Anderson, J. L. (2009) Ionic liquids containing the tris(pentafluoroethyl) trifluorophosphate anion: a new class of highly selective and ultra hydrophobic solvents for the extraction of polycyclic aromatic hydrocarbons using single drop microextraction. *Analytical Chemistry* 81 (12), 5054-5063.

Zhao, Q.; Eichhorn, J.; Pitner, W. R. & Anderson, J. L. (2009) Using the solvation parameter model to characterize functionalized ionic liquids containing the tris(pentafluoroethyl)trifluorophosphate (FAP) anion. *Analytical and Bioanalytical Chemistry* 395 (1), 225-234.

Zhao, F.-Q.; Li, J.; Zeng, B.-Z. (2008) Coupling of ionic liquid-based headspace single-drop microextraction with GC for sensitive detection of phenols. *Journal of Separation Science* 31 (16-17), 3045-3049.

Zhao, F.; Lu, S.; Du, W. & Zeng, B. (2009) Ionic liquid-based headspace single-drop microextraction coupled to gas chromatography for the determination of chlorobenzene derivatives. *Microchimica Acta* 165(1-2), 29-33.

Zhu, J.; Chen, J. & Fei, W. (2004) Separation of aromatic hydrocarbons or olefins from paraffins using new ionic liquids. *Huagong Xuebao* (Chinese Edition), 55, 2091-2094.

Zissimos, A. M.; Abraham, M. H.; Barker, M. C.; Box, K. J.; Tam, K. Y. (2002a) Calculation of Abraham descriptors from solvent-water partition coefficients in four different systems; evaluation of different methods of calculation. *Journal of the Chemical Society, Perkin Transactions 2* (3), 470-477.

Zissimos, A. M.; Abraham, M. H.; Du, C. M.; Valko, K.; Bevan, C.; Reynolds, D.; Wood, J.; Tam, K. Y. (2002b) Calculation of Abraham descriptors from experimental data from seven HPLC systems; evaluation of five different methods of calculation. *Journal of the Chemical Society, Perkin Transactions 2* (12), 2001-2010.

14 Chapter

DBU-DERIVED IONIC LIQUIDS AND THEIR UTILIZATION IN ORGANIC SYNTHESIS REACTIONS

1. INTRODUCTION: AN OVERVIEW

In light of the ongoing depletion of natural resources and increasing environmental consciousness, present and future chemists are being educated to devise synthetic pathways that are economically advantageous. A technique to implement green processes in organic synthesis is to substitute harmful volatile organic solvents with more ecologically benign reaction media. Ionic liquids (ILs) have emerged as a prominent category of novel green solvents in green chemistry over the past decade, attributed to their exceptional chemical and thermal properties, including excellent thermal stability, negligible vapor pressure, ease of handling, recyclability, and superior coordinating and dissolving capabilities. Consequently, ionic liquids were extensively employed in numerous organic transformations (Miao & Chan, 2006; Ying et al., 2008; Ying et al., 2008). Nevertheless, the majority of the ionic liquids examined by chemists are structurally derived from imidazole, which is unreactive in organic processes. The necessity for the advancement of novel task-specific ionic liquids persists.

Conversely, 1,8-diazabicyclo[5.4.0]undec-7-ene (2,3,4,6,7,8,9,10-octahydropyrimido [1,2-α]azepine, DBU) shown superiority over other tertiary amines in its roles as a catalyst, base, or promoter. For instance, the utilization of DBU as a catalyst in the Balylis-Hillman reaction and aza-Michael addition was documented by Aggarwal and Kim, respectively (Aggarwal & Mereu, 1999; Yeom et al., 2007). The DBU-mediated CO2-fixation reaction (Yoshida et al., 2008), DBU-promoted S-N-type Smiles and Ireland-Claisen rearrangement reactions (Ma et al., 2007; Li et al., 2007), DBU-catalyzed addition reactions of sulfonylimidates (Matsubara & Kobayashi, 2008), DBU-promoted chemoselective cleavage of the acetylenic TMS group (Yeom et al., 2008), and DBU-assisted unusual dehydrogenation (Kim et al., 2008) were all demonstrated to be highly efficient. In light of the unique function of DBU in organic synthesis, ionic liquids containing DBU moisture were formulated

(Tolstikoua & Shainyan, 2006). The substantial size, low symmetry of the cationic moiety, and charge delocalization across the N-C=N triad contribute to the low melting temperatures of these ionic liquids.

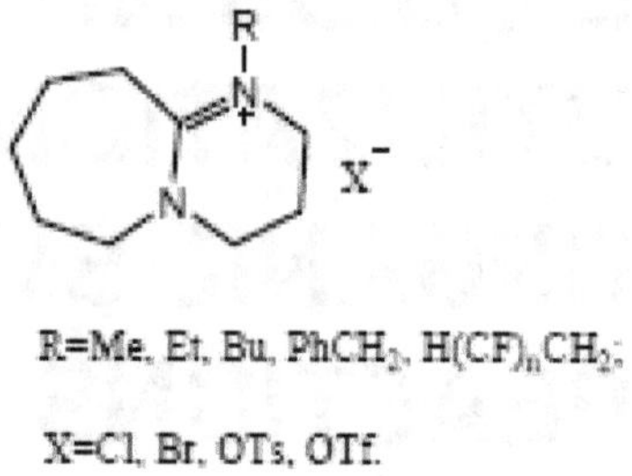

Scheme 1

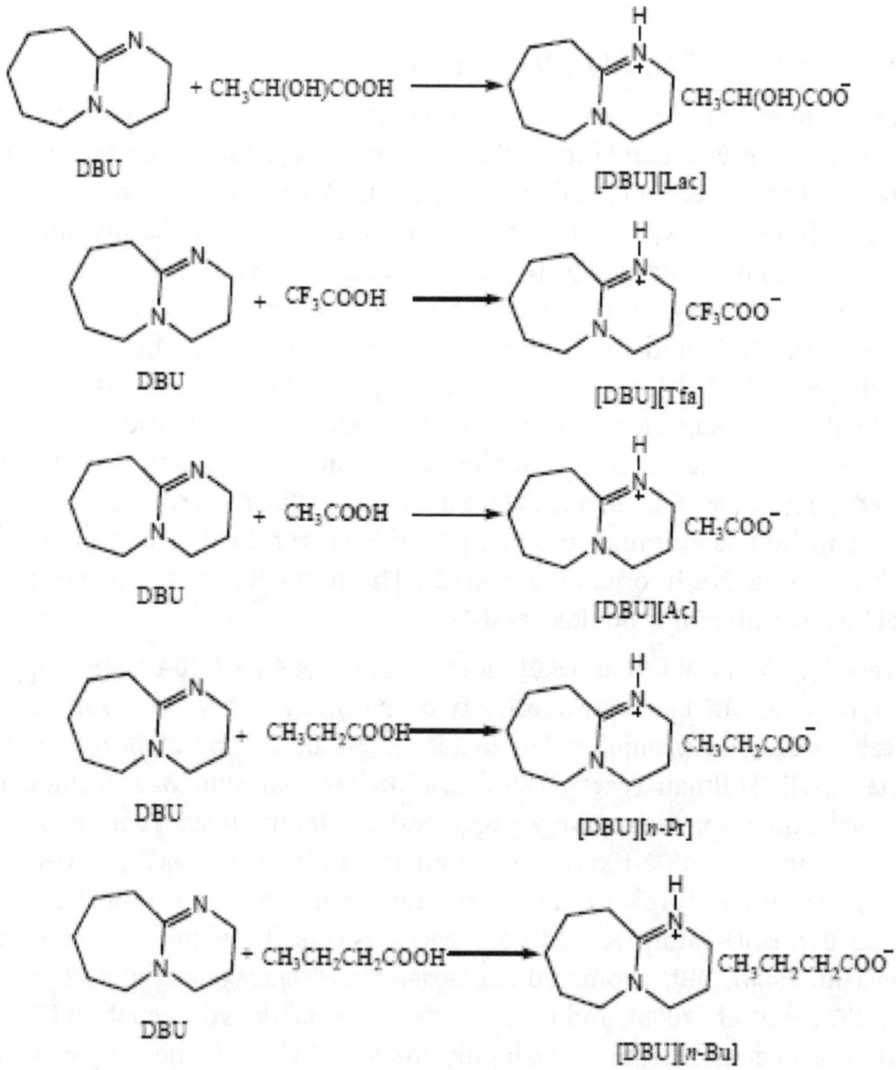

Scheme 2

However, the salts in Scheme 1 are inert and can only been used as reaction solvents. Then, we attempted to develop a new class of task-specific DBU-derived ionic liquids which could not only serve as reaction medium but also as catalyst or promotor. The preparation of the five novel ILs can be simply realized by neutralization reaction of DBU and the corresponding carboxyl acids, including acetic acid, lactic acid, propionic acid and butyric acid (Scheme2)(Ying et al., 2009; Ying et al., 2010; Ying, 2010). Among these ILs, [DBU][Tfa] is liquid with weak acidicity at room temperature, while the other four ionic liquids, [DBU][Ac], [DBU][Lac], [DBU][n-Pr] and [DBU][n-Bu] show property of weak basicity[11].

Subsequently, we shall examine the uses of the innovative DBU-based ionic liquids in Michael addition and Knoevenagel condensation. This chapter will summarize the use of ionic liquids immobilized on magnetic nanoparticles (MNP) to enhance recycling capabilities.

2. APPLICATION IN MICHAEL ADDITION

2.1 Michael addition (aliphatic amines as Michael donors)

Carbon-nitrogen bond forming reaction is one of the most important methodologies in synthetic organic chemistry for preparation of β-aminocarbonyl compounds, which not only constitute a component of biologically active natural products but also serve as a key intermediate for the syntheisis of β-aminoalcohols, β-lactams and β-amino acids(Kleinmann, 1991; Georg, 1993; Corey, 1989). Manich reaction between enolates and imines provide a classic route for the construction of β-amino carbonyl compounds(Arend et al., 1998; Kobayashi & Ishitani, 1999). However, this type of reaction often requires harsh reaction conditions and long reaction time. Compared with the Manich reaction, the aza-Michael addition of amines to electron deficient alkenes has attracted considerable attention as an alternative protocol for C-N bond formation due to its atomic economy and operational simplicity. Most aza-Michael reactions are usually carried out with a strong acid or a base, which would lead to by-products or undesired harmful residues. Thus, milder Lewis acidic catalysts such as $LiClO_4$(Azizi & Said, 2004), $Yb(OTf_3)$(Jenner, 1995; Matsubara et al., 1994), $Bi(NO_3)$(Srivastava & Banik, 2003), $FeCl_3.6H_2O$(Xu et al., 2004), $CeCl_3.7H_2O$(Bartoli et al., 2001), $InCl_3$(Loh & Wei, 1998), SmI_2(Reboule et al., 2005), $Cu(OTf)_2$(Xu et al., 2005), and so on are employed in the Michael protocol. Recently, some novel reagents used as catalyst or promoter in conjugate addition have been reported, including β-cyclodextrin(Surendra et al., 2006), bromodimethylsulfonium bromide(Khan et al., 2007), boric acid in water(Chaudhuri et al., 2005), $ZrOCl_2.8H_2O$ on montmorillonite K10(Hashemi et al., 2006), imidazoliHum-based polymer supported CuI(Alleti et al., 2008),

KF/Al_2O_3(Kantam et al., 2008), $[HP(HNCH_2CH_2)_3N]NO_3$(Yang et al., 2005), etc. However, many of the above methods suffered from some drawbacks, such as the requirement for a large excess of reagents, substrate-selective for some catalysts, and often involvement of some toxic solvents such as 1,2-dichloroethane or acetonitrile.

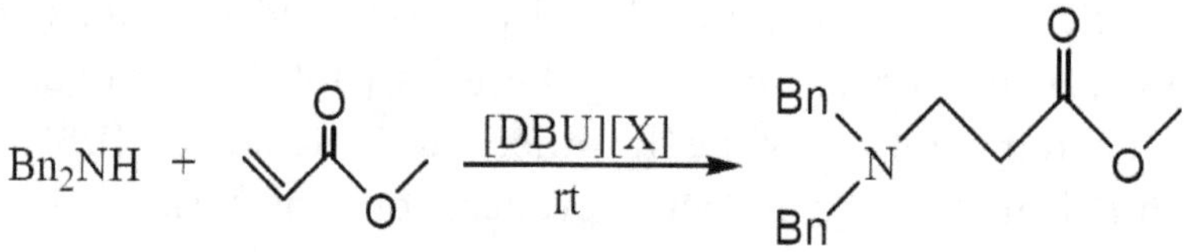

Among the five ionic liquids, [DBU][Ac] showed slightly higher catalytic activity than the other ionic liquids(Table 1, entry 7 and entries 10-13). For the comparison with DBU as promoter for aza-Michael addition reported by Kim et al., 50 mol % of the ionic liquid [DBU][Ac] was employed for the model reaction between dibenzylamine and methyl acrylate in acetonitrile at room temperature(Yeom et al., 2007). The reaction catalyzed by [DBU][Ac] was slightly faster than that promoted by its parent base DBU (Table 1, entries 1- 2), indicating the rationality of [DBU][Ac] as catalyst for the reaction. Other solvents such as methanol, toluene, and dichloromethane were also tested and they were all effective.

Table 1. Aza-Michael reaction of dibenzylamine (1 mmol) with methyl acrylate (1.3 mmol) under various reaction conditions

Entry	Catalyst	Solvent	Catalyst amount (mmol)	Time (h)	Yield (%)
1b	DBU	Acetonitrile	0.5	6	95
2	[DBU][Ac]	Acetonitrile	0.5	5	94
3	[DBU][Ac]	Methanol	0.5	5	86
4	[DBU][Ac]	Toluene	0.5	5	89
5	[DBU][Ac]	Dichloromethane	0.5	5	90
6	[DBU][Ac]	Solvent-free	0.5	4.5	96
7	[DBU][Ac]	Solvent-free	0.3	4.5	95
8	[DBU][Ac]	Solvent-free	0.1	4.5	72
9	[DBU][Ac]	Solvent-free	0.01	4.5	43
10	[DBU][Tfa]	Solvent-free	0.3	4.5	83
11	[DBU][Lac]	Solvent-free	0.3	4.5	92
12	[DBU][*n*-Pr]	Solvent-free	0.3	4.5	89
13	[DBU][*n*-Bu]	Solvent-free	0.3	4.5	83

[a] Isolated yields. [b] reference (Yeom et al., 2007): using DBU as catalyst.

The main aim of the work is to examine if the non-solvent addition of aliphatic amines to electron deficient olefins can proceed smoothly with DBU based task-specific ionic liquids as catalysts.

Reaction medium for the model reaction (entries 3-5). However, because of the toxicity of organic solvents and our pursuit for the establishment of the environmentally benign process for organic transformations, we attempted to conduct the reaction of dibenzylamine and methyl acrylate under solvent-free condition and to our delight, good yield (96 %) was obtained within 4.5 h (Table 1, entry 6). So we chose solvent-free conditions for further study. Next, the amount of catalyst [DBU][Ac] was reduced to 0.3 equiv and almost no decrease of yield was observed (entry 7). However, the product yield decreased obviously when the amount was further reduced to 0.01 equiv (Table 1, entries 8-9). As a result, we adopted 0.3 equiv of [DBU][Ac] under solvent-free condition at room temperature for the following investigations.

With the optimal catalytic system in hand, we investigated the suitability of a wide range of nitrogen nucleophiles for the [DBU][Ac] catalyzed aza-Michael reactions using methyl acrylate as substrate, aliphatic secondary amines such as morpholine, piperidine, 1- methylpiperazine, 1-ethylpiperazine and pyrazole underwent conjugate reaction with methyl acrylate favourably and excellent yields of Michael adducts were obtained at ambient temperature under solvent-free conditions within short reaction time (Table 2, entries 1-4). Morpholine, piperidine and pyarzole reacted faster and afforded higher products yields in the presence of ionic liquid [DBU][Ac] than those promoted by DBU (entries1, 4 and 5). It is worthy to note that when imidazole were treated as Michael donor with methyl acrylate, it disppeared in 2 hours and no Michael product was detected with the formation of acylation product. Judging from the disappearance of OCH3 group of products from ^{1}H NMR spectrum, it may provides a possible method for acylation reaction of Michael acceptors. However, 2-isopropylimidazole could react with methyl acrylate and 93 % Michael adduct yield was obtained in 10 hours. From the above, steric bulkiness of reagent can assist the formation of Michael adduct and is detrimental to acylation process. Primary amine, benzylamine, were also treated with methyl acrylate and 20 % yield of di- substituted product was formed which was the same with that catalyzed by DBU(Yeom et al., 2007) (Table 2, entry 8).

Having obtained favourable results with methyl acrylate, we studied the solventless reaction of morpholine with other electron-deficient olefins under the same conditions. Other various vinyl esters, acrylonitrile and acrylamide were effective substrates to give desired products in good to excellent yields (Table 2, entries 9-11 and 13-14). Among the acrylate esters, the increasing of carbon number acrylates would lead to decreased reactivity (entries 9 and 10), as reported by Lin et al.(2007) α-Methyl and β-bezene substituted ester, ethyl cinnamate were also tested as acceptors respectively and were found that the former gave the rational yield in short time and the latter exhibited relatively inert reaction activity due to steric hindrance of benzene at β position (entries 11 and 12).

Table 2. Results of aza-Michael addition of various aliphatic amines to electron-deficit alkenes using [DBU][Ac] as catalyst under solvent-free conditions[a]

Entry	Nitrogen nucleophile	Michael acceptor	Reaction time/ Ref. 19 (h)	Product	Yield[b]/Ref. 19 (%)
1		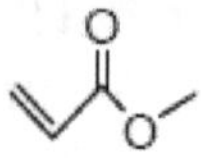	1.5/4	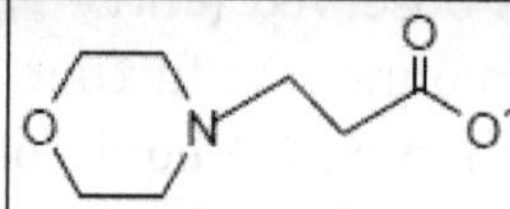	93/90
2	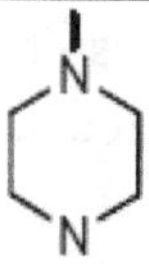	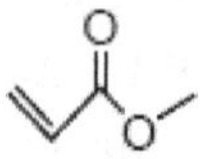	1.5	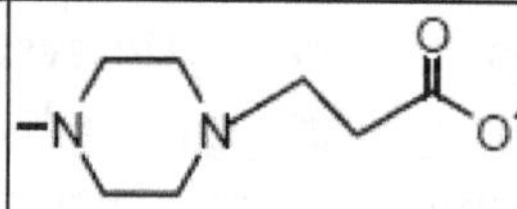	90
3	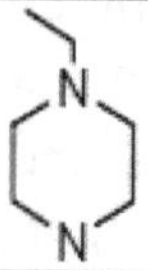	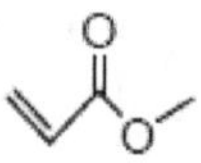	1.5	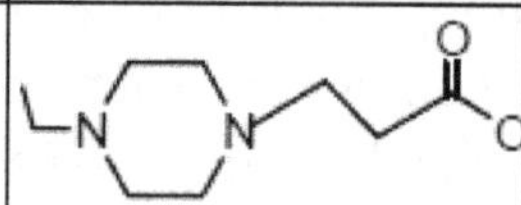	87
4	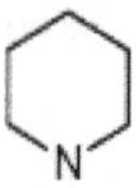	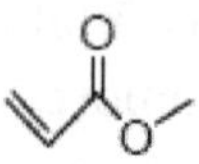	1.5/4	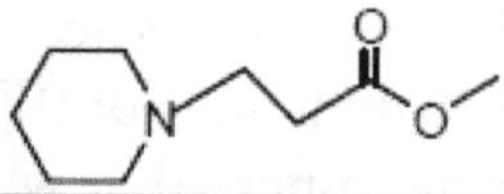	92/88
5	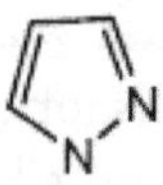	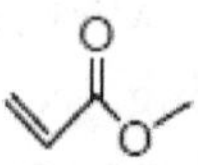	5/14	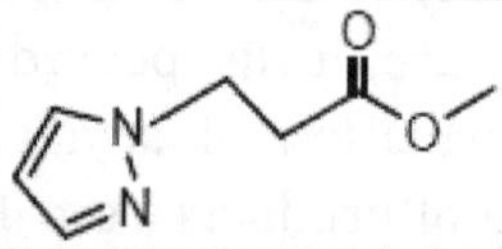	93/95
6	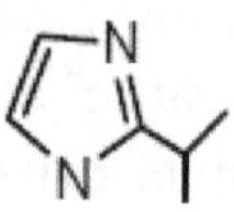	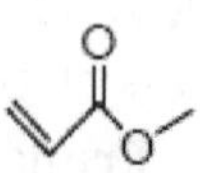	10	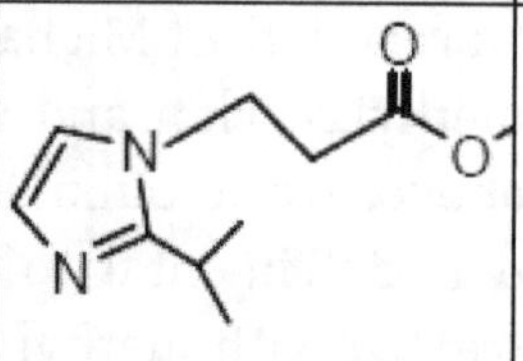	93
7	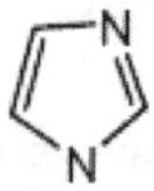	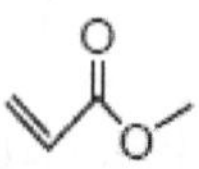	2/14	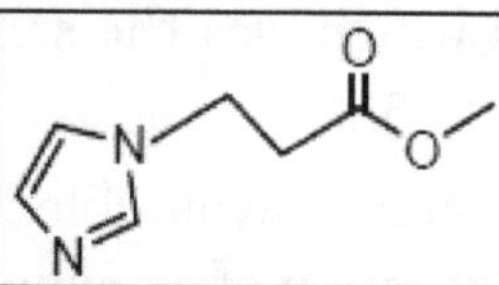	0/95

8	BnNH2	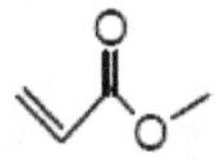	3/3	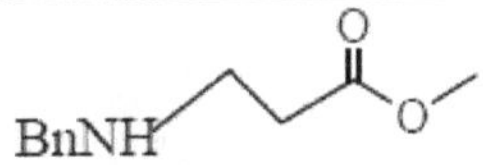	80/75
9		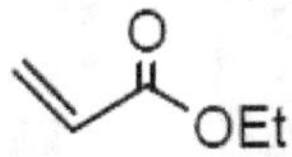	3	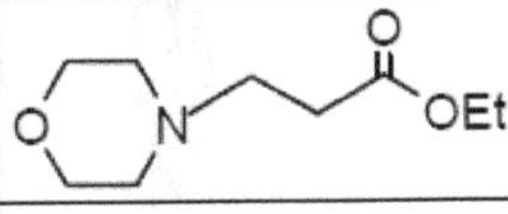	90
10	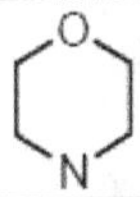	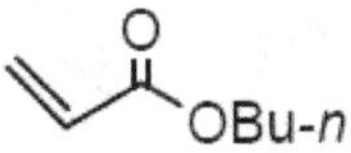	5	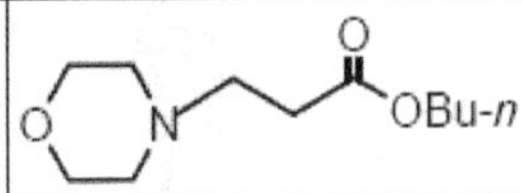	93
11		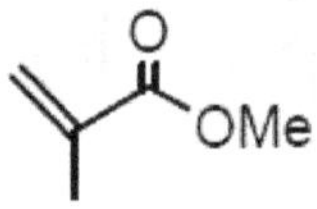	4	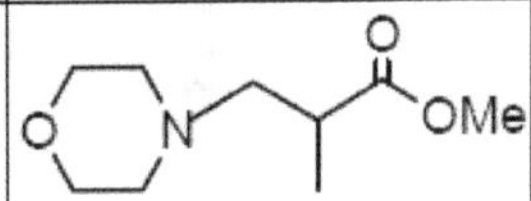	90
12	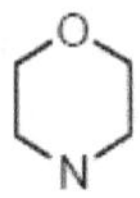	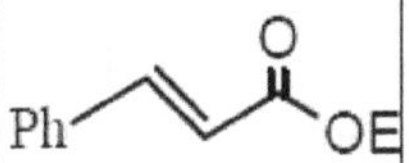	15	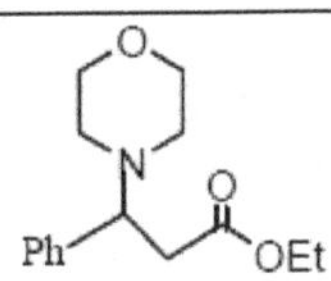	65
13	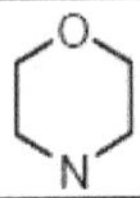	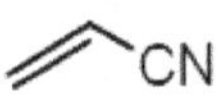	3	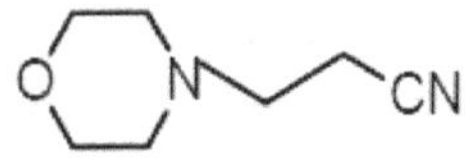	88
14	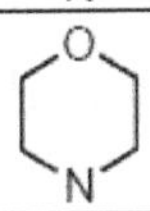	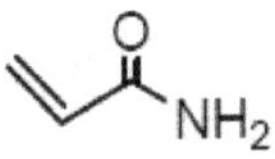	3	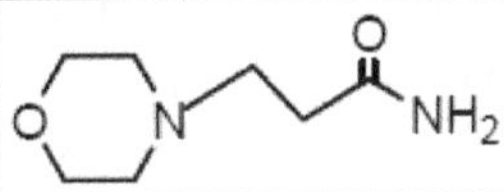	93

[a]Reactions were carried out on 1.0 mmol scale of substrate with 1.3 equiv of α,β-unsaturated compounds in the presence of 0.3 equiv ionic liquid at room temperature;

[b]Yields of isolated products.

In order to demonstrate the industrial applicability of this methodology, the solvent-free aza-Michael condensation of piperidine and methyl acrylate was carried out on a larger scale (100 mmol). The reaction was completed in 2 hours, excellent yield of 94% for the conjugate product was obtained. On the same scale, the recyclability of catalytic system was investigated using the same reaction as model reaction. Upon the completion of the reaction, the product was isolated through vacuum distillation while the residue ionic liquid was dried to remove water at 60 °C under vacuum. The recovered ionic liquid was reused in subsequent reactions. As shown in Figure 1, the ionic liquid [DBU][Ac] can be recycled for six times without considerable decrease of activity and the used ionic liquid remained intact (^{1}H NMR).

Although [DBU][Ac] has weaker basicity than DBU, its catalytic property for aza-Michael reaction is better than that of DBU. The reason, we speculated, is that amines exhibited higher nucleophilicity in the presence of ionic liquids than in organic solvents(Xu et al., 2007; Kim et al., 2002; Crowhurst et al., 2006; Meciarova et al., 2006).

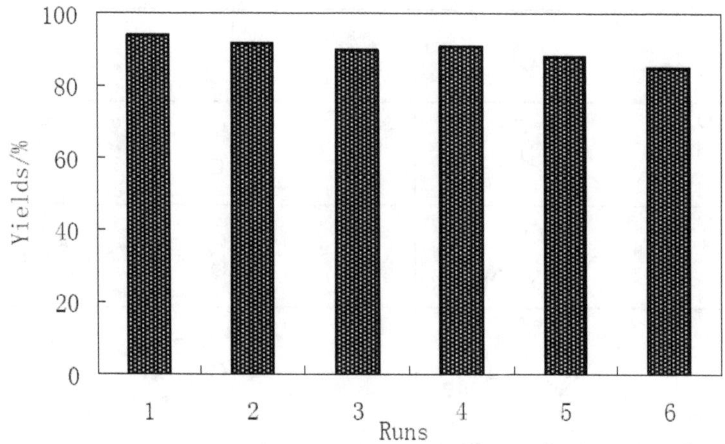

Fig. 1. Reuse of ionic liquid for aza-Michael reaction between piperidine (100 mmol) and methyl acrylate under solvent-free conditions

In conclusion, we have developed a mild, simple and efficient methodology using a new basic ionic liquid [DBU][Ac] as catalyst for the conjugate reaction of various aliphatic and aromatic amines with a variety of structurally diverse olefins. The reactions are conducted at room temperature under solvent-free conditions and in most cases, from high to excellent yields of the desired 1,4-adducts were obtained. Also, the protocol could be scaled up to 100 mmol and proceeded smoothly, showing the potential for industrial applicability. Upon completion of the reaction, the catalyst [DBU][Ac] could be recovered by drying under vacuum at 60 °C and reused for six times without significant loss of activity. The applicability of the task-specified ionic liquid [DBU][Ac] in other fields of organic transformation are underway in our laboratory.

2.2 Michael addition (aromatic amines as Michael donors)

Because of the inertness of arylamines compared with aliphatic amines, most of the available methods suitable for Michael addition of aliphatic amines are not successful with aromatic amines. Duan and his co-workers used CAN (ceric ammonium nitrate) as effective promoter for aza-Michael addition of aromatic and aliphatic amines to α,β-unsaturated electrophiles in absence of solvent under ultrasound irradiation(Duan et al., 2006). Very recentzly, Bhanage et al. reported $Y(NO_3)_3 \cdot 6H_2O$ catalyzed aza-conjugate reaction between weakly nucleophilic aromatic amines and various Michael acceptors such as esters, nitriles and amides under solvent-free conditions(Bhanushali et al., 2008). However, both of the two procedures utilized transition metal catalysts, which limited their application from economic and environmental viewpoints, and more importantly, they have not been involved in the aza-Michael reaction with α,β-unsaturated ketones as Michael acceptors. Basic ionic liquid [bmim]

OH (1-butyl-3-methylimidazolium hydroxide) was first introduced as catalyst for the conjugate reaction of aromatic amines and to our disappointment, however, relatively long reaction times were required(Yang et al., 2006). Thus, the development of the efficient and green protocol for aza-Michael addition of aromatic amines to electron-deficient ketones still remains a challenging task and is highly desirable. Herein we wish to used our recently developed task-specific ionic liquids rather than the reported inert ionic liquids(Tolstikova & Shainyan, 2006) derived from DBU for aza-Michael addition of aromatic amines to α,β-unsaturated ketones at room temperature under solvent-free conditions.

The reaction of aniline and ethyl vinyl ketone (EVK) was selected as model to optimize the reaction conditions. Firstly, using [bmim]OH which is highly efficient in Michael reaction of carbon nucleophiles and aliphatic amines as catalyst(Yang et al., 2007), 8 h was required for consumption of the starting material (Table 3, entry 6)(Yang et al., 2007). Inert ionic liquid [bmim]BF_4 (1-butyl-3-methylimidazolium tetrafluoroborate) was also tested as promoter of the reaction and only 55 % yield of the desired product was obtained in 40 min (Table 3, entry 7). For comparison, some Lewis bases such as DMAP, PPh_3, and DBU were subjected to the model reaction and bis-adduct was formed thereby decreasing the product yield (Table 3, entries 8-10). Especially, with DBU, the bis-addition product was produced as much as 32 % yield while using its corresponding ionic liquids [DBU][Ac] , [DBU][Lac], [DBU][n-Pr], [DBU][n-Bu] and [DBU][Tfa] no bis-addition reaction was observed and 89-99% yields were afforded within the same period, demonstrating good chemo- and regio-selectivity of these ionic liquids in the reaction of aniline and ethyl vinyl ketone (Table 3, entries 1-5). Among the three ionic liquids, [DBU][Lac] performed better than the other ionic liquids, [DBU][Ac], [DBU][n-Pr], [DBU][n-Bu] and [DBU][Tfa] (entries 1-5). So we chose [DBU][Lac] as catalyst for further investigations. A blank experiment was also carried out to demonstrate the catalytic activity of the DBU-based task-specific ionic liquids (Table 3, entry 11). To optimize the details of the reaction conditions, solvent effect was studied in the same model reaction between aniline and EVK (Table 3, entry 1). Toluene, CH_2Cl_2, CH_3OH, and CH3CN were all found to be effective media for the reaction and relatively lower reactant concentration in organic solvent than that under solvent-free conditions decreased the reaction rate. We then selected the solvent-free conditions rather than using organic solvents on consideration of the environmental effect and reaction rate.

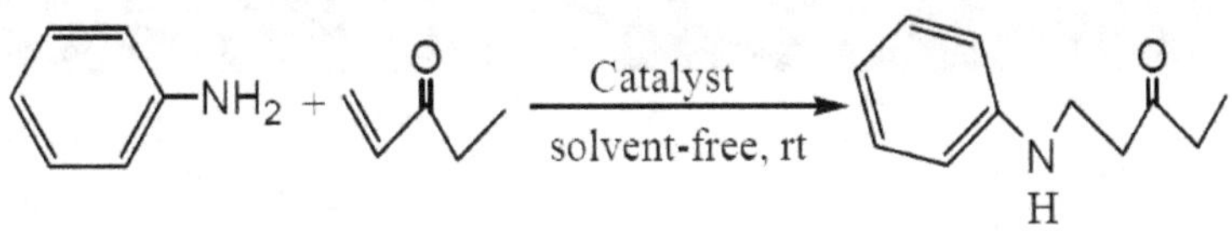

Table 3. Aza-Michael Reaction of Aniline with Ethyl Vinyl Ketone under Various Reaction Conditions[a]

Entry	Catalyst	Time (min)	Yield[b] (%)
1	[DBU][Lac]	30	99 (82[c], 75[d], 76[e], 70[f])
2	[DBU][Ac]	40	96
3	[DBU][n-Pr]	40	90
4	[DBU][n-Bu]		89
5	[DBU][Tfa]	40	96
6[g]	[bmim]OH	480	9823
7	[bmim]BF4	40	55
8	DMAP	40	64
9	PPh3	40	59
10	DBU	40	43
11	-	40	37

[a] Reactions conditions: aniline (1 mmol), ethyl vinyl ketone (1.5 mmol), catalyst (30 mol %), solvent-free conditions, rt. [b] GC yield. [c] Toluene. [d] CH_2Cl_2. [e] CH_3OH. [f] CH_3CN.

[g] Reference: Yang et al., 2007.

With the efficient catalytic system in hand, EVK was treated with other aromatic amines. The results are summarized in Table 4 (entries 1-5). Aromatic amines with electron-donating group at bezene ring were effective nucleophiles to react smoothly with EVK and excellent yields were afforded in very short reaction time (Table 4, entries 1-2), while 2-Methyl substituted amine, due to its steric hindrance, reacted relatively slowly (entry 3). p-NO_2 substituted amine could not react all (entry 5). Arylamine with weakly electron-drawing substitution Cl was also tested with EVK under the above reaction conditions and 95 % GC yield was obtained in 1 h (entry 4).

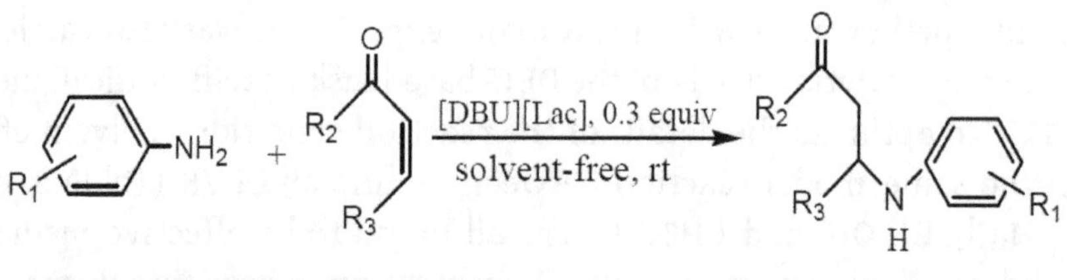

Table 4. [DBU][Lac] catalyzed aza-Michael reactions of various aromatic amines with α,β- unsaturated ketones at room temperature[a]

Entry	R1	R2	R3	Time (h)	GC yield (%)	Isolated yield (%)
1	4-MeO	Et	H	0.5	98	95
2	4-Me	Et	H	0.1/8[c]	97/98[c]	90
3	2-Me	Et	H	0.8	92	85

4	4-Cl	Et	H	1/8[c]	95/97[c]	90
5	2-NO2	Et	H	2	NR	NR
6	4-H	$-(CH_2)_3-$		5	89	82
7	4-Cl	$-(CH_2)_3-$		6	90	79
8[b]	3-Cl	$-(CH_2)_3-$		5	75	68
9	2-Me	$-(CH_2)_3-$		8	85	77
10	4-Me	$-(CH_2)_3-$		5/24[c]	95/71[c]	92
11[b]	2-Cl	$-(CH_2)_2-$		5/24[c]	95/71[c]	92
12[b]	2-Br	$-(CH_2)_2-$		8	75	69
13[b]	2-Me	$-(CH_2)_2-$		8	86	78
14[d]	4-H	Ph	Ph	5	92	85
15[d]	4-Cl	Ph	Ph	12	75	72
16[d]	4-MeO	Ph	Ph	8	88	83
17[d]	4-Me	Ph	Ph	8	91	87

[a] Reaction conditions: aromatic amines (1 mmol), α,β-unsaturated ketones (1.5 mmol), [DBU][Lac] (30 mol %), solvent-free conditions, rt.

[b] Reaction conducted at 60 °C.

[c] Results obtained from reference: Yang et al., 2007, reaction conditions: [bmim]OH (0.3 equiv.), aromatic amines (1.0 mmol), acceptors (1.5 mmol), rt, solvent-free. [d] 1.0 Equiv of [DBU][Lac] required for dissolving chalcone and aromatic amines.

In order to evaluate the generality of the ionic liquid [DBU][Lac] as catalyst for aza-Michael reaction of aromatic amines, 2-cyclohexen-1-one, 2-cyclopenten-1-one and chalcone were tested under the same conditions (Table 4, entries 6-17). To our delight, 2-cyclohexen-1-one and chalcone were effective Michael acceptors to react with various arylamines, giving good to excellent isolated yields (entries, 6-10, 14-17). However, 2-cyclopenten-1-one showed relatively lower reactivity. Accordingly, higher temperature (60 °C) was required (entries, 11-13). For further comparison with [bmim]OH, [DBU][Lac] was used as promoter for the additions of p-methyl aniline to EVK and p-methyl aniline to 2-cyclohexen-1-one respectively (entries, 2 and 10). With comparable or higher products yields obtained, the two reactions catalyzed by [DBU][Lac] proceeded much faster than those promoted by [bmim]OH. Moreover, chalcone reacted smoothly with various aromatic amines using [DBU][Lac] as catalyst (entries, 14-17) while no reaction occurred in the presence of the ionic liquid [bmim]OH(Yang et al., 2007). All cases summarized in Table 4 obviously demonstrate that the novel ionic liquid

[DBU][Lac] has excellent catalytic activity for aza-Michael reaction of aromatic amines and α,β-unsaturated ketones.

The recyclability of the ionic liquid [DBU][Lac] was then studied using the reaction of aniline and 2-cyclohexen-1-one as model. The results are shown in Table 5. Upon the completion of the reaction, the reaction solution was extracted with ethyl acetate and purified by flash chromatography. The addition product was identified by ^{1}H NMR, ^{13}C NMR and MS. The residual ionic liquid was washed with ethyl ether, dried under vacuum at 60 °C for 2 h and reused for subsequent reactions. As shown in Table 5, the recovered ionic liquid could be used for 8 times without obvious loss of catalytic activity. It is worthwhile to note that the ionic liquid [DBU][Lac] used for 8 runs remained intact, judging from its ^{1}H NMR spectrum.

Table 5. Recycling and Reusability of the catalyst [DBU][Lac] in the Reaction between aniline and 2-cyclohexen-1-one[a]

Cycle	t (h)	GC yield[b] (%)
1	5	89
2	5	90
3	5	88
4	5	89
5	5	83
6	5	83
7	6	90
8	6	90

[a] Reaction conditions: 1 mmol of aniline, 1.5 mmol of 2-cyclohexen-1-one, 0.3 mmol of [DBU][Lac] without solvents at rt. [b] GC yields.

As for the role of DBU-based ionic liquids for the activation in the aza-Michael addition of aromatic amines to α,β-unsaturated ketones, [DBU][Lac], [DBU][Ac], [DBU][*n*-Pr] and [DBU][*n*-Bu] were regarded as Brønsted bases to promote the addition and [DBU][Tfa] played the role as Brønsted acid for the this type of aza-Michael reaction. The reason for relatively lower catalytic activity of [DBU][*n*-Pr] and [DBU][*n*-Bu] might be their much higher hindrance of anions. The reason for slightly higher catalytic activity of [DBU][Lac] than those of the other four ionic liquids may be the activation of carbonyl group of Michael acceptors by the hydroxyl group at the α-position of the carboxyl in lactate anion.

In conclusion, we have developed five task-specific ionic liquids, [DBU][Lac], [DBU][Ac], [DBU][*n*-Pr], [DBU][*n*-Bu] and [DBU][Tfa]. The ionic liquids were then used as catalysts for aza-Michael addition of aniline to EVK. Among the five ionic liquids, [DBU][Lac] exhibited the best catalytic activity. Using [DBU][Lac] as catalyst, various aromatic amines were subjected to 2-cyclohexen-1-one,

2-cyclopenten-1-one and chalcone, from good to excellent yields were obtained. This protocol also has some advantages, such as readily work-up of the reactions, excellent reaction selectivity and good recyclability of the ionic liquid (reused for 8 times without significant loss of activity).

3. APPLICATION IN THE KNOEVENAGEL CONDENSATION

3.1 Condensation under solvent-free conditions

Knoevenagel condensation is one of the most important methodologies for carbon-carbon double bond formation in synthetic chemistry(Freeman, 1980; Tietze, 1996). The condensation products are the key intermediates for synthesis of natural and therapeutic drugs, polymer, cosmetics and perfumes(Tietze, 2004; Yu et al., 2000). Generally, Knoevenagel reactions are carried out by condensation of active methylene compounds with aldehydes using some organic bases with their salts as catalysts, including dimethylamino pyridine, piperidine, guanidine, ethylenediamine(Narsaiah et al., 2004; Han et al., 2008), and so on. Also, alternative protocols for Knoevenagel condensations catalyzed by Lewis acids such as $ZnCl_2$(Shanthan Rao & Venkataratnam, 1991), $CuCl_2$(Attanasi et al., 1983), $TiCl_4$(Green et al., 1985), $LaCl_3$(Narsaiah & Nagaiah, 2003), $Mg(ClO_4)_2$(Bartoli et al., 2008) and various heterogeneous solid bases, including Zeolites(Saravanamurugan et al., 2006; Martns et al., 2008), sulfate-ion promoted Zirconia(Reddy et al., 2006), clay(Bigi et al., 1999), and layered double hydroxides (LDHs)(Kantam et al., 2006) have been reported in literatures. However, many of those procedures required the use of large amount of organic solvents as reaction medium, long reaction time, harsh reaction conditions and had difficulties in the reuse of catalysts, which prompt chemical researchers for further development of more environmentally benign, efficient and operationally simple Knoevenagel protocols.

Ionic liquids, due to their unique properties such as good solvating ability, negligible vapor pressure, variable polarity, nonflammability and recyclability have been widely used as catalyst as well as reaction medium in Knoevenagel condensations with more or less success(Paun et al., 2007; Hu et al., 2005; Gao et al., 2007; Yue et al., 2008; Santamarta et al., 2008; Cai et al., 2006; Yeom et al., 2007). However, the costs of those methods increased because large excess amount of ionic liquids were required, which would greatly limit their industrial applications. From both environmental and economical points of view, Knoevenagel reaction promoted by catalytic amount of ionic liquid without solvent may be an attractive catalytic reaction system with many advantages such as lower cost, reduced pollution and simple operation.

Very recently, Kakade et al. investigated the use of DBU as catalyst for Knoevenagel reaction of 2-chloroquinoline-3-carbaldehyde and ethyl cyanoacetate under ultrasonic irradiation(Kakade et al., 2008). Encouraged by this, we tested the reaction of benzaldehyde with ethyl cyanocetate catalyzed by 20 mol % amount of DBU without solvent (Table 6, entry 10) and were pleased to found that 91 % yield of condensation product was obtained within 20 min. However, using DBU as catalyst has some problems that it could not be reused and has unplesant flavour during the operational process which also exists in other reaction with organic bases as promoters for synthetic transformation. In order to overcome these problems and promped by our wish for invention of new task-specified ionic liquid used for organic transformations, we would like to report a novel, basic, task specific ionic liquids, [DBU][X](X=Tfa, Lac, Ac, *n*-Pr, *n*-Bu) and their use as catalysts for Knoevenagel reactions between aromatic aldehydes and active methylene ingredients.

Then we examined novel DBU derived ionic liquids used as catalyst for the reaction of benzaldehyde with ethyl cyanoacetate under solvent-free conditions at room temperature. The results shows that [DBU][Lac] demonstrates the most excellent catalytic activity(Table 6, entry 3 and entries 6-9). A remarkable yield of desired product was obtained (Table 6, entry 3). Some literatures reported that water could promote the Knoevenagel coupling(Shaabani et al., 2007; EI-Rahman et al., 2007; Deb & Bhuyan, 2005). Thus, we carried out the reaction in water for the purpose of comparison with that under solvent-free conditions and found that solvent-free condensation is much faster than aqueous reaction (Table 6, entries 3, 4).

A blank experiment to demonstrate the catalytic ability of [DBU][Lac] was carried out. Only 37 % of the desired product was obtained in absence of [DBU][Lac] even with long reaction time (Table 6, entry 11). This result clearly indicated the effective catalytic role of the new ionic liquid in Knoevenagel condensation. To find a optimal amount of catalyst used for the model reaction of benzaldehyde and ethyl cyanoacetate, the amount of [DBU][Lac] was reduced from 0.5 equiv to 0.01 equiv. The results are collected in Table 5. The reactivity for lower loading of [DBU][Lac] (0.01 equiv) decreased obviously. Increasing the amount of [DBU][Lac] from 20 mol % to 50 mol %, expected improvement in yield of electrophilic alkene has not been observed (Table 6, entries 1-4). Thus, 0.2 equiv was the optimal amount of catalyst for further experiments.

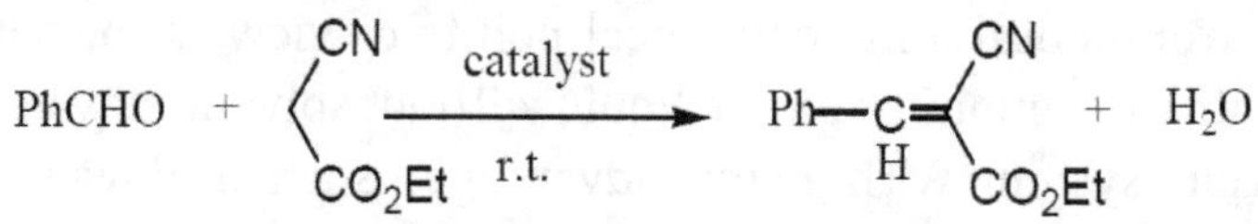

Table 6. Results of varying the amounts of [DBU][Lac] in the solvent-free Knoevenagel condensation of benzaldehyde and ethyl cyanoacetate at room temperature

Entry	Catalysts (mol %)	Time (min)	Yieldsa (%)
1	[DBU][Lac] (0.1)	30	59
2	[DBU][Lac] (1)	30	73
3	[DBU][Lac] (20)	20	93
4b	[DBU][Lac] (20)	120	95
5	[DBU][Lac] (50)	20	94
6	[DBU][Ac] (20)	20	90
7	[DBU][n-Pr] (20)	20	85
8	[DBU][n-Bu] (20)	20	83
9	[DBU][Tfa] (20)	20	78
10	DBU (20)	20	91
11	-	10	37

[a] Isolated yields of products [b] Reaction in water.

With the best catalytic system in hand, we investigated the Knoevenagel reaction of various aromatic aldehydes with ethyl cyanoacetate and malononitrile. The results are shown in Table 7. A variety of structurally diverse aromatic aldehydes reacted favorably with active nucleophilic reagents to give the desired products in good to excellent yields. The products could be isolated by simple filtration. Some crude products need recrystallization in ethanol. All products were verified by their melting point, ^{1}H NMR and ^{13}C NMR spectroscopy which were in good agreement with literatures.

Aromatic aldehydes bearing withdrawing groups reacted with ethyl cyanoacetate, affording lower yields of desired products, while the reaction time is shorter than those with electron-donating substituents (Table 7, entries 1-6). Such results are mainly attributed to the strong activities of electro-deficient aromatic aldehydes, which led to some side reactions and thus resulted in lower yield. When malononitrile was treated with aromatic aldehydes carrying electron donating or withdrawing groups, the reactions proceeded within short time to achieve very high isolated yields (Table 7, entries 1-5). It is notable that electron-donating or electron-withdrawing substituents on aromatic rings have less effect on Knoevenagel reaction presumably due to the strong activity of acidic malononitrile. It can also been observed from Table 7 that more sluggish methylene compounds, acetylacetone can not converse completely and moderate yield afforded (Table 7, entry 17). The order of reaction rate as well as efficiency of Knoevenagel reaction is as follows: malononitrile>ethyl cyanoacetate> acetylacetone, which is accordance with the acidic activity of three nucleophiles. In addition, Knoevenagel condensation of hetero aromatic aldehydes such as 2-furaldehyde and 3- pyridinecarboxaldehyde with active methylene compounds also underwent smoothly at room temperature (Table 7, entries 7, 8, 14, 15). It is worthy to note that all products obtained are E-geometry exclusively and no subsequent Michael adduct is detected. Unfortunately, almost no desired condensation product was found when aliphatic aldehydes and ketones were used as the substrates under the above reaction conditions due to the inertness of these two types of compounds.

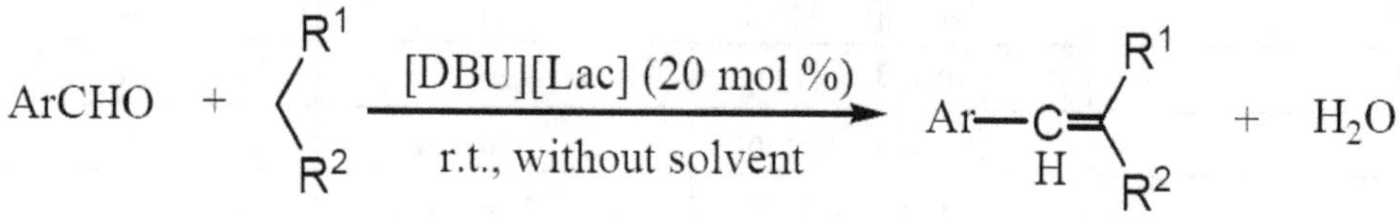

Table 7. Results of Knoevenagel condensation between various aromatic aldehydes and methylene active compounds at room temperature under solvent-free conditions[b]

Entry	Ar	R1	R2	Time (min)	Products	Yields a (%)
1	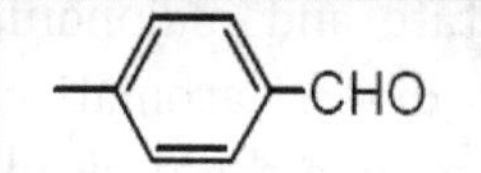	CN	CO2Et	30	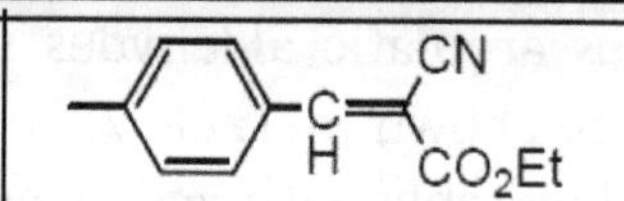	95
2	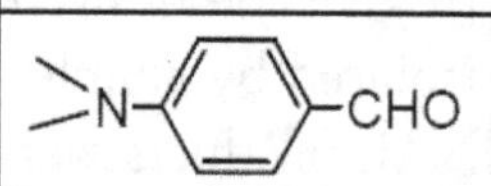	CN	CO2Et	60	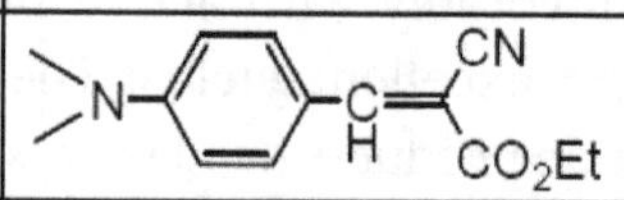	95,
3	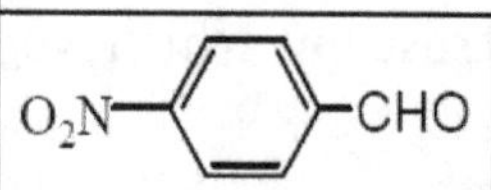	CN	CO2Et	3	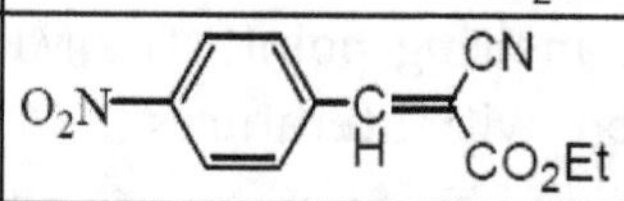	80
4	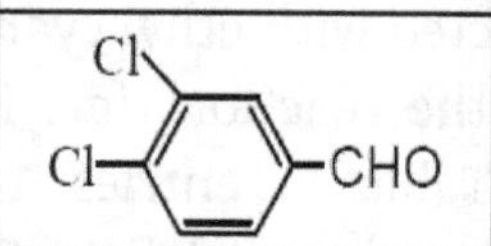	CN	CO2Et	3	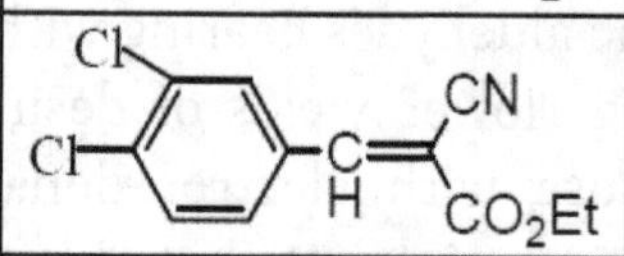	83
5	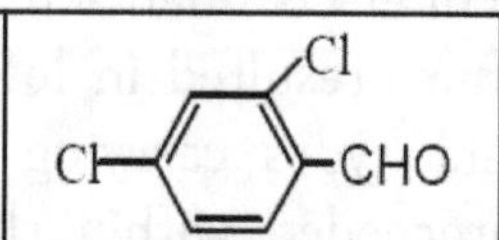	CN	CO2Et	10	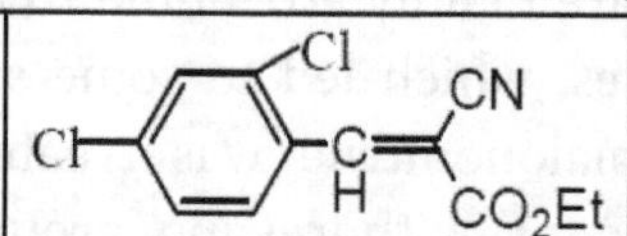	86
6	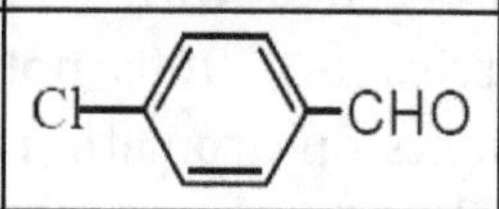	CN	CO2Et	5	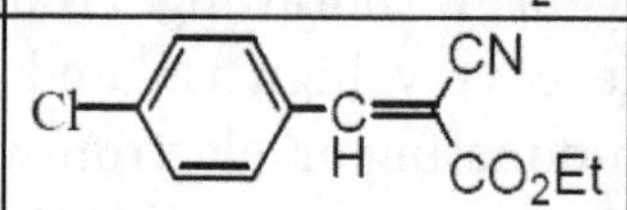	93
7	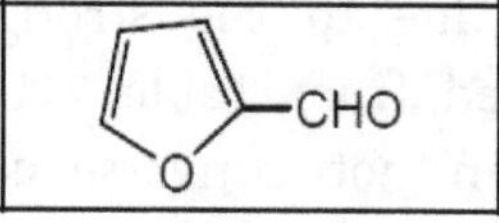	CN	CO2Et	1	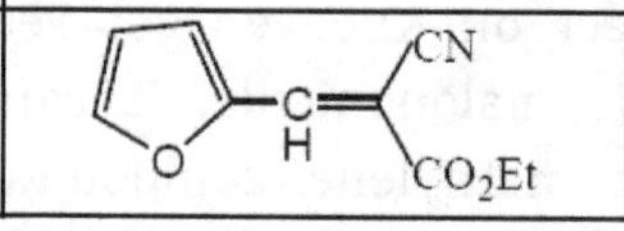	90

8	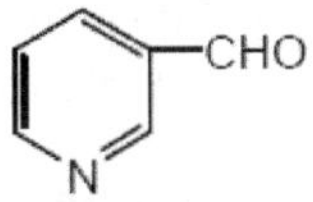	CN	CO2Et	20	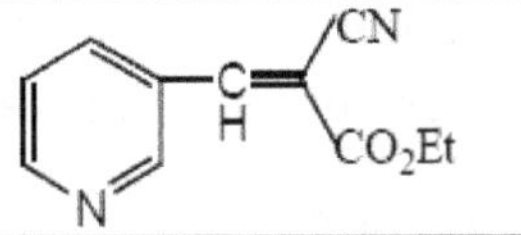	75
9	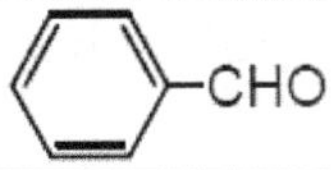	CN	CN	2	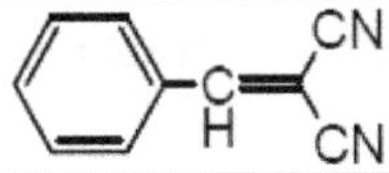	93
10	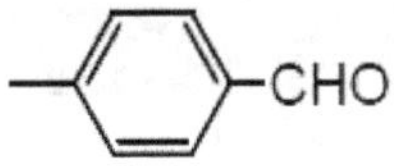	CN	CN	1	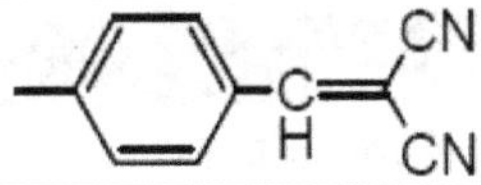	92
11	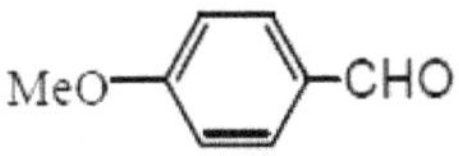	CN	CN	2	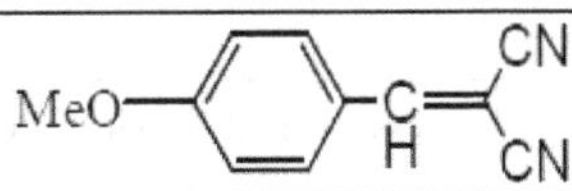	90
12	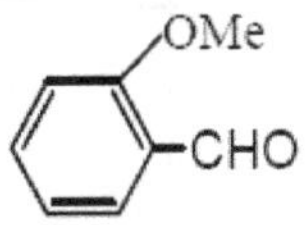	CN	CN	5	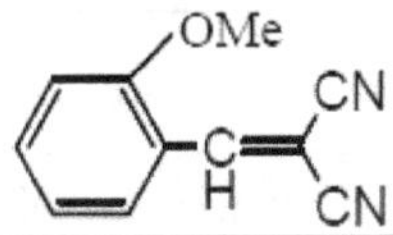	95
13	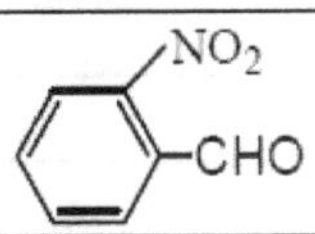	CN	CN	2	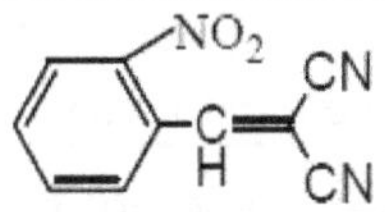	86
14	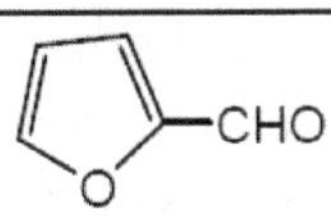	CN	CN	5	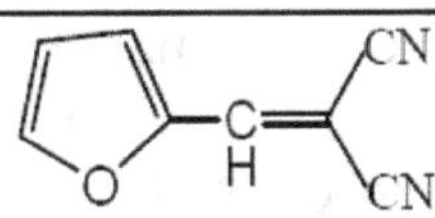	90
15	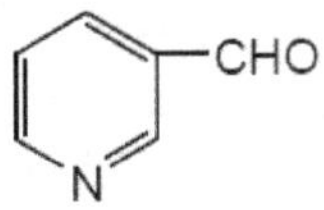	CN	CN	10	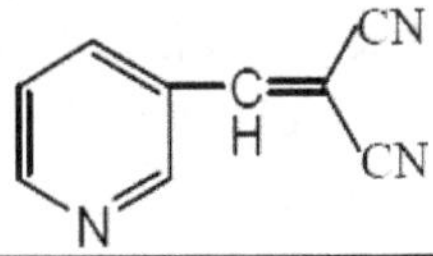	85
16	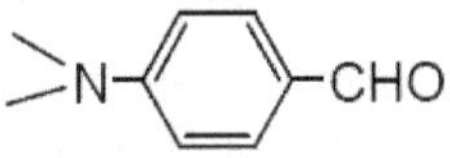	CN	CN	3	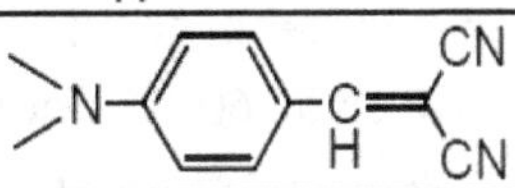	93
17c	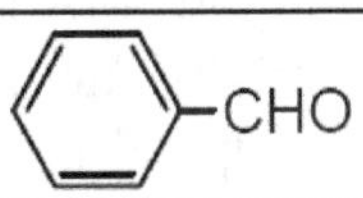	COMe	COMe	300	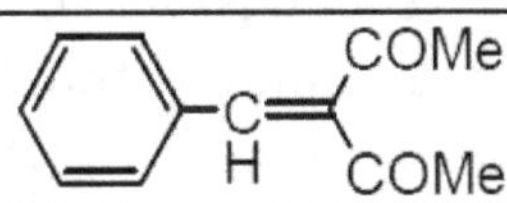	65

[a] Isolated yields of desired products; [b] Reaction conditions: aromatic aldehydes (1 mmol), active methylene compounds (1mmol) stired with 20 mol % [DBU][Lac] as catalyst without solvent at room temperature; [c] Reactions at 60°C;

For the purpose of comparison with other methodologies on the catalytic efficiency, we carried out the reaction of slugguish substrate 4-(dimethylamino) benzaldehyde with ethyl cyanoacetate. As shown in Table 8, only reaction in guanidinium lactate provided the comparable yield (97.2 %) while it required 2.5 equiv amount of ionic liquid, which would limit its scale-up (Table 8, entry 4-5). There was a problem about recyclability of catalytic system in the condensation reaction promoted by phase-transfer catalyst TEBA (Table 8, entry 3). Other methods suffered from the longer reaction times as well as lower yields of products

(Table 8, entries 1-2). All the results show that the present catalytic system (Table 8, entry 5) is very efficient and economic for Knoevenagel condensation.

Table 8. Cmparison of the present catalytic system with other reported protocols in the model reaction between 4-(dimethylamino) benzaldehyde with ethyl cyanoacetate.

Entry	Reaction conditions	Yields (%)	Ref.
1	Perfluoroalkylated pyridine catalyzed in n-octane, at 80°C for 10 h.	78	Yi & Cai, 2008
2	PEG as solvent, catalyzed by L-proline, reaction time: 210 min	81.2	Liu et al., 2008
3	Catalyst: TEBA, reaction time: 15 min	86	Rong et al., 2006
4	2.5 equiv amount of cyclic guanidinium lactate as solvent, reaction time: 3 min	97.2	Liang et al., 2008
5	60 min for reaction, 0.2 equiv [DBU][Lac] as catalyst	95	-

Finally, in order to demonstrate the industrial applicability of this methodology, the solvent- free Knoevenagel condensation of benzaldehyde and ethyl cyanoacetate was carried out on a larger scale (100 mmol). The reaction completed in 20 min., excellent yield of 94 % for the condensation product was achieved. On the same scale, the recyclibility of catalytic system was investigated using the same reaction as model reaction. Upon the completion of the reaction, the product was isolated via readily filtration, washed with water and dried under vacuum without further separation, while the filtrate containing [DBU][Lac] was dried to remove water at 60 °C under vacuum. The recycled ionic liquid was reused in subsequent reactions. As shown in Figure 2, the ionic liquid [DBU][Lac] can be recycled for six times without considerable decrease of activity and the used ionic liuid remained intact (^{1}H NMR). As for the role of [DBU][Lac] in Knoevenagel condensation, we speculate that the basicity of the ionic liquid can attract the proton of the nucleophile to form carbon anion firstly, and secondly, catalytic amount of ionic liquid can improve the nucleophilicity of carbon anion to facilitate the condensation reaction, as same as the behavior observed in larger amount of ionic liquid(Kim et al., 2002; Kim et al., 2003; Liang et al., 2008). However, more work need to be done to clarify the exact mechanism of the [DBU][Lac] mediated Knoevenagel reaction.

In conclusion, we have established a simple, mild and efficient methodology for solvent-free Knoevenagel condensation between aromatic aldehydes and methylene compounds at ambient temperature with ionic liquid [DBU][Lac] as catalyst. This protocol has notable advantages, such as high product yields, reuse of ionic liquid on large scale.

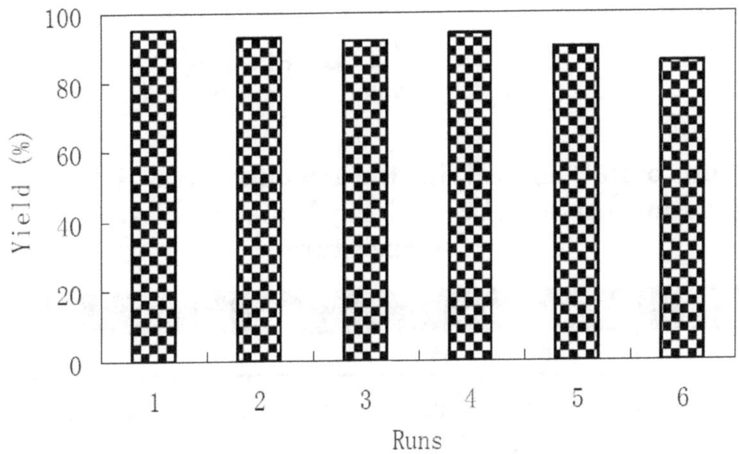

Fig. 2. Reuse of catalyst for Knoevenagel condensation between benzaldehyde and ethyl cyanoacetate (100 mmol) under solvent-free conditions

3.2 Condensation reaction in water

In the last decades, there has been incredible growth in research involving the use of water as a green, environmentally benign replacement for a wide range of processes that currently rely on conventional organic solvents from both environmental and economical points of view(Breslow, 1991; Li, 1993; Ribe & Wipf, 2001). More recently, Song and co-workers employed water as reaction solvent for Knoevenagle condensation between aromatic aldehydes and malononitirle(Cai et al., 2006). Compared with organic solvents media, the advantage of water is very obvious from points of reaction efficiency or environmental "greeness".

Then we examined DBU based ionic liquids as catalysts for the reaction of benzaldehyde with ethyl cyanoacetate in water at room temperature. Among the five ionic liquids, [DBU][Ac] has the best efficient catalytic activity(Table 9, entries 3 and 5-8). An excellent yield of desired product was obtained (Table 9, entry 3). The result is comparable with that catalyzed by parent catalyst DBU. Bhuyan et al. publicated a method for Knoevenagel reaction in aqueous medium without any catalyst(Deb & Bhuyan, 2005). However, the method was substrate-selective and the aqueous reaction of benzaldehyde with ethyl cyanoacetate in absence of [DBU][Ac] is very slow with low yield obtained (Table 9, entry 9). The result indicated that the ionic liquid [DBU][Ac] played an important role as catalyst during the reaction process.

To find a optimal loading amount of catalyst for the model reaction of benzaldehyde and ethyl cyanoacetate, the amount of [DBU][Ac] was reduced from 0.5 equiv to 0.01 equiv. The results are collected in Table 9. Considering both catalytic activity and the cost of catalyst, we chose 0.2 equiv as optimal amount of catalyst for further examinations.

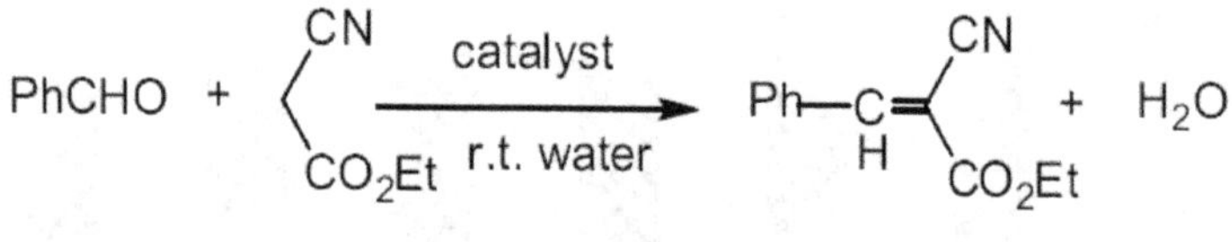

Table 9. Results of DBU and varying the amounts of [DBU][Ac] in the aqueous Knoevenagel condensation of benzaldehyde and ethyl cyanoacetate at room temperature

Entry	Catalysts (mol %)	Time (h)	Yieldsa (%)
1	[DBU][Ac] (1)	3	59
2	[DBU][Ac] (5)	3	73
3	[DBU][Ac] (20)	2	95
4	[DBU][Ac] (50)	2	94
5	[DBU][Lac] (20)	2	92
6	[DBU][n-Pr] (20)	2	90
7	[DBU][n-Bu] (20)	2	85
8	[DBU][Tfa] (20)	2	85
9	-	10	36
10	DBU (20)	2	93

[a] Isolated yields of products

With the rational catalytic system in hand, we studied the Knoevenagel reaction of various aromatic aldehydes with malononitrile and ethyl cyanoacetate. The results are shown in Table 10. A variety of structurally diverse aromatic aldehydes reacted favorably with active nucleophilic reagents, giving desired products in good to excellent yields. The products could be isolated by simple filtration without further tedious purification process. All products were verified by their melting point, ^{1}H NMR and ^{13}C NMR spectroscopy which were in good agreement with the data in other published literatures.

In order to demonstrate the industrial applicability of this methodology, the Knoevenagel condensation of benzaldehyde and ethyl cyanoacetate was carried out on a larger scale (100 mmol) in water (100 mmL). The reaction completed in 2 h, excellent yield of 97% for the condensational product was achieved. On the same scale, the recyclability of catalytic system was investigated using the same reaction as model. Upon the completion of the reaction, the high pure product was afforded by readily filtration without further separation, while the remaining aqueous medium containing [DBU][Ac] was reused directly without additional recovery. As shown in Figure 3, the catalytic system of [DBU][Ac] in water can be reused for ten times without noticeable loss of activity.

In summary, we have established a simple, mild and efficient methodology for Knoevenagel condensation between aromatic aldehydes and methylene compounds at ambient temperature in water with ionic liquid [DBU][Ac] as catalyst. This protocol has notable advantages, such as high product yields, aqueous reaction medium, ease of work-up, reuse of ionic liquid on large scale, which make it more efficient and cleaner with industrial potential.

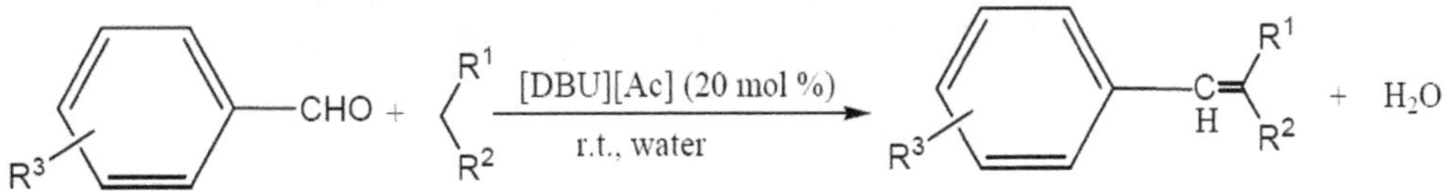

Table 10. Results of Knoevenagel condensation between various aromatic aldehydes and methylene active compounds at room temperature in aqueous media[c]

Entry	R3	R1	R2	Time (h)	Yieldsa (%)
1	H	CN	CN	0.3	93
2	4-Me	CN	CN	0.3	94
3	4-MeO	CN	CN	0.3	94
4	2-MeO	CN	CN	0.3	90
5	2-NO2	CN	CN	0.2	88
6d	4-Dimethyl	CN	CO2Et	6	96
7	4-Me	CN	CO2Et	4	95
8	4-NO2	CN	CO2Et	1	90
9	3,4-Dichloro	CN	CO2Et	3	82
10	2,4-Dichloro	CN	CO2Et	2	75
11	4-Cl	CN	CO2Et	1.5	95
12d	H	COMe	COMe	10	63b
13	2-Furaldehyde	CN	CN	1	86
14	3-Pyridinecarboxaldehyde	CN	CN	1	96
15	2-Furaldehyde	CN	CO2Et	3	92
16	3-Pyridinecarboxaldehyde	CN	CO2Et	3	94

[a] Isolated yields of desired products

[b] Starting materials detected

[c] Reaction conditions: aromatic aldehydes (1 mmol), active methylene compounds (1mmol) stired in water (1mL) with 20 mol % [DBU][Ac] as catalyst at room temperature

[d] Reactions at 60°C

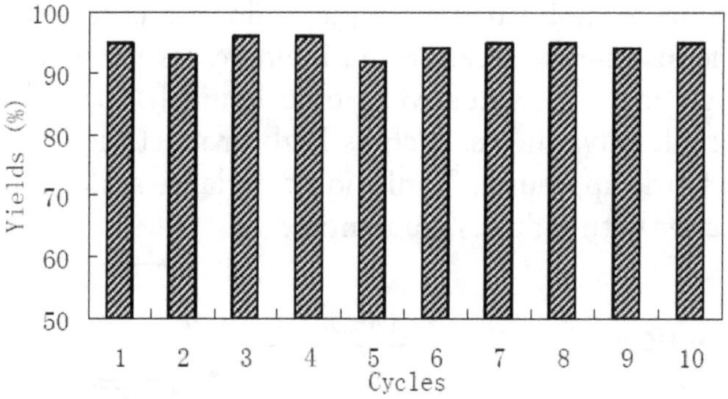

Fig. 3. Reuse of catalyst for Knoevenagel condensation between benzaldehyde and ethyl cyanoacetate (100 mmol) in 100 mL water

4.IONICLIQUIDSIMMOBILIZEDONMNPANDTHEIRAPPLICATIONS IN ORGANIC SYNTHESIS

In order to improve the recycling and reuse property of ionic liquids, immobilization of ionic liquids on polymers is frequently employed(Welton, 2004). However, the reduced activities of these type of ionic liquids catalysts are often observed because of poor dispersion of supported ionic liquids in the reaction system. Thus, novel supports that ensure both good recyclability and catalytic activity are still desirable.

MNP have recently appeared as a new type of catalyst support because of their easy preparation and functionalization, large surface area ratio, facile separation via magnetic force as well as low toxicity and price(Yoon et al., 2003; Stevens et al., 2005; Lee et al., 2006; Abu-Reziq et al., 2006; Luo et al., 2008). These features have made MNP a promising alternate of porous/mesoporous catalyst supports.

Combined advantages of ionic liquids with those of MNP, MNP-supported ionic liquid catalysts (MNP-ILs) were developed by some famous research groups. Luo and his co- workers(Zheng et al., 2009) developed a series of novel MNP-ILs and successfully utilized these catalyst in CO2 cycloaddition reactions (Scheme 3). The activity of the supported catalyst is comparable with that of the free ILs catalysts for this reaction. The catalysts could be easily recycled using a magnetic force and reused for up to 11 times with essentially no loss of activity.

Pd-NHC-ionic liquid matrix was immobilized into ionic liquid layers coated on the surface of Fe_3O_4(Taher et al., 2009). The catalyst 2 has shown excellent catalytic activity and high stability for the Suzuki coupling reaction in water (Scheme 4). This heterogeneous catalyst can be recycled for five times without significant loss of the catalytic activity. Furthermore, recovery of the catalyst by an external permanent magnet is facile and efficient.

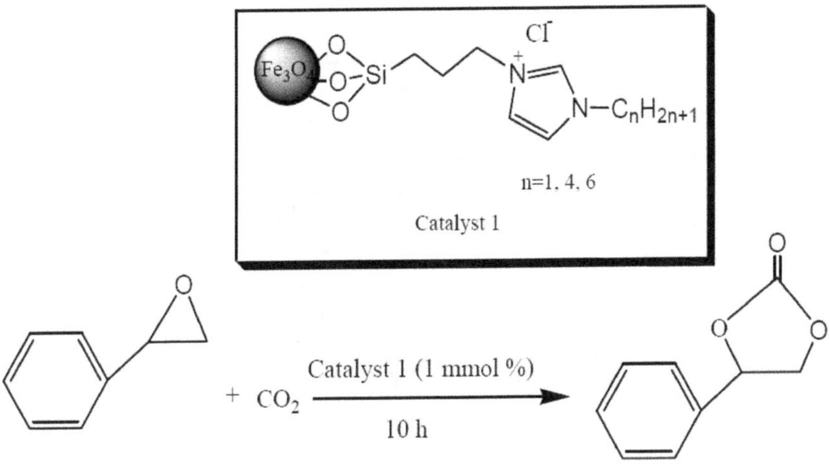

Scheme 3

Shan and his co-workers(Wang et al., 2010) prepared a novel magnetic nano-solid acid catalyst, which includes grafting ionic liquid onto Fe_3O_4 nanoparticles, followed by the sulfonation of phenyls groups in the ionic liquid. The catalyst 3 shows an excellent performance in the acetalization of the carbonyl under mild reaction conditions (Scheme 5), and it can be recycled without obvious loss of catalytic activity.

In summery, MNP-supported ionic liquid catalysts (MNP-ILs) have been developed, which overcome the disadvantages of both magnetic nano-particles and ionic liquids. The three examples have showed the excellent catalytic activities and facile work-up of the MNP-ILs.

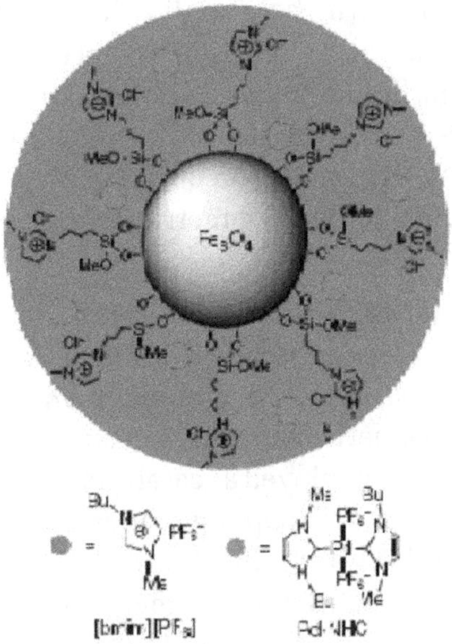

Catalyst 2

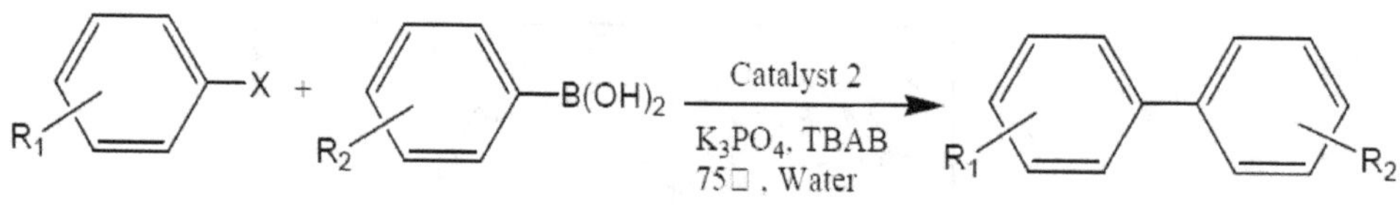

Scheme 4

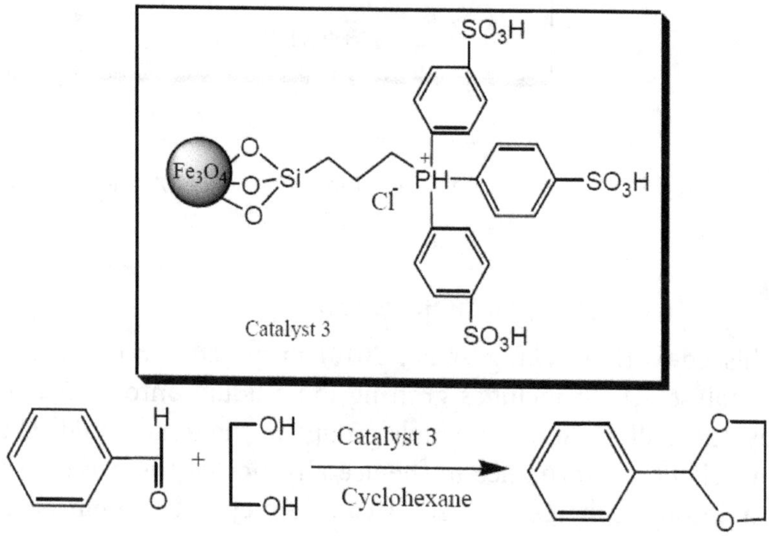

Scheme 5

5. CONCISE SUMMARY

This chapter primarily delineates our recently synthesized task-specific ionic liquids generated from DBU, namely [DBU][Tfa], [DBU][Lac], [DBU][Ac], [DBU][n-Pr], and [DBU][n-Bu], along with their applications in aza-Michael addition and Knoevenagel condensation. In comparison to certain traditional catalysts, the innovative DBU-based ionic liquid catalysts exhibit significantly enhanced catalytic activity, superior chemo- and stereoselectivity, facile recovery, and exceptional recyclability. Nonetheless, the utilization of DBU-based ionic liquids in additional organic transformations, including asymmetric organic synthesis, requires further advancement.

Magnetic nanoparticles (MNP) attract significant interest from researchers due to their excellent stability, straightforward synthesis and functionalization, high surface area, and ease of separation. In light of the characteristics of ionic liquids, chemists have created magnetic nanoparticle-supported ionic liquid catalysts (MNP-ILs). Recently, MNP-ILs employed as catalysts have been effectively exploited in cycloaddition reactions, Suzuki coupling reactions, and acetalization reactions. The results indicate that MNP-ILs can efficiently catalyze reactions, be easily recovered from the reaction solution using magnetic force, and exhibit excellent recyclability and stability. This review aims to catalyze the pertinent research on MNP-ILs.

REFERENCES

Abu-Reziq, R.; Alper, H.; Wang, D. & Post, M. L. (2006). Metal supported on dendronized magnetic nanoparticles: highly selective hydroformylation catalysts *J. Am. Chem. Soc.* Vol. (128) : 5279-5282.

Aggarwal, V. K. & Mereu, A. (1999). Superior amine catalysts for the Baylis-Hillman reaction : the use of DBU and its implication *Chem. Commun.* 2311-2312.

Alleti, R.; Woon, S. O.; Perambuduru, M.; Ramena, C. V. & Reddy, V. P. (2008). Imidazolium-based polymer supported gadolinium triflate as a heterogeneous recyclable Lewis acid catalyst for Michael additions *Tetrahedron Lett.* Vol.(49):3466- 3470.

Arend, M.; Westermann, B. & Risch, N. (1998). Modern variants of the Mannich reactions Angew. *Chem. Int. Ed.* Vol.(37):1044-1070.

Attanasi, O; Fillippone, P & Mei A. (1983). Effect of metal ions in organic synthesis. Part XVI. Knoevenagel condensations of aldehydes and tosylhydrazones with 2,4- pentanedione by copper(II) chloride-catalyzed reaction *Synth Commun.* Vol. (13):1203-1208.

Azizi, N. & Said, M. R. (2004). LiClO4 Accelerated Michael addition of amines to α,β- unsaturated olefins under solvent-free conditions *Tetrahedron* Vol.(60):383-387.

Bartoli, G.; Bosco, M. & Carlone A. (2008). Magnesium perchlorate as efficient Lewis acid for the Knoevenagel condensation between β-diketones and aldehydes *Tertrahedron Lett.* Vol. (49):2555-2557.

Bartoli, G.; Bosco, M.; Marcantoni, E.; Petrini, M.; Sanbri, L. & Torregiani, E. (2001). Conjugate Addition of Amines to α,β-Enones Promoted by CeCl3·7H2O-NaI System Supported in Silica Gel *J. Org. Chem.* Vol.(66):9052-9055.

Bhanushali, M. J.; Nandurkar, N. S.; Jagtap, S. R. & Bhanage, B. M. (2008). Y(NO3)3·6H2O catalyzed aza-Michael addition of aromatic/hetero-aromatic amines under solvent- free conditions *Catal. Commun.* Vol. (9):1189-1195.

Bigi, F.; Chesini, L. & Maggi R. (1999). Montmorillonite KSF as an Inorganic, Water Stable, and Reusable Catalyst for the Knoevenagel Synthesis of Coumarin-3-carboxylic Acids *J. Org. Chem.* Vol. (64):1033-1035.

Breslow, R.(1991). Hydrophobic effects on simple organic reactions in water *Acc. Chem. Res.* Vol. (24):159-164

Cai, Y. Q.; Peng, Y. Q. & Song, G. H. (2006). Amino-functionalized ionic liquid as an efficient and recyclable catalyst for Knoevenagel reactions in water *Catal. Lett.* Vol. (109):61-64.

Chaudhuri, M. K.; Hussain, S.; Kantam, M. L. & Neelima, B. (2005). Boric acid: a novel and safe catalyst for aza-Michael reactions in water *Tetrahedron Lett.* Vol.(46):8329-8331.

Corey, E. J. & Reichard, G. A. (1989). Enantioselective and practical syntheses of R- and S- fluoxetines *Tertrahedron Lett.* Vol.(30):5207-5210.

Crowhurst, L.; Lancaster, N. L.; Arlandis, J. M. P. & Welton, T. (2006). Manipulating Solute Nucleophilicity with Room Temperature Ionic Liquids *J. Am. Chem. Soc.* Vol. (126):11549-11555.

Deb, M. L. & Bhuyan, P. J. (2005). Uncatalyzed Knoevenagel condensation in aqueous medium at room temperature *Tetrahedron Lett.* Vol. (46):6453-6456.

Duan, Z.; Xuan, X.-J.; Li, T.; Yang, C.-F. & Wu, Y.-J. (2006). Cerium(IV) ammonium nitrate (CAN) catalyzed aza-Michael addition of amines to α,β-unsaturated electrophiles *Tetrahedron Lett.* Vol. (47):5433-5436.

El-Rahman, N. M. A. ; El-Kateb, A. A. & Mady, M. F. (2007). Simplified approach to the uncatalyzed Knoevenagel condensation and Michael addition reactions in water using microwave irradiation *Synth. Commun.* Vol. (37):3961-3970.

Freeman, F. (1980). Properties and reactions of ylidenemalononitriles *Chem. Rev.* Vol. (80):329-350.

Gao, G. H.; Lu, L.; Zou, T., et al. (2007). Basic ionic liquid: A reusable catalyst for Knoevenagel condensation in aqueous media *Chem. Res Chinese Univ.* Vol. (23):169- 172.

Georg, G. I. (1993). *The Organic Chemistry of β-Lactams VCH*, New York.

Green, B.; Grane, R. I. & Khaidem I. S. (1985). Synthesis of steroidal 16,17-fused unsaturated δ-lactones *J. Org. Chem.* Vol. (50):640-644.

Han, J. J.; Xu Y. F. & Su Y. P. (2008). Effect of hydrophobic and hydrophilic properties of solid acid catalysts on the esterification of maleic anhydride with ethanol *Catal. Commun.* Vol. (9):2077-2079.

Hashemi, M. M.; Eftekhari-Sis, B.; Abdollahifar, A. & Khalili, B. (2006). ZrOCl2·8H2O on montmorillonite K10 accelerated conjugate addition of amines to α,β-unsaturated alkenes under solvent-free conditions *Tetrahedron* Vol.(62):672-677.

Hu, Y.; Chen, J.; Le Z. G., et al. (2005). Organic reactions in ionic liquids. Ionic liquids ethylammonium nitrate-promoted Knoevenagel condensation of aromatic aldehydes with active methylene compounds *Synth. Commun.* Vol. (35):739-744.

Jenner, G. (1995). Catalytic high pressure synthesis of hindered β-amino ester *Tetrahedron Lett.* Vol. (36):233-236.

Kakade, G. K.; Madje, B. R.; Pokalwar, R. U., et al. (2008). DBU-catalyzed Knoevenagel condensation under ultrasonic irradiation *Indian J. Het. Chem.* Vol. (17):379-380.

Kantam, M. L.; Ravindra, A. & Reddy C. V. (2006). Layered double hydroxides-supported diisopropylamide: synthesis, characterization and application in organic reactions *Adv. Synth. Catal.* Vol. (348):569-578.

Kantam, M. L.; Roy, M.; Roy, S.; Sreedhar, B. & De, R. L. (2008). Polyaniline supported CuI: An efficient catalyst for C-N bond formation by N-arylation of N(H)-heterocycles and benzylamines with aryl halides and arylboronic acids, and aza-Michael reactions of amines with activated alkenes *Catal. Commun.* Vol.(9):2226-2230.

Khan, A. T.; Parvin, T.; Gazi, S. & Choudhury, L. H. (2007). Bromodimethylsulfonium bromide-mediated Michael addition of amines to electron deficient alkenes *Tetrahedron Lett.* Vol.(48):3805-3808.

Kim, D. W.; Song, C. E. & Chi, D. Y. (2003). Significantly Enhanced Reactivities of the Nucleophilic Substitution Reactions in Ionic Liquid *J. Org. Chem.* Vol. (68):4281- 4285.

Kim, D. W.; Song, C. E. & Chi, D. Y. (2002). New method of fluorination using potassium fluoride in ionic liquid: significantly enhanced reactivity of fluoride and improved selectivity *J. Am. Chem. Soc.* Vol. (124):10278-10279.

Kim, S. C.; Lee, K. Y.; Lee, H. S. & Kim, J. N. (2008). Synthesis of poly-substituted benzenes starting from Baylis-Hillman adducts: DBU-assisted unusual dehydrogenation *Tetrahedron* Vol.(64):103-109.

Kleinmann, E. F. (1991). *Comprehensive Organic Synthesis*; Pergamon: New York, NY. Kobayashi, S. & Ishitani, H. (1999). Catalytic Enantioselective Addition to Imines *Chem. Rev.* Vol.(99):1069-1094.

Lee, D.; Lee, J.; Lee, H.; Jin, S.; Hyeon, T. & Kim, B. M. (2006). Filtration-free recyclable catalytic asymmetric dihydroxylation using a ligand immobilized on magnetic mesocellular mesoporous silica *Adv. Synth. Catal.* Vol. (348):41-46.

Li, C. J. (1993). Organic reactions in aqueous media - with a focus on carbon-carbon bond formation *Chem. Rev.* Vol.(93):2023-2035.

Li ,C. J. & Chen, T. H. (1997). *Organic Reacions in Aqueous Media*, Wiley, New York.

Li, P.; Wang, L.; Zhang, Y. & Wang, G. (2008). Silica gel supported pyrrolidine-based chiral ionic liquid as recyclable organocatalyst for asymmetric Michael addition to nitrostyrenes *Tetrahedron* Vol. (64):7633-7638.

Li, Y. X.; Goeke, A.; Wang, R. Y.; Wang, Q. R. & Fráter, G. (2007). DBU-mediated Ireland- Claisen rearrangement of allyl alk-3-enoates :an efficient synthesis of 2-ethylidene- γ,δ-unsatuated carboxylic acids *Tetrahedron* Vol.(63):9605-9613.

Liang, D. P.; Xin, X.; Duan, H. F., et al. (2008). Tetramethylguanidium lactate ionic liquid: catalyst-free medium for Michael addition reaction *Chem. Res. Chinese Univ.* Vol. (24):36-41.

Liu, Y.; Liang, J.; Liu, X. H., et al. (2008). Polyethylene glycol (PEG) as a benign solvent for Knoevenagel condensation *Chinese Chem. Lett.* Vol. (19):1043-1046.

Loh, T. P. & Wei, L. L. (1998). Indium trichloride-catalyzed conjugate addition of amines to α,β-ethylenic compounds in water *Synlett* 975-976.

Luo, S. Z.; Zheng, X. X. & Cheng, J. P. (2008). Asymmetric bifunctional primary aminocatalysis on magnetic nanoparticles *Chem. Commun.* 5719-5721.

Ma, C.; Zhang, Q.; Ding, K.; Xin, L. & Zhang, D. J. (2007). A novel DBU-promoted S-N-type Smiles rearrangement reaction under controlled microwave heating *Tetrahedron Lett.* Vol.(48):7476-7479.

Martns, L.; Vieira K. M. & Rios L. M. (2008). Basic catalyzed Knoevenagel condensation by FAU zeolites exchanged with alkylammonium cations *Catal. Today* Vol. (706):133- 135.

Matsubara, R. & Kobayashi, S. (2008). DBU-catalyzed addition reactions of sulfonylimidates *Synthesis* 3009-3011.

Matsubara, S.; Yoshiyoka, M. & Utimoto, K. (1994). Lanthanoid triflate catalyzed conjugate addition of amines to α ,β-unsaturated esters. A facile route to optically active β- lactam *Chem. Lett.* 827-830.

Meciarova, M.; Toma, S. & Kotrusz, P. (2006). Michael addition of thiols to □ -enones in ionic liquids with and without organocatalysts *Org. Biomol. Chem.* Vol. (4):1420-1424.

Miao, W. W. & Chan T. H. (2006). Ionic-Liquid-Supported Synthesis: A Novel Liquid-Phase Strategy for Organic Synthesis *Acc. Chem. Res.* Vol.(39):897-908.

Narsaiah A.V.; Basak A.K. & Visali B. (2004). An eco-friendly synthesis of electrophilic alkenes catalyzed by dimethylaminopyridine under solvent-free conditions *Synth. Commun.* Vol. (16):2893-2901.

Narsaiah, A. V. & Nagaiah, K. (2003). An efficient Knoevenagel condensation catalyzed by LaCl3.7H2O in heterogeneous medium *Synth. Commun.* Vol. (33):3825-3832.

Paun, C.; Barklie, J. ; Goodrich P. et al. (2007). Supported and liquid phase task specific ionic liquids for base catalyzed Knoevenagel reactions *J. Mol. Catal. A: Chem.* Vol. (269):64-71.

Reboule, I.; Gil, R. & Collin, J. (2005). Aza-Michael reactions catalyzed by samarium diiodide *Tetrahedron Lett.* Vol.(46):7761-7764.

Reddy, B. M.; Patil, M. K. & Rao K. N. (2006). An easy-to-use heterogeneous promoted zirconia catalyst for Knoevenagel condensation in liquid phase under solvent-free conditions *J. Mol. Catal. A: Chem.* Vol. (258):302-307.

Ribe, S. & Wipf, P. (2001). Water-accelerated organic transformations *Chem. Commun.* 299- 307.

Rong, Y. C.; Li, X. Y.; Wang, H. Y., et al. (2006). Efficient green procedure for the Knoevenagel condensation under solvent-free conditions *Synth. Commun.* Vol. (36):2407-2412.

Santamarta, F.; Verdía, P. & Tojo E. (2008). A simple, efficient and green procedure for Knoevenagel reaction in [MMIm][MSO4] ionic liquid *Catal. Commun.* Vol. (9):1779- 1781.

Saravanamurugan, S.; Palanichamy, M. & Hartmann M. (2006). Knoevenagel condensation over □ and Y zeolites in liquid phase under solvent free conditions *Appl. Catal. A: General*, Vol. (298):8-15.

Shaabani, A.; Samadi, S. & Rahmati A. (2007). One-pot, three-component condensation reaction in water. An efficient and improved procedure for the synthesis of pyran annulated heterocyclic systems *Synth. Commun.* Vol. (37):491-499.

Shanthan Rao P. & Venkataratnam R.V. (1991). Zinc chloride as a new catalyst for Knoevenagel condensation *Tetrahedron Lett.* Vol. (32):5821-5822.

Srivastava, N. & Banik, B. K. (2003). Bismuth nitrate-catalyzed versatile Michael reactions *J. Org. Chem.* Vol.(68):2109-2114.

Stevens, P. D.; Fan, J.; Gardimalla, H. M. R.; Yen, M. & Gao, Y. (2005). Superparamagnetic Nanoparticle-Supported Catalysis of Suzuki Cross-Coupling Reactions *Org. Lett.* Vol. (7):2085-2088.

Surendra, K.; Krishnaveni, N. S.; Sridhar, R. & Rao, K.R. (2006). β-Cyclodextrin promoted aza-Michael addition of amines to conjugated alkenes in water *Tetrahedron Lett.* Vol.(47):2125-2127.

Taher, A.; Kim, J. B.; Jung, J. Y.; Ahn, W. S. & Jin, M. J. (2009). Highly Active and Magnetically Recoverable Pd-NHC Catalyst Immobilized on Fe3O4 Nanoparticle- Ionic Liquid Matrix for Suzuki Reaction in Water *Synlett* 2477-2482.

Tietze, L.F. (1996). Domino Reactions in Organic Synthesis *Chem. Rev.* Vol. (96):115-136.

Tietze, L. F. (2004). Domino reactions in the synthesis of heterocyclic natural products and analogs *Pure. Appl. Chem.* Vol. (76):1967-1983.

Tolstikoua, L. L. & Shainyan, B. A. (2006). Ionic liquids on the basis of 2,3,4,6,7,8,9,10 - octahydropyrimido-[1,2-a]azepine (1,8-diazabicyclo[5.4.0]undec-7-ene) *Russian J. Org. Chem.* Vol.(42):1068-1074.

Wang, P.; Kong, A. G.; Wang, W. J.; Zhu, H. Y. & Shan, Y. K. (2010). Facile Preparation of Ionic Liquid Functionalized Magnetic Nano-Solid Acid Catalysts for Acetalization Reaction *Catal. Lett.* Vol. (135):159-164.

Welton, T. Coord. (2004). Ionic liquids in catalysis *Chem. Rev.* Vol. (248):2459-2477.

Xu, D. Q.; Luo, S. P.; Wang, Y. F.; Xia, A. B.; Yue, H. D.; Wang, L. P. & Xu, Z. Y. (2007). Organocatalysts wrapped around by poly(ethylene glycol)s (PEGs): A unique host- guest system for asymmetric Michael addition reactions *Chem. Commun.* 4393-4395.

Xu, J. M.; Qian, C.; Liu, B. K.; Wu, Q. & Lin. X. F. (2007). A fast and highly efficient protocol for Michael addition of N-heterocycles to α,β-unsaturated compound using basic ionic liquid [bmIm]OH as catalyst and green solvent *Tetrahedron* Vol.(63):986-990.

Xu, L. W.; Li, L. & Xia, C. G. (2004). Transition-metal-based Lewis acid catalysis of aza-type Michael additions of amines to α,β-unsaturated electrophiles in water *Helv. Chim. Acta.* Vol.(87):1522-1526.

Xu, L. W.; Li, J. W.; Xia, C. G.; Zhou, S. L. & Hu, X. X. (2005). Efficient copper-catalyzed chemoselective conjugate addition of aliphatic amines to α,β-unsaturated compounds in water *Synlett* 2425-2427.

Yang, L.; Xu, L. W. & Xia, C. G. (2005). Highly efficient KF/Al2O3-catalyzed versatile hetero-Michael addition of nitrogen, oxygen, and sulfur nucleophiles to α,β - ethylenic compounds *Tetrahedron Lett.* Vol.(46):3279-3282.

Yang, L.; Xu, L.-W.; Zhou, W.; Li, L. & Xia, C.-G. (2006). Highly efficient aza-Michael reactions of aromatic amines and N-heterocycles catalyzed by a basic ionic liquid under solvent-free conditions *Tetrahedron Lett.* Vol. (47):7723-7726.

Yeom, C. E.; Kim, M. J.; Choi, W. & Kim, B. M. (2008). DBU-promoted facile, chemoselective cleavage of acetyletynic TMS group *Synlett.* 565-568.

Yeom, C. E.; Kim, M. J. & Kim, M. B. (2007). 1,8-Diazabicyclo[5.4.0]undec-7-ene(DBU) promoted efficient and versatile aza-Michael addition *Tetrahedron* Vol.(63):904-909.

Yi ,W. B. & Cai C. (2008). Perfluoroalkylated pyridine catalyzed Knoevenagel condensation: an important complement of fluorous catalysis without fluorous solvent *Catal. Commun.* Vol. (9):1291-1296.

Ying, A. G. (2010). Preparation of DBU derived task-specific ionic liquids and the study of their applications in the Michael additions and Knoevenagel condensations; Dissertation for Doctor degree, Zhejiang university, pp. 42-50.

Ying, A. G.; Chen, X. Z.; Ye, W. D.; Zhang, D. F.; Liu, L. & Chen, J. H. (2008). Application of ionic liquids in organic synthesis promoted by microwave irradiation *Prog. Chem.* Vol.(20):1642-1650.

Ying, A. G.; Liu, L.; Wu, G. F.; Chen, G.; Chen, X. Z. & Ye, W. D. (2009). Aza-Michael addition aliphatic or aromatic amines to 〈,β-unsaturated compounds catalyzed by a DBU-derived ionic liquid under solvent-free conditions *Tetrahedron Lett.* Vol.(50):1653-1657.

Ying, A. G.; Liu, L.; Wu, G. F.; Chen, X. Z.; Ye, W. D.; Chen, J. H. & Zhang, K. Y. (2009). Knoevenagel condensation catalyzed by DBU brönsted ionic liquid without solvent *Chem. Res. Chinese Universities* Vol.(25):876-881.

Ying, A. G.; Wang, L. M.; Deng, H. X.; Chen, J. H.; Chen, X. Z. & Ye, W. D. (2009). Green and efficient aza-Michael additions of aromatic amines to 〈,β-unsaturated ketones catalyzed by DBU based task-specific ionic liquids without solvent *Arkivoc,* Vol.(XI):288-298.

Ying, A. G.; Wang, L. M.; Wang, L. L.; Chen, X. Z. & Ye, W. D. (2010). Green and efficient Knoevenagel condensation catalysed by a DBU based ionic liquid in water *J. Chem. Res.* 30-33.

Ying, A. G.; Ye, W. D.; Liu, L.; Wu, G. F.; Chen, X. Z.; Qian, S. & Zhang, Q. P. (2008). Progress in the applicatioin of ionic liquids to organic synthesis *Chinese J. Org. Chem.* Vol.(28):2081-2094.

Yoon, T. J.; Lee, W.; Oh, Y. S. & Lee, J. K. (2003). Magnetic nanoparticles as a catalyst vehicle for simple and easy recycling *New J. Chem.* Vol. (27):227-229.

Yoshida, M.; Komatsuzaki, Y. & Ihara, M. (2008). Synthesis of 5-vinylideneoxazolidin-2-ones by DBU-mediated CO_2 fixation reaction of 4-(benzylamino)-2-butynyl carbonates benzoates *Org. Lett.* Vol. (10):2083-2086.

Yu. N.; Aramini J. M. & Germann M. W. (2000). Reactions of salicylaldehydes with alkyl cyanoacetates on the surface of solid catalysts: syntheses of 4H-chromene derivatives *Tertrahedron Lett.* Vol. (41):6993-6996.

Yue, C. B.; Mao, A. Q.; Wei Y. Y., et al. (2008). Knoevenagel condensation reaction catalyzed by task-specific ionic liquid under solvent-free conditions *Catal. Commun.* Vol. (9):1571-1574.

Zheng, X. X.; Luo, S. Z.; Zhang, L. & Cheng, J. P. (2009). Magnetic nanoparticle supported ionic liquid catalysts for CO2 cycloaddition reactions *Green Chem.* Vol. (11):455-458.

INDEX

D

E

F

G

H

www.ingramcontent.com/pod-product-compliance
Lightning Source LLC
LaVergne TN
LVHW081940200926
848484LV00009B/2477